Intersections of Conservation Biogeography and Wildlife Management

Kaushalendra Kumar Jha ·
Michael O'Neal Campbell
Editors

Intersections of Conservation Biogeography and Wildlife Management

Navigating the Anthropocene Era

Springer

Editors
Kaushalendra Kumar Jha
Indian Institute of Forest Management
Bhopal, Madhya Pradesh, India

Michael O'Neal Campbell
Simon Fraser University
Burnaby, BC, Canada

ISBN 978-981-95-0120-5 ISBN 978-981-95-0121-2 (eBook)
https://doi.org/10.1007/978-981-95-0121-2

This Springer imprint is published by the registered company Springer Nature Singapore Pte Ltd.
The registered company address is: 152 Beach Road, #21-01/04 Gateway East, Singapore 189721, Singapore

Preface

This book examines the complex science of conservation biogeography from a novel perspective, examining applications to wildlife management issues, and in relation to human relations with carnivores and conservation policy management. These critical topics are currently developing within the mindset of conservation biogeography, and new methodological and technological developments. Human relations with medium to large, especially predatory mammals and birds, and the perception of how this relationship may be managed from a political, economic sand political perspective has gained traction over the twentieth and twenty-first centuries. The central theme of the book is that conservation biogeography, which is the science of biological and ecological distribution and environmental relationships, from a conservation perspective, has the tools necessary to critically assess current wildlife issues at global, regional and local levels. The emergence of the conservation movement in the twentieth century and the associated science of conservation biology, linked to biogeography as both a biological and geographical discipline, influenced the development of the modern science of conservation biogeography, which also has roots in cultural biogeography, conservation psychology, and political ecology. Current contributions come from the areas of computerization and the topics of the social discourses. The topics covered in this book reflect the background education and knowledge of the editors and authors, all of whom work within the multidisciplinary and interdisciplinary sciences.

This book is divided into 2 parts and 9 chapters. Part I, titled "Conservation Biogeography and Human-Carnivore Relations", includes 4 chapters that examine the development of conservation biogeography and applications, with case studies applied to the conservation of big cats in Eurasia, Africa and the Americas, and the political ecology and conservation psychology of these same species, and human-animal relations with case studies from India that examine indigenous communities in tiger reserves. Chapter 1, titled "Conservation Biogeography, Political Ecology, and Conservation Psychology", examines these big cats from a more social science perspective, and argues that human dimensions of conservation ecology include political, socioeconomic, cultural and psychological issues, which provide the background for political ecology, defined as political economy applied to the environment, and conservation and/environmental psychology, which explains human attitudes toward wildlife and habitats. These topics complement the biogeographical questions in conservation biogeography, presented in Chap. 7 in relation to the ecology of carnivores. Particularly important are the political power differences between different levels of stakeholders and the mental constructs and attitudes of local people in relation to their environment. It is argued that few studies have examined the triangular relationship between conservation biogeography, psychology, and political ecology, especially in the context of human relations with large carnivores.

Chapter 2, titled "Conservation Biogeography of Big Cats in Eurasia, Africa, and the Americas" examines the species of the Genera Panthera and Puma, in order of size: tigers (Panthera tigris Linnaeus, 1758), lions (Panthera leo Linnaeus, 1758), jaguars (Panthera onca Linnaeus, 1758), cougars (Puma concolor Linnaeus, 1771), leopards (Panthera pardus Linnaeus, 1758) and snow leopards (Panthera uncia Schreber, 1775) from an ecological and biogeographical perspective. The selection of these species as case studies for conservation biogeography is justified because they are dangerous, charismatic and even charming animals to humans, and create strong landuse conflicts with humans in land occupancy. It is noted that the global conservation status of big cats depends mostly on human conservation efforts, with variable outcomes.

Chapter 3, titled "Voluntary Village Relocation and Sustainability: A Case of Ecological Component in Kanha Tiger Reserve, Central India" examines tiger-human relations in a populus tiger reserve. Here tiger extinction has been avoided by focussing on human population, tiger habitat and the exclusion of anthropogenic activities. Human populations in some cases were relocated, and the study examines the pre and post relocation situation in terms of the three constituents of sustainability, which comprise environmental/ecological, social, and economic parameters. The results of the study suggest that relocated human populations were satisfied with the relocation, which was assisted by the Forest Department, and generated maps suggested that carnivores and herbivores entered the vacated areas. Relocation of human populations, resulted in vegetation change, declines in grazing cattle, reduced firewood extraction, and fewer human-wildlife conflicts, such as cattle depredation. It is argued that these findings support policy makers and conservationists involved in tiger conservation and inform the broader issues of wildlife conservation elsewhere.

Chapter 4, titled "Rise of Tigers from a Steep Fall: Panna Tiger Reserve in Case" examines the tigers' fall from a large population to near extirpation and a later increase in population. In Panna Tiger Reserve (PTR) in central-north-east India, the population recovered from a low point. The chapter notes that tigers were formerly numerous in their range countries, but there was a sharp decline in their populations, after which there was a marginal recovery due to conservation efforts. The Indian tiger populations declined but recovered under conservation interventions at the federal government level: forming Project Tiger institution and promulgating Wildlife (Protection) Act, about 50 years ago. For PTR, the situation was different, as the tiger population declined to total extirpation and then recovered rapidly due to tiger relocation-based natural breeding management and strict monitoring. It is argued that the success of this management intervention could be repeated with success.

Part II, titled "Human-Wildlife Conflicts, Methods and Developments" includes 5 chapters that examine human relations with mesopredators (smaller or medium sized carnivores), an issue generally neglected in published studies, compared with human relationships with the larger, more charismatic apex predators; new methods for the study of human-wildlife conflict, with the suggestion of the concept of environmental stewardship; the role of trophy hunting in ecosystem management and conservation; and a study of the vegetation in the diet of some Indian wild ungulates. These topics are selected as themes generally neglected in the published literature but are nevertheless critical for conservation knowledge. Chapter 5, titled "Human-Wildlife Conflict: Can *Loco Parentis* Help?" by Arvind Kumar Jha, examines the concept of the idea of "in loco parentis" or "guardian", questions the idea of human dominance, emphasizes a symmetrical view of the natural world in relation to the rights of human and non-human species, and recognizes the need to balance human and non-human interests and protect rights to create harmony. This concept has been applied to the context of India, where the growing population is heavily dependent on forests and there are land shortages, unemployment issues in communities that were previously hunters, illegal wildlife trade and legal interference leading to the decentralization of powers leading to a reduction in the Forest Ministry mandate for protection and enforcement. These lead to a continuation and aggravation of disputes, which can be avoided by changing problem-solving methods. It is argued that the idea of "in loco parentis" or "guardian" challenges the idea of human dominance, emphasizes a symmetrical view of the natural world in relation to the rights of human and non-human species, and recognizes the need for human and non-human animals to be protected in their interests and rights in order to create harmony. The reason for institutionalizing the Loco Parentis paradigm was to reduce conflict, promote coexistence, and achieve a broader strategy for sustainable conservation. This is a departure from the anthropocentric approach currently used. It is suggested that this strategy would support effective, efficient and sustainable production.

Chapter 6, titled "Ecosystem Management and Trophy Hunting," by Shahbaz Ahmad, examines the current emphasis on the idea of ecosystem services. The ecosystem services of the entire biosphere are worth far more than the global GDP (gross domestic product). With this realization, the unplanned exploitation of natural resources gave way to their conservation and sustainable use. Numerous laws and agreements also ensured the protection of wild animals. In the twentieth century, tourist trophy hunting began in Kenya and eventually spread throughout Africa. There is no evidence that it reduces human-wildlife conflict or that it is a necessary management strategy. General opinion is against trophy hunting as awareness of the practice grows. However, since trophy hunting is a pastime of wealthy people, the professionals in this business offer many explanations and defenses that are not supported by data or empirical research. To stop trophy hunting, which is rooted in a colonial mentality, governments of various nations must develop an exit strategy. The chapter argues that managing existing forests and natural ecosystems from a global perspective is necessary to prevent climate change and ensure the continued existence of ecosystem services for the benefit of humanity. An ecosystem manager is responsible for a variety of tasks, such as maintaining ecosystem services, engaging people in the sustainable use of usufruct rights, managing human-wildlife conflicts through integrated strategies, and raising awareness of the natural world through more sustainable management. Tourism initiatives, conducting campaigns and events to provide livelihood and other benefits to locals, promote harmony among wildlife, etc. are important. Trophy hunting is incompatible with ecologically sound management practices.

Chapter 7, titled "Human Relations with Mesopredators" by Michael O'Neal Campbell, examines mesopredators as smaller, less dangerous and charismatic to humans, and less researched in terms of conservation compared with large predators. It is nevertheless argued that mesopredators are critical components of ecosystems and may have vital relationships with humans and their life spaces and associated animals. Important issues examined include the tendency of mesopredators to assume the status of apex predators with the local extinction of the latter, disputes as to the size distinction between mammalian mesopredators and apex predators, the ecological status of avian apex predators, which function at the same level as mammalian mesopredators (for example, bobcats *Lynx rufus* Schreber, 1777 and golden eagles *Aquila chrysaetos* Linnaeus, 1758, which may take similar prey), and the less intense human attitudes toward mesopredators compared with apex predators, based on smaller size and less dangerous habits. The evidence suggests that mesopredators are less studied than apex predators, a relevant omission for ecosystem studies. The social science perspective, particularly that of conservation psychology, is important for the study of conservation biogeography and human-wildlife relationships.

Chapter 8, titled "Composition and Plant Species Diversity in the Diet of Some Wild Ungulates in India: A Systematic Review", by Rohit Chaudhary, Kavya Shah, Taksh Kelaiya, Aadil Kazi, Prakhar Sharma, Mohmad Navaz Dahya, and Priyank Kapdi examines the case that wild ungulates can influence the structure and function of their resident and transitory ecosystems through their feeding habits. Consequently, an analysis of the feeding habits of ungulates can support efforts to protect ungulates, including the conservation of their wider habitats. The dietary habits of a case study of nine wild ungulate species of different sizes in India are examined through the study of literature published from 1990 to 2020. The species studied were chital (Axis axis), sambar (Rusa unicolor), nilgai (Boselaphus tragocamelus) and blackbuck (Antilope cervicapra), gaur (Bos gaurus), swamp deer (Rucervus duvaucelii), chinkara (Gazella bennetii), chowsingha (Tetracerus quadicornis) and hog deer (Axis porcinus). The aim of the chapter was to evaluate the techniques used in ungulate nutrition studies to capture the diet of ungulates and any similarities between species. To evaluate the diet of ungulates, the number of plant species, life forms (trees, shrubs, herbs, grasses, sedges and climbing plants), the diversity of life forms and the ratio of tree/grass species were used. The results of this study suggest that the diversity of plant species that ungulate species eat is decreasing as they become increasingly specialized in the use of their habitat.

Chapter 9, titled "Geomatics Applications to Conservation and Social Developments" by Michael O'Neal Campbell, examines the computer-aided systems, referred to as geomatics in North America and geoinformatics in Europe and Asia, which are applied to conservation biogeography topics, as well as to the environmental and spatial aspects of economics, anthropology, sociology, human geography, psychology and political science. Geomatics is a very broad field, which includes remote sensing, geographic information science or geographic information systems (GIS), global positioning systems (GPS), geodesy, and cartography and mapping. The current state of geomatics in conservation biogeography is assessed, through the assessment of the research results in the social sciences, conservation biology and biogeography. Complexities based on the multidisciplinary background of conservation biogeography are also examined. Current evidence suggests that geomatics-based systems enable more reliable and precise calculations for environmental research with more applicable results. Geomatics-based techniques easily measure land cover changes that can impact conservation and biogeography. However, predictive modeling of human activities in complex environments are more problematic.

These chapters serve as an example of the wide range of topics covered in conservation biogeography. They also shed light on issues related to political ecology, such as political power, access to resources, and cultural differences, and conservation psychology, which includes the various perspectives that participants and wildlife have of the environment. Novel approaches, including research methodologies, are needed to address the political, social, economic, and ecological concerns under discussion. It is the hope of the editors and authors that this book's significant topics will increase readers' interest in the critical and fascinating field of conservation biogeography.

Bhopal, India — Kaushalendra Kumar Jha
Burnaby, Canada — Michael O'Neal Campbell

Contents

About the Editors

Kaushalendra Kumar Jha is a forester with high administrative and academic skills in forest management and environmental science. He holds degrees, research, and teaching position from Indian Universities and Management Institute (botany and forestry: aquatic, afforestation, avian, and raptor ecology; forest produce management; climate change and international conventions). He acquired skill-upgradation degree and trainings from European and North American Universities (sustainable tropical forest management with specialization in environment, tree improvement, and geoinformatics). He has considerable number of publications.

Michael O'Neal Campbell is an environmental scientist with expertise in applied ecology and biochemical relations in environmental changes. He holds degrees, research, and teaching positions in applied biogeography, geomatics, conservation biology, and landscape ecology, from the Universities of London (Imperial College), Manchester Metropolitan, Leicester, and Lancaster (all in the UK), and Carleton University in Canada. He has published widely in the environmental sciences.

List of Figures

Part I
Conservation Biogeography and Human-Carnivore Relations

This part includes four chapters:

Chapter 1: Conservation Biogeography, Political Ecology and Conservation Psychology

Chapter 2: Conservation Biogeography of Big Cats in Eurasia, Africa and the Americas

Chapter 3: Voluntary Village Relocation and Sustainability of Eco-system: A Case Study of Kanha Tiger Reserve

Chapter 4: Rise of Tigers from a Steep Fall: Panna Tiger Reserve in Case

These chapters examine the relationship between the complex associations of conservation biogeography and scaled human relations, at national, regional, local and individual levels. Conservation biogeography is considered, based on the ecological relations of the organisms in the case studies. Political ecology, defined as the political economy of the human-environment relations, studies political, economic and social groups and their power relations, these groups including international governmental and non-governmental organizations, national, regional and local governments, and local rural stakeholders of varying power structures and participatory development. Conservation psychology, the study of human perceptions of environmental phenomena, is mainly concerned with individual level perceptions of people, concerning agricultural and non-agricultural environmental activities, including medicinal plant extraction and hunting. The first chapter introduces conservation biogeography and explores its links with the social, interdisciplinary and multidisciplinary sciences of political ecology and conservation psychology, taking the example of human relations with large carnivores. The second chapter, takes the conservation biogeography of big cats as a case study. The third chapter examines a case study of Kanha Tiger Reserve in India, based on voluntary village relocation and the sustainability of the relevant ecosystem. The final chapter examines the natural and social context of the Panna Tiger Reserve in India.

Chapter 1
Conservation Biogeography, Political Ecology, and Conservation Psychology

Michael O'Neal Campbell

Abstract The factors that enable or constrain the features of effective conservation biogeography increasingly include political, socioeconomic, cultural, and psychological aspects, which may vary at continental, national, regional, and local scales. Political ecology, broadly defined as political economy applied to the environment, is a crucial background for conservation biogeography because political and economic structures and developments can influence or even determine the existence or extinction, spread or decline, composition and characteristics, and human dimensions of species. Psychology, particularly conservation and/environmental psychology, is closely related to this dynamic as it can describe or explain human attitudes toward wildlife and habitats. The central themes include the political power differences between different levels of actors and stakeholders as well as the mental constructs and attitude structures of local people in relation to their environment. These topics complement the biological and biogeographical questions that form the basis of conservation biogeography and become more important as human engagement with the natural environment develops. However, few studies have examined the triangular relationship between conservation biogeography, psychology, and political ecology. Such a study contributes to the development of effective conservation management and policy.

Keywords Conservation biogeography · Conservation psychology · Environmental Psychology · Carnivores · Political Ecology · Big cats

Introduction

Conservation biogeography relates to the links between the natural and human worlds, and how the relationship between humans, animals, and their shared environment may be conserved for environmental sustainability (Campbell, 2017). Such studies are based on components of human psychology, political relationships,

M. O. Campbell (✉)
Simon Fraser University, Burnaby, BC, Canada
e-mail: ecol55@hotmail.com

K. K. Jha and M. O. Campbell (eds.), *Intersections of Conservation Biogeography and Wildlife Management*, https://doi.org/10.1007/978-981-95-0121-2_1

economic power, and their interaction with connected animal ecologies (Perreault et al., 2015). The word "conservation biogeography" refers to the study of biological conservation from a distributional, spatial, and integrated geographic standpoint (Ladle et al., 2016; Whittaker & Ladle, 2011). Whittaker and Ladle (2011) and Whittaker et al. (2005) defined conservation biogeography as a field of conservation biology that includes "the application of biogeographical principles, theories, and analyses, being those concerned with the distributional dynamics of taxa individually and collectively, to problems concerning the conservation of biodiversity." This illustrates the link between conservation biogeography and conservation biology. Yong (2018) adds that "fusing the traditions of biogeography with the concerns of conservation biology, conservation biogeography is a relatively new academic endeavor that develops conservation and applied concerns. The spatial organization of biological diversity is the subject of the well-established scientific field of biogeography, in contrast".

The well-established scientific field of biogeography studies how biological diversity is arranged spatially. According to Ripple et al. (2014), conservation biogeography, based on a biogeographical model, has gained strength since 2005 based on attempts to conceptually and empirically define the possibly neglected spatial aspects of conservation biology. Conservation biogeography was envisioned as a distinct applied and interdisciplinary, yet spatially oriented science concerned with the conservation of nature (Riddle, 2021; see also Richardson, 2005; Franklin et al., 1999; Campbell, 2017, 2018, 2019, 2022; Campbell & Jha, 2021; Jha & Campbell, 2021, 2022). Therefore, conservation biogeography is a relatively new and possibly under-researched field that needs to be evaluated against three powerful and established trends: biogeography, conservation science, and biological background themes. As Young (2021) pointed out, there are few general reviews because conservation biogeography is still in its infancy, and the majority of reviews can be found in the biogeography literature. However, as demonstrated by journal articles and departmental research at universities, conservation biogeography is progressively changing within academic circles. These advancements are consistent with the understanding of the broader context of biogeographic change and the multiplicity of academic fields with shared interests in this field (Campbell, 2017, 2018). These encompass natural science and more human-oriented topics (Poloni et al., 2022; Dansereau, 2025).

In conservation biogeography, the social sciences play a crucial role that cannot be overstated (Campbell & Jha, 2021; Jha & Campbell, 2022). According to Ferraro and Pressey (2015), reducing human pressure in areas that support biodiversity and ecosystem services is essential to the success of conservation efforts. Bennett et al. (2018, p. 104) contend that in order to make effective conservation decisions in planning, implementation, and management, the conservation social sciences are a necessary complement to the natural sciences, and that the best available data and relevant conceptual frameworks should and can serve as the basis for conservation policy and practice. In the past, information used to direct conservation efforts has usually come from the natural sciences alone or as the main source. But for a long time, a number of significant conservation scientists have understood how crucial social and natural factors are to conservation (Bennett et al., 2018).

Bennett et al. (2018, p. 94) note that "the social sciences have also been used to research conservation and environmental management at all scales from local to regional and global" (see Bennett et al., 2018; Pietri et al., 2015; Fleischman et al., 2014). Applied research and practical applications include avian conservation, wildlife, fisheries, agricultural landscape, protected area and forest management, marine conservation planning, and water governance. Bennett et al. (2018) also give lists of the "classic conservation social sciences," the "applied conservation social sciences" and the "interdisciplinary conservation social sciences." The main conservation social sciences include environmental specializations within anthropology, economics, geography, conservation and history, political science and psychology, environmental sociology, and philosophy and ethics. Applied conservation social sciences include conservation and development studies, environmental and conservation law, environmental and conservation education, conservation marketing, and human dimensions of natural resource management. The interdisciplinary conservation social sciences include political ecology, science and technology studies, and environmental humanities (Bennett et al., 2018).

The justification of the social science approaches into conservation biogeography is justified on the grounds that the issues are not limited to the biological approach (Campbell, 2017). Adams (2014, p. 275) notes that "this much we know: most people active in conservation are trained as biologists, yet most of the problems of conservation are to do with people." Coghlan (2008) argues that "interdisciplinary teams are now commonplace in natural resource research centers. The evolution from pure biologists to a systems approach with ecologists and later on, social scientists and geographers reflects our understanding that no natural environment is exempt from human impacts; to protect the natural environment we must incorporate the human dimensions within the environment." Fox et al. (2006, p. 1817) add that "the ecology of threatened species and ecosystems is the foundation of conservation biology, but ecological knowledge is not sufficient for conservation success. Conservation actions are ultimately human behaviors, and it is vital to understand how social factors (e.g., markets, cultural beliefs and values, laws and policies, demographic change) shape human interactions with the environment and choices to exploit or conserve biodiversity."

Sandbrook et al. (2013, p. 1487) note that "there is now widespread agreement that social research is relevant to conservation. Extensive efforts have been made to improve communication between natural and social scientists interested in conservation. However, this communication is challenging personally and professionally." Additionally, "despite these efforts, we believe that the role and place of social research in conservation remains a major source of misunderstanding, miscommunication, and contention among conservation researchers. There are problems of method (e.g., use of both qualitative and quantitative methods in social research), of epistemology (e.g., positivist versus postpositivist, and problem solving versus critical approaches), of understanding (it takes time to become expert in any discipline), and of language (terminology and writing styles can make publications effectively incomprehensible, or at least deeply unattractive and difficult, for people trained in a different discipline)" (Sandbrook et al. 2013, p. 1487).

These issues are particularly relevant in the case of the relations between people and large carnivores in terms of shared forest and other natural habitats, attacks on people and associated animals, and the agricultural and non-agricultural requirements of people with such shared landscapes (Bennett et al., 2018). For example, Bombieri et al. (2023) note that "large carnivores have long fascinated human societies and have profound influences on ecosystems. However, their conservation represents one of the greatest challenges of our time, particularly where attacks on humans occur. Where human recreational and/or livelihood activities overlap with large carnivore ranges, conflicts can become particularly serious." This position is supported by Lorand et al. (2022) who argue that "human-wildlife conflicts are associated with a threat to large carnivores, as well as with economic and social costs, thus challenging conservation management around the world" and also note that "the literature on conflict management is often imprecise and lacks consistency between studies or situations, which generally makes comparisons difficult." The study of these dynamics, especially when considered in spatial perspective, forms the basis for the political, economic, and social/psychological foundations of conservation biogeography (Bennett et al., 2018; Bombieri et al., 2023; Jha & Campbell, 2021, 2022; Lorand et al., 2022).

Bennett et al. (2018, p. 93) emphasize the importance of social sciences in conservation and define conservation social sciences as a subset of traditional and applied social science fields with a focus on environmental management or conservation. Environmental anthropology, environmental psychology, environmental economics, environmental history, human-environmental geography, environmental sociology, and environmental governance are some examples of the traditional conservation social sciences, which are defined as those that emerged from specific disciplines. The applied conservation social sciences include many concepts from the social science areas, such as science and technology studies, conservation education, conservation and development, conservation communication studies and conservation law. There are also numerous interdisciplinary areas, including human ecology, political ecology, ethnoecology, ecological economics, social-ecological systems, and the human aspects of conservation. These include concepts from multiple areas of the natural and/or social sciences (Bennett et al., 2018, p. 93).

The political issues of power, social stratification, income inequality, land-use marginalization, and local participation, and the psychological issues of perceptions of the environment and components such as habitat type land cover and wildlife are major contemporary issues for conservation (Robbins, 2012). Concerning the issue of power, for example, Green (2016, p. 88) cites "the politics and power dynamics of natural resource management" and argues that the devolution of authority to the local level for natural resource management is the center of community-based natural resource management, and a wealth of literature exists to comprehend the ways in which power is transferred, the limitations imposed on that transfer, and the realities of management at the community level. Micropolitics at the local level, particularly in relation to power dynamics and the advantages of community-based natural resource management, as well as the socio-political-economic context of power transfer in community-based natural resource management, are noteworthy research areas that have emerged from this literature (Green, 2016, p. 88).

Concerning the issue of social inequality in land access, Ergas et al. (2021, p. 15) cite the importance of this field for current environmental studies, noting that environmental sociologists are using intersectional analyses more frequently, which is vital given the scope, depth, and duration of developments in intersectionality scholarship and its relevance to themes of environmental inequality. Recent empirical research has indicated links between ecological degradation and social inequality, highlighting the significance of approaching environmental issues from an intersectional perspective. Researchers have recently started to integrate the issues of environmental justice and gender-and-environment frameworks to create critically intersectional socioecological theory. It is argued that from an integrated perspective, these varied studies highlight the possibilities inherent in the study of the links between social position, privilege, and disadvantage interact with environmental access and exploitation, to produce disparate impacts on and perceptions of the natural world in society (Ergas et al., 2021, p. 15).

Concerning the issues of psychology and cooperation with the political issues of the environment, the first issue concerns the role of psychological research in environmental understanding (Campbell, 2017; Cranston & Khalil, 2022). According to Saunders (2003, p. 137), psychologists who specialize in human behavior have offered a variety of methods for understanding the attitudes, motives, beliefs, values, and behaviors related to conservation issues. The importance of psychology in promoting environmental protection and environmental sustainability has recently received greater attention. Psychology has addressed conservation issues from several research perspectives, but potential possibilities exist. Dominant discourses in psychology may focus on the human relationships rather than socio-environmental relations, hence the latter cluster of studies may be marginalized. Saunders (2003, p. 137) further argues that for the development of understandings of the experiences of human nature and what drives people to maintain these relationships, psychologists, and the research frameworks that they utilize may make a strong contribution. There is insufficient cooperation between psychology and other disciplines on this topic of human dimensions of environmental and conservation issues, partly due to the disciplinary boundaries encountered, that may impede multidisciplinary research attempts. Therefore, psychologists must cooperate with other disciplines that approach conservation issues and apply its concepts and techniques critically (Saunders, 2003, p. 137).

The rest of this chapter will examine these concepts and applications within the field of conservation biogeography. The concepts, development, and applications of political ecology and conservation/environmental psychology are examined, as is their role in the understanding of the issues of conservation biogeography. These issues are then applied to the shared relationships between people and wildlife, taking the example of big cats (lions, tigers, leopards, jaguars, snow leopards) and cougars, the justification for this selection being the extreme tensions unique to human-big cat relations, including fear and danger for people, companion animals and livestock; contested habitats, which cannot be shared due proximity dangers; the charismatic status of these felines, their rarity and substantial habitat requirements. The central argument is that political ecology plays an important role, as such problems may only be solved by human participation regardless of power, class, and habitat, while ameliorating the issue of nonparticipation that may bedevil solutions to conservation issues at international, national, regional, and local levels. The inclusion of the psychological analysis is also important, and human viewpoints, emotions, and resultant perspectives, at individual and group levels, and across power, economic, social, and cultural groups are vital for the successful participation and sustainability. This is especially the case for big cat conservation, as local proximity to perceived and real dangers, and marginalization of less powerful people into habitats frequented by big cats may increase the problems inherent in big cat conservation.

Political Ecology

The term "political ecology" was formulated by Thone (1935) and Wolf (1972). It is defined as a subdiscipline at the intersection of applied ecology and social sciences, with a primary focus on power relations, social group or structural differences, resource access, and the sociopolitical or socioeconomic impacts of access to environmental structures (Robbins, 2012). Definitions of political ecology vary in terms of emphases on geography, politics, power relations, anthropology, culture, and applications (Baer, 1996; Minch, 2011). Some of these definitions are listed in Table 1.1. A central theme is its breadth: Political ecology is an "overarching framework of inquiry for exploring the politics of access and use of natural resources at multiple scales over time" (Wilshusen, 2003, p. 41). This complex and chaotic nature is also discussed in some definitions, as it is "far from a coherent theory of the complex metabolism between nature and society" (Watts, 2009, p. 545). This chaotic nature is partly due to the great complexity of the field of study; political ecology "seeks to understand the complex relationships between nature and society through a careful analysis of what might be called the forms of access and control over resources and their impacts on environmental health and sustainable livelihoods" (Watts, 2000, p. 257).

These definitions point to the important role of politics and power, which affect and may define access to environmental resources, with social and environmental consequences. Some definitions also emphasize the conflicted nature of political relations. Bryant (1998, p. 79), for example, defines political ecology as the study of "political dynamics surrounding material and discursive struggles over the environment in the Third World." This definition is significant because it recognizes the importance of two main areas of environmental study: discursive and material struggles, which may be post-structural, which focuses on knowledge and ideas, and structural, which focuses on the interrelationship between ecology and political economy"(also described by Blaikie & Brookfield, 1987, p. 17; Blaikie, 1999; Watts, 2000). This has to do with the disciplinary focus that has become pluralized in various political ecologies alongside the study of numerous interdisciplinary relationships (Watts, 2009,

Table 1.1 Definitions of political ecology

Author	Definitions of political ecology
Neumann (2009)	"An interdisciplinary field that analyzed environmental problems using the concepts and methods of political economy… ecological change cannot be understood without consideration of the political and economic structures and institutions within which it is embedded. The nature–society dialectic is the fundamental focus of analysis." also, post-structural social theory, nonequilibrium ecology, urban political ecology, environmental security theory, ethics, environment and identity
Benjaminsen and Svarstad (2019)	"A field within environmental studies focusing on power relations as well as the coproduction of nature and society. Theoretical inspirations are taken from different sources such as political economy, poststructuralism, and peasant studies." the "place and role of "ecology" within political ecology remains an on-going discussion." focus on natural science, or "more social science-based theoretical debates where "ecology" refers to the environment more broadly"
McMahan and Nichter (2011)	"Applies a system's perspective to understanding the ecological ramifications and sequelae of political economic inspired practices and policy on both macro- and microenvironments. It looks upstream at the actors and actions responsible for setting environmental changes in motion, and downstream at those subject to such changes." Integrates local and global environments, and natural and cultural, and "the ecological and the sociopolitical as coextensive"
de Micheaux and Jenia (2021)	"A field that critically interrogates the nature–society relations" especially the power relations that "intersect and affect access to natural resources, in order to reveal disparities and injustices in the distribution of costs and benefits" and land degradation and development/environmental issues, political economy and ecology. It "borrows its methods" from "geography, anthropology, political economy, political sciences, social sciences, history, or environmental sciences"
Shrestha and Ojha (2017)	"It is an outgrowth of ecological and social science that combines social and political investigation with environmental processes. It combines the concerns of ecology with a broadly defined political economy. Together this encompasses the constantly shifting dialectic between society and land-based resources, and also within classes and groups within society itself." Topics include discursive socioenvironmental relations between humans and decentralization of power
Minch (2011)	"The study of the intersection and relationship between the political, broadly understood, and environmental and ecological phenomena. Political, economic, social, and cultural forces affect, and are affected by, ecological and environmental trends…there exists no agreed-upon definition or conception of the field" … it is interdisciplinary, concerning "political economy, cultural ecology, social ecology, green socialism, environmental sociology, development ecology, anthropology
Roberts (2023)	"A critical research field within anthropology, geography and related disciplines"… "that examines how and why economic structures and power relations drive environmental change in an increasingly interconnected world." initially studied "the practices and impacts of large-scale resource development projects in subsistence-oriented communities in the global south" … "now including environmental politics and socio-ecological degradation in urbanized settings

Sources Neumann (2009), McMahan and Nichter (2011), Minch (2011), Shrestha and Ojha (2017), Benjaminsen and Svarstad (2019), de Micheaux et al. (2021), Roberts (2023)

p. 545). Science studies (Demeritt, 1998), Marxism (Leff, 1995; O'Connor, 1998), environmental history (Grove, 1995), actor-network theory (Braun & Castree, 1998), gender theory (Agrawal, 1998), and discourse analysis (Escobar, 1995) are some of them. Political ecology is currently considered the leading paradigm in the study of the relationships between nature and society and has its roots in political economy, cultural ecology, geography, and sociology. The basic idea behind this approach is that all changes at the human/environment interface involve factors and influences that span time, national borders, and political, economic, and social contexts.

Political ecology is applicable to the intersections of socioeconomic marginalization; political conflicts; environmental and social movements; and environmental degradation and conservation (Robbins, 2004). Having developed first as a structuralist and then as a more flexible paradigmatic framework, political ecology has enabled discourses neglected by its predecessor, cultural ecology. These discourses recognized the broader political and economic forces affecting local human-environment interactions (Robbins, 2012). Political ecology went further than cultural ecology by examining the sociopolitical issues that determine, influence, enable, and constrain behavior at the socio-ecological interface, while cultural ecology focused on the direct relationship between culture and ecology (Walker, 2005).

Bryant and Bailey (1997) note that there are three key assumptions that increase the relevance of political ecology. Firstly, due to socioeconomic and political differences and inequalities, there is an unequal distribution of the costs and benefits associated with environmental change, which requires an examination of these socioeconomic and political differences and inequalities. Secondly, conversely, existing inequalities and differences are increased or reduced by the unequal distribution of costs and benefits; therefore, environmental changes affect the political and economic system. Thirdly, changing power relations result as external effects both from the unequal distribution of costs and benefits and from the associated change in already existing inequalities. Political ecology can provide guidance in analyzing and resolving these problems: essentially, a "normative understanding that there are most likely better, less coercive, less exploitative, and more sustainable ways of doing things"; potentially more effective environmental policies; more competent environmental decision-making; and understanding different socio-economic and political structures and inequalities (Robbins, 2004, p. 12).

Political ecology has increasing links with social theory (Pomeroy, 2012). As a complement, social theory has contributed to the development of a more socially sensitive political ecology with a deeper view of actor orientation (Walker, 2005, 2006). The overly structuralist character of early political ecology was similar to structuralist views within sociology (Perry, 2003; Ritzer, 2008) and anthropology (Wolf, 1997). Furthermore, while there has generally been more attention to actor-level details in cultural ecology, there has also been a perceived neglect of interactions between socioeconomic and environmental variables (Perry, 2003). The most common applications of studies referred to as political ecology include: the political, economic, and social relationships behind sociopolitical movements; scaled resource utilization policy; conservation of natural resources; marginalization of power groups; consumption and production policy; environmental conflicts; identities and claims; externalities of government incompetence; and property access and rights (McCarthy, 2002, 2005, 2006; Pomeroy, 2012; Robbins, 2004; Walker, 2005, 2006).

Proponents of political ecology primarily sought to reevaluate the methods of inquiry within the field of geographical perspectives on conservation. They sought to "conceptualize and articulate the relationships between structural processes and local contexts, and clearly establish which methods and data may be best used to get at which aspects of these relationships" (Elwood, 2010, p. 104). Some have argued that, in order to avoid structuralist bias, a more actor-oriented focus should be included in political ecology research on socio-environmental, participation, and policy issues related to conservation (Dove & Carpenter, 2008; Robbins, 2012). Thus, structuration theory serves as a paradigm that is complementary (Pomeroy, 2012; Zimmerer, 1996). Within the framework of what is known as a duality, structuration theory encompasses both structures and actors (Craib, 1992). According to Giddens (1979, 1984), Craib (1992), Marshall (1994), a structure is any recurrent pattern of social behavior, including kinship, religious, political, and economic institutions, as well as norms, values, and social roles. The primary sociological theoretical framework prior to the late 1960s, known as structural functionalism, defined structures as having various functions that increase in quantity and complexity in tandem with the growth and development of society (Craib, 1992). In contrast, an actor is a unique individual within society, and an actor-oriented approach looks at individual experiences from everyday life.

Through the integration of an actor-oriented approach and a structuralist framework, structuration theory asserts that structures and agency are one and the same (a duality) instead of two distinct entities (a dualism). Actors are influenced, constrained, and enabled by the social structures they and other actors develop. Actors also create, enable, reinforce, maintain, and constrain these social structures. The purpose of structuration is to integrate and transcend the structure/action polarity, as stated by Benton and Redclift (1994, p. 19) and Redclift and Woodgate (1993). Furthermore, the environment may be assessed as a structure or even a social construction, that both permits and restricts action and can be altered by social action, and hence lends itself easily to the application of the structuration theory (Benton & Redclift, 1994, p. 19). A few examples of the structures that actors in agricultural development create and are influenced by are the household, socio-cultural norms, the state-farmer relationship, and patterns of livelihood activity (Hakeem & Vemuri, 1997; Woodgate & Redclift, 1998). These relationships are important, as are the more general political, economic, and social concerns of decision-makers, international trade blocs, global economic systems, and related environmental change.

Current environmental problems that can be assessed using the political ecology approach include environmental degradation and declines in ecosystem productivity and wildlife numbers and habitats (Miller Jr & Hackett, 2011). The political, economic, and social connections with the environment mostly originate in countries with a significantly different perception of the environment than those in which the actual exploitation took place. Therefore, the dominant "core" nations of the

North (in North America, Western Europe, and Japan) controlled a global system of resource exploitation that was largely dependent on the environment of the nations of the South. The rapidly growing population of the countries of the South has exacerbated the resulting environmental problems. Events of recent decades with current relevance include the development of the United Nations' first development decade in the 1960s and 1970s, in which development was defined as the promotion of economic growth, measurable in terms of per capita income and consumption, gross domestic product, and of cultivated land, to the neglect of social issues (such as poverty, inequality, unemployment and gender disparities) and environmental issues (such as deforestation, pollution, biodiversity loss and soil erosion) have been marginalized or ignored. This method was considered a failure because it did not bring about real development in the global south. The second development decade evolved into a shift in thinking in the 1970s, with an emphasis on reducing or preventing poverty, inequality, and unemployment, and greater attention to the environment (Hope, 1984). This approach recognized that agricultural development policies and the exploitation of resources for economic growth were the main factors in the pronounced environmental degradation (e.g., famine and desertification, mainly due to deforestation and soil erosion) that dominated development issues in the 1990s (Elliot, 1994).

Social activism also emerged, referred to as environmentalism, which suggests that human activities are the major causal and contributing factors to environmental degradation. Terms such as deep ecology, political ecology, radical ecology, ecologism, and ecofeminism referred to a new politically and ecologically based environmental awareness. The concepts discussed also included concepts of intra- and intergenerational justice. Further developments included the Stockholm Conference of 1972 and the subsequent establishment of numerous environmental departments, agencies and/or committees and non-governmental organizations in various countries. Sustainable development was first set out in the World Conservation Strategy, which came into vogue after the World Commission on Environment and Development (the "Brundtland Commission 1987") expressed it as development that meets the needs of current generations without compromising the needs of future generations. Conflicts arose between the goals of these organizations and those of policymakers committed to industrial development. Disciplinary contributions to the idea of sustainable development, included those of economists focused on natural capital stocks, ecologists concerned with biodiversity conservation, and political "green" activists who advocated limited or zero growth to alleviate environmental degradation. Broader perspectives emerged on issues such as agricultural development, requiring examination of historical, political, and economic aspects (Hajian & Kashani, 2021).

Conservation Psychology

The main connection between political ecology and psychology concerns the fields of general psychology and a range of environmental issues, and more specifically conservation psychology and environmental psychology (Elmhirst, 2011; González-Hidalgo & Zografos, 2020; Uzzell & Moser, 2009). Conservation psychology, an important, emerging, and evolving branch of environmental psychology, is at the forefront of psychological research due to the current importance of ecological change and the critical role of public attitudes toward conservation policy and management. This is a comparatively new field with numerous definitions. The focus is on people's attitudes and actions toward the environment and on answering questions about this connection. It exists largely in the older field of environmental psychology, an interdisciplinary field that studies the human mind and its perception of its environment (Adams, 2014; Asah & Blahna, 2013; Campbell, 2018; Fielding et al., 2014; Osbaldiston, 2013; Pearson, 2013; Pearson et al., 2014; Plesa, 2019). Recent massive advances in social media technology and reach through platforms such as Facebook, Instagram, Twitter, YouTube, and various variants of email have enabled a new global debate and discussion about the human dimensions of wildlife conservation, with potential implications for the public, and the perception and application of nature conservation policy. Large carnivores, a particularly fearsome and charismatic group, feature prominently in this complex (Bennett et al., 2018; Bombieri et al., 2023; Lorand et al., 2022).

The term "environmental psychology" refers to a broad category of psychological disciplines concerned with the interaction between the environment and human behavior (De Young, 2013). "An interdisciplinary and international field that views people and their physical environment as interdependent," is how Nasar (2015, p. 384) defines environmental psychology. To study these human-environment relationships, social science methods are used, and the importance of a multi-level, multidisciplinary, social-ecological approach to such problems is recognized (González-Hidalgo & Zografos, 2020). The connections between the environment (at different scales, from a room to a city) and the range of human responses addressed in this field are among the many facets of environmental psychology, as Nasar (2015, p. 384) further notes. These connections include "environmental perception and cognition; environmental attitudes and evaluations; environmental stress, noise and crowding; Responses to disasters, environments, personal space, territoriality and privacy; crime and fear of crime; behavior changes; home, neighborhood, work and educational environments, and facility planning and evaluation."

As Uzzell and Moser (2009, p. 307) point out, environmental psychology can be distinguished from other subfields or topics of psychology by emphasizing context. The underlying premise is that every physical environment acquires social meaning through culture and, similarly, every social environment is affected in some way by its physical context (Uzzell & Moser, 2009, p. 307). Environmental psychology focuses on environmental context, and environmental context also needs to convince proponents of other psychological subfields that environmental context is an integral part of the narrative and not just a neutral background for human action. A mental model is an oversimplified version of reality that people store in their minds and use to make sense of things, to plan, to evaluate possibilities, in short, to function throughout the day (De Young, 2013, p. 21). Formal education, direct and indirect experience, and "moving around a space or concept to learn more about it" are some of the ways environmental psychology can use mental models in this endeavor (De Young, 2013,p. 21). Some concepts that emerge from these ideas are place identities and attachments (based on areas that can foster mental attachments due to memorable events, perceptions, and/or ideas), personal space, and territory (the area perceived by the human mind as pleasant and unwelcoming is delimited) and environmental awareness (the response of the human mind and physiology to environmental stimuli) (Bell et al., 2001; Cantrill, 1998; De Young, 2013; Proshansky et al., 1983; Smaldone, 2007).

Several studies that seek to support the validity of this approach justify it. For example, Kaplan and Kaplan (2009, p. 329) point out that human perception, behavior, and well-being are broadly significantly influenced by the environment. If this is true, environmental psychology has much greater potential to benefit humanity than is usually realized. According to this assessment, an approach that views the environment as fulfilling human information needs and potentially facilitating human development—supported by strong connections to other areas of psychology—could be the key to the success of environmental psychology (Kaplan & Kaplan, 2009, p. 329). As these authors explore new avenues of relevance to human development, they build the "reasonable person model" as an example or perspective that leverages current knowledge about human-environment connections (Kaplan & Kaplan, 2009, p. 329).

Conservation psychology is a subfield of environmental psychology, which is a broader field that is continually changing and redefining its boundaries (Saunders et al., 2006). According to Saunders (2003), conservation psychology is the study of applying conservation theory to the broader topic of connections between human development and environment and mind. In particular:

> Conservation psychology is the scientific study of the reciprocal relationships between humans and the rest of nature, with a particular focus on how to encourage conservation of the natural world. Conservation psychology is an applied field that uses psychological principles, theories, or methods to understand and solve issues related to human aspects of conservation. It has a strong mission focus in that it is motivated by the need to encourage people to care about and take care of the natural world. As well as a field of study, conservation psychology is also the actual network of researchers and practitioners who work together to understand and promote a sustainable and harmonious relationship between people and the natural environment (Saunders, 2003, p. 137).

The underlying theory of this subdiscipline is that it is more than just a subset of psychology; it has the capacity to broaden its scope and foster the exchange of ideas and contributions among scholars of other pertinent fields, all with the goal of improving our understanding of how environment and human behavior interact. Because of this, conservation psychology primarily belongs in the field of environmental psychology, though it also extends beyond it by incorporating information from other academic fields to broaden its core and related paradigms (De Young, 2013; Saunders, 2003).

The advancement of conservation psychology, influenced by academics from a wide range of fields, occasionally calls for an ambitious goal: the promotion of conservation policy at different levels (Clayton & Brook, 2005). Nonetheless, some writers contend that by modeling the "social psychology of conservation behavior, which focuses on situational context, existing schemas, and personal motives," it is possible to better understand a field that is "largely unfamiliar to many psychologists and to those working in the environmental field" (Clayton & Brook, 2005, p. 87). Several related studies have adopted a similar approach (Abell, 2012; Adams, 2014; Asah & Blahna, 2013; Fielding et al., 2014; Fischer & Young, 2007; Johnson et al., 2009; Koger & Scott, 2007; Osbaldiston, 2013; Pearson, 2013; Pearson et al., 2014; Plesa, 2019; Uzzell & Räthzel, 2009). Saunders (2003, p. 137) made the case for the development of conservation psychology by citing a natural evolution from academic, policy, and social events, noting that the increasing understanding that the social sciences are essential to conservation necessitates more effective means of collaboration toward a shared goal. To strengthen ties between research and practice, between psychology and other social sciences, and between the natural and social sciences, a new field of conservation psychology is required. Directly oriented psychological research into environmental sustainability is the aim of this kind of network (Saunders, 2003, p. 137).

In addition, Saunders (2003) contends that comprehending conservation psychology necessitates evaluating its connections to disciplines like environmental psychology and conservation biology (Saunders, 2003, p. 137). Severe environmental issues are becoming more prevalent globally. These issues provide the social sciences—psychology specifically—with new opportunities as well as challenges. The discipline's research topics include those that inspire concern for nature and human role in it, as well as those that drive individuals to act in a more environmentally friendly manner. Individual and group analyses of these themes are conducted (Saunders, 2003, p. 137). Psychology's methods for comprehending the thoughts, attitudes, motives, beliefs, values, and behaviors connected to conservation issues are among its advantages as a field studying conservation (Saunders, 2003, p. 137). As a result of these problems being acknowledged, its role in advancing environmental sustainability and protection has grown. Saunders (2003) also lists works on human behavior connected to conservation issues that are helpful for learning the fundamentals of conservation psychology (Bechtel & Churchman, 2002; Cvetkovich & Werner, 1994; Gardner & Stern, 2002; Nickerson, 2003; Winter & Koger, 2003) and those who link environmental issues and psychology (e.g., Oskamp, 2000; Zelezny & Schultz, 2000). According to Saunders (2003, p. 137), one significant limitation is the inclination of psychologists to focus on social relations as opposed to the socio-environmental relationships that shape social relations.

Conservation psychology has developed in response to shifting interfaces between individuals, social groups, and nations as well as theoretical academic questions. The modern social media that is based on the internet and other online communication systems has gradually replaced the earlier "mail" and non-digital media (letters, telegrams, radio, and television) as a means of improving social contacts between individuals and groups. Academic disciplines have new opportunities within this faster and far more extensive communication system in terms of research reach, results dissemination, and application insights (Cranston & Khalil, 2022). As a result, the public is equipped with a far wider range of information sources and presentations from which to make decisions and win over the public on matters pertaining to research findings and their practical applications. It is in this context that the evolution of conservation psychology needs to be assessed.

The development of conservation psychology is linked not only to theoretical academic issues, but to the changing interfaces between people as individuals, social groups, and nations. The gradual improvement in social contact between people and groups, from the former "snail mail" and non-digital media (letters and telegrams, radio and television) has slowly given way to the modern social media based on internet-based online communication systems. Within this faster and far more voluminous communication system, academic subjects have new possibilities in terms of research reach, dissemination of results, and application findings (Bergman et al., 2022). The human mind has thus acquired far wider information bases to inform decision-making and to assemble public support for issues emanating from research findings and their application. The evolution of conservation psychology must therefore be evaluated in this regard (American Press Institute, 2015; Newman, 2009; Norton, 2017; Perloff, 2008; Yong, 2018).

Accordingly, social media refers to web-based services that make it easy for people, groups, and organizations to share and engage with each other by allowing them to co-create, modify, and create user-generated content (Bergman et al., 2022; McCay-Peet & Quan-Haase, 2017). Due to the vast scope of this technology, social media has a growing impact on people's thoughts and on issues related to the relationship between humans and their environment. These relationships and interactions, including human ways of thinking and interactions between individuals and their surroundings, are relevant and conditional (McKendree et al., 2014; Moss, 2019; Roth, 2019). Because social media has influenced human perspectives to such an extent that it may be impossible to separate their influence from mental views, it is becoming problematic to conduct psychological research without acknowledging it (Kaplan & Kaplan, 2009; Uzzell & Moser, 2009).

Applied to environmental and conservation issues, these developments are increasingly relevant. As noted by Toivonen et al. (2019, 298), through the use of cutting-edge data sources like social media and other sizable datasets, the field of conservation science known as "digital conservation" aims to comprehend and possibly alleviate the biodiversity crisis. It may be possible to monitor biodiversity more affordably by using user-generated big data. Significantly, they facilitate the examination of human-nature interactions at a never-before-seen magnitude. Social media is a valuable resource for researching people's outdoor activities and for comprehending online debates or discussions about conservation, among other big data sources (Bergman et al., 2022).

The application of social media studies to the study of charismatic animals like large carnivores has supported the development of conservation psychology, which is undoubtedly an exciting and highly relevant field of study (Bergman et al., 2022; Lorand et al., 2022). Insufficient research has been conducted in this field; however, this could be partially attributed to the fact that online social media is still a relatively new area of interaction. Its future is dependent on technological advancements, expanded audience reach, improved social management, and user education to improve information accuracy. Three main areas of research appear to be supported by the evidence. Social media users should be interacted with and respected for their knowledge, political inclinations, access to information, and technological advancements. Technology's usability, privacy access, post security, competition, and demographic reach are further issues. People in developed nations, for instance, have far greater access to social media platforms than people in developing nations, despite the fact that many conservation issues are either worldwide in scope or present in developing nations (Campbell, 2019). Another problem concerns the information generated; In particular, is this information sufficiently detailed, reliable, and measurable information available on these social media platforms to support conservation psychology research? This is data and information that can be used to learn more about people's thought processes, cultural influences and patterns, and changing viewpoints on environmental issues (Bergman et al., 2022; Lorand et al., 2022).

Social media has developed new methods for the study and practice of the relationships between nature and society (Arts et al., 2015; Bergman et al., 2022; Di Minin et al., 2015; Ladle et al., 2016). Social media posts can provide information about visitation patterns and activities in protected areas (Toivonen et al., 2019). The result of this shift is a much larger communications audience, reaching billions of contact addresses (blog audiences, contacts, friends, chat buddies, etc.) who, when connected to Internet search engines, have near-universal access to the relevant information, which may drive forces and even contribute to important outcomes (Campbell, 2019). CAISE Admin (2016) notes that the way an animal and conservation issues are portrayed in the media can influence an individual's views, although television stories about animals with sometimes contradictory information can create familiarity among the public, which in turn creates more positive feelings. In addition, children may have an inaccurate idea of how common certain behaviors (e.g., hunting) occur because of the emphasis these behaviors are in films and television, and they may view animals on television or in films as having them "seen" in their real lives (CAISE Admin, 2016). Several researchers working on both older, more traditional media and modern online media technologies are cited to support the thesis that media representation is one of the most important factors in psychological perspectives on wildlife (Bouse, 2003; Ross et al., 2008, 2011; Schroepfer et al., 2011; Serpell, 2004; Snaddon et al., 2008).

Another issue concerns the benefits of social media for environmental protection. Negative impacts exist. For example, Puentes (2021) argues that "it appears as though the inclination for people to take to social media to capture and address environmental problems isn't going away anytime soon. Social media undoubtedly presents a unique set of dangers, from misinformation to mental health impacts, that can undermine environmental protection efforts and degrade trust." Puentes (2021) also points out that the increasing use of social media and the amount of information consumed by users has resulted in a reduction of the global attention span for news and trends. This is particularly problematic for environmental issues, because images and videos of climate disasters and environmental destruction are so easily accessible that they can seem normal, which can lead to a feeling of fatalism among audiences. This may reduce the influence of such media on stakeholders' determination to take action to prevent recurring incidents (Puentes, 2021).

There may be two sides to public awareness-raising, according to available data. Kidd et al. (2018, p. 106) found that unpopular and uncharismatic species receive less conservation support, which may have implications for their long-term survival. As for the connection to nature, it is about the extent to which social media can draw attention to nature, leading to an increase in tourism (both excessive visits and disturbance to wildlife and environmental degradation, as well as spending on visits that finances many projects) and more scientific studies (some spurred by public support after watching videos on YouTube or images on Flickr) and an overall increase in public sympathy for the environment (Campbell, 2019). For natural disasters, online resources such as Twitter and Facebook allow for rapid dissemination of details about wildlife suffering from events such as hurricanes and volcanoes, enabling rapid funding, information, and awareness. This can help raise funds through online payment methods for wildlife organizations (Bergman et al., 2022; Campbell, 2019; Lorand et al., 2022).

The type and utility of data and information that can be presented on social media are also important. The decisive factors are the sources, the form of presentation and the user-friendliness of the data in social media. As Toivonen et al. (2019, p. 298), collecting data can be a major problem due to temporal, spatial and resource issues and requires more efficient methods. One answer has been the development of big data, described as important because vast amounts of available digital information offer new research opportunities in various areas of science in virtual networks (Toivonen et al., 2019, p. 298; see also Castells, 2010; Boyd & Crawford, 2012; Waldron et al., 2013; Ruths & Pfeffer, 2014). Information shared via social media can include text and visual (photo and video) data (Tenkanen et al., 2017).

Another issue is the development of geotagged data and information. Geotagging is the process of adding geographic identification metadata to social media, including photos, videos, or websites. This can be in the form of geographical coordinates (latitude and longitude, UTM) and/or altitude, distance, and place names (Sui & Goodchild, 2011). Older studies may use manual content and/or rudimentary spatial analysis. Recent studies have attempted to automatically analyze social media content to track illegal wildlife trade (Di Minin et al., 2018; Di Minin et al., 2019) and human attitudes toward wildlife (Becken et al., 2017). Social media data can also be analyzed using various methods relevant to research and development. For example, application programming interfaces (APIs) enable interactions with social media platforms, allowing third parties to interact with the platform in research mode (Toivonen et al., 2019).

Political Ecology, Psychology, and Large Carnivores

Large carnivores and humans have always had a complicated relationship (Lozano et al., 2019; Fernández-Sepúlveda & Martín, 2022, p. 1911). They impact on human activities because they may endanger farm animals and even human lives. Figures 1.1, 1.2, 1.3, 1.4, 1.5, and 1.6 show the main domestic animals killed by the big cats. Although large carnivores are admired and even revered for their strength, beauty, and dominant predatory behavior, they are also feared and persecuted, and over the past 200 years their numbers have declined (Bruskotter et al., 2017; Crooks, 2002; Fernández-Sepúlveda & Martín, 2022; Linnell et al., 2001; Ripple et al., 2014). Numerous carnivore species are threatened with extinction or may soon become extinct (Gittleman et al., 2001). Fernández-Sepúlveda & Martín (2022: p. 1911) wrote that "carnivores are more threatened than mammals in general (26.9% of endangered species vs. 22.7%) and have a significantly higher proportion of species with declining populations (48.3% vs. 31.9%)." The animal groups with the lowest conservation status include big cats and bears, making them more vulnerable to human threats than other animal species, which has led to significant global population declines in many of these species (Bergman et al., 2022; Lorand et al., 2022). The main threats to the carnivores are land cover change under human action, consequent habitat losses through deforestation and agricultural expansion, and hunting and trapping of carnivores and prey species. Carnivore conservation should be a major concern for mammals, and initiatives directed at this group should be undertaken or increased as a matter of urgency (Fernández-Sepúlveda & Martín, 2022, p. 1911).

Human relations with such carnivores are generally regionally uneven. North American large carnivores, namely brown and black bears, pumas and jaguars, are generally less dangerous to humans than large carnivores in Africa and Asia, which include lions, tigers, and leopards, based on the number of fatal attacks. This is attributed to the gentler behavior of the former species and less contact with humans on the less densely populated North American continent (Bombieri et al., 2023). Human responses to such large carnivores can include conservation, tolerance, and hostility (Balme et al., 2010). Inskip et al. (2016, p. 1) point out that tolerance of wildlife is a central component of conservation strategies for many endangered species and an ongoing global challenge to effective conservation and that intolerance includes attitudes toward a species and/or conservation judgements (Inskip et al., 2016).

Both political ecology and conservation psychology attempt to reach a larger audience, concerning large carnivores, as shown in the figures. Political ecology emphasizes the importance of assessing the majority of poorer, less empowered, generally rural actors, who experience at first hand the impacts and dynamics of conservation policies designed in some cases by elite policymakers out of touch with the local context (Bombieri et al., 2023). Conservation psychology, in some cases through the vast social media networks seeks to reach a large audience, possibly comprised of more affluent people with technological access, who nevertheless may

Fig. 1.1 Indian cattle are frequently killed by tigers, and sometimes by leopards. *Source* Zaneta Misutova-Pexels

Fig. 1.2 African cattle are frequently killed by lions. *Source* Abubakar Balogun-Unsplash

Fig. 1.3 Both jaguars and cougars are capable of killing cattle, but jaguars are generally blamed for killing larger cattle. *Source* Michael O'Neal Campbell, Central America

Fig. 1.4 Herd of cattle in Central America. Cougars are generally considered to take smaller cattle and other livestock. *Source* Michael O'Neal Campbell, Central America

Fig. 1.5 Small African livestock, such as these in Arusha, Tanzania are frequently the prey of leopards. *Source* Kureng Workx-Pexels

Fig. 1.6 Domestic Yak are killed by snow leopards. *Source* Suman Thapa-Pexels

through media outlets experience environmental conservation issues at first hand, and through the political power of voting and consumer pressure impact conservation policy (Campbell, 2019). Both approaches therefore emphasize the power of large pressure groups, the necessity of bottom-up approaches, the sharing of information at all levels of human societies, and the need for consultation between decision-making elites and less empowered people (Bauer et al., 2021; Gebo et al., 2022).

Large carnivore conservation is perhaps the most profound test of the tools of political ecology and conservation psychology (Ripple et al., 2014). Carnivores are considered more threatening than herbivores due to their ability to kill humans and domestic animals, which can lead to their extinction (Kansky et al., 2014). Commentators in political ecology and conservation psychology emphasize the need to examine human attitudes toward carnivores and to shift the goals of carnivore management from those based on fear and narrow economic interests to those based on a better understanding of ecosystem function and adaptive management based (Treves & Karanth, 2003, p. 1491). It is recognized that the conflict between conservation and security concerns is embedded in discussions of conservation policy and management (Campbell, 2012; Campbell & Lancaster, 2010; Holthe & Baldus, 2013; Kellert, 1994; Schaltegger & Beständig, 2012).

For political ecology, a major concern is the stronger impacts of conservation policies on local disempowered actors, who experience more acutely the conflicts with large carnivores that the policy elites may seek to protect. For example, Margulies and Karanth (2018) look at human-wildlife conflict, cited as "a political animal geography of encounter." The dynamic in the case study of the surroundings of Bandipur National Park in Karnataka, India, was of how changes in a local rural economy have created new areas where farmers, their animals, and carnivores interact and state discourses about conflict between humans and wildlife in addition to rural agricultural communities' viewpoints regarding shifts in these relationships. The authors argue that "Political ecology, as a shared field of practice engaged in studying the social and political dimensions of human relations within their lived environments, is a well-suited lens for unpacking this discourse of crisis" (Margulies & Karanth, 2018). The methods of political ecology are applied to address political issues, closely considering the interactions of people and animals in disputed areas. The impact of carnivore predation on community livelihoods in a high-conservation agricultural landscape may be relevant in areas where animal and population densities are comparable and protected areas are nearby (see also Bombieri et al., 2023).

In addition, Margulies & Karanth et al. (2018) demonstrate that the narrative advanced by Indian Forest Ministry actors about declining tolerance for coexistence with large carnivores obscures both the creation of new geographies encounters with wild animals—the spaces in which species meet—and the fractures in a regional agrarian economy. The study demonstrated how state narratives about conflicts between humans and wildlife obscure deeper, fundamental shifts in agricultural villages' livelihood strategies in response to park management and regional economic shifts, and how these shifts are intrinsically linked to the creation of new human-wildlife encounter geographies. The findings of the study indicated that a shift in the livestock industry's economic structure and the adoption of more assertive protected area management techniques have resulted in a decreasing tolerance for the harm and death caused by carnivores to cattle, and demonstrates possibilities for the role of political ecology in the awareness of animals as political actors and social justice (Margulies & Karanth, 2018).

In the context of conservation psychology, issues pertaining to large carnivore conservation include reaching out to a wider audience—particularly younger, social media-savvy individuals—in order to raise awareness of the problems (Bauer et al., 2021). Because there are more people involved in environmental issues through social media, the younger generation may be more likely to engage with them and the results could be more significant (Fry et al., 2018; Thomas, 2016). Younger people may be more activist and knowledgeable about these issues than older people, with the former holding stronger views on the need to protect wildlife and, in certain cases, reintroducing potentially dangerous species like wolves (*Canis lupus*, Linnaeus, 1758), cougars (*Puma concolor*, Linnaeus, 1771), jaguars (*Panthera onca*, Linnaeus, 1758), and tigers (*Panthera tigris*, Linnaeus, 1758).

Large carnivores such as tigers, lions (Panthera leo, Linnaeus, 1758), leopards (*Panthera pardus*, Linnaeus, 1758), cougars, bears (family *Ursidae*, G. Fischer de Waldheim, 1817) and wolves are particularly targeted due to their large size, charisma, character (real or imagined), beauty, mystery, and visibility (Bombieri et al., 2023). Large carnivores appear as cartoons (e.g., Yogi Bear, the Lion King), central film characters, storylines (e.g., Shere Khan and Bagheera in the Jungle Book), advertising stars (e.g., the ESSO tiger), and news reports (e.g., Cecil the Lion). These visualizations can influence the way people perceive such animals, potentially creating contradictory images based entirely on reality or on constructed characterizations that can replace the more realistic depictions from media outlets such as National Geographic, Time, and the British Broadcasting Corporation. The result, in many cases, has been the development of sympathetic media forces that attack violations of "animal rights" (Campbell, 2017).

A liberal, socially constructed narrative has emerged based on such visualizations of large carnivores (Campbell, 2019). Based on some social media users' opinions that large carnivores are ornamental or enhance human quality of life, Daly (2019) highlights some points pertinent to human conservation cognition of large carnivores. Events such as people taking selfies in front of wild animal enclosures and people approaching potentially lethal animals in zoos and parks are based on these perceptions. "What is the psychology behind the impulse?" (Daly, 2019). These contentious actions might support the myth that these big predators aren't as dangerous as people think (Daly, 2019). Susan Clayton, a psychology professor, is cited by Daly in support of her claim that because people no longer associate animals with danger, they have lost some of their implicit reality. Social media, as the main face of conservation psychology, is blamed for these developments. Channels, like YouTube, Dailymotion, Facebook, may promote exciting animal encounters and provide instant gratification (psychologically, such posts can provide a temporary ego boost that can be compulsive or addictive). A psychological argument is that children's education and the zoo environment can encourage the perception of wild animals as toys, which is currently reinforced by social media pressures (Daly, 2019).

A counter narrative is generally enabled and supported by the local people who participate in activities near large carnivore habitation, and are much more likely to encounter their predation on people and associated animals (Bauer et al., 2021). Meena et al. (2014) comment that the viability of the large predator population is further threatened by interactions with humans in the border areas surrounding protected areas. Concerns about human safety and livestock predation make large carnivores vulnerable to retaliation. Therefore, conflicts between humans and carnivores along the boundaries of protected areas represent a serious problem for both species conservation and human livelihoods. The extent of this threat depends on the availability of wild prey to carnivores, on land-use practices outside protected areas, on the perception of carnivores by humans and the extent of damage to human resources and security. Resolving the human-carnivore conflict is a global concern and is particularly relevant for endangered and isolated carnivore species (Meena et al., 2014; See also Woodroffe & Ginsberg, 1998; Sillero-Zubiri & Laurenson, 2001; Ogada et al., 2003; Woodroffe & Frank, 2005; Treves et al., 2006, 2009; Loveridge et al., 2010; Macdonald et al., 2010; MacDonald, 2019). The following sections examine these issues, from the perspectives of political ecology and conservation psychology, applied to human interactions with the large cat species, frequently termed big cats or great cats (lions, tigers, leopards, jaguars, snow leopards, and cougars).

Political Ecology, Conservation Psychology, and the Lion

Political ecology and conservation psychology applied to lion conservation concern mainly the political power relations between elites and local actors, and how these relate to narratives constructed as a result of individual peoples' fear of their predation and the possible danger to humans, which leads to indiscriminate killing (mainly as a result of retaliatory or pre-emptive killing to protect human life and livestock) and the depletion of the prey base in local contexts, and other narratives created by those not as affected in by lion presence, especially in unrelated urban contexts, who perceive lions from the viewpoints of charisma, conservation justice and rights and even anthropomorphization (Campbell, 2018).

Concerning political ecology, two important issues concern the main threats to lion populations, and the conservation actions that may be pursued to ameliorate these effects. The main factors for lion number decline, and local to regional extirpation, are killing in defense of human life and livestock, habitat loss, and prey base depletion (including poaching and the bushmeat trade), with more recent, lesser effects from trade in bones and body parts for traditional medicine, in Africa and in Asia, and trophy hunting (IUCN, 2006a, 2006b; Riggio et al., 2012; Bauer et al., 2016; see also Becker et al., 2013; Rosenblatt et al., 2014). Bauer et al. (2016) note that "while attention is currently focused on lion hunting reforms to ensure sustainability, the leading causes of population decline are more difficult to address and are likely to continue. The observed and inferred reductions are based on direct observation;

appropriate indices of abundance; declines in area of occupancy, extent of occurrence and habitat quality; and actual and potential levels of exploitation."

In terms of human casualties during the twentieth century, Löe and Röskaft (2004) placed lions third, after tigers and leopards. West Africa had fewer human attacks from lions than the rest of the continent, possibly as a result of Tanzania's sizable lion population. Between 1990 and 2006, lions there killed 563 people and injured 308 others (Packer et al., 2005, 2006). Between 1923 and 1994, lions attacked 275 individuals in Uganda (Uganda Game Department archives, 1923–1994). Although some admit that isolated, unreported attacks may occur, the lower number of reported killings in West Africa may be due to the region's smaller lion population and potentially less trustworthy records (Chardonnet et al., 2010).

The bushmeat trade, engaged in by poor local stakeholders, is an important economic activity across sub-Saharan Africa, which has major effects on the prey base of lions (Campbell, 2019). These species, which are mostly antelopes, are becoming more vulnerable to the unsustainable and commercialized bushmeat trade, which is causing prey populations in many savanna regions of Africa to collapse (Lindsey, Balme, Becker, et al., 2013). Time series data on the primary Lion prey species from 78 herbivore populations monitored in West, Eastern, and Southern Africa between 1970 and 2005 closely match the regional trends in lion population sizes reported in this assessment; herbivore population sizes increased by 24% in Southern Africa, declined by 52% in Eastern Africa, and increased by 85% in West Africa (Craigie et al., 2010).

In East Africa, the largest lion populations exist (Campbell, 2018). Hazzah et al. (2017) point out that "East Africa supports nearly 60% of the remaining lion population and indiscriminate killing poses the greatest threat in this region. Many lions are illegally speared, shot, and poisoned annually across East Africa. High levels of killing, even at a local level, can have far-reaching impacts on lion populations." However, the ranges of lions in these areas are highly variable. In East Africa, lion conservation is problematic, largely due to proximity and conflicts with human stakeholders (Dolrenry et al., 2014).

In Tazania, lion population declines have been noted, although this country is "home to 40% of the world's remaining wild lions" and Tanzania "has long been considered a leader in lion conservation" (WildAid, 2019). Here, "the lion's survival is threatened by a number of factors—60% of them live outside of protected areas, where they are vulnerable to habitat loss, conflict with humans, and the bushmeat trade" (WildAid, 2019). Conservation areas are vital in Tanzania; Schooler et al. (2022) notes in a study of the Serengeti and surrounding areas, that "lions demonstrated lower probability of use of unprotected areas as compared to core protected areas and strongly protected buffer areas, which confirms the importance of protected areas to conserve lion habitat." Despite the large lion population in Tanzania, serious conflicts and exploitation issues have been recorded: Brink et al. (2016) write that concerning the killing of lions, "over-utilization is a particular concern for lion conservation in Tanzania, as the country supports between a quarter and half of the remaining free-ranging lions in the world. In addition, Tanzania is the most important destination for sport hunting of lions, exporting an average of 243 wild lion trophies per year between 1996 and 2006, compared to 96/year from Zimbabwe, and 55/year from Zambia, while no other country exported more than 20/year" (see also Packer et al., 2009; Packer et al., 2011; Riggio et al., 2012).

Lion predation on rural people is also recorded in certain regions of southeast Tanzania. Up to 400 recorded human fatalities from lion attacks occurred between 1997 and 2007 (Ikanda, 2007). Lion killings of livestock can be an important economic loss. For example, Patterson et al. (2004) calculated that the annual livestock loss caused by each lion to livestock farmers in Kenya adjacent to Tsavo East National Park was US$290 (see Bauer et al., 2016). Similarly, annual livestock losses to lions in Cameroonian areas bordering Waza National Park accounted for only about 31.1% of all livestock losses but were estimated to account for over 22% of financial losses, or about $370 per owner (Bauer et al., 2015). As a result, lions face severe persecution in livestock areas in many African nations. Their scavenging habits make them particularly vulnerable to poisoned carcasses that are discarded to kill predators. Although this is believed to be the greatest threat to the survival of lions outside protected areas, there is little actual data on the number of lions killed by locals as problem animals. One of the main strategies for resolving human-lion conflicts is the implementation of appropriate livestock management measures in conjunction with problem animal control measures and livestock loss compensation mechanisms (Bauer et al., 2015, 2016; Campbell, 2019; Frank et al., 2006; Hazzah et al., 2014).

In Kenya, four categories of threats have been documented as facing lion populations; "geopolitical, national policy, landscape management and proximate threats" and "human-wildlife conflict, habitat loss and conversion, wildlife disease, infrastructure development, and poor conservation management are ranked as the top five threats to lion and spotted hyena conservation" (Kenya Wildlife Service, 2020). Gneezy (2023) argues that "over the past few decades, economic development has drastically reduced Kenya's lion population, and the Maasai lion-killing tradition hardly helped matters." Hazzah et al. (2014, p. 851) point out that the problem of lion conservation and human-lion conflicts is "particularly acute in southern Kenya, where Maasai pastoralists have been spearing and poisoning lions at a rate that will ensure near term local extinction." Cooperation with local residents, including Maasai stakeholders is therefore crucial for the amelioration of these effects. Hazzah et al. (2017) note that "despite legal protection, deliberate killing by local people is one of the major threats to the conservation of lions and other large carnivores in Africa." The findings indicated that lion killing was mainly determined by general attitudes toward lions, which were the strongest predictor of lion killing behavior. "Other factors were engagement in traditional customs, lion killing intentions to defend property, and socio-cultural killing intentions" (Hazzah et al., 2017).

In Uganda, Omoya et al. (2014) give a study of lions (and hyenas), noting that "despite >60 years of conservation in Uganda's national parks the populations of lions and spotted hyenas in these areas have never been estimated using a census method." The study was based on the three main savannah national parks: Kidepo Valley National Park, Murchison Falls Conservation Area, and the Queen Elizabeth Protected Area (the last including the Queen Elizabeth National Park and the linked wildlife reserves of Kigezi and Kyambura). Small numbers of lions have also been recorded in Toro-Semliki Wildlife Reserve, Lake Mburo National Park, and Kabwoya Wildlife Reserve. This study gives a chronology of lion numbers in Uganda: late 1970s, 400 in the Queen Elizabeth Protected Area (Din, 1978; Van Ordsol, 1982) late 1990s, early 2000s, 185 and 206 in the same area (Driciru, 1999; Driciru et al., 2005, p. 200); and mid-1990s, total for Uganda 345–745, (Omoya et al. 2014). Omoya et al. (2014, p. 395) point out that "such a wide range indicates the uncertainty of the species' status in the country in the 1990s." They conclude that "studies of the lion in Uganda have been neither consistent nor exhaustive" and the Queen Elizabeth Protected Area is the most studied area (see Din, 1978; Driciru, 1999; Van Ordsol, 1982), and their findings show that lion numbers have been decreasing in the Queen Elizabeth Protected Area and Murchison Falls National Park but have been increasing in Kidepo Valley National Park (Omoya et al., 2014, p. 400).

Kiggwe (2022) describe the problems facing "unique tree-climbing lions" in Uganda, these including the illegal wildlife trade, the recent killing and mutilation of six lions in the Ishasha sector of Uganda's Queen Elizabeth National Park, habitat degradation, human-wildlife conflict, illegal wildlife trade, snaring, and lion body part trafficking. Human wildlife conflicts include retaliations for livestock predation, and local beliefs that lion body parts may have medicinal value and may be sources of power and wealth. Hall and Bale (2018) also report the fatal poisoning of eight young lions. The World Wildlife Fund (2018) notes that "lions in Uganda have continued to lose their lives as the population throws and feeds them on meat laced with poison. Between May 2006 and July 2007, 15 lions were killed in the area in attacks blamed on landless herdsmen defending their cattle. In May 2010, five lions were killed in the park in another possible poisoning case." Poisoning may be based on aldicarb, an insecticide also known as Temik. This is a crucial issue because, based on tourism, "a WCS assessment in 2006 showed that each individual lion in Queen Elizabeth National Park generated about $13,500 USD per year for the national economy in terms of the revenue it brought into the country. An influencing factor was that tourists are willing to stay longer just to see lions" (World Wildlife Fund, 2018).

In Ethiopia, Yirga et al. (2021) there is "an extensive extant lion range providing connectivity between East and Central Africa" (see also EWCA, 2012), and "Ethiopia's lions are under increased pressure and are declining rapidly due to conflict with humans and habitat loss (Gebresenbet et al., 2009). Traditional lion hunting for prestige and as a rite of passage also persists in places like Awash and Mago National Parks (NP) among the Mursi, Hamar, Tsemai and Ari ethnic groups." Yirga et al. (2014) reported livestock conflict as a major cause of decline in large carnivore populations in Nechisar, with Guji herders persecuting large carnivores. Anthropogenic factors led to lion extinction in Nechisar; reportedly, any lion that ventures near livestock is shot before any contact (Yirga et al., 2014). In Ethiopia, lions are systematically killed because they prey on livestock. Other threats likely also contributed, particularly lack of prey and habitat degradation. However, there is a lack of information about the extent to which lions are affected by various threats.

Human-lion conflict is certainly one of the main causes of the decline in lion populations throughout Ethiopia, including the areas where they are officially protected (EWCA, 2012).

The lion population in Ethiopia is currently experiencing a sharp decline due to anthropogenic factors. Although the exact range and population size of lions in Ethiopia are still unknown, the rate of decline in lion populations is alarming. Lions are viewed by pastoralists as a threat to human security and livestock. People are in favor of well-organized regional lion conservation, especially in small protected areas. There are those who could live with lions in their midst. Cattle-killing cases are known to influence opinion. Following predation by livestock, communities are less likely to accept carnivores, more likely to kill them in retaliation, and are resistant to the possibility of growing carnivore populations (see also Baral & Henien, 2007; Chess, 2000; Dar et al., 2009; Ericsson & Heberlein, 2003; Hazzah et al., 2009; Kissui, 2008; Naughton-Treves et al., 2003; Palmeira & Barrella, 2007; Røskaft et al., 2007; Sogbohossou et al., 2011; Zimmermann et al., 2005).

In West Africa, lions are in a particularly dire situation because conservation efforts are minimal and human land use is increasing. Riggio et al. (2016, p. E107) note that the extinction threat is exaggerated, because "of nine West and Central African lion populations, the largest are stable (Bénoué, 200) or increasing (Pendjari, 100) …The declining populations are the smallest, with starting populations of fewer than 70." However, Henschel et al. (2014) note that the countries with native lion populations include Benin, Burkina Faso, Niger, Nigeria, Senegal. In other countries, Ghana and Guinea Côte d'Ivoire, Gambia, Guinea-Bissau, Mali, Mauritania, Sierra Leone, and Togo lions are extremely rare or may be extinct (see also Bauer et al., 2015). Henschel et al. (2014) contend that the lion subpopulation in West Africa is thought to number only 400 animals at time of writing, and this subpopulation is declining. Henschel et al. (2014) suggested that there may be only between 269 and 583 lions in West Africa, of these 121–375 being mature individuals, with "continuing calamitous declines in prey populations are almost certainly causing concomitant declines in lions" (Henschel et al., 2014). Howard (2014) argues that "lions may soon disappear entirely from West Africa unless conservation efforts improve, a new study predicts" and cites studies that "found an estimated total of only 250 adult lions occupying less than 1% of that historic range. The lions form four isolated populations: one in Senegal; two in Nigeria; and a fourth on the borders of Benin, Niger, and Burkina Faso. Only that last population has more than 50 lions."

The habitat in the Sahel and Sudan savanna may still be suitable for lions, but the lion populations are still declining or locally extinct. An important concern is that in West Africa, the range of the lion is generally composed of isolated patches with few usable corridors (Angelici et al., 2015; Henschel et al., 2014). Suitable lion habitats include the WAP-Complex, which includes the W-Arly-Pendjari National Parks in Burkina Faso, Benin, and Niger. There is also the Niokolo-Koba National Park in Senegal and the Kainji Lake National Park and the Yankari Game Reserve in Nigeria (Henschel et al., 2010, 2014). Works by the IUCN and Henschel et al. (2014) on 17 protected areas, mostly using tracks for identification, found lions were present in three of the surveyed protected areas; in Senegal (Niokolo-Koba National Park, with fewer than 60 lions), in Benin, Burkina Faso, and Niger (the tri-national W-Arly-Pendjari Complex, with 246 to 466 lions), and in Nigeria (the Kainji Lake National Park, 23–63) (Henschel et al., 2014). There was little evidence of lion presence in Mole National Park in Ghana and in Guinea's Haut Niger and Kankan Reserves, and in Togo, Guinea-Bissau, Mali, and Côte d'Ivoire.

The greatest threats to West African lions from urbanization are the ever-shrinking spaces between urban centers, the massive development of connectivity networks in formerly rural areas, the increasing exploitation of land cover and herbivorous wildlife, and the overall increased proximity of human presence. West Africa is one of the more urbanized regions of Africa. The common solution of establishing national parks is being questioned, partly because of land-use conflicts with urban expansion and partly because of costs (Campbell, 2012). As argued by Kinzig and McShane (2015), land cover and habitat conservation generally involves the exclusion of pre-existing settlers from required areas. These people may have limited understanding of the purpose of conservation areas and may derive little or no benefit from conservation policies. As a result, locally excluded peoples have no trust in the authorities that establish or maintain the protected areas, in part because these authorities, supported by conservationists, have traditionally paid no attention to the connection between ecology and wildlife survival and the livelihoods of displaced people (Kinzig & Mcshane, 2015; see also McShane & Wells, 2004).

Agricultural expansion is another threat (Schwartz, 2017). Killing lions in retaliation for livestock depredation has become a major factor in lion extinction in land areas dominated or even visited by humans. There is the increasing importance of cattle in the income of rural actors in West Africa, particularly as cattle are not only an immediate source of meat but also long-term sources of status and capital (Campbell, 2012). This highlights the problem of land sharing between humans (in this case farmers) and large predators. Chardonnet et al. (2010, p. 12) also note that in Mali, for example, lion predation on cattle is one of the main reasons for the drastic decline in lion numbers. In the national parks in the north of the Central African Republic, the decline in lion numbers is largely due to systematic shooting by herders during the dry season. Stephens (2015, p. 14754) argues that in many parts of the world it is probably not possible to conserve extensive areas, so there is a choice coexistence (sharing land with large mammals) and confining species to areas which are too small for them to play a signficant ecological role. In the case of lions, any attack on humans or livestock reduces coexistence, especially given the lack of funding, the increasing land dependence of the human population, and the vast areas required for lions to forage (Di Minin et al., 2021).

Another threat is ineffective conservation area management, especially at the edges of these areas. For example, Tumenta et al. (2010) examined the Waza National Park in northern Cameroon, to assess trends in local lion populations. In this study, two out of four collared lions were killed by illegal retaliation within a year, and two additional males and one additional female were also likely killed during this period. The lion population appears to have declined over the past 5 years. Six lions were killed annually over 5 years, which was a much higher rate than in previous decades (Tumenta et al., 2010). Pressure between humans and livestock increased enormously during this time, leading to frequent conflicts between humans and lions. Although lions are most often killed in retaliation for livestock theft, poaching also appears to occur. The carcasses of both lions were never recovered. The communities surrounding the park place socio-economic value on various parts of the Lion and there is evidence of trade in these parts (Tumenta et al., 2010).

For the South African lion, DeMarco (2015) notes that lions in southern Africa were doing comparatively well and were even increasing in some areas. Factors included the competent management of South African populations based on fences, as opposed to the open-range system of East Africa. An important development for lion conservation in South Africa is the use of translocation, which concerns introducing lions into existing populations, or into areas where lions have been extirpated (Becker et al., 2022). Becker et al. (2022) note that "South Africa was pioneering in this regard and has an established metapopulation of lions in small, fenced reserves that rely on translocations to mimic natural systems within a metapopulation, including dispersal and gene flow. Translocations since the early 1990s have reestablished wild lions in >50 sites in South Africa (see also Funston, 2008; Miller et al., 2013, 2015; Miller & Funston, 2014; Slotow & Hunter, 2009).

For the Asiatic lion, now limited in range to western India, Jhala et al. (2019) note that "Asiatic lions typify most challenges faced by large carnivores: single population, historical bottlenecks, habitat loss, poaching, and conflict with humans. Their recovery from <50 in a few hundred km^2 to >500 occupying 13,000 km^2 of agro-pastoral Saurashtra landscape, Gujarat, India is an enigma." Meena et al. (2014, p. 120) also note the recovery of the Asiatic lion, as "the last population of free-ranging Asiatic lion (Panthera leo persica) has been steadily increasing in response to successful conservation initiatives spanning five decades. With the increasing population size, lions have expanded their range and move throughout the Greater Gir Landscape in Western India." However, like elsewhere a major issue for lion conservation is the rising human-lion conflicts as lions intrude into former habitats now included in human landscapes of agriculture, transportation, urbanization, and/or residence. It is noted that the landscape around the Gir Protected Area (GPA) "includes agro-pastoral villages with large human communities and vast numbers of livestock" (Meena et al. (2014, p. 120).

The findings of this study, using "analysis of the Gujarat State Forest Department monetary compensation records, and human perceptions of forested habitat, lion conservation and wildlife damage mitigation, via interview surveys" indicated the population of lions in the GPA was stable, "the number and severity of livestock depredations has increased over time" especially near villages around the GPA (Meena et al., 2014, p. 120). The authors comment that "these hot spots of depredation are associated with movement paths of dispersing lions out of GPA" and that based on the interview surveys human livelihood dependency on forest was described as "minimal, while economic losses associated with crop-raiding wild herbivores and livestock-depredation by lions were higher." Lion conservation in the Greater Gir landscape is still currently moderately successful, but it is crucial to manage conflict and promote a positive public, mostly local perception of lion presence. Currently, the agro-pastoral economy, land use, and cultural tolerance appear to be favorable to the survival of lions in the GPA region. This population has recovered as a result of conservation efforts over the course of nearly 50 years, upgrading the Asiatic lion's conservation status from Critically Endangered to Endangered in the IUCN's assessment. Due to the lions'dispersal and utilization of their habitat within an agro-pastoral landscape matrix, there is a greater chance of human-lion interaction and conflict. The lions'recent range expansion has included human spaces among densely populated villages and intensively utilized land cover. In these villages, people frequently come into contact with lions, which raises worries about human safety and financial loss due to livestock theft. These issues have major effects on lion survival in this ecosystem (Breitenmoser et al., 2008; Meena et al., 2014).

Across all the regions occupied by lions, important aspect of the political ecology and conservation psychology of lion-human relations is the perception that lion body parts fulfill human health functions, including disease treatment. Regional lion conservation strategies call on countries to prohibit and control the trade in lion bones and other parts (IUCN, 2006a, 2006b). It is argued that the illegal trade in lion body parts for medicinal purposes threatens subpopulations of Asiatic and African lions. Sogbohossou (2006) discovered that there are numerous reports of the use of lion skins and bones in Benin (fat and bone being the most commonly used: N'Diaye, 2014) and that over 22 lion parts have medicinal value in the Yankari National Park in Nigeria (Born Free Foundation, 2008).

The use of African lion bones appears to be becoming increasingly popular in Asia. Although the use of lion bones in traditional medicine is not documented there, China has approved the use of lion bones from captive animals to make medicinal wines previously made from tiger bones. Large exports of lion bones from South Africa to China, Lao People's Democratic Republic, and Vietnam have been reported. Illegally acquired parts of lions (and other big cats) can be concealed by the legitimate global trade in lion bones from captive-bred animals (Nowell & Pervushina, 2014). There are also concerns that the massive illegal wildlife trade in elephant ivory to Asia could affect wild lion parts from eastern and southern Africa. Conservation actions, developed from international regional power centers with varied empowerment and involvement of local groups may be evolved to contend with these developments (Campbell, 2019). For example, wildlife tourism is a major economic incentive for lion conservation in East and Southern Africa, with powerful narratives derived from political power in advanced countries (IUCN, 2006a, 2006b). These regional strategies have been used in many countries to develop Lion Conservation Action Plans. It is nevertheless argued that the ongoing decrease in the range and population of lions indicates that political priority and funding are insufficient, despite awareness of the threats and recognition of solutions, and local involvement may be insufficient, due to the unwillingness central power elites to fully engage with involved local, generally peripheral groups (Bauer et al., 2016; see also Packer et al., 2013).

Political Ecology Conservation Psychology, and the Tiger

Political ecology and conservation psychology inform the narrative of tiger conservation throughout the tiger's range in Asia, as the tiger is considered the continent's dominant and most charismatic predator, the most feared, historically a leading human killer, with vast habitat and prey requirements and a dominant tourist attraction (Campbell, 2022). Policy actors involved in this dynamic include North American, European, and Asian governments, international business and conservation organizations, smaller nongovernmental organizations, and local land users, including farmers, developers, tourism workers, and local residents (Campbell, 2022). Tigers are considered one of the most popular wild mammals in the United States, where there are more tigers in various degrees of captivity than in the wild worldwide (Hannaford, 2019). World Wildlife Fund (2014) contends that "one of the world's largest populations of tigers exists not in the wild—but in captivity in the United States. With an estimated 5000 tigers, the U.S. captive tiger population exceeds the approximately 3200 tigers in the wild." Therefore, the political ecology of tiger conservation is a dominant referential and comparative narrative in studies of all large carnivores (Campbell, 2022).

Various political strands have been applied to the study of tiger conservation, at international to local and individual stakeholder levels. For example, at the international level, Slappendel (2022) points at the failures of international tiger conservation efforts, arguing that "with so many countries, organizations and industries involved, tiger conservation has strayed far from the initial goals and into politics over the decades." Also cited was the 2010 St. Petersburg Tiger Summit, where the heads of the 13 Tiger Range Countries made declarations about the conservation of tigers, and there was the adoption of the National Tiger Recovery Program (GTRP), in which the tiger range countries acknowledged cooperation other governments and international organizations (Slappendel, 2022). There is also the fourth Asia Ministerial Conference (AMC) on Tiger Conservation which ended on January 21, 2022, with the adoption of the Kuala Lumpur Joint Statement, which noted the "vital need to address the trade in tigers across their range by strengthening law enforcement against tiger crimes, including online trade, improving transboundary collaboration and intelligence sharing, and phasing out tiger farms, as well as halting the demand for tiger parts and products which drives the trade and poaching" (World Wildlife Fund, 2022).

At the national level, Margulies (2019, p. 150) examined the "practice and politics of classifying a tiger as a "man-eater" in South India to explore what doing so reveals more broadly about the relations between animal life and the kinds of human life marked as expendable by the state." This refers to the killing of people by tigers around conservation areas, as an example of the politics of human-animal encounters in shared spaces, and how the power structures manage these situations. This is placed within the framework of necropolitics, which is the "politics mediating death" (Margulies, 2019: p. 150). Margulies (2019, p. 150) cites a study around how the Indian State "goes about reclassifying tigers from a strictly protected endangered species to killable—the process of making the "man-eater"—in relation to how the state both values and devalues human and non-human life as a process rooted in colonial histories of accumulation by dispossession." The actors in this case are tigers that are situated and protected within conservation areas, but may also move into densely human-occupied areas outside the protected area, and people who are both marginalized by their vulnerable political and economic situation, and require protection from danger including that of tigers, by the state. The author notes that "in bringing animals to the center of this discussion of violent more-than-human politics, the interdisciplinary field of political ecology, being broadly concerned with issues of justice in the co-production of environments and access to natural resources, has increasingly come to engage with more-than-human geography and post-humanist theory on the one hand, and the politics of animals on the other" (Margulies, 2019, p. 151).

Another study by Menon and Rai (2019, pp. 652–665) focuses on the issues of political, economic and spatial marginalization of less powerful people in the context of contested land-use and tiger conservation, based on a case study of the Biligiri Rangaswamy Temple Hills of Karnataka. The authors argue that "the Indian state has conserved tigers by establishing reserves that are governed as a form of fortress conservation. Residence and local uses in these tiger reserves are often criminalized. It is in this context that we critique recent neoliberal attempts to estimate the economic value of ecosystem services from tiger reserves. Proponents of valuation argue that it will not only provide a justification for the reserves, but also recognize the importance of ecosystem services for human well-being" (Menon & Rai, 2019, p. 652). The study makes the case—based on a political ecology approach—that institutional frameworks that prioritize some actors over others are the larger context in which economic valuation operates, and that it is not necessarily a neutral instrument (Menon & Rai, 2019, p. 652). In India, protected areas are argued to be valued, even if the people who live there are displaced and their use of the forest is restricted. The authors conclude that economic valuation can obscure complex relationships between humans and nature and undermine different ways of understanding local problems and landscapes, including the issues of land access, social marginalization, local economic development and human-animal relations (Menon & Rai, 2019, p. 652).

The conservation psychology of the tiger is mainly focused on human fear of the big cats, which are generally perceived as larger, more powerful and more dangerous than sympatric predators such as leopards and bear species (Campbell, 2022). The psychology of human-tiger relations may be related to economic, social, cultural and/or religious structures, as well as individual volition (Campbell, 2017, 2019). For example, McKay et al. (2018) cite a case study of the Kerinci-Seblat region, Indonesia, and conducted an investigation into the differences in tiger beliefs among the ethnic groups (Minangkabau, Melayu, Kerincinese) residing close to a sizable protected area with Sumatra's largest tiger population. The objective of the study was to assess the potential role that cultural tolerance may play in tiger survival (McKay et al., 2018). The findings indicated that the majority of respondents (73%) who identified as Minangkabau, along with 52% of Melayu and 44% of Kerincinese, claimed that their ethnic group had customary laws regarding tigers. Most participants, regardless of ethnicity, did not perceive any correlation between Islam and tigers (McKay et al., 2018). Two groups (Minangkabau and Kerincinese) considered that spirit tigers existed. An unprovoked tiger attack in the village would create the strongest calls for the tiger to be killed, compared with other scenarios of human-tiger conflict, regardless of how serious other attacks were to livestock or human lives (McKay et al., 2018). However, the majority of respondents from all ethnic groups stated they would not take action if a tiger killed a villager or family member while they were hunting in the forest. Based on beliefs about the tiger‘s role as a moral law enforcer, this reaction became more common when a tiger killed someone in the village who had committed adultery (McKay et al., 2018).

Struebig et al. (2018) describe a case study of Sumatran tigers in Indonesia, noting that tigers are threatened with extinction due to deforestation and persecution, but Sumatran tigers range close to human-occupied landscapes, and such occurrences provide investigative opportunities, the results of which may be considered for studies in other parts of the tiger's range. The findings indicated that populated villages adjacent to forests or rivers that connect tiger habitat had a higher chance of encountering tigers; geographic profiles further refined these predictions into three important areas. The underlying attitudes, feelings, social mores, and spiritual convictions of people were associated with their tolerance for tigers. For example, it is argued that some of the social support for tiger protections is based on the acceptance of Islam, which forbids the eating of animals that hunt with claws, along with other indigenous farming communities' belief systems, has been used to explain the tolerance shown toward Sumatran tigers. The Kerincinese and Minangkabau people are described as having historical, spiritual ties to animals, and some believe that tigers, who protect customary laws, are embodied souls of their ancestors. Despite this, some have lost their kinship with tigers through time. Some people also think that people have the ability to assume the shape of a "weretiger," which can have disastrous effects.

Struebig et al. (2018) comment that these spiritual belief systems offer a rare chance to investigate how these elements might encourage tolerance and coexistence with dangerous wildlife, especially in light of the ongoing monitoring of tigers and their interactions with humans (Struebig et al., 2018: see also McNeeley & Sochaczewski, 1988; Mangunjaya & McKay, 2012; Bakels, 2013; Boomgaard, 2013; McKay et al., 2018). Struebig et al. (2018) commented on the effectiveness of their methdology, contending that tolerance predictions using socio-ecological models combined with broader data outperformed models based solely on social predictors by a factor of 32. Interestingly, these socio-ecological predictions could have prevented up to 51% of attacks on humans and livestock and saved 15 tigers through preventive measures (Struebig et al., 2018). The authors conclude that current findings suggest that the application of socio-ecological models such as theirs may be positive, but such models are recently emerging. In addition, the results suggest that "the prevailing anthropological view that attitudes toward wildlife in Southeast Asia are driven by spiritual beliefs unique to specific indigenous ethnic groups could be oversimplified. Rather, our findings indicate that tolerance toward tigers, in Kerinci-Seblat at least, is driven by a number of complex factors that include spirituality, as well as risk of actual attack" and the people under study would be "more likely to support an increase in tiger populations if they held beliefs concerning the importance of the animal for spiritual well-being, but this inclination appeared broadly prevalent among respondents and independent of their ethnicity" (Struebig et al., 2018; see also Kartomi, 1977; McNeeley & Sochaczewski, 1988; Mangunjaya & McKay, 2012; Bakels, 2013; Boomgaard, 2013; McKay et al., 2018).

Political Ecology, Conservation Psychology, and the Jaguar

Political ecology and conservation psychology influence jaguar conservation Latin America, concerning human actions, which can be influenced by power relations between international, national, regional, and local actors, as well as by the psychology of these actors toward the large carnivore presence (Campbell, 2017). The prevailing dynamics in this regard concern the power relations and outcomes between conservation actors, economic policymakers (concerning forestry, agriculture, urban, and infrastructural development), and local land users (including farmers and hunters) (Campbell, 2022). Conservation psychology can be influenced by local traditional religious and socio-cultural perspectives and norms, physical and economic outcomes of encounters between big cats and humans/livestock, and more generally by the possibilities of "modern" socioeconomic issues, such as tourism (Campbell, 2017).

The main political ecology issues concern high-level governmental and non-governmental actions for international jaguar conservation, and how these are able to connect with, and involve local, largely disempowered stakeholders (Campbell, 2017, 2022). For example, the World Bank (2023) cites International Jaguar Day on November 29th, which was designed to increase awareness of the issues facing jaguars, and support "efforts stretch from the local, national and regional levels to high-level international initiatives seeking to protect the jaguar and its habitat" and the GEF-funded Global Wildlife Program, which "works in three jaguar-range countries Belize, Ecuador, and Panama to address multiple threats such as habitat loss and fragmentation, human-jaguar conflict, and the trafficking of jaguar parts." Developments included the Jaguar 2030 New York Statement, which acknowledged the relevance of jaguar conservation to the global Sustainable Development Goals, and involved range country governments, and was co-hosted by, among others the Permanent Missions of Mexico and Colombia to the United Nations, and co-organized by the United Nations Development Programme, and non-governmental organizations such as Panthera, Conservation International, Wildlife Conservation Society and the World Wildlife Fund. Related projects and actions involve landowners in Belize, cattle ranchers in Panama, and local actors and law enforcement in solutions to wildlife crime, including hunting, trafficking, and trade in jaguars and jaguar parts in the Ecuadorian Amazon (World Bank, 2023).

For conservation psychology local actor issues assume prominence, which may influence the grassroots implementation of the policy actions promoted at policy-making levels. For example, Caruso et al. (2020, p. 678) cite a study of local perceptions of jaguars by people living in and around protected areas in northern Argentina. The study is justified on the grounds that "in Argentina little is known about the socio-economic factors that influence people's perceptions of and attitudes toward jaguars" and "the attitudes and perceptions of local people toward jaguars have not been studied in the northern region of Argentina, hindering the promotion of coexistence between people and jaguars" despite scattered studies showing both positive and negative attitudes to the big cats (Caruso et al., 2020, p. 679; see also Altrichter et al., 2006; Payán, Castaño-Uribe, et al., 2013; Caruso et al., 2017). Important psychological themes covered in the research included support for jaguar conservation, the identification of the jaguar's ecological role, and feelings about jaguar extinction and personal encounters jaguars, and general socio-economic factors influencing perceptions of and attitudes toward jaguars. The findings of this study indicated positive attitudes and perceptions of jaguars were linked to potential economic gains for humans from the species' presence, such as increased tourism income (Caruso et al., 2020, p. 678). Surprisingly, more positive attitudes and perceptions were not linked to higher levels of formal education. Fear was found to be the driving force behind negative attitudes and perceptions toward the species, as people generally saw jaguars as a physically threatening. The degree of tolerance for jaguars was influenced by socioeconomic factors that go beyond financial losses (Caruso et al., 2020, p. 678).

In terms of support for jaguar conservation, a higher proportion of support was found among elementary school graduates over those without such education, older people over younger people, and people involved in tourism over those not involved in tourism, such as cattle herders and farmers (Caruso et al., 2020). In terms of the identification of the important ecological role of the jaguar, elementary and college educated people were more supportive than those with less formal education, but those with secondary education levels did not have higher recognition of the ecological role of the jaguar role than those with no education (Caruso et al., 2020). Most young people and students recognized the ecological role of the jaguar, but most older people did not. These findings were similar for opinions of jaguar extinction, as younger and more educated people, and those engaged in tourism, being more concerned about jaguar extinction than older and less educated people, and those not engaged in tourism activities, such as cattle herders and farmers (Caruso et al., 2020). Concerning the physical threat from the jaguar, most people regardless of age, education, or occupation, cited fear, with the college educated least likely to cite fear (Caruso et al., 2020).

Manchini (2010) cites a similar study from Amazonia and the Pantanal in Brazil, where conservationists and policymakers have succeeded in promoting conservation management for jaguar survival, by both protecting jaguars and attempting to reduce jaguar-human conflict. Manchini (2010) examined a human psychological perspective, writing that the hypothesis of the study was that "the ultimate motivation for persecution was not the loss of livestock to jaguars (or attacks on humans—which are very rare), but rather the perceived potential for these threats and for their impact on human livelihoods" and "jaguar killing was not strictly retaliatory, and that the perceptions that determine the human behavior of killing jaguars were amenable to change through education and communication." This conservation psychology approach found that "perceptions behind the conflicts between people and jaguars in Brazil have more to do with social and psychological factors than previously considered, and therefore may be more amenable to change through education and communication than generally believed" (Manchini, 2010). The findings noted psychological factors, including fear, as important, and these varied regionally and by income of the respondents. Importantly, livestock predation by jaguars was not the factor for jaguar killings, but "it acts in combination with attitudes, knowledge and perceptions of the economic situation to determine how people perceive the impact jaguars have on their livelihoods. Attitudes and knowledge, in turn, are influenced by age, gender and whether respondents live in urban or rural areas" (Manchini, 2010). School-based education on conservation was found to influence students, and importantly this was transferred to family members, through student homework and communication. The authors conclude that "if jaguar killing in rural Brazil is to be prevented, conservationists will need to find and support ways to make jaguar killing not only illegal and unprofitable, but also socially and personally unacceptable" (Manchini, 2010).

Political Ecology, Conservation Psychology, and the Cougar

Political ecology and conservation psychology influence the conservation of the cougar, as this predator is a prime example of an animal that is considered both dangerous and essential to conservation policy in North America. For many years, cougar presence and survival have been largely determined by human fears, feelings of protection, dangers to livestock, and the cougar may be thought to be more dangerous than actual records indicate (Campbell, 2012; Kertson & Keren, 2022). For instance, cougars have a generally negative social history in the United States: "A disruption to urban life, a nuisance to society, or a threat to humans" was how the cougar was described in old records (Wolch et al., 1997, p. 106, 109). Serial killers are those who commit crimes with premeditation, and the identification of the cougar as such took advantage of public concerns about increasing crime and lawlessness (Wolch et al., 1997, p. 106, 109). Some claimed that murderers should not be allowed to walk free in any civilization and made a comparison between cougars and human criminals (Perry, 2003). Some people speculated that urban cougars might develop a lazy streak and choose to consume humans and companion animals over their natural prey. This may have touched on persistent beliefs regarding the worth of labor, the immorality of individuals believed to be avoiding employment, and their standing as unworthy of assistance from the public or protection (Wolch et al., 1997, p. 110).

Current psychological attitudes toward cougars may vary depending on the age and gender of the human evaluator (Campbell, 2012; Campbell & Lancaster, 2010; Campbell & Torres Alvarado, 2011). Using Canadian case studies, Campbell (2012) and Campbell and Lancaster (2010) hypothesized that women and older people would be more afraid of cougars and that women and younger people would be more supportive of cougar conservation. These hypotheses were consistent with the results of the relevant literature (Hayward et al., 2007; Thorn et al., 2013; Campbell, 2017). However, Campbell (2012) found that more women than men believed that cougars improved human quality of life. The majority of men argued that cougars were detrimental to human quality of life, regardless of the age of the respondent (Campbell, 2012). Campbell and Lancaster (2010) found that regardless of the gender or age of respondents, people did not believe that cougars impact quality of life and argued that cougars are less likely than bears to invade and/or cause harm to human homes, parks, or other property (Campbell & Lancaster, 2010).

Campbell and Torres Alvarado (2011), in a study of El Salvador, Central America, tested the hypothesis that older people would be less tolerant of wildlife than younger people, women would be more tolerant than men, and women would be more afraid than men concerning the reintroduction of jaguars and cougars, and be more concerned about the danger to children and are more afraid of larger jaguars than cougars (Campbell & Torres Alvarado, 2011). The results showed that these hypotheses were either refuted or provided weak support for these assumptions, although there was evidence that younger people were more likely than older people to access knowledge from online social media. Men were more tolerant of large carnivores than women and were also more likely than women to argue that large carnivores improved human life (Campbell & Torres Alvarado, 2011). More women than men were afraid of cougars and more women than men perceived a danger to pets, livestock, and people and a negative impact on quality of life. More men than women supported tolerance of cougars, with no gender differences in trapping, shooting, protection, and reintroductions. These depart from the general assumptions based on ecofeminist, women and the environment, and political ecology literature that women have greater understanding of the environment and/or wildlife conservation due to cultural conditioning and/or natural cognition (Campbell & Torres Alvarado, 2011, see also Campbell & Lancaster, 2010).

These findings (Campbell & Torres Alvarado, 2011) might be connected to the "rational short-term interest" stance taken by some writers in the literature on women and the environment, according to which fear or other rational considerations may take precedence over any moral convictions women may have regarding the preservation of nature (Campbell, 2017). The results of Campbell and Torres Alvarado (2011) did not show that women believe there is a greater risk to children, but they did support the idea that women are more afraid of wildlife. These results confirmed those of other studies conducted in other nations, which found that people's negative perceptions of pumas and jaguars varied according to gender (Conforti & Azevedo, 2003). In terms of tolerance, trapping and removal, shooting, protection, reintroductions, impacts on human life, and danger to cattle, adults, and children, cougars and jaguars were thought to be equivalent, despite the fact that jaguars are typically larger and stronger than cougars. Both species were thought to be capable of killing people, defying the popular belief that the cougar is an elusive and fearful animal, and the jaguar is a more potent killer of cattle and people (Conforti & Azevedo, 2003; Palmeira et al., 2008).

Additionally, Campbell and Torres Alvarado (2011) found that whereas younger people were more in favor of reintroducing cats into unrestricted rural areas, older people were more likely to support culling the cats, but young and older people had similar attitudes to reintroductions to unrestricted rural areas and zoos, more likely to support reintroducing them into special areas (Campbell & Torres Alvarado, 2011). Older adults were much less likely to get their information from social media; instead, they were more likely to rely on personal experiences, information passed down from earlier generations, or hearsay and gossip. Compared to younger people, older people were less concerned about the extinction of cats and more concerned about the threat to cattle (perhaps because they owned more cattle), presumably because younger people were better informed and had easier access to pertinent topics on the Internet). These findings aligned with previous research on human attitudes to the conservation of large carnivores (Andersone & Ozolins, 2004).

Public perceptions of large carnivores have been reported in a more informal manner on social media than in the previously mentioned academic literature. For example, Heidenreich (2017) cites the case of Steve Ecklund, a television host, in Global News, noting that "a Canadian host of outdoor TV show The Edge continues to face attacks from animal rights activists on social media after posting photos of himself with a giant cougar he met in northern Alberta earlier this month." Numerous online posts contributed to the backlash, which received even more attention after several UK media outlets published reports about the cougar killing (Heidenreich, 2017). The report also noted that cougar hunting is sparking debate among scientists and nearly 20,000 were involved, including hunters and activists. Derworiz (2018) argued that a cultural divide still exists regarding the hunting of large carnivores, that appeared to be based on the value of the individual animal rather than the value of the population as a whole. This public pressure contributed to the repealing of some hunting laws.

Another social media post showed a cougar in a SkyTrain station in British Columbia (Luymes, 2017). The video was published by The Buzzer, TransLink's free newsletter. As of April 21, 2017, 6:04 p.m., the Vancouver Sun website reported that "199 people are talking about it." The report states that "a pair of cougars strolled through the Inlet Center SkyTrain station in Port Moody early Saturday morning."...and "the video shows the two cats strolling along the tracks through the abandoned station of the new Evergreen Line In addition, the cougars triggered a burglar alarm around 4:15 a.m., but it is uncertain how they managed to access or exit the tracks. An official search did not detect the two intruders."

A social media development concerned the Cougar Network, which is a non-profit research organization focused on the study of cougar habitat and the role of cougars in ecosystems, and since 2015 has posted a popular weekly feature called #CougarOrNot on Twitter. The polling-style game involves posting a photo each Friday and then inviting participants to identify the beast as a cougar... or not a cougar (Crosby, 2018). There is also the World Wildlife Day 3 March also posted a "social media kit for World Wildlife Day 2018. The theme was Big Cats—predators under threat and the main hashtags were #WorldWildlifeDay, #PredatorsUnderThreat, #WWD2018. Supplementary hashtags were #DoOneThingToday and #iProtectBigCats. The main Twitter (now X) handles were @wildlifeday, @CITES, @UNDP @UNEP, @UNODC, @WWF, @WCS, @PantheraCats and @AfricanParks and @youth4wildlife. The Twitter hashtags and messages placed online by the World Wildlife Day organizers included: #DidYouKnow, #BigCats, #WorldWildlifeDay, #PredatorsUnderThreat, #iProtectBigCats, #WWD2018, #HabitatLoss, #HumanWildlifeConflict, #poaching, #wildlifetrafficking & #ClimateChange, #pumas, #CITES, #SnowLeopard, #DoOneThingToday, and #tiger.

Attempts have also been made to engage the public in cougar identification through social media (Campbell, 2019). Crosby (2018) notes that people often misidentify other mammals in images, such as bobcats, domestic cats, dogs, and other species, as cougars. Additionally, online games such as #CougarOrNot are also being considered as possible means of promoting public identification Cougars discussed. Scientific researchers and organizations, including The Cougar Network, report numerous reports of alleged sightings, which can include photographs, tracks, scat, and visual sightings. The visual sightings are considered less reliable because they usually cannot be verified. The questions concern the accuracy of the visual sightings and the possibility that the identifier's demographic information, including location, education, and age is relevant. Surveys of #CougarOrNot participants are used to determine the accuracy of photos and link participant demographics to the accuracy of sightings, and to improve the accuracy of #CougarOrNot responses (Crosby, 2018).

Political Ecology, Conservation Psychology, and the Leopard

From the socio-economic and political perspective, for leopards, as for other big cats, one of the biggest survival problems is the human perception, whether real or imagined, that they pose a threat to humans and/or livestock and pets, and the resulting killing of the big cats by humans. This combines with other more ecological factors, such as habitat losses to agricultural, urban and infrastructural developments, and hunting of leopards for body parts (Athreya et al., 2011; Breitenmoser et al., 2007; Campbell, 2022; Datta et al., 2008; Kaboodvandpour et al., 2021; Raza et al., 2012; Spalton et al., 2006; Swanepoel et al., 2015). The political ecology of leopard conservation across the leopard's range includes the effective management of human-leopard conflict, which is encouraged by livestock predation, and in rarer cases killing of people, and trade in leopard parts (Stein et al., 2023).

For example, in Asia, Farhadinia et al. (2017, p. 452) write that in continental Asia there are six leopard subspecies in continental Asia, four of which are threatened ranging through several countries: the Persian *Panthera pardus saxicolor*, Indochinese *P. pardus delacouri*, Arabian *P. pardus nimr*, and Amur *P. pardus orientalis* leopards, and "a relatively high proportion (23–100%) of the ranges of each leopard subspecies occurs in borderlands." Farhadinia et al. (2017, p. 455) list the main political issues in this cross-boundary scenario as the varying levels of legal protection and management across national jurisdictions, military conflicts, poaching of leopards from international trade, and infrastructure developments. For example, it is argued that for the Persian leopard, the political leaders in Turkmenistan, Kazakhstan, and Afghanistan have not developed adequate conservation protections, which may affect the conservation actions in states with shared borders, contrast to Oman, where protections are stronger (Farhadinia et al., 2017). The same situation exists for the Indochinese leopard in Cambodia, China, Myanmar, and Laos, which may have insufficient protections (Farhadinia et al., 2017).

It is also argued that military conflicts, political unrest, poor border security, and insufficient control of illegal animal trade have weakened political supports for conservation, especially with the territories and border zones of Yemen and Saudi Arabia (armed conflict), Armenia and Azerbaijan (Nagorno-Karabakh-armed conflict), northern Iraq, western Iran and Turkey (political instability, armed conflict), Afghanistan and Pakistan (armed conflict), and Myanmar, and Thailand (political instability and armed conflict) (Farhadinia et al., 2017; Oswell, 2010). There is also weak governmental control over illegal trade in big cat parts in the border areas of Thailand, Laos, China, and India, including those of Thailand–Myanmar, Iran–Iraq and Armenia–Azerbaijan borders) (Farhadinia et al., 2017; Oswell, 2010).

The political ecology of leopard conservation in Asia is also influenced by the local peoples cooperation with governing elites, especially when leopard conservation may imply increased big cat populations and related threats to people and wildlife (Campbell, 2022). For example, Naha et al. (2018) contend that "human killing is the decisive and most critical expression of human-leopard conflict and needs to be addressed sensitively to maintain local support for leopard conservation in India," using a case study of Pauri Garhwal and North Bengal in the Indian Himalayan region (IHR). According to the findings, human injuries occurred in roughly 97% of leopard attacks in North Bengal (mostly on tea estate workers, in tea estates) and 60% of leopard attacks in Pauri (mostly on children and people, in dense scrub) (Naha et al., 2018). The proportion of people who died due to leopard attacks was higher in Pauri (40%) than in North Bengal (3%). However, despite these occurrences, 75% of respondents in North Bengal and 41% in Pauri supported the conservation of leopards (Naha et al., 2018).

For the African leopard, similar problems emerge. For example, Viollaz et al. (2021) present a case study of human-leopard relations in a small village in South Africa, where people were killing leopards. Cited reasons included the relationship between the government authorities and their policies, and the local stakeholders (Viollaz et al., 2021). The government's slow and unwilling response to the human-leopard conflict, the use of ineffective methods in that response, the lack of resources to address the killings, and the absence of laws or their enforcement were the four main issues that respondents identified as driving these killings. A variety of creative tactics were offered by local partners to lessen hostility and conflict between humans and leopards (Viollaz et al., 2021). Even though each party had a different explanation for why these solutions worked or failed, their conclusions were frequently in agreement. The parties were unable to acknowledge or accept their similarities because of their mutual mistrust (Viollaz et al., 2021). Viollaz et al. (2021) cite "distrust in the government." As an important factor cited by farmers, noting that "respondents expressed feelings of anger from the perceived economic hardships imposed by leopards, ongoing land tenure and land-use issues, and the lack of government support. Distrust also emerged as an overarching emotion expressed by farmers and government officials."

Another study of human-leopard conflict is a case study in the Blouberg Mountain Range of South Africa (Constant, 2015). According to Constant (2015), disputes over the use of land and natural resources, in particular, have a significant impact on people's opinions of conservation. These disputes also have an impact on people's historical and present relationships with protected areas (Constant, 2015). Disempowerment, marginalization, and stigmatization are legacy issues that show up in social conflicts, conservation discourses, and opposition to the creation of protected areas—a process that is often regarded as traumatic (Constant, 2015). Human-leopard conflicts are exacerbated by traumatizing environments, which lowers local stakeholder support for wildlife conservation and foster mistrust of government and wildlife authorities among local people (Constant, 2015).

Constant (2015) also cites the political issues of declining traditional livestock sharing systems and the absence of substitute livelihood options, which place subsistence communal farmers at a significant financial disadvantage when livestock is killed by predators. Predation on livestock may have negative effects on local resident's well-being, social capital, material and functional benefits, and spiritual resources, with perceived cultural decline. Farming communities may support or oppose the conservation of leopards by attributing different environmental values to the leopard (Constant, 2015). Constant (2015) argues that by developing stronger working relationships with farming communities, environmental institutions can gain the trust and legitimacy of the local population and respond more effectively to reports of conflict between humans and leopards. Additionally, reducing leopard predation on livestock requires giving local government actors more control over human-leopard conflicts and giving farmers more accountability for enhancing livestock husbandry techniques (Constant, 2015). Programs for education and incentives may also be important to lowering the use of lethal control methods, increasing predator tolerance, and promoting the conservation of leopards (Constant, 2015).

These political issues, which situate local stakeholders in scenarios of conflict and possible cooperation, may also be explained at the individual and group level by conservation psychology (Campbell, 2019). Fear of big cats may exist simultaneously with the political dynamics between government and non-governmental elites, and local, mostly disempowered stakeholders (Campbell, 2017). This fear is generally real, but mindsets may emerge and dominate human-carnivore relations, and lead to inter-species generalizations, where in some cases one carnivore species is blamed for the activities of another carnivore (Kaltenborn, Bjerke, & Nyahongo, 2006; Kaltenborn, Bjerke, Nyahongo, & Williams, 2006). In some cases, a general mindset against carnivores may emerge, which fails to correctly identify the carnivore responsible for livestock losses (Campbell, 2022).

For example, Farhadinia et al. (2017) present a case of Persian leopards and wolves in a West Asian mountainous landscape with a high density of wild ungulates and livestock. In this study, wolf predation on livestock was considered to be fivefold more common per herd than that of leopard predation had positive attitudes toward leopards, but those who reported more wolf attacks tended to have more negative views of leopards (Farhadinia et al., 2017). The authors conclude that in ecosystems with multiple predators, people's perceptions of the activities of other predators may impact perceptions of each species, and this acknowledgment is crucial for preserving public tolerance, particularly with regard to large carnivores that are endangered, such as the Persian leopard, which may occur with more resilient species, in this case wolves (Farhadinia et al., 2017).

Consolee et al. (2020) present a case of the north China leopard leopard in and around Tieqiaoshan Provincial Nature Reserve in Shanxi Province, China. The leopard caused significant economic losses to the locals by raiding livestock, causing them to engage in serious conflict. Locals wanted to lessen or even eradicate the leopard because the species posed a serious threat to their livestock and cost them money. The opinions of the locals differed based on their economic pursuits. About 76% of farmers who raised livestock and 8% of farmers who raised crops believed that leopards were "very dangerous" (Consolee et al., 2020). The authors came to the conclusion that the Tieqiaoshan Province Nature Reserve was the main area where the North China leopard preyed on cattle. This is probably because the cattle were left in the wild, which made it easier for the nocturnal leopard to hunt them (Consolee et al., 2020).

Another study by Gebresenbet et al. (2018) examined the social dynamics surrounding livestock losses to lions and leopards in the Adiyo and Gimbo districts in Kafa Biosphere Reserve, Ethiopia, noting that from 2009 to 2013 there were 350 attacks by lions and 62 by leopards, despite these numbers, tolerance toward lions was higher because lions had more respect in the local culture. Protecting lions and leopards was a positive cultural practice, but respect for lions persisted even during livestock attacks (Gebresenbet et al., 2018). There was also fear of legal action, as killing lions is banned by law in Ethiopia. In some cases, lion predation preceded significant increases in livestock numbers, therefore respondents saw the loss of livestock to lions as a sign of good fortune prosperity. When lion attacks increased, some local elders started holding ceremonies to protect the animals. The majority of people supported conservation efforts in the reserve but avoided the fields out of fear of lions(Gebresenbet et al., 2018). Although there was still more tolerance for lions than leopards, lions posed greater problems in the reserve. The authors argued that retaliatory killings may increase because attitudes toward leopards were less favorable. This is despite the leopard's more adaptable diet and secretive nature compared to the lion, which enables the former to coexist with humans (Gebresenbet et al., 2018; see also Hayward et al., 2006; Odden et al., 2014; Athreya et al., 2016a, for leopard-human interactions).

Kichloo et al. (2023) present another study from the Kishtwar National Park in India, which is cited as an under-researched area in the Greater Himalayan region of

India. In this area, populated by agro-pastoralists and transhumant pastoralists, leopards are responsible for high livestock predation, cited as 71 incidents over 2 years, with households with larger livestock holdings attracting more predation. The findings of the study indicated that the respondents' age and period of activity in the Park significantly influenced their opinions regarding leopards, with increased livestock predation contributing to more negative attitudes toward leopards (Kichloo et al., 2023). The authors conclude that the existence of leopards near human settlements has often resulted in negative human-leopard interactions, leading to the creation of a negative narrative about these interactions. Livestock predation is a major cause of economic loss in affected communities and leads to negative perceptions of predators. This case study showed that financial compensation for livestock losses is a key factor in improving human-leopard coexistence. A comprehensive, cross-sector collaborative approach would help improve conflict resolution and promote positive attitudes toward these predators (Kichloo et al., 2023; see also Karanth et al., 2018; Ankit et al., 2021).

Another topic that remains poorly understood but is nevertheless important to the study of human-leopard relationships concerns leopards that invade human spaces and feed almost exclusively on domestic animals as food, sometimes more on domestic animals than on livestock. For example, Athreya et al. (2016b, p. 152) cite the case of Ahmednagar district, Maharashtra, India, where there was a "almost complete dependence of leopards Panthera pardus on domestic animals as prey in the crop lands" and "87% of the leopard's prey biomass consisted of domestic animals, with 39% consisting of domestic dogs... alone." Despite the greater density goats and cattle, dogs and cats were over-represented in the leopard's diet, and "the standing biomass of dogs and cats alone was sufficient to sustain the high density of carnivores at the study site" (Athreya et al., 2016a, p. 152). The authors conclude that there is a possibility of increased conflicts between people and leopards, and negative views of leopards, due to the high abundance of domestic prey biomass in leopard diets, and the high density of leopards in human-dominated spaces (Athreya et al., 2016b, p. 152).

Political Ecology, Conservation Psychology, and the Snow Leopard

Political ecology and conservation psychology influence the human relationship with the snow leopard, which is facilitated by the national and local political ecology (power structures including state and non-state elites, less powerful stakeholders) and the latter's social psychology, which may be based on their self-interest, economic survival, and personal security, as well as their positive and negative perceptions of big cats (World Wildlife Fund, 2023). World Wildlife Fund (2023) notes that humans are the sole predators of the snow leopard and lists current threats, including human-caused problems such as poaching, habitat loss, declines in natural prey species, and retaliatory killings resulting from human-wildlife conflict. Snow leopards are increasingly poached for their fur, bones and other body parts used in traditional medicine, and are often killed by local farmers and herders for stealing livestock such as sheep, goats, horses and yak calves. Their natural prey, such as the argali and the blue sheep, are hunted by the local population, and their habitat as a food source for livestock is becoming increasingly rare. As their natural prey and habitat become increasingly rare, snow leopards are killing livestock in areas disturbed by humans and increasingly used for grazing (World Wildlife Fund, 2023).

Numerous case studies illustrate the points above and show the importance of political ecology and the issues of conservation psychology for snow leopard status and conservation. For example, Suryawanshi et al. (2024, pp. 149–156) argue that investigations into human perceptions, attitudes, and behavior toward snow leopards have been primarily prompted by widespread snow leopard livestock predation and retaliatory killing of snow leopards by herders and other stakeholders. Research findings indicate that attitudinal drivers can be diverse and context-specific, and factors such as human individuals age, gender and education are important in attitudes toward snow leopards, and current understanding of the relationship between human attitudes and behavior toward snow leopards remains inadequate (Suryawanshi et al., 2024, p. 149).

Hanson et al. (2019) observe that the snow leopard is found in mountainous areas of 12 Asian countries and faces a variety of overlapping human-induced threats, such as the illegal wildlife trade, killings by herders in retaliation for livestock losses, habitat and prey loss, infrastructure, energy and mining developments and the alteration of climate. Crucially, the snow leopard's range includes areas outside protected areas, where conflicts occur with human land users and human attitudes may evolve according to the conflicts with these cats. China is an instructive case study, because it has the largest population and range of snow leopards (Alexander et al., 2016; Munkhtsog et al., 2024). Alexander et al. (2016, p. 308) point out that China's strong economic growth increases both threats to snow leopards and possibilities for their conservation. The results of their literature-based survey (1950–2014) focused on four "critical and interrelated topics underpinning snow leopard ecology and conservation: habitat (distribution and protected area coverage); prey (distribution and abundance, predator-prey relationships); human interactions (hunting and trade, livestock interactions and conflicts); and the underlying policy context," and the study concluded "the modest quantity and limited scope of published research on the snow leopard in China calls for a continued and intensified effort to inform and support national conservation policies" (Alexander et al., 2016, p. 308).

In another related case study of human attitudes to snow leopards in China, Alexander, Gopalaswamy, et al. (2015, pp. 1–9) assessed "local perceptions toward snow leopards in Northwest China using a framework depicting key conflict domains." This study compared human attitudes to the threats to humans and livestock of snow leopards and compared these threats and human attitudes to those posed by other large carnivores (grey wolf (Canis lupus)), Eurasian lynx (Lynx lynx) and brown bear (Ursus arctos) in rural areas near the Qilianshan National Nature

Reserve, Gansu Province, China (Alexander, Gopalaswamy, et al., 2015). The findings showed that herders of yak, sheep, and goats reported that the commonest factor for livestock losses was to predation by the four carnivore species, with such 3.6% of the livestock population lost during the previous year. The lynx was blamed for most kills (53%), well above that of snow leopards (7.8%) (Alexander, Gopalaswamy, et al., 2015). However, the findings also indicated positive attitudes to snow leopards and possible protective actions for their conservation, and minimal impacts of snow leopard presence on herding activities. However, lynx, wolves, and bears were perceived negatively, in terms of their threat to people and livestock. The authors conclude that "understanding community perceptions of threats posed by wildlife is vital for gaining community support for, and engagement in, conflict mitigation" (Alexander, Gopalaswamy, et al., 2015).

Local people's culture and religious beliefs may also influence their attitudes to environmental conservation (Campbell, 2022). For example, in another study of snow leopards in China, Hacker et al. (2021, p. 783) examine agro-pastoralists, in Yushu Prefecture in Qinghai Province, which "provides some of the largest known stretches of habitat for the Vulnerable snow leopard in China." The findings indicated that "herders had higher agreement with positive than with negative statements about snow leopards despite nearly half reporting livestock loss to snow leopards within the last 5 years. No retaliatory killing was reported. Herders with more years of formal education and fewer livestock losses were more likely to have positive attitudes whereas those with lower importance of snow leopards to their religion, fewer livestock losses, and fewer years of education were more likely to have negative attitudes" (Hacker et al., 2021, p. 783). The authors argue that these findings support the "importance of increased education and adherence to Tibetan beliefs in promoting conservation tolerance toward snow leopards in Qinghai Province, but also indicate a need for further research into the impact of livestock loss" (Hacker et al., 2021, p. 783).

The findings of Hacker et al. (2021) were consistent with other China-based studies (e.g., Alexander, Chen, et al., 2015; Li et al., 2014; Li et al., 2015), which found that positive public attitudes were associated with more years of formal education and fewer livestock losses positive attitudes among younger respondents, possibly due to more education and a reduced nomadic lifestyle of the younger generation. Students were more likely than their elders to be taught about conservation and environmental issues, as well as the potential ecological benefits of the presence of carnivores (Hacker et al., 2021). This result was also consistent with a study of human-snow leopard-wolf relations in Spiti Valley in the Indian Trans-Himalaya, which found that "factors significant at the individual level were gender, education, and age of the respondent (for wolves and snow leopards), number of income sources in the family (wolves), agricultural production, and large-bodied livestock holdings (snow leopards). At the community level, the significant factors included the number of smaller-bodied herded livestock killed by wolves and mean agricultural production (wolves) and village size and large livestock holdings (snow leopards)" (Suryawanshi et al., 2014, p. 1657).

Hacker et al. (2021) also found that livestock losses contributed to more negative attitudes toward snow leopards (unlike the Indian study just cited, Suryawanshi et al. (2014) and that livestock losses were positively correlated with number of livestock owned, indicating that herders might all experience similar rates of losses to predation. Tibetan Buddhist beliefs were also important, snow leopards were regarded as religiously important by over 90 of the responding herders, and Tibetan Buddhism "influences c. 80% of the geographical regions that overlap with snow leopard habitat" (Hacker et al., 2021, see also Li et al., 2014). Despite this, the respondents considered predation as considered the primary threat to livestock, ahead of disease and grassland degradation respectively (Hacker et al., 2021).

Another study conducted in China is that of Li et al. (2014), which also pointed out the importance of Tibetan Buddhism for snow leopard conservation. Li et al. (2014, p. 87) note that "current conservation strategies, including nature reserves and incentive programs, have limited capacities to protect snow leopards" and their study "investigated the role of Tibetan Buddhist monasteries in snow leopard conservation in the Sanjiangyuan region in China's Qinghai Province on the Tibetan Plateau." The study showed that 7674 km^2 of the 89,602 km^2 snow leopard habitat in the Sanjiangyuan region was within the core zones of the Sanjiangyuan Nature Reserve. An important finding was the spatial relation between snow leopard habitat and Buddhist monasteries, 46% of monasteries were in snow leopard habitat and another 90% were within 5 km of snow leopard habitat. By enforcing social norms and conducting proactive patrols, the 336 monasteries in the Sanjiangyuan region may be able to protect more habitat for snow leopards than the central regions of the reserve. The majority of the local herders denied ever killing any wildlife, and 42% claimed they never did so because doing so is considered a Buddhist sin. These findings demonstrated the significance of monasteries in the preservation of snow leopards. Other Tibetan Buddhist regions that are within the snow leopard's habitat may benefit from monastic protection of the species (Li et al., 2014).

Another study conducted in China is that of Hacker et al. (2020), which also pointed out the importance of local knowledge for snow leopard conservation, arguing that "pastoralists are key stakeholders in environmental management decisions in China," their "involvement in wildlife conservation and research is imperative for the success of long-term initiatives" and "despite the many opportunities for herders to be included in these efforts, biases have hindered knowledge exchange and collaborative outcomes" (Hacker et al., 2020). Hacker et al. (2020) note that pastoralists residing in the habitats of snow leopards on China's Qinghai-Tibet Plateau possess an intricate and profound understanding of their surroundings, making them valuable allies in conservation efforts and active participants in scientific investigations. Research findings included an increased emphasis on conservation issues, a better integration of Western science and indigenous ecological knowledge, more income for the community, improved trust and communication among conservation actors, more flexibility in research, and more venues for community-based conservation.

It is also argued that through active participation in research and conservation initiatives, local residents can help improve assessments of snow leopard ecology and status (Hacker et al., 2020). The interdependence of these parties may also offer supplementary and alternative revenue streams unrelated to animal husbandry. The engagement of young residents may provide a catalyst for some of them to develop into conservation participants. Through these initiatives, the added value of the natural resources that pastoralists damage be developed, and open channels of communication and information may be provided to improve conservation outcomes and applications that align with traditional values. These may create opportunities to increase understanding of snow leopard ecology and status (Hacker et al., 2020).

Another study conducted in Mongolia, noted that "Mongolia is second only to China in the number of wild snow leopards and extent of snow leopard habitat" (Munkhtsog et al., 2024, p. 555). Increased research is noted, including "the most extensive radio-collaring study of the species anywhere, and one of the more successful and novel community-based conservation programs" and "several camera-trapping surveys… allowing for rigorous estimation of the snow leopard population size in the country and a better understanding of the species ecology" (Munkhtsog et al., 2024, p. 555). However, the authors argue that although people's attitudes toward snow leopards have generally improved across much of their range in the country, snow leopards still face various threats, including poaching for trade and retaliatory killings for livestock predation. Prey animals such as mountain ungulates also compete with livestock for land space and are therefore decreasing in number (Munkhtsog et al., 2024, p. 555).

Several studies have also been conducted on snow leopard interactions in Nepal. For example, Hanson et al. (2019) present a case study of the Nepalese Himalayas, including Sagarmatha National Park, which has a less decentralized management model, and Annapurna Conservation Area, which has a more decentralized model. They analyzed human attitudes toward snow leopards and snow leopard conservation, the reasons behind these attitudes, and the specific factors that best explain them. The findings demonstrated that local attitudes toward snow leopards were most strongly influenced by attitudes toward snow leopard conservation and vice versa (Hanson et al., 2019). This applied to both protected areas, despite the disparities in their management systems. Human age, the number of animals kept, years of education, and household income were other significant explanatory variables. Furthermore, the most often mentioned factor by respondents to justify their opinions regarding snow leopards and

Another Nepalese study is that of Chetri et al., 2020, which also examined local attitudes to snow leopards and compared these with attitudes to wolves. The findings demonstrated that people's tolerance for snow leopards was significantly higher than that of wolves. Remarkably, perceptions of carnivores were not significantly affected by past livestock losses (Chetri et al., 2020). Occupation had a significant impact on respondents' opinions about snow leopards but not wolves. Age and literacy alone had little bearing on how snow leopards were perceived, but their combination had a significant impact. For older respondents, the negative effects of illiteracy were greater. The most significant predictors of wolf perceptions among the various factors were gender and the quantity of animals kept (Chetri et al. (2020). More negative attitudes toward wolves were found among those who kept larger herds of livestock. In terms of gender, males showed greater acceptance of wolves than females, but snow leopards showed no such tendency. The isolation of the villages was also associated with people's unfavorable attitudes regarding wolves. The study argued that it was impossible to generalize factors influencing people's perceptions across the two species (Chetri et al. (2020).

Another study, also taken for the Annapurna Conservation Area, Nepal, is that of Tiwari et al. (2020), who examined the patterns and factors of livestock depredation by snow leopards, and herders and public attitudes to these occurrences. The findings indicated that during the 2 years (2017–2019) of the study period, 74% of the respondent households lost livestock to snow leopards, with an annual average loss of two livestock per household. The losses were mostly yak (55.4%) followed goats (31.7%), sheep (6.8%), horses (3.2%), and cattle (2.8%) (Tiwari et al., 2020). The number of livestock killings depended on the number of livestock kept and the number of livestock species with larger body sizes. Many herders (41%) believed the snow leopard's preference for domestic livestock was the main predation factor, and 5% blamed poor husbandry practices for livestock losses. The majority of herders (71%) viewed better herding as a solution to snow leopard predation on livestock, while some (15%) viewed total snow leopard eradication as a solution (Tiwari et al., 2020). In addition, herders' tolerance and awareness of snow leopard conservation increased. This was attributed to the Buddhist religion and law enforcement within the sanctuary. The authors recommended a community-based livestock insurance system to offset economic losses in livestock, improved husbandry practices for livestock safety, and sustainable protection of snow leopards (Tiwari et al., 2020).

Another Nepalese study is that of Ikeda, 2004, which considered the economic impacts of livestock killing by the snow leopard in the Kanchenjunga Conservation Area. The study noted impacts of livestock predation on herders, the retaliatory killings of snow leopards, and conservation efforts by conservation stakeholders. The authors presented that for the study period, the average annual damage did "not appear to have adversely affected fundamental livelihoods in households with an average herd size (36.6 head)," but in "the worst scenario of livestock depredation, households with medium or small-sized herds (<40 head) might risk their living conditions becoming unsustainable or having to withdraw from yak pastoralism" (Ikeda (2004). The authors noted that the majority of herders had a negative attitude toward the snow leopard conservation policy, with the exception of those who had no opinion at all regarding the regional conservation and development program. For the snow leopard conservation program to be successful, compensation for herders' households for damage to livestock would be necessary.

Conclusions

Political and economic structures and developments have the power to influence or even determine the existence or extinction, the expansion or decline, the composition and characteristics, and the human dimension of species. For this reason, political ecology, broadly defined as political economy applied to the environment, is an essential background for conservation biogeography. This dynamic is closely related to psychology, particularly environmental and/or conservation psychology, which can characterize or explain how people think about wildlife and their habitats. Yet little research has addressed the triangle that exists between political ecology, psychology, and conservation biogeography.

The development of political ecology, its relationship to wildlife ecology and conservation biogeography, and the role that related streams of psychology play in these dynamics have been discussed in this chapter. Aspects such as conservation and protected area developments, social conservation movements, the effects of social power and stratification, government policy and participation, the changing influence of stakeholder interactions, the role of human-wildlife conflict, population growth and land use, and animal adaptation are also considered applications of the paradigms of political ecology and conservation psychology. These species pose a real or imagined threat to human life as they attack and even eat humans, large pets, and livestock. They are also considered charismatic and powerful in the human psyche, are heavily involved in human-animal conflicts, and require large, pristine habitats. In studying such animals, political ecology and conservation psychology provide an explanatory framework for understanding human interactions and perceptions behind conservation decisions in conservation biogeography. This contributes to knowledge about protected area management and policies related to conservation.

References

Abell, J. (2012). Volunteering to help conserve endangered species: An identity approach to human–animal relationships. *Journal of Community & Applied Social Psychology, 23*(2), 157–170. https://doi.org/10.1002/casp.2114

Adams, M. (2014). Approaching nature, 'sustainability' and ecological crises from a critical social psychological perspective. *Social and Personality Psychology Compass, 8*(6), 251–262. https://doi.org/10.1111/spc3.12104

Agrawal, B. (1998). The gender and environment debate. In R. Keil, D. V. J. Bell, P. Penz, & L. Fawcett (Eds.), *Political ecology* (pp. 193–220). Routledge.

Alexander, J., Chen, P., Damerell, P., Youkui, W., Hughes, J., Shi, K., et al. (2015). Human wildlife conflict involving large carnivores in Qilianshan, China and the minimal paw-print of snow leopards. *Biological Conservation, 187*, 1–9. https://doi.org/10.1016/j.biocon.2015.04.002

Alexander, J. S., Gopalaswamy, A. M., Shi, K., & Riordan, P. (2015). Face value. Towards robust estimates of snow leopard densities. *PLoS One, 10*(8), e0134815. https://doi.org/10.1371/journal.pone.0134815

Alexander, J. S., Zhang, C., Shi, K., & Riordan, P. (2016). A spotlight on snow leopard conservation in China. *Integrative Zoology, 11*(4), 308–321. https://doi.org/10.1111/1749-4877.1220

Altrichter, M., Boaglio, G., & Perovic, P. (2006). The decline of jaguars Panthera onca in the Argentine Chaco. *Oryx, 40*, 302–309.

American Press Institute. (2015). *How millennials use and control social media*. Retrieved from https://www.americanpressinstitute.org/publications/reports/survey-research/millennials-social-media/

Andersone, Z., & Ozolins, J. (2004). Public perception of large carnivores in Latvia. *Ursus, 15*, 181–187. https://doi.org/10.2192/1537-176(2004)015<0181:PPOLCI>2.0.CO;2

Angelici, F. M., Mahama, A., & Rossi, L. (2015). The lion in Ghana: Its historical and current status. *Animal Biodiversity and Conservation, 38*(2), 151–162.

Ankit, K., Ghanekar, R., Morey, B., Mondal, I., Khandekar, V., Jayramegowda, R., Mondol, S., Nigam, P., & Habib, B. (2021). Inhabiting terra incognita: Two-decadal patterns of negative human–leopard interactions in human-dominating landscape of Maharashtra, India. *Global Ecology and Conservation, 29*, e01740. https://doi.org/10.1016/j.gecco.2021.e01740

Arts, K., van der Wal, R., & Adams, W. M. (2015). Digital technology and the conservation of nature. *Ambio, 44*, 661–673. https://doi.org/10.1007/s13280-015-0705-1

Asah, S. T., & Blahna, D. J. (2013). Practical implications of understanding the influence of motivations on commitment to voluntary urban conservation stewardship. *Conservation Biology, 27*(4), 866–875. https://doi.org/10.1111/cobi.12058

Athreya, V., Odden, M., Linnell, J. D., & Karanth, K. U. (2011). Translocation as a tool for mitigating conflict with leopards in human-dominated landscapes of India. *Conservation Biology: The Journal of the Society for Conservation Biology, 25*(1), 133–141. https://doi.org/10.1111/j.1523-1739.2010.01599.x

Athreya, V., Odden, M., Linnell, J. D. C., Krishnaswamy, J., & Karanth, K. U. (2016a). A cat among the dogs: Leopard Panthera pardus diet in a human-dominated landscape in western Maharashtra, India. *Oryx, 50*(1), 156–162. https://doi.org/10.1017/S0030605314000106

Athreya, V., Odden, M., Linnell, J. D. C., Krishnaswamy, J., & Karanth, K. U. (2016b). A cat among the dogs: Leopard Panthera pardus diet in a human-dominated landscape in western Maharashtra, India. *Oryx, 60*, 156–162.

Baer, H. A. (1996). Bringing political ecology into critical medical anthropology: A challenge to biocultural approaches. *Medical Anthropology, 17*(2), 129–141. https://doi.org/10.1080/01459740.1996.9966132

Bakels, J. (2013). In R. Corbey & A. Lanjouw (Eds.), *The politics of species: Reshaping our relationships with other animals*. Cambridge University Press.

Balme, G. A., Slotow, R., & Hunter, L. T. B. (2009). Impact of conservation interventions on the dynamics and persistence of a persecuted leopard (Panthera pardus) population. *Biological Conservation, 142*(11), 2681–2690.

Balme, G. A., Slotow, R., & Hunter, L. T. B. (2010). Edge effects and the impact of nonprotected areas in carnivore conservation: leopards in the Phinda–Mkhuze Complex, South Africa. *Animal Conservation, 13*, 315–323. https://doi.org/10.1111/j.1469-1795.2009.00342.x

Baral, N., & Henien, J. T. (2007). Resources use, conservation attitudes, management intervention and park people relations in the Western Terai landscape of Nepal. *Environmental Conservation, 34*(1), 64–72. https://doi.org/10.1017/S0376892907003670

Bauer, H., Chapron, G., Nowell, K., Henschel, P., Funston, P., Hunter, L. T., MacDonald, D. W., & Packer, C. (2015). Lion (Panthera leo) populations are declining rapidly across Africa, except in intensively managed areas. *Proceedings of the National Academy of Sciences of the United States of America, 112*(48), 14894–14899. https://doi.org/10.1073/pnas.1500664112

Bauer, H., Chardonnet, B., Scholte, P., Kamgang, S. A., Tiomoko, D. A., Tehou, A. C., Sinsin, B., Gebresenbet, F., Asefa, A., Bobo, K. S., Garba, H., Abagana, A. L., Diouck, D., Mohammed, A. A., & Sillero-Zubiri, C. (2021). Consider divergent regional perspectives to enhance wildlife conservation across Africa. *Nature Ecology & Evolution, 5*(2), 149–152. https://doi.org/10.1038/s41559-020-01343-6

Bauer, H., Packer, C., Funston, P.F., Henschel, P., & Nowell, K. (2016). Panthera leo (errata version published in 2017). The IUCN red list of threatened species 2016. https://doi.org/10.2305/IUCN.UK.2016-3.RLTS.T15951A107265605.en

Bechtel, R. B., & Churchman, A. (2002). *Handbook of environmental psychology*. John Wiley and Sons.

Becken, S., Stantic, B., Chen, J., Alaei, A. R., & Connolly, R. M. (2017). Monitoring the environment and human sentiment on the great barrier reef: Assessing the potential of collective sensing.

Journal of Environmental Management, 203, 87–97. https://doi.org/10.1016/j.jenvman.2017.07.007

Becker, M., McRob, R., Watson, F., Droge, E., Kanyembo, B., Murdoch, J., & Kakumbi, C. (2013). Evaluating wire-snare poaching trends and the impacts of by-catch on elephants and large carnivores. *Conservation Biology, 158*, 26–36. https://doi.org/10.1016/j.biocon.2012.08.017

Becker, M. S., Almeida, J., Begg, C. M., Bertola, L. D., Breitenmoser, C., Breitenmoser, U., Coals, P., Funston, P. J., Gaylard, A., Groom, R. J., Henschel, P., Ikanda, D., Jorge, A. A., Kruger, J., Lindsey, P. A., Maimbo, H., Mandisodza-Chikerema, R., Maude, G., Mbizah, M. M., Miller, S. M., Mudongo, E. I., Mwape, H., Mweetwa, T., Naude, V. N., Nyirenda, V. R., Parker, A., Parker, D. M., Reid, C., Robson, A., Sayer, E., Selier, S. A., Sichande, M., Simukonda, C., Uiseb, K., Williams, V. L., Zimba, D., & Hunter, L. T. (2022). Guidelines for evaluating the conservation value of African lion (Panthera leo) translocations. *Frontiers in Conservation Science, 3*, 963961. https://doi.org/10.3389/fcosc.2022.963961

Bell, P. A., Greene, T. C., Fisher, J. D., & Baum, A. (Eds.). (2001). *Environmental psychology* (5th ed.). Harcourt.

Benjaminsen, T. A., & Svarstad, H. (2019). Political ecology a critical engagement with global environmental issues. In B. Fath (Ed.), *Encyclopedia of ecology* (Vol. 4, 2nd ed., pp. 391–396). Palgrave Macmillan. https://doi.org/10.1007/978-3-030-56036-2

Bennett, N. J., Roth, R., Klain, S. C., Chan, K. M., Christie, P., Clark, D. A., Cullman, G., Curran, D., Durbin, T. J., Epstein, G., Greenberg, A., Nelson, M. P., Sandlos, J., Stedman, R. C., Teel, T. L., Thomas, R. E., Veríssimo, D., & Wyborn, C. (2018). Conservation social science: Understanding and integrating human dimensions to improve conservation. *Biological Conservation, 205*, 93–108. https://doi.org/10.1016/j.biocon.2016.10.006

Benton, T., & Redclift, M. (1994). Introduction. In M. Redclift & T. Benton (Eds.), *Social theory and the global environment* (pp. 1–27). Routledge.

Bergman, J. N., Buxton, R. T., Lin, H., Lenda, M., Attinello, K., Hajdasz, A. C., Rivest, S. A., Tran Nguyen, T., Cooke, S. J., & Bennett, J. R. (2022). Evaluating the benefits and risks of social media for wildlife conservation. *FACETS, 7*, 360–339. https://doi.org/10.1139/facets-2021-0112

Blaikie, P. (1999). A review of political ecology: Issues, epistemology and analytical narratives. *Zeitschrift fur Wirtschaftsgeographie, 43*(3–4), 131–147. https://doi.org/10.1515/zfw.1999.0009

Blaikie, P., & Brookfield, H. (1987). *Land degradation and society*. Methuen.

Bombieri, G., Penteriani, V., Almasieh, K., Ambarlı, H., Ashrafzadeh, M. R., Das, C. S., Dharaiya, N., Hoogesteijn, R., Hoogesteijn, A., Ikanda, D., Jędrzejewski, W., Kaboli, M., Kirilyuk, A., Jangid, A. K., Sharma, R. K., Kushnir, H., Lamichhane, B. R., Mohammadi, A., Monroy-Vilchis, O., Mukeka, J. M., & Delgado, M. D. M. (2023). A worldwide perspective on large carnivore attacks on humans. *PLoS Biology, 21*(1), e3001946. https://doi.org/10.1371/journal.pbio.3001946

Boomgaard, P. (2013). *Frontiers of fear: Tigers and people in the Malay world, 1600–1950* (p. 322). Yale University Press.

Born Free Foundation. (2008). *Too much pressure to handle? Lion derivatives used in traditional medicine in Nigeria, West Africa*. Born Free Foundation.

Bouse, D. (2003). False intimacy: Close-ups and viewer involvement in wildlife films. *Visual Studies, 18*(2), 123–132. https://doi.org/10.1080/14725860310001631994

Boyd, D., & Crawford, K. (2012). Critical questions for big data. *Information, Communication & Society, 15*(5), 662–679. https://doi.org/10.1080/1369118X.2012.678878

Braun, B., & Castree, N. (1998). *Remaking reality. Nature at the millennium*. Routledge. https://doi.org/10.4324/9780203983966

Breitenmoser, C., Breitenmoser, U., Mallon, D., & Zazanashvili, N. (2007). *Strategy for the conservation of the leopard in the Caucasus ecoregion*. IUCN.

Breitenmoser, U., Mallon, D.P., Ahmad Khan, J., & Driscoll, C. (2008). Panthera leo ssp. persica. IUCN red List of threatened species. Version 2013.2.

Brink, H., Smith, R. J., Skinner, K., & Leader-Williams, N. (2016). Sustainability and long term tenure: Lion trophy hunting in Tanzania. *PLoS One, 11*(9), e0162610. https://doi.org/10.1371/journal.pone.0162610

Bruskotter, J. T., Vucetich, J. A., Manfredo, M. J., Karns, G. R., Wolf, C., Ard, K., Carter, N. H., López-Bao, J. V., Chapron, G., Gehrt, S. D., & Ripple, W. (2017). Modernization, risk, and conservation of the world's largest carnivores. *Bioscience, 67*(7), 646–655. https://doi.org/10.1093/biosci/bix049

Bryant, R. L. (1998). Power, knowledge and political ecology in the third world: A review. *Progress in Physical Geography: Earth and Environment, 22*(1), 79–94. https://doi.org/10.1177/030913339802200104

Bryant, R. L., & Bailey, S. (1997). *Third world political ecology*. Routledge.

CAISE Admin. (2016). *Impact of media on public perceptions of wildlife*. Retrieved from http://www.informalscience.org/news-views/impactmedia-public-perceptions-wildlife

Campbell, M. (2012). The relevance of age and gender for public attitudes to Brown bears (Ursus arctos), black bears (Ursus americanus), and cougars (Puma concolor) in Kamloops, British Columbia. *Society & Animals, 21*(2013), 341–359. https://doi.org/10.1163/15685306-12341260

Campbell, M. (2017). *Biological conservation in the twenty first century: A conservation biology of large wildlife*. Nova Science Publishers.

Campbell, M. (2018). The ecology and conservation of the West African lion. In O. P. Jenkins (Ed.), *Advances in animal science and zoology* (pp. 73–110). Nova Science Publishers.

Campbell, M. (2019). Conservation psychology, social media, and the conservation of large carnivores. In A. Colombus (Ed.), *Advances in psychology research, 138* (pp. 1–48). Nova Science Publishers.

Campbell, M. (2022). The two tigers: The jaguar ('El Tigre') of the Americas and the Tiger of Asia. In K. K. Jha & M. Campbell (Eds.), *Dynamics and interrelations between nature, science, and society* (pp. 83–123). Nova Science Publishers.

Campbell, M., & Jha, K. K. (2021). *Critical research techniques in animal and habitat ecology international examples*. Nova Science Publishers.

Campbell, M., & Lancaster, B. L. (2010). Public attitudes towards black bears Ursus americanus and cougars Puma concolor on Vancouver Island. *Journal of Society and Animals, 10*(1), 47–51. https://doi.org/10.1163/106311110X12586086158448

Campbell, M., & Torres Alvarado, M. E. (2011). Public perceptions of jaguars Panthera onca, pumas Puma concolor and coyotes Canis latrans in El Salvador. *Area, 43*(3), 250–256. https://doi.org/10.1111/j.1475-4762.2011.00996.x

Cantrill, J. G. (1998). The environmental self and a sense of place: Communication foundations for regional ecosystem management. *Journal of Applied Communication Research, 26*(3), 301–318. https://doi.org/10.1080/00909889809365509

Caruso, F., Perovic, P. G., & Altrichter, M. (2017). Actitudes y percepciones sociales ante el jaguar (Panthera onca) en el noroeste argentino. In C. Castaño-Uribe, C. A. Lasso, R. Hoogesteijn, & E. Payán-Garrido (Eds.), *Conflicto entre Felinos y Humanos en América Latina* (pp. 349–361). Serie Editorial Fauna Silvestre.

Caruso, F., Perovic, P. G., Tálamo, A., Trigo, C. B., Andrade-Díaz, M. S., Marás, G. A., & Altrichter, M. (2020). People and jaguars: New insights into the role of social factors in an old conflict. *Oryx, 54*(5), 678–686. https://doi.org/10.1017/S0030605318001552

Castells, M. (2010). *The rise of the network society*. John Wiley Sons.

Chardonnet, P., Soto, B., Fritz, H., Crosmary, W., Drouet-Hoguet, N., Mesochina, P., Pellerin, M., Mallon, D., Bakker, L., Boulet, H., & Lamarque, F. (2010). Managing the conflicts between people and lions. In *Wildlife management working paper 13*. FAO.

Chess, C. (2000). Evaluating environmental public participation: Methodological questions. *Journal of Environmental Planning and Management, 43*, 769–784. https://doi.org/10.1080/09640560020001674

Chetri, M., Odden, M., Devineau, O., McCarthy, T., & Wegge, P. (2020). Multiple factors influence local perceptions of snow leopards and Himalayan wolves in the Central Himalayas, Nepal. *PeerJ, 8*, e10108. https://doi.org/10.7717/peerj.10108

Clayton, S., & Brook, A. (2005). Can psychology save the world? A model for conservation psychology. *Analyses of Social Issues and Public Policy, 5*(1), 87–102. https://doi.org/10.1111/j.1530-2415.2005.00057.x

Coghlan, A. (2008). *Towards a close integration of social and natural sciences*. https://research-repository.griffith.edu.au/items/b3d56f1f-32f4-50f8-99fa-0d36955c843b

Conforti, V. L., & Azevedo, F. C. C. (2003). Local perceptions of jaguars Panthera once and pumas Puma concolor in the Iguacu National Park area South Brazil. *Biological Conservation, 111*, 215–221. https://doi.org/10.1016/S0006-3207(02)00277-X

Consolee, K. T., Gao, C., Vitekere, K., Li, C., Yan, H., & Jiang, G. (2020). Human-leopard conflict: An emerging issue of North China leopard conservation in Tieqiaoshan provincial nature Reserve in Shanxi Province, China. *Animals, 10*(6), 996. https://doi.org/10.3390/ani10060996

Constant, N. (2015). *A social-ecological approach towards understanding conflict between leopards (Panthera pardus) and humans: Implications for leopard conservation and farming livelihoods*. Unpublished PhD thesis. Durham University, UK.

Craib, I. (1992). *Anthony giddens*. Routledge.

Craigie, I. D., Baillie, J. E. M., Balmford, A., Carbone, C., Collen, B., Green, R. E., & Hutton, J. M. (2010). Large mammal population declines in Africa's protected areas. *Biological Conservation, 143*(9), 2221–2228.

Cranston, K. A., & Khalil, K. (2022). Conservation psychology: The future of conservation built upon a legacy. *Zoo Biology, 41*(5), 377–378. https://doi.org/10.1002/zoo.21740

Crooks, K. R. (2002). Relative sensitivities of mammalian carnivores to habitat fragmentation. *Conservation Biology, 16*(2), 488–502. https://doi.org/10.1046/j.1523-1739.2002.00386.x

Crosby, T. (2018). *Cougar, or not? Researchers use social media to learn how accurately the public IDs the big cat as it expands its range*. https://blog.news.siu.edu/20181025-cougar-or-not-researchers-use-social-media-game-to-learn-how-accurately-the-public-can-identify-the-big-cat-as-it-expands-its-range/

Cvetkovich, G. T., & Werner, R. (1994). How can psychology help save the planet? A research agenda on environmental problems. In *Statement distributed by the APA taskforce on psychology and environmental problems*. American Psychological Association.

Daly, N. (2019). *Why people risk their lives for the ultimate animal selfie*. National Geographic. Retrieved from https://www.nationalgeographic.com/animals/2019/03/psychology-of-why-people-enter-wild-animalenclosures-at-zoos/

Dansereau, G., Braga, J., Ficetola, G. F., Galiana, N., Gravel, D., Maiorano, L., Montoya, J. M., O'Connor, L., Pollock, L., Thuiller, W., Poisot, T., & Barros, C. (2025). Overcoming the disconnect between species interaction networks and biodiversity conservation. *Trends in ecology & evolution*, S0169-5347(25)00156-9. Advance online publication. https://doi.org/10.1016/j.tree.2025.05.010

Dar, N. I., Minhas, R. A., Zaman, Q., & Linkie, M. (2009). Predicting the patterns, perceptions and causes of human-carnivore conflict in and around Machiara National Park, Pakistan. *Biological Conservation, 142*, 2076–2082. https://doi.org/10.1016/j.biocon.2009.04.003

Datta, A., Anand, M. O., & Naniwadekar, R. (2008). Empty forests: Large carnivore and prey abundance in Namdapha National Park, north-East India. *Biological Conservation, 141*(5), 1429–1435. https://doi.org/10.1016/j.biocon.2008.02.022

De Micheaux, F. L., & Jenia, M. (2021). Groundwater and society: Enmeshed issues, interdisciplinary approaches. In *Global groundwater source, scarcity, sustainability, security, and solutions* (pp. 359–369). Elsevier. https://doi.org/10.1016/B978-0-12-818172-0.00026-8

De Young, R. (2013). Environmental psychology overview. In A. H. Huffman & S. Klein (Eds.), *Green organizations: Driving change with IO psychology* (pp. 17–33). Routledge.

DeMarco, E. (2015). *East, West Africa could lose 50% of their lions by 2035*. Retrieved from http://www.sciencemag.org/news/2015/10/east-west-africacould-lose-50-their-lions-2035

Demeritt, D. (1998). Science, social constructivism, and nature. In B. Braun & N. Castree (Eds.), *Remaking reality: Nature at the millennium* (pp. 177–197). Routledge.
Derworiz, C. (2018, January 7). Cougar hunt in Alberta sparks debate among scientists, hunters and activists. CBC. https://www.cbc.ca/news/canada/edmonton/cougar-hunt-alberta-1.4476780
Di Minin, E., Fink, C., Hiippala, T., & Tenkanen, H. (2019). A framework for investigating illegal wildlife trade on social media with machine learning. *Conservation Biology, 33*, 210–213. https://doi.org/10.1111/cobi.13104
Di Minin, E., Fink, C., Tenkanen, H., & Hiippala, T. (2018). Machine learning for tracking illegal wildlife trade on social media. *Nature Ecology & Evolution, 2*, 406–407. https://doi.org/10.1038/s41559-018-0466-x
Di Minin, E., Slotow, R., Fink, C., et al. (2021). A pan-African spatial assessment of human conflicts with lions and elephants. *Nature Communications, 12*, 2978. https://doi.org/10.1038/s41467-021-23283-w
Di Minin, E., Tenkanen, H., & Toivonen, T. (2015). Prospects and challenges for social media data in conservation science. *Frontiers in Environmental Science, 3*, 1–6. https://doi.org/10.3389/fenvs.2015.00063
Din, N. A. (1978). Notes on two lion prides near Mweya in the Ruwenzori National Park, Uganda. *Pakistan Journal of Zoology, 10*(2), 133–138.
Dolrenry, S., Stenglein, J., Hazzah, L., Lutz, R. S., & Frank, L. (2014). A Metapopulation approach to African lion (Panthera leo) conservation. *PLoS One, 9*(2), e88081. https://doi.org/10.1371/journal.pone.0088081
Dove, M. R., & Carpenter, C. (Eds.). (2008). *Environmental anthropology: A historical reader*. Blackwell Publishing.
Driciru, M. (1999). *Lions of queen Elizabeth, their population and health status*. Makerere University.
Driciru, M., Siefert, L., & Mapesa, M. (2005). Status of lions in Uganda. *Unpublished Report to Uganda Wildlife Authority*.
Elliot, J. (1994). *An introduction to sustainable development in the developing world*. Routledge.
Elmhirst, R. (2011). Introducing new feminist political ecologies. *Geoforum, 42*(2), 129–132. https://doi.org/10.1016/j.geoforum.2011.01.006
Elwood, S. (2010). Mixed methods: Thinking, doing, and asking in multiple ways. In D. Delyser, S. Herbert, S. Aitken, M. Crang, & L. McDowell (Eds.), *Qualitative geography* (pp. 94–113). Sage.
Ergas, C., McKinney, L., & Bell, S. E. (2021). Intersectionality and the environment. In B. Schaefer Caniglia, A. Jorgenson, S. A. Malin, L. Peek, D. N. Pellow, & X. Huang (Eds.), *Handbook of environmental sociology. Handbooks of sociology and social research* (pp. 15–34). Springer. https://doi.org/10.1007/978-3-030-77712-8_2
Ericsson, G., & Heberlein, T. A. (2003). Attitudes of hunters, locals, and the general public in Sweden now that the wolves are back. *Biological Conservation, 111*(2), 149–159. https://doi.org/10.1016/S0006-3207(02)00258-6
Escobar, A. (1995). *Encountering development: The making and unmaking of the third world*. Princeton University Press.
EWCA. (2012). *National Action Plan for the conservation of lions in Ethiopia*. Addis Ababa.
Fleischman, F. D., Ban, N. C., Evans, L. S., Epstein, G., Garcia-Lopez, G., & Villamayor-Tomas, S. (2014). Governing large-scale social-ecological systems: Lessons from five cases. *International Journal of the Commons, 8*(2), 428. https://doi.org/10.18352/ijc.416
Farhadinia, M. S., Johnson, P. J., Hunter, L. T. B., & Macdonald, D. W. (2017). Wolves can suppress goodwill for leopards: Patterns of human-predator coexistence in northeastern Iran. *Biological Conservation, 213*, 210–217. https://doi.org/10.1016/j.biocon.2017.07.011
Fernández-Sepúlveda, J., & Martín, C. A. (2022). Conservation status of the world's carnivorous mammals (order Carnivora). *Mammalian Biology, 102*, 1911–1925. https://doi.org/10.1007/s42991-022-00305-8

Ferraro, P. J., & Pressey, R. L. (2015). Measuring the difference made by conservation initiatives: protected areas and their environmental and social impacts. Philosophical transactions of the Royal Society of London. *Series B, Biological Sciences, 370*(1681), 20140270. https://doi.org/10.1098/rstb.2014.0270

Fielding, K. S., Hornsey, M. J., & Swim, J. K. (2014). Developing a social psychology of climate change. *European Journal Social Psychology, 44*, 413–420. https://doi.org/10.1002/ejsp.2058

Fischer, A., & Young, J. C. (2007). Understanding mental constructs of biodiversity: Implications for biodiversity management and conservation. *Biological Conservation, 136*(2), 271–282. https://doi.org/10.1016/j.biocon.2006.11.024

Fox, H. E., Christian, C., Nordby, J. C., Pergams, O. R. W., Peterson, G. D., & Pyke, C. R. (2006). Perceived barriers to integrating social science and conservation. *Conservation Biology, 20*(6), 1817–1820. https://doi.org/10.1111/j.1523-1739.2006.00598.x

Frank, L., Hemson, G., Kushnir, H., & Packer, C. (2006). Lions, conflict and conservation. Background paper for the east and southern African lion conservation workshop Johannesburg, South Africa, 8–13 January 2006.

Franklin, W. L., Johnson, W. E., Sarno, R. J., & Iriarte, J. A. (1999). Ecology of the Patagonia puma Felis concolor patagonica in southern Chile. *Biological Conservation, 90*, 33–40.

Fry, R., Igielnik, R., & Patten, E. (2018). *How millennials today compare with their grandparents 50 years ago*. Pew Research Center. Retrieved from https://www.pewresearch.org/fact-tank/2018/03/16/how-millennialscompare-with-their-grandparents/

Funston, P. J. (2008). Conservation and management of lions in southern Africa: Status, threats, utilization, and the restoration option. In B. Croes, R. Buij, H. De Longh, & H. Bauer (Eds.), *Management and conservation of large carnivores in West and Central Africa* (pp. 109–131). Institute of Environmental Sciences. Retrieved from https://leofoundation.org/wp-content/uploads/2017/01/W-C-Africa_large-carnivores_seminar-2006.pdf

Gardner, G. T., & Stern, P. C. (2002). *Environmental problems and human behavior* (2nd ed.). Pearson Custom Publishing.

Gebo, B., Kabtehyimer, S., & Shibru, S. (2022). Knowledge, attitude and practice of the local people towards human–carnivore coexistence in Faragosa–Fura Landscape, Gamo Zone, southern Ethiopia. *Wildlife Biology*. https://doi.org/10.1002/wlb3.01018

Gebresenbet, F., Baraki, B., Yirga, G., Sillero-Zubiri, C., & Bauer, H. (2018). A culture of tolerance: Coexisting with large carnivores in the Kafa highlands, Ethiopia. *Oryx, 52*(4), 751–760. https://doi.org/10.1017/S0030605316001356

Gebresenbet, F., Bauer, H., Hunter, L., & Gebretensae, K. (2009). *Proceedings of the national lion conservation workshop*. EWCA.

Giddens, A. (1979). *Central problems in social theory*. Macmillan.

Giddens, A. (1984). *The constitution of society*. Polity Press.

Gittleman, J., Funk, S., MacDonald, D., & Wayne, R. (2001). *Carnivore conservation*. Cambridge University Press.

Gneezy, U. (2023). From lion killers to lion savers. *Stanford Social Innovation Review*. https://doi.org/10.48558/YW6C-ZN62

González-Hidalgo, M., & Zografos, C. (2020). Emotions, power, and environmental conflict: Expanding the 'emotional turn' in political ecology. *Progress in Human Geography, 44*(2), 235–255. https://doi.org/10.1177/0309132518824644

Green, K. E. (2016). A political ecology of scaling: Struggles over power, land and authority. *Geoforum, 74*, 88–97. https://doi.org/10.1016/j.geoforum.2016.05.007

Grove, R. (1995). *Green imperialism*. Cambridge University Press.

Hacker, C., Dai, Y., Cheng, Y., Zhang, Y., Zhang, Y., Miller, L., & Janecka, J. (2021). Determinants of herder attitudes towards the vulnerable snow leopard Panthera uncia in Yushu prefecture, China. *Oryx, 55*(5), 783–790. https://doi.org/10.1017/S0030605319001315

Hacker, C. E., Janecka, J. E., Jackson, R. M., Li, D., & Zhang, Y. (2020). Pastoralism partnerships: Recognizing the value of local involvement in China's snow leopard conservation efforts. *Sustainability, 12*(16), 6491. https://doi.org/10.3390/su12166491

Hajian, M., & Kashani, S. J. (2021). Evolution of the concept of sustainability. From Brundtland Report to sustainable development goals. In *Sustainable resource management: Modern approaches and contexts* (pp. 1–24). Elsevier. https://doi.org/10.1016/B978-0-12-824342-8.00018-3
Hakeem, S. A., & Vemuri, S. R. (1997). Structural adaption of human settlements to climate for sustainable agriculture. *Journal of Sustainable Agriculture, 10*(1), 63–74. https://doi.org/10.1300/J064v10n01_07
Hall, J., & Bale, R. (2018). *8 lion cubs killed in suspected poison attack*. Retrieved from https://www.nationalgeographic.com/animals/article/wildlife-watch-lions-poisoned-uganda-cattle-retaliation
Hannaford, A. (2019). *The tiger next door: America's backyard big cats*. Retrieved from https://www.theguardian.com/global/2019/nov/10/the-tiger-next-door-americas-backyard-big-cats
Hanson, J. H., Schutgens, M., & Leader-Williams, N. (2019). What factors best explain attitudes to snow leopards in the Nepal Himalayas? *PLoS One, 14*(10), e0223565. https://doi.org/10.1371/journal.pone.0223565
Hayward, M. W., Henschel, P., O'Brien, J., Hofmeyr, M., Balme, G., & Kerley, G. I. (2006). Prey preferences of the leopard (Panthera pardus). *Journal of Zoology, 270*(2), 298–313.
Hayward, M. W., O'Brien, J., & Kerley, G. I. H. (2007). Carrying capacity of large African predators: Predictions and tests. *Biological Conservation, 139*, 219–229.
Hazzah, L., Bath, A., Dolrenry, S., Dickman, A., & Frank, L. (2017). From attitudes to actions: Predictors of lion killing by Maasai warriors. *PLoS One, 12*(1), e0170796. https://doi.org/10.1371/journal.pone.0170796
Hazzah, L., Dolrenry, S., Naughton-Treves, L., Edwards, C. T., Mwebi, O., Kearney, F., & Frank, L. (2014). Efficacy of two lion conservation programs in Maasailand, Kenya. *Conservation Biology: The Journal of the Society for Conservation Biology, 28*(3), 851–860. https://doi.org/10.1111/cobi.12244
Hazzah, L., Mulder, M. B., & Frank, L. (2009). Lions and warriors: Social factors underlying declining African lion populations and the effect of incentive-based management in Kenya. *Biological Conservation, 142*, 2428–2437. https://doi.org/10.1016/j.biocon.2009.06.006
Heidenreich, P. (2017). *TV host faces social media backlash after posting photos with cougar he killed in Alberta*. Retrieved from https://globalnews.ca/news/3928191/tv-host-faces-social-media-backlashafter-posting-photos-with-cougar-he-killed-in-alberta/
Henschel, P., Azani, D., Burton, C., Malanda, G., Saidu, Y., Sam, M., & Hunter, L. (2010). Lion status updates from five range countries in West and Central Africa. *Cat News, 52*, 34–39.
Henschel, P., Bauer, H., Sogbohoussou, E., & Nowell, K. (2015). Panthera leo (West Africa subpopulation). The IUCN Red List of Threatened Species 2015. https://doi.org/10.2305/IUCN.UK.2015-2.RLTS.T68933833A54067639.en
Henschel, P., Coad, L., Burton, C., Chataigner, B., Dunn, A., MacDonald, D., et al. (2014). The lion in West Africa is critically endangered. *PLoS One, 9*(1), e83500. https://doi.org/10.1371/journal.pone.0083500
Holthe, V., & Baldus, R. D. (2013). *CIC position paper: Management of large carnivores in Europe*. International Council for Game and Wildlife Conservation (CIC).
Hope, K. B. (1984). *The dynamics of development and development administration*. Greenwood Publishing Group.
Howard, B. C. (2014). *Lions approach extinction in West Africa*. http://news.nationalgeographic.com/news/2014/01/140108-west-africanlions-endangered-conservation-science/
Ikanda, D. K. (2007). Assessment of man-eating outbreaks by African lions Panthera leo in south-eastern Tanzania. In J. Hughes & R. Mercer (Eds.), *Felid biology and conservation conference 17–20 September: Abstracts* (p. 53). WildCRU.
Ikeda, N. (2004). Economic impacts of livestock depredation by snow leopard Uncia uncia in the Kanchenjunga conservation area, Nepal Himalaya. *Environmental Conservation, 31*(4), 322–330. https://doi.org/10.1017/S0376892904001778

Inskip, C., Carter, N., Riley, S., Roberts, T., & MacMillan, D. (2016). Toward human-carnivore coexistence: Understanding tolerance for tigers in Bangladesh. *PLoS One, 11*(1), e0145913. https://doi.org/10.1371/journal.pone.0145913

IUCN. (2006a). *Conservation strategy for the lion in west and Central Africa.* IUCN SSC Cat Specialist Group.

IUCN. (2006b). *Conservation strategy for the lion in eastern and southern Africa.* IUCN SSC Cat Specialist Group.

Jacobson, A. P., Gerngross, P., Lemeris, J. R., Jr., Schoonoer, R. F., Anco, C., Breitenmoser-Würsten, C., Durant, S. M., Farhadinia, M. S., Henschel, P., Kamler, J. F., Laguardia, A., Rostro-García, S., Stein, A. B., & Dollar, L. (2016). Leopard (Panthera pardus) status, distribution, and the research efforts across its range. *PeerJ, 4*, e1974. https://doi.org/10.7717/peerj.1974

Jha, K. K., & Campbell, M. (2021). *Critical research techniques in animal and habitat ecology: Examples from India.* Nova Science Publishers.

Jha, K. K., & Campbell, M. (2022). *Dynamics and interrelations between nature, science and society.* Nova Science Publishers.

Jhala, Y. V., Banerjee, K., Chakrabarti, S., Basu, P., Singh, K., Dave, C., & Gogoi, K. (2019). Asiatic lion: Ecology, economics, and politics of conservation. *Frontiers in Ecology and Evolution, 7*, 312. https://doi.org/10.3389/fevo.2019.00312

Johnson, L. R., Johnson-Pynn, J. S., Sweeney, S. S., & Williams, C. T. (2009). Youth civic action: Going green, going global. *Ecopsychology, 1*(2), 75. https://doi.org/10.1089/eco.2009.0007

Kaboodvandpour, S., Almasieh, K., & Zamani, N. (2021). Habitat suitability and connectivity implications for the conservation of the Persian leopard along the Iran-Iraq border. *Ecology and Evolution, 11*(19), 13464–13474. https://doi.org/10.1002/ece3.8069

Kaltenborn, B. P., Bjerke, T., & Nyahongo, J. (2006). Living with problem animals—Self-reported fear of potentially dangerous species in the Serengeti region, Tanzania. *Human Dimensions of Wildlife, 11*(6), 397–409. https://doi.org/10.1080/10871200600984323

Kaltenborn, B. P., Bjerke, T., Nyahongo, J. W., & Williams, D. R. (2006). Animal preferences and acceptability of wildlife management actions around Serengeti National Park, Tanzania. *Biodiversity and Conservation, 15*, 4633–4649. https://doi.org/10.1007/s10531-005-6196-9

Kansky, R., Kidd, M., & Knight, A. T. (2014). Meta-analysis of attitudes toward damage causing mammalian wildlife. *Conservation Biology, 28*, 924–938. https://doi.org/10.1111/cobi.12275

Kaplan, S., & Kaplan, R. (2009). Creating a larger role for environmental psychology: The reasonable person model as an integrative framework. *Journal of Environmental Psychology, 29*(3), 329–339. https://doi.org/10.1016/j.jenvp.2008.10.005

Karanth, K. K., Gupta, S., & Vanamamalai, A. (2018). Compensation payments, procedures and policies towards human–wildlife conflict management: Insights from India. *Biological Conservation, 227*, 383–389. https://doi.org/10.1016/j.biocon.2018.07.006

Kartomi, M. J. (1977). Tigers into kittens. *Hemisphere, 22*, 23–29.

Kellert, S. R. (1994). Public attitudes toward bears and their conservation. *Bears: Their Biology and Management, 9*, 43–50. https://doi.org/10.2307/3872683

Kenya Wildlife Service. (2020). *National Recovery and action plan for lion and spotted hyena in Kenya (2020–2030).* Kenya Wildlife Service. Ministry of Tourism and Wildlife.

Kertson, B. N., & Keren, I. N. (2022). Cougar use of residential areas and interactions with people in periods of population stability and growth. *Journal of Mammalogy, 103*(2), 347–360. https://doi.org/10.1093/jmammal/gyab145

Kichloo, M. A., Sohil, A., & Sharma, N. (2023). Living with leopards: An assessment of conflict and people's attitudes towards the common leopard Panthera pardus in a protected area in the Indian Himalayan region. *Oryx.* https://doi.org/10.1017/S0030605323001278

Kidd, L. R., Gregg, E., Bekessy, S., Garrard, G., & Robinson, J. (2018). Tweeting for their lives: Visibility of threatened species on twitter. *Journal for Nature Conservation, 46*, 106–109. https://doi.org/10.1016/j.jnc.2018.10.001

Kiggwe, L. (2022). *Unique tree-climbing lions roar again in Uganda (Op-Ed).* Retrieved from https://www.livescience.com/ishasha-lions-uganda-conservation

Kinzig, A. P., & McShane, T. O. (2015). Conservation in Africa: Exploring the impact of social, economic and political drivers on conservation outcomes. *Environmental Research Letters, 10*, 090201. https://doi.org/10.1088/1748-9326/10/9/090201/pdf

Kissui, B. (2008). Livestock predation by lions, leopards, spotted hyenas, and their vulnerability to retaliatory killing in the Maasai steppe, Tanzania. *Animal Conservation, 11*, 422–432. https://doi.org/10.1111/j.1469-1795.2008.00199.x

Koger, S. M., & Scott, B. A. (2007). Psychology and environmental sustainability: A call for integration. *Teaching of Psychology, 34*(1), 10. https://doi.org/10.1080/00986280709336642

Ladle, R. J., Correia, R. A., Do, Y., Joo, G.-J., Malhado, A. C. M., Proulx, R., Roberge, J.-M., & Jepson, P. (2016). Conservation culturomics. *Frontiers in Ecology and the Environment, 14*, 269–275. https://doi.org/10.1002/fee.1260

Leff, E. (1995). *Green production.* Guilford Press.

Li, C., Jiang, Z., Li, C., Tang, S., Li, F., Luo, Z., Ping, X., Liu, Z., Chen, J., & Fang, H. (2015). Livestock depredations and attitudes of local pastoralists toward carnivores in the Qinghai Lake region, China. *Wildlife Biology, 21*(4), 204–212. https://doi.org/10.2981/wlb.00083

Li, J., Wang, D., Yin, H., Zhaxi, D., Jiagong, Z., Schaller, G. B., Mishra, C., McCarthy, T. M., Wang, H., Wu, L., Xiao, L., Basang, L., Zhang, Y., Zhou, Y., & Lu, Z. (2014). Role of Tibetan Buddhist monasteries in snow leopard conservation. *Conservation Biology: The Journal of the Society for Conservation Biology, 28*(1), 87–94. https://doi.org/10.1111/cobi.12135

Lindsey, P. A., Balme, G., Becker, M., Begg, C., Bento, C., Bocchino, C., Dickman, A., Diggle, R. W., Eves, H., Henschel, P., Lewis, D., Marnewick, K., Mattheus, J., McNutt, J. W., McRobb, R., Midlane, N., Milanzi, J., Morley, R., Murphree, M., Opyene, V., Phadima, J., Purchase, G., Rentsch, D., Roche, C., Shaw, J., Van der Westhuizen, H., Van Vliet, N., & Zisadza-Gandiwa, P. (2013). The bushmeat trade in African savannas: Impacts, drivers, and possible solutions. *Biological Conservation, 160*, 80–96. https://doi.org/10.1016/j.biocon.2012.12.020

Linnell, J. D. C., Swenson, J. E., & Andersen, R. (2001). Predators and people: Conservation of large carnivores is possible at high human densities if management policy is favourable. *Animal Conservation, 4*(4), 345–349. https://doi.org/10.1017/S1367943001001408

Löe, J., & Röskaft, E. (2004). Large carnivores and human safety: A review. *Ambio, 33*(6), 283–288. https://doi.org/10.1579/0044-7447-33.6.283

Lorand, C., Robert, A., Gastineau, A., Mihoub, J. B., & Bessa-Gomes, C. (2022). Effectiveness of interventions for managing human-large carnivore conflicts worldwide: Scare them off, don't remove them. *The Science of the Total Environment, 838*(Pt 2), 156195. https://doi.org/10.1016/j.scitotenv.2022.156195

Loveridge, A. J., Wang, S. W., Frank, L. G., & Seidensticker, J. (2010). People and wild felids: Conservation of cast and management of conflict. In D. W. MacDonald & A. J. Loveridge (Eds.), *Biology and conservation of wild felids* (pp. 161–195). Oxford University Press.

Lozano, J., Olszańska, A., Morales-Reyes, Z., Castro, A. A., Malo, A. F., Moleón, M., & Martín-López, B. (2019). Human-carnivore relations: A systematic review. *Biological Conservation, 237*, 480–492. https://doi.org/10.1016/j.biocon.2019.07.002

Luymes, G. (2017). *Cougar commute: Big cats caught strolling through SkyTrain station (with video).* Retrieved from https://vancouversun.com/news/local-news/cougar-commute-big-cats-caught-strollingthrough-skytrain-station-with-video

Ma, Y. Q. (1998). Panthera pardus. In S. Wang (Ed.), *China red data book of endangered animals: Mammalia* (pp. 116–118). Science Press.

MacDonald, D. W. (2019). Mammal conservation: Old problems, new perspectives, transdisciplinarity, and the coming of age of conservation geopolitics. *Annual Review of Environment and Resources, 44*(1), 61–88. https://doi.org/10.1146/annurev-environ-101718-033039

Macdonald, D. W., Loveridge, A. J., & Rabinowitz, A. (2010). Felid futures: Crossing disciplines, borders, and generations. In D. W. MacDonald & A. J. Loveridge (Eds.), *Biology and conservation of wild felids*. Oxford University Press.

Manchini, S. (2010). *Human dimensions of the conflicts between people and jaguars (Panthera Onca) in Brazil.* University of Oxford.

Mangunjaya, F., & McKay, J. (2012). Reviving an Islamic approach for environmental conservation in Indonesia. *Worldviews, 16*, 286–305.
Margulies, J. D. (2019). Making the 'man-eater': Tiger conservation as necropolitics. *Political Geography, 69*, 150–161. https://doi.org/10.1016/j.polgeo.2018.12.011
Margulies, J. D., & Karanth, K. K. (2018). The production of human-wildlife conflict: A political animal geography of encounter. *Geoforum, 95*, 153–164. https://doi.org/10.1016/j.geoforum.2018.06.011
Marshall, G. (1994). *The concise Oxford dictionary of sociology*. Oxford University Press.
McCarthy, J. F. (2002). First world political ecology: Lessons from the wise use movement. *Environment and Planning A, 34*(7), 1281–1302. https://doi.org/10.1068/a3526
McCarthy, J. F. (2005). Guest editorial. *Environment and Planning A, 37*(6), 953–958. https://doi.org/10.1068/a3813
McCarthy, J. F. (2006). *The fourth circle. A political ecology of Sumatra's rainforest frontier*. Stanford University Press.
McCay-Peet, L., & Quan-Haase, A. (2017). What is social media and what questions can social media research help us answer. In L. Sloan & A. Quan-Haase (Eds.), *The SAGE handbook of social media research methods* (pp. 13–26). SAGE.
McKay, J. E., John, F. A. V., Harihar, A., Martyr, D., Leader-Williams, N., Milliyanawati, B., et al. (2018). Tolerating tigers: Gaining local and spiritual perspectives on human-tiger interactions in Sumatra through rural community interviews. *PLoS One, 13*(11), e0201447. https://doi.org/10.1371/journal.pone.0201447
McKendree, M. G., Croney, C. C., & Windmar, N. J. (2014). Effects of demographic factors and information sources on United States consumer perceptions of animal welfare. *Journal of Animal Science, 92*, 3161–3173. https://doi.org/10.2527/jas.2014-6874
McMahan, B., & Nichter, M. (2011). Medical anthropology. In *Encyclopedia of environmental health* (pp. 274–281). Elsevier. https://doi.org/10.1016/B978-0-444-63951-6.00541-6
McNeeley, J. A., & Sochaczewski, P. S. (1988). *Soul of the Tiger: Searching for nature's answers in Southeast Asia 399*. University of Hawaii Press.
McShane, T. O., & Wells, M. P. (Eds.). (2004). *Getting biodiversity projects to work: Toward more effective conservation and development*. Columbia University Press.
Meena, V., Macdonald, D. W., & Montgomery, R. A. (2014). Managing success: Asiatic lion conservation, interface problems and peoples' perceptions in the Gir protected area. *Biological Conservation, 174*, 120–126. https://doi.org/10.1016/j.biocon.2014.03.025
Menon, A., & Rai, N. D. (2019). The mismeasure of nature: The political ecology of economic valuation of Tiger reserves in India. *Journal of Political Ecology, 26*(1), 652–665. https://doi.org/10.2458/v26i1.23194
Miller, G. T., Jr., & Hackett, D. (2011). *Living in the environment* (2nd ed.). Nelson Education.
Miller, S. M., Bissett, C., Parker, D. M., Burger, A., Courtenay, B., Dickerson, T., et al. (2013). Management of reintroduced lions in small, fenced reserves in South Africa: An assessment and guidelines. *South African Journal of Wildlife Research, 43*(2), 138–154. https://doi.org/10.3957/056.043.020
Miller, S. M., & Funston, P. J. (2014). Rapid growth rates of lion (Panthera leo) populations in small, fenced reserves in South Africa: A management dilemma. *South African Journal of Wildlife Research, 44*(1), 43–55. https://doi.org/10.3957/056.044.0107
Miller, S. M., Harper, C. K., Bloomer, P., Hofmeyr, J., & Funston, P. J. (2015). Fenced and fragmented: Conservation value of managed metapopulations. *PLoS One, 10*(12), e0144605. https://doi.org/10.1371/journal.pone.0144605
Minch, M. (2011). Political ecology. In D. K. Chatterjee (Ed.), *Encyclopedia of global justice*. Springer. https://doi.org/10.1007/978-1-4020-9160-5_119
Mora, J. M., Polisar, J., Portillo, H., & Franklin, C. (2016). Estado del Jaguar en Honduras. In R. A. Medellín, J. A. de la Torre, C. Chávez, H. Zarza, & G. Ceballos (Eds.), *El Jaguar en el Siglo XXI: La Perspectica Continental, Fondo de Cultura Económica*. Universidad Nacional Autónoma de México, Ciudad de México.

Moreno, R., Bustamante, A., Méndez-Carvajal, P., & Moreno, J. (2016). Jaguares (Panthera onca) en Panamá, Estado Actual y Conservación. In R. A. Medellín, J. A. de la Torre, C. Chávez, H. Zarza, & G. Ceballos (Eds.), *El Jaguar en el Siglo XXI: La Perspectiva Continental, Fondo de Cultura Económica*. Universidad Nacional Autónoma de México, Ciudad de México.

Moss, S. (2019). *How social media is inspiring children to save the natural world*. Retrieved from https://www.theguardian.com/environment/2019/jan/25/social-media-inspiring-children-to-save-natural-world

Munkhtsog, B., Augugliaro, C., Bayrakcismith, R., Munkhtsog, B., & McCarthy, T. (2024). Current status and conservation of snow leopards in Mongolia. In D. Mallon & T. McCarthy (Eds.), *Snow leopards* (2nd ed., pp. 555–564). Academic Press. https://doi.org/10.1016/B978-0-323-85775-8.00005-4

N'Diaye, C. (2014). Utilisation et Commerce des sous-produits de grands carnivores dans les communes périphériques de la Réserve de Biosphère de la Pendjari. *Ecole Polytechnique D'Abomey-Calavi*.

Naha, D., Sathyakumar, S., & Rawat, G. S. (2018). Understanding drivers of human-leopard conflicts in the Indian Himalayan region: Spatio-temporal patterns of conflicts and perception of local communities towards conserving large carnivores. *PLoS One, 13*(10), e0204528. https://doi.org/10.1371/journal.pone.0204528

Nasar, J. L. (2015). Advances in environmental psychology. *Behavioral Sciences, 5*(3), 384–387. https://doi.org/10.3390/bs5030384

Naughton-Treves, L., Grossberg, R., & Treves, A. (2003). Paying for tolerance. Rural citizens' attitudes toward wolf depredation and compensation. *Conservation Biology, 17*, 1500–1511. https://doi.org/10.1111/j.1523-1739.2003.00060.x

Neumann, R. P. (2009). Biodiversity. In R. Kitchin & N. Thrift (Eds.), *International encyclopedia of human geography* (pp. 228–233). Elsevier. https://doi.org/10.1016/B978-008044910-4.00580-0

Newman, N. (2009). The rise of social media and its impact on mainstream journalism. *Reuters Institute for the Study of Journalism, 8*(2), 1–5.

Nickerson, R. S. (2003). *Psychology and environmental change*. Lawrence Erlbaum Associates.

Norton, O. (2017). *Animal welfare frames: How social media messages bridge the gap between the protein industry and consumers*. Retrieved from http://scholarworks.uark.edu/etd/2562

Nowell, K., & Pervushina, N. (2014). *Review of implementation of resolution Conf. 12.5 (rev. CoP16) on conservation and trade in tigers and other appendix-I Asian big cats*. IUCN and TRAFFIC report prepared for the CITES Secretariat, 65th meeting of the CITES standing committee, Geneva, 7-11 July. Geneva.

O'Connor, J. (1998). *Natural causes: Essays in ecological marxism*. Guilford Press.

Odden, M., Athreya, V., Rattan, S., & Linnell, J. D. (2014). Adaptable neighbours: Movement patterns of GPS-collared leopards in human dominated landscapes in India. *PLoS One, 9*(11), e112044.

Ogada, M., Woodroffe, R., Oguge, N., & Frank, G. (2003). Limiting depredation by African carnivores: The role of livestock husbandry. *Conservation Biology, 17*(6), 1521–1530. https://doi.org/10.1111/j.1523-1739.2003.00061.x

Omoya, E. O., Mudumba, T., Buckland, S. T., Mulondo, P., & Plumptre, A. J. (2014). Estimating population sizes of lions Panthera leo and spotted hyaenas Crocuta crocuta in Uganda's savannah parks, using lure count methods. *Oryx, 48*, 394–401. https://doi.org/10.1017/S0030605313000112

Osbaldiston, R. (2013). Synthesizing the experiments and theories of conservation psychology. *Sustainability, 5*(6), 2770–2795. https://doi.org/10.3390/su5062770

Oskamp, S. (2000). A sustainable future for humanity? How can psychology help? *American Psychologist, 55*(5), 496–508. https://doi.org/10.1037/0003-066X.55.5.496

Oswell, A. (2010). *The big cat trade in Myanmar and Thailand*. TRAFFIC Southeast Asia.

Packer, C., Brink, H., Kissui, B. M., Maliti, H., Kushnir, H., & Caro, T. (2011). Effects of trophy hunting on lion and leopard populations in Tanzania. *Conservation Biology, 25*(1), 142–153. https://doi.org/10.1111/j.1523-1739.2010.01576.x

Packer, C., Canney, S., Loveridge, A., Garnett, S. T., Zander, K. K., Balme, G., Bauer, H., Begg, C., Begg, K., Bhalla, S., Bonham, R., Brink, H., Burton, C., Caro, T. M., Clegg, B., Dloniak, S., Frank, L., Funston, P., Groom, R., Heath, B., Hill, T., Hunter, L., DeIongh, H. H., Joubert, D., Kissui, B., Knocker, W., Leatham, B., Lindsey, P. A., Maclennan, S. D., MacNutt, T., Nicholls, K., Patterson, B., Plumptre, A., Salerno, J., Slotow, R., Sogbohossou, E., Stratford, K., Winterbach, C., Winterbach, H., & Polasky, S. (2013). Conserving large carnivores: Dollars and fence. *Ecology Letters, 16*(5), 635–641. https://doi.org/10.1111/ele.12091

Packer, C., Ikanda, D., Kissui, B., & Kushnir, H. (2005). Lion attacks on humans in Tanzania. *Nature, 436*, 927–928.

Packer, C., Ikanda, D., Kissui, B., & Kushnir, H. (2006). The ecology of man-eating lions in Tanzania. In *Nature & faune*. FAO.

Packer, C., Kosmala, M., Cooley, H. S., Brink, H., Pintea, L., Garshelis, D., et al. (2009). Sport hunting, predator control and conservation of large carnivores. *PLoS One, 4*, 5941. https://doi.org/10.1371/journal.pone.0005941

Palmeira, F. B., Crawshaw, P. G., Haddad, C. M., Ferraz, K. M. P., & Verdade, L. M. (2008). Cattle depredation by puma (Puma concolor) and jaguar (Panthera onca) in Central-Western Brazil. *Biological Conservation, 141*, 118–125. https://doi.org/10.1016/j.biocon.2007.09.015

Palmeira, F. B. L., & Barrella, W. (2007). Conflitos causados pela predaçao de rebanhos domésticos por grandes felines em comunidades quilombolas na Mata Atlântica. *Biota Neotropica, 7*, 119–128. https://doi.org/10.1590/S1676-06032007000100017

Patterson, B. D., Kasiki, S. M., Selempo, E., & Kays, R. W. (2004). Livestock predation by lions (Panthera leo) and other carnivores on ranches neighboring Tsavo National Parks, Kenya. *Biological Conservation, 119*, 507–516.

Payán, E., Castaño-Uribe, C., Gonzalez-Maya, J. F., Valderrama, C., Ruiz-Garcia, M., & Soto, C. (2013). Distribución y estado de conservación del Jaguar en Colombia. In E. Payán & C. Castaño-Uribe (Eds.), *Grandes Felinos de Colombia* (pp. 23–36). Fundación Herencia Ambiental Caribe, Conservación Internacional Colombia y Cat Specialist Group UICN/SSC.

Pearson, E. (2013). Conservation psychology: A gap in current Australian undergraduate psychology education? *Sustainability, 5*(12), 1266–1281. https://doi.org/10.3390/su5031266

Pearson, E. L., Lowry, R., Dorrian, J., & Litchfield, C. A. (2014). Evaluating the conservation impact of an innovative zoo-based educational campaign: 'Don't palm us off' for orangutan conservation. *Zoo Biology, 33*(3), 184–196.

Perloff, R. M. (2008). *The dynamics of persuasion: Communication and attitudes in the 21st century* (3rd ed.). Lawrence Erlbaum Associates.

Perreault, T., Bridge, G., & McCarthy, J. (2015). *Routledge handbook of political ecology*. Routledge.

Perry, R. J. (2003). *Five key concepts in anthropological thinking*. Prentice Hall.

Plesa, P. (2019). A theoretical foundation for ecopsychology: Looking at ecofeminist epistemology. *New Ideas in Psychology, 52*, 18–25. https://doi.org/10.1016/j.newideapsych.2018.10.002.pone.0026048

Poloni, R., Iannella, M., Fusco, G., & Fattorini, S. (2022). Conservation biogeography of high-altitude longhorn beetles under climate change. *Insect Conservation and Diversity, 15*(4), 429–444. https://doi.org/10.1111/icad.12570

Pomeroy, J. Y. (2012). *Analyzing municipal annexations: Case studies of Frederick and Caroline counties of Maryland 1990–2010*. University of Maryland.

Proshansky, H., Fabian, A., & Kaminoff, R. (1983). Place identity: Physical world socialization of the self. *Journal of Environmental Psychology, 3*(1), 57–83. https://doi.org/10.1016/S0272-4944(83)80021-8

Puentes, C. (2021). *Social media and environmental activism: An evolving relationship*. University of Washington.

Raza, R. H., Chauhan, D. S., Pasha, M. K. S., & Sinha, S. (2012). *Illuminating the blind spot: A study on illegal trade in leopard parts in India (2001–2010)*. Traffic International.
Raza, R. H., Chauhan, D. S., Pasha, M. K. S., & Sinha, S. (2012b). *Illuminating the blind spot: A study on illegal trade in leopard parts in India (2001–2010)*. TRAFFIC India/WWF India.
Redclift, M., & Woodgate, G. W. (1993). *Concepts of the environment in the social sciences*. In Wye college external programme - M.Sc. In Enviromental management, course file, units 6 and 10.
Richardson, D. M. (2005). Diversity, distributions and conservation biogeography. *Diversity and Distributions, 11*(1), 1–2. https://doi.org/10.1111/j.1366-9516.2005.00157.x
Riddle, B. R. (2021). Conservation biogeography: Our place in the world. *Biogeography*, 321–333. https://doi.org/10.1002/9781119882381.ch13
Riggio, J., Caro, T., Dollar, L., Durant, S. M., Jacobson, A. P., Kiffner, C., Pimm, S. L., & van Aard, L. J. (2016). Lion populations may be declining in Africa but not as Bauer et al. *Proceedings of the National Academy of Sciences of the United States of America, 113*(2), E107.
Riggio, J., Jacobson, A., Dollar, L., Bauer, H., Becker, M., Dickman, A., et al. (2012). The size of savannah Africa: A lion's (Panthera leo) view. *Biodiversity and Conservation, 22*, 17–35. https://doi.org/10.1007/s10531-012-0381-4
Ripple, W. J., Estes, J. A., Beschta, R. L., Wilmers, C. C., Ritchie, E. G., Hebblewhite, M., Berger, J., Elmhagen, B., Letnic, M., Nelson, M. P., Schmitz, O. J., Smith, D. W., Wallach, A. D., & Wirsing, A. J. (2014). Status and ecological effects of the world's largest carnivores. *Science, 343*, 1241484. https://doi.org/10.1126/science.1241484
Ritzer, G. (2008). *Modern sociological theory*. McGraw-Hill.
Robbins, P. (2004). *Political ecology: A critical introduction*. Blackwell Publishing.
Robbins, P. (2012). *Political ecology: A critical introduction* (2nd ed.).
Roberts, J. (2023). Political ecology. In F. Stein (Ed.), *Encyclopedia of anthropology*. Elsevier.
Rosenblatt, E., Becker, M. S., Creel, S., Droge, I., Mweetwa, T., Schuette, P. A., Watson, F., Merkle, J., & Mwape, H. (2014). Detecting declines of apex carnivores and evaluating their causes: An example with Zambian lions. *Biological Conservation, 180*, 176–186. https://doi.org/10.1016/j.biocon.2014.10.006
Røskaft, E., Händel, B., Bjerke, T., & Kaltenborn, B. P. (2007). Human attitudes towards large carnivores in Norway. *Wildlife Biology, 13*, 172–185. https://doi.org/10.2981/0909-6396(2007)13[172:HATLCI]2.0.CO;2
Ross, S. R., Lukas, K. E., Lonsdorf, E. V., Stoinski, T. S., Hare, B., Shumaker, R., & Goodall, J. (2008). Inappropriate use and portrayal of chimpanzees. *Science, 319*, 1487. https://doi.org/10.1126/science.1154490
Ross, S. R., Vreeman, V. M., & Lonsdorf, E. V. (2011). Specific image characteristics influence attitudes about chimpanzee conservation and use as pets. *PLoS One, 6*(7), e22050. https://doi.org/10.1371/journal.pone.0022050
Roth, A. (2019). *Don't be fooled by social media—wild animals make terrible pets*. Retrieved from https://www.nationalgeographic.com/animals/2019/01/cute-wild-animals-make-terrible-pets/
Ruths, D., & Pfeffer, J. (2014). Social media for large studies of behavior. *Science, 346*(5213), 1063–1064. https://doi.org/10.1126/science.346.6213.1063
Sandbrook, C., Adams, W. M., Büscher, B., & Vira, B. (2013). Social research and biodiversity conservation. *Conservation biology: The journal of the Society for Conservation Biology, 27*(6), 1487–1490. https://doi.org/10.1111/cobi.12141
Sanei, A., & Zakaria, M. (2011). Distribution pattern of the Persian leopard (Panthera pardus saxicolor) in Iran. *Asia Life Sciences, 7*(Supplement), 7–18.
Saunders, C. D. (2003). The emerging field of conservation psychology. *Human Ecology Review, 10*(2), 137–149. Retrieved from http://www.jstor.org/stable/24706965
Saunders, C. D., Brook, A. T., & Myers, O. E. (2006). Using psychology to save biodiversity and human well-being. *Conservation Biology, 20*(3), 702–705. https://doi.org/10.1111/j.1523-1739.2006.00435.x

Schaltegger, S., & Beständig, U. (2012). *Corporate biodiversity management handbook: A guide for practical implementation.* Federal Ministry of the Environment.

Schooler, S. L., Finnegan, S. P., Fowler, N. L., et al. (2022). Factors influencing lion movements and habitat use in the western Serengeti ecosystem, Tanzania. *Scientific Reports, 12*, 18890. https://doi.org/10.1038/s41598-022-22053-y

Schroepfer, K. K., Rosati, A. G., Chartrand, T., & Hare, B. (2011). Use of "entertainment" chimpanzees in commercials distorts public perception regarding their conservation status. *PLoS One, 6*(10), e26048.

Schwartz, M. (2017). *The lion and the cow: Conservation, pastoralism, and conflict.* Cat Watch. Retrieved from https://voices.nationalgeographic.org/2017/04/28/the-lion-and-the-cow-conservation-pastoralism-and-conflict/

Serpell, J. A. (2004). Factors influencing human attitudes to animals and their welfare. *Animal Welfare, 13*(S1), S145–S151. https://doi.org/10.1017/S0962728600014500

Shrestha, K. K., & Ojha, H. R. (2017). Theoretical advances in community-based natural resources management: Ostrom and beyond. In *Redefining diversity & dynamics of natural resources management in Asia, sustainable natural resources management in dynamic Asia* (pp. 13–40). Elsevier. https://doi.org/10.1016/B978-0-12-805454-3.00002-5

Sillero-Zubiri, C., & Laurenson, M. (2001). *Interactions between carnivores and local communities: conflict or co-existence?* (Vol. 5, pp. 282–312). Cambridge University Press.

Slappendel, C. (2022). *Tiger politics and tiger conservation: Where the stakeholders are going wrong.* Mongabay. Retrieved from https://news.mongabay.com/2022/04/tiger-politics-and-tiger-conservation-where-the-stakeholders-are-going-wrong-commentary/

Slotow, R., & Hunter, L. T. B. (2009). Reintroduction decisions taken at the incorrect social scale devalue their conservation contribution: The African lion in South Africa. In M. W. Hayward & M. J. Somers (Eds.), *Reintroduction of top-order predators* (pp. 41–73). Blackwell Publishing Ltd. https://doi.org/10.1002/9781444312034.ch3

Smaldone, D. (2007). The role of time in place attachment. In *Proceedings of the 2006 Northeastern Recreation Research Symposium GTR-NRSP-14* (pp. 47–56). Retrieved from https://www.nrs.fs.usda.gov/pubs/gtr/gtr_nrs-p-14/7-smaldone-p-14.pdf

Snaddon, J., Turner, E., & Foster, W. (2008). Children's perceptions of rainforest biodiversity: Which animals have the lions share of environmental awareness? *PLoS One, 3*(7), e2579.

Sogbohossou, E. A. (2006). *Conservation des grands carnivores Afrique de L'ouest: Perception par les populations et commerce des sous-produits.* Wildlife Conservation Society.

Sogbohossou, E. A., De Iongh, H. H., Brice, S., Snoo, G. R., & Funstone, P. (2011). Human–carnivore conflicts around Pendjari biosphere reserve, northern Benin. *Oryx, 45*(4), 569–578. https://doi.org/10.1017/S0030605310001109

Spalton, J. A., Al Hikmani, H. M., Willis, D., & Said, A. S. B. (2006). Critically endangered Arabian leopards Panthera pardus nimr persist in the Jabal Samhan nature reserve, Oman. *Oryx, 40*(3), 287–294. https://doi.org/10.1017/S0030605306000743

Stein, A. B., Athreya, V., Gerngross, P., Balme, G., Henschel, P., Karanth, U., Miquelle, D., Rostro-García, S., Kamler, J. F., Laguardia, A., Khorozyan, I., & Ghoddousi, A. (2023). Panthera pardus. The IUCN red list of threatened species 2023. https://doi.org/10.2305/IUCN.UK.2023-1.RLTS.T15954A215195554.en

Stephens, P. A. (2015). Land sparing, land sharing, and the fate of Africa's lions. *PNAS, 112*(48), 14753–14754.

Struebig, M. J., Linkie, M., Deere, N. J., Martyr, D. J., Millyanawati, B., Faulkner, S. C., Le Comber, S. C., Mangunjaya, F. M., Leader-Williams, N., McKay, J. E., & St John, F. A. V. (2018). Addressing human-tiger conflict using socio-ecological information on tolerance and risk. *Nature Communications, 9*(1), 3455. https://doi.org/10.1038/s41467-018-05983-y

Sui, D., & Goodchild, M. (2011). The convergence of GIS and social media: Challenges for GIScience. *International Journal of Geographical Information Science, 25*, 1737–1748. https://doi.org/10.1080/13658816.2011.604636

Suryawanshi, K. R., Bhatia, S., Bhatnagar, Y. V., Redpath, S., & Mishra, C. (2014). Multiscale factors affecting human attitudes toward snow leopards and wolves. *Conservation Biology, 28*(6), 1657–1666. https://doi.org/10.1111/cobi.12320

Suryawanshi, K. R., Suresh, S., Young, J., Bhatia, S., & Mishra, C. (2024). Promoting coexistence through improved understanding of human perceptions, attitudes, and behavior toward snow leopards. In D. Mallon & T. McCarthy (Eds.), *Snow leopards* (2nd ed., pp. 149–156). Academic Press. https://doi.org/10.1016/B978-0-323-85775-8.00053-4

Swanepoel, L. H., Somers, M. J., Van Hoven, W., Schiess-Meier, M., Owen, C., Snyman, A., & Dalerum, F. (2015). Survival rates and causes of mortality of leopards Panthera pardus in southern Africa. *Oryx, 49*(4), 595–603. https://doi.org/10.1017/S0030605313001282

Tenkanen, H., Di Minin, E., Heikinheimo, V., Hausmann, A., Herbst, M., Kajala, L., & Toivonen, T. (2017). Instagram, Flickr, or twitter: Assessing the usability of social media data for visitor monitoring in protected areas. *Scientific Reports, 7*, 17615. https://doi.org/10.1038/s41598-017-18007-4

Thomas, S. (2016). *A virtual life: How social media changes our perceptions.* The Chicago School of Professional Psychology. Retrieved from https://www.thechicagoschool.edu/insight/a-virtual-life/

Thone, F. (1935). Nature ramblings: We fight for grass. *The Science News-Letter, 27*(717), 14–14.

Thorn, M., Green, M., Scott, D., & Marnewick, K. (2013). Characteristics and determinants of human-carnivore conflict in South African farmland. *Biodiversity and Conservation, 22*, 1715–1730.

Tiwari, M. P., Devkota, B. P., Jackson, R. M., Chhetri, B. B. K., & Bagale, S. (2020). What factors predispose households in trans-Himalaya (Central Nepal) to livestock predation by snow leopards? *Animals, 10*(11), 2187. https://doi.org/10.3390/ani10112187

Toivonen, T., Heikinheimo, V., Finka, C., Hausmann, A., Hiippala, T., Järv, O., Tenkanen, H., & Di Minin, E. (2019). Social media data for conservation science: A methodological overview. *Biological Conservation, 233*, 298–315. https://doi.org/10.1016/j.biocon.2019.01.023

Treves, A., & Karanth, K. U. (2003). Human-carnivore conflicts and perspectives on carnivore management worldwide. *Conservation Biology, 17*(6), 1491–1499. https://doi.org/10.1111/j.1523-1739.2003.00059.x

Treves, A., Wallace, R. B., Naughton-Treves, L., & Morales, A. (2006). Co-managing human–wildlife conflicts: A review. *Human Dimensions of Wildlife, 11*(6), 383–396. https://doi.org/10.1080/10871200600984265

Treves, A., Wallace, R. B., & White, S. (2009). Participatory planning of interventions to mitigate Humanâ wildlife conflicts. *Conservation Biology, 23*(6), 1577–1587. https://doi.org/10.1111/j.1523-1739.2009.01242.x

Tumenta, P. N., Kok, S., Rijssel, J. C., Buij, R., Croes, B. M., Funston, P., & Udo De Haes, H. A. (2010). Threats of rapid extermination of the lion in a former West-African stronghold; case of the National Park, Cameroon. *African Journal of Ecology, 48*(4), 888–894. https://doi.org/10.1111/j.1365-2028.2009.01181.x

Uzzell, D., & Moser, G. (2009). Introduction: Environmental psychology on the move. *Journal of Environmental Psychology, 29*(3), 307–308. https://doi.org/10.1016/j.jenvp.2009.08.003

Uzzell, D., & Räthzel, N. (2009). Transforming environmental psychology. *Journal of Environmental Psychology, 29*(3), 340. https://doi.org/10.1016/j.jenvp.2008.11.005

Van Ordsol, K. G. (1982). *Lion predation in Rwenzori National Park, Uganda.* University of Cambridge.

Viollaz, J. S., Thompson, S. T., & Petrossian, G. A. (2021). When human-wildlife conflict turns deadly: Comparing the situational factors that drive retaliatory leopard killings in South Africa. *Animals, 11*(11), 3281. https://doi.org/10.3390/ani11113281

Waldron, A., Mooers, A. O., Miller, D. C., Nibbelink, N., Redding, D., Kuhn, T. S., Roberts, J. T., & Gittleman, J. L. (2013). Targeting global conservation funding to limit immediate biodiversity declines. *Proceedings of the National Academy of Sciences of the United States of America, 110*(29), 12144–12148. https://doi.org/10.1073/pnas.1221370110

Walker, P. A. (2005). Political ecology: Where is the ecology? *Progress in Human Geography, 29*(1), 73–82. https://doi.org/10.1191/0309132505ph530pr

Walker, P. A. (2006). Political ecology: Where is the policy? *Progress in Human Geography, 30*(3), 382–395. https://doi.org/10.1191/0309132506ph613pr

Watts, M. (2000). Political ecology. In E. Sheppard & T. Barnes (Eds.), *A companion to economic geography* (pp. 257–274). Blackwell Publishing.

Watts, M. (2009). *The dictionary of human geography*. John Wiley and Sons.

Whittaker, R. J., Araújo, M. B., Jepson, P., Ladle, R. J., Watson, J. E. M., & Willis, K. J. (2005). Conservation biogeography: Assessment and prospect. *Diversity and Distributions, 11*(1), 3–23. https://doi.org/10.1111/j.1366-9516.2005.00143.x

Whittaker, R. J., & Ladle, R. J. (2011). The Roots of Conservation Biogeography. 1–12. https://doi.org/10.1002/9781444390001.ch1

WildAid. (2019). *Tanzanians rally to protect world's largest population of lions*. Retrieved from https://wildaid.org/tanzanians-rally-to-protect-worlds-largest-population-of-lions/

Wilshusen, P. R. (2003). Exploring the political contours of conservation: A conceptual view of power in practice. In S. R. Brechin, P. R. Wilshusen, C. L. Fortwangler, & P. C. West (Eds.), *Contested nature* (pp. 41–57). University of New York Press.

Winter, D. D., & Koger, S. (2003). *Psychology of environmental problems*. Lawrence Erlbaum.

Wolch, J., Gullo, A., & Lassiter, U. (1997). Changing attitudes toward California's cougars. *Society and Animals, 5*(2), 95–116. https://doi.org/10.1163/156853097X00015

Wolf, E. (1972). Ownership and political ecology. *Anthropological Quarterly, 45*(3), 201–205.

Woodgate, G., & Redclift, M. (1998). From a 'sociology of nature' to environmental sociology: Beyond social Construction. *Environmental Values, 7*(1), 3–24. https://doi.org/10.1177/096327199800700102

Woodroffe, R., & Frank, L. G. (2005). Lethal control of African lions (Panthera leo): Local and regional population impacts. *Animal Conservation, 8*(1), 91–98. https://doi.org/10.1017/s1367943004001829

Woodroffe, R., & Ginsberg, J. R. (1998). Edge effects and the extinction of populations inside protected areas. *Science, 280*(5372), 2126–2128. https://doi.org/10.1126/science.280.5372.2126

Wolf, E. (1972). Ownership and political ecology. *Anthropological Quarterly, 45*(3), 201. https://doi.org/10.2307/3316532

World Bank. (2023). *The 3 C's vital for jaguar conservation: Coordination, connectivity, and coexistence*. Retrieved from https://www.worldbank.org/en/news/feature/2021/n11/25/the-3-c-s-vital-for-jaguar-conservation-coordination-connectivity-and-coexistence

World Wildlife Fund. (2014). *More Tigers in American backyards than in the wild*. Retrieved from https://www.worldwildlife.org/stories/more-tigers-in-american-backyards-than-in-the-wild

World Wildlife Fund. (2018). *The Ugandan lions: Don't let them die*. Retrieved from https://wwf.panda.org/wwf_news/?332812/The-Ugandan-Lions-Dont-Let-Them-Die/

World Wildlife Fund. (2022). *The Fourth Asia Tiger ministerial conference concludes in Malaysia*. Retrieved from https://tigers.panda.org/?5804966/Forth-Asia-Tiger-Ministerial-Conference-concludes

World Wildlife Fund. (2023). *Snow leopard*. Retrieved from https://www.worldwildlife.org/species/snow-leopard

Yirga, G., Amare, S., Gebresenbet, F., De Iongh, H. H., Vos, M., Sillero-Zubiri, C., & Bauer, H. (2021). Lion (Panthera leo) ecology and survival in protected areas of Ethiopia. *Mammalian Biology, 101*, 791–801. https://doi.org/10.1007/s42991-021-00137-y

Yirga, G., Gebresenbet, F., Deckers, J., & Bauer, H. (2014). Status of lion (Panthera leo) and spotted hyena (Crocuta crocuta) in Nechisar National Park, Ethiopia. *Momona Ethiopian Journal of Science, 6*, 127–137. https://doi.org/10.4314/mejs.v6i2.109714

Young, K. R. (2014). *Conservation biogeography*. Oxford Bibliographies Online Datasets. https://doi.org/10.1093/obo/9780199874002-0043

Yong, E. (2018). *Wait, have we really wiped out 60 percent of animals?* The Atlantic. Retrieved from https://www.theatlantic.com/science/archive/2018/10/have-we-really-killed-60-percent-animals-1970/574549/

Zelezny, L. C., & Schultz, P. W. (2000). Promoting environmentalism. *Journal of Social Issues, 56*(3), 365–578. https://doi.org/10.1111/0022-4537.00172

Zimmerer, K. (1996). Ecology as cornerstone and chimera in human geography. In C. Earle, K. Mathewson, & M. Kenzer (Eds.), *Concepts in human geography* (pp. 161–188). Rowman and Littlefield Publishers.

Zimmermann, A., Walpole, M. J., & Leader-Williams, N. (2005). Cattle ranchers attitudes to conflicts with jaguar Panthera onca in the Pantanal of Brazil. *Oryx, 39*, 406–412. https://doi.org/10.1017/S0030605305000992

Chapter 2
Conservation Biogeography of Big Cats in Eurasia, Africa, and the Americas

Michael O'Neal Campbell

Abstract Big cats of the Genera *Panthera* and *Puma*, in order of size tigers (*Panthera tigris* Linnaeus, 1758), lions (*Panthera leo* Linnaeus, 1758), jaguars (*Panthera onca* Linnaeus, 1758), cougars (*Puma concolor* Linnaeus, 1771), leopards (*Panthera pardus* Linnaeus, 1758), and snow leopards (*Panthera uncia* Schreber, 1775) are large obligate carnivores, perceived as potentially dangerous (to humans, livestock and companion animals), charismatic, and even charming animals. These species are considered in this study due to their powerful, charismatic status in the human psyche, their strong involvement in human-animal conflicts, their need for large, pristine habitats, their recent adaptive abilities leading to their incursions into human habitats, and in contrast to large, carnivorous birds and large mammal herbivores, their threat to human life through attacks and even consumption of humans, large domestic animals and livestock. Currently, the global status of big cats depends almost entirely on human conservation efforts. There are areas of strong protection, rarity, controversial settlement by local and more powerful human interests, and ongoing local extirpation or even national extinction of big cats. Understanding the status of big cats from the flexible perspective of conservation biogeography, which combines natural ecological and human dimensions as well as spatial and temporal aspects, is crucial for formulating conservation policy and management. This chapter provides a background for the next chapter by examining the issues of ecological conservation, the 'biogeography' in conservation biogeography, while the next chapter examines the human dimensions of conservation biogeography. It is concluded that the ecological and conservation biogeographic background questions are important for the further study of political ecology and psychology and that these approaches provide a concise explanatory framework for understanding the human dimensions of conservation biogeography. This contributes to conservation policy and protected area management.

Keywords Tiger · Lion · Leopard · Jaguar · Cougar · Snow leopard

M. O. Campbell (✉)
Simon Fraser University, Burnaby, BC, Canada
e-mail: ecol55@hotmail.com

K. K. Jha and M. O. Campbell (eds.), *Intersections of Conservation Biogeography and Wildlife Management*, https://doi.org/10.1007/978-981-95-0121-2_2

Introduction

The conservation biogeography of big cats of the Genera *Panthera* (Oken, 1816) and *Puma* (Jardine, 1834; Figs. 2.1, 2.2, 2.3, 2.4, 2.5, 2.6, 2.7, 2.8, 2.9, 2.10, 2.11, 2.12, 2.13, 2.14, 2.15, 2.16, 2.17, 2.18, 2.19, 2.20, 2.21, 2.22, 2.23, 2.24, 2.25, 2.26, 2.27, 2.28, 2.29, 2.30, and 2.31), in order of size tigers (*Panthera tigris* Linnaeus, 1758), lions (*Panthera leo* Linnaeus, 1758), jaguars (*Panthera onca* Linnaeus, 1758), cougars (*Puma concolor* Linnaeus, 1771), leopards *(Panthera pardus* Linnaeus, 1758), and snow leopards (*Panthera uncia* Schreber, 1775) is justified based on the extreme tensions that characterize relationships between humans and these big cats, including fear and danger to humans, pets and livestock; contested habitats that cannot be shared due to conflict; the charismatic status of these cats, their rarity and their considerable habitat requirements. Large carnivores and humans have always had a complicated relationship (Campbell, 2017; Lozano et al., 2019; Fernández-Sepúlveda & Martín, 2022, p. 1911). As noted by Athreya et al., 2013), "charismatic predatory species have long held a central place in global conservation strategies, both in terms of attracting the public's attention and serving as a focus for research and conservation effort." They impact human activities as they can endanger livestock and even human lives. Although large carnivores are admired and even revered for their strength, beauty, and dominant predatory behavior, they are also feared and persecuted, and over the past 200 years, their numbers have declined (Bruskotter et al., 2017; Crooks, 2002; Fernández-Sepúlveda & Martín, 2022; Linnell et al., 2001; Ripple et al., 2014).

Numerous predator species are threatened with extinction (Campbell, 2017; Gittleman et al., 2001). Fernández-Sepúlveda & Martín (2022, p. 1911) wrote

Fig. 2.1 Indian tiger. *Source* Shravan Khare-Pexels

Fig. 2.2 Young Indian tiger. *Source* Vatsal Parekh-Pexels

Fig. 2.3 Indian tiger in water. *Source* Yash Bohra-Pexels

that "carnivores are more threatened than mammals in general (26.9% of endangered species vs. 22.7%) and have a significantly higher proportion of species with declining populations (48 0.3% vs. 31.9%)." Thompson and Jenks (2010, p. 3457) write that "large terrestrial carnivores have undergone some of the largest population declines and range reductions of any species, which is of concern as they can have large effects on ecosystem dynamics and function." The animal groups with

Fig. 2.4 Tiger habitat. *Source* Keyur Nandaniya—Unsplash

Fig. 2.5 Sambar, a common tiger prey species. *Source* Anupa Uthsara-Pexels

the worst conservation status include big cats and bears, making them more vulnerable to human threats than other animal species, which has led to significant global population declines for many of these species (Campbell, 2017, 2018, 2022). The main threats to carnivores are land cover changes due to human activity, resulting in habitat loss due to deforestation and agricultural expansion, and hunting and trapping of carnivores and prey. The protection of carnivores should be a primary

Fig. 2.6 Water buffalo, a common tiger prey species. *Source* Erik Ding-Pexels

Fig. 2.7 Male and female lion resting. *Source* H. Bieser-Pixabay

concern for mammals, and initiatives targeting this group should be urgently adopted or strengthened (Fernández-Sepúlveda & Martín, 2022, p. 1911).

Human relationships with such carnivores are generally regionally uneven. North American large carnivores, namely brown and black bears, pumas, and jaguars, are generally less dangerous to humans than large carnivores in Africa and Asia, which include lions, tigers, and leopards, in terms of the number of fatal attacks. This is attributed to the gentler behavior of the former species and less contact with humans

Fig. 2.8 Male lion. *Source* Xi Serge-Pixabay

Fig. 2.9 Zebras and wildebeest, common lion species. *Source* Twilight Kenya-Pexels

on the less densely populated North American continent (Campbell, 2017). Human responses to such large carnivores can include protectiveness, tolerance, and hostility (Balme et al., 2010). Inskip et al. (2016, p. 1) point out that tolerance of wildlife is a central component of conservation strategies for many endangered species and an ongoing global challenge to effective conservation. Inskip et al. (2016) point out that intolerance includes attitudes toward a species and/or conservation judgments, particularly toward a species and its behavior. The rest of this chapter will examine the conservation status of each big cats according to their size. Figures 2.1, 2.2,

Fig. 2.10 Wildebeest, a common lion species. *Source* Jan McCarthy-Pixabay

Fig. 2.11 Jaguar Stalking Prey. *Source* PublicDomainPictures-Pixabay

2.3, 2.4, 2.5, 2.6, 2.7, 2.8, 2.9, 2.10, 2.11, 2.12, 2.13, 2.14, 2.15, 2.16, 2.17, 2.18, 2.19, 2.20, 2.21, 2.22, 2.23, 2.24, 2.25, 2.26, 2.27, 2.28, 2.29, 2.30, and 2.31 show illustrations of these species, their main prey species and habitats.

Fig. 2.12 Head of jaguar. *Source* Mike van den Bos-Unsplash

Fig. 2.13 Black (melanistic) jaguar resting. *Source* Juan Camilo Guarin P- Unsplash

Conservation Biogeography of the Tiger

The conservation of the tiger (Panthera tigris), another dominant apex predator, is an ecologically important species and popular wildlife symbol (Boomgaard, 2013). The current conservation status of the tiger is shown in the Table 2.1, as vulnerable based on the most recent assessment of the tiger in 2020 (Goodrich et al., 2022). Recent classifications of tigers are disputed. For example, Wozencraft (2005a) describe *Panthera*

Fig. 2.14 Jaguar habitat in Amazon, Guaporé, Brazil. *Source* deltreehd-Pixabay

Fig. 2.15 Capybara, a common jaguar prey, Águas de Lindoia, SP, Brazil. *Source* Rafael Rodrigues-Pexels

tigris (Linnaeus, 1758) with 8 offspring: the Siberian tiger *Panthera tigris altaica*, the South China tiger *Panthera tigris amoyensis*, the Bali tiger *Panthera tigris* balica, the Indo-Chinese tiger *Panthera tigris corbetti*, the Sunda Island tiger *Panthera tigris sondaica*, the Sumatran tiger *Panthera tigris sumatrae,* and the Caspian tiger *Panthera tigris virgata.* Kitchener (1999) notes that due to the similarity of many tiger subspecies, eventually the number of subspecies was reduced two subspecies;

Fig. 2.16 Cougar in rocky terrain. *Source* Priscilla Du Preez-Unsplash

Fig. 2.17 Head of cougar. *Source* Wildphoto1-Pixabay

Panthera tigris tigris on the Asian mainland, and *Panthera tigris sondaica* in the Greater Sunda Islands.

Wilting et al. (2015) point out that in their study of tiger genetics, the analyses showed that putative subspecies exhibited high trait overlap and minimal variation, while molecular data indicated very low diversity as a result of a significant population decline in the Late Pleistocene. In addition "although significantly more money is spent on the conservation of tigers than on any other threatened species, today

Fig. 2.18 Young cougar. *Source* LapineBlanche-Pixabay

Fig. 2.19 Common cougar Habitat, Squamish, Canada. *Source* Bryce Evans-Unsplash

only 3200–3600 tigers roam the forests of Asia, occupying only 7% of their historical range. Despite the global significance of and interest in tiger conservation, global approaches to plan tiger recovery are partly impeded by the lack of a consensus on the number of tiger subspecies or management units, because a comprehensive analysis of tiger variation is lacking" (Wilting et al., 2015). Therefore, the recommendation is that only two subspecies be recognized: the Sunda tiger *Panthera tigris sondaica*, and the continental tiger *Panthera tigris tigris*, the latter divided into northern and

Fig. 2.20 White-tailed deer, a common cougar prey species in North America. *Source* Thomas Shockey-Pexels

Fig. 2.21 Leopard walking. *Source* Rajit Galaiya-Unsplash

the southern administrative units (Wilting et al., 2015). These classifications are important, because noted by Wilting et al. (2015) programs for conservation management, including captive breeding, reintroduction efforts, and transboundary projects, depend on a stable, consistent definition of subspecies as taxonomic units, which is established by solid evidence from multiple scientific lines of inquiry, as opposed to single characteristics or haphazard descriptions of a single or small number of specimens.

Fig. 2.22 Black leopard resting. *Source* Denishan Joseph-Pexels

Fig. 2.23 Leopard in savanna habitat. *Source* Vik Joshi-Pexels

Goodrich et al. (2022) justify the classification of the tiger as endangered by the IUCN, which is "based on the declines of tiger populations and ranges in most of the 13 countries that record tiger populations: However, we recognize that there are several uncertainties in the available data from the past and that the classification of Endangered as opposed to Vulnerable is marginal." Population estimates include those of Nowell and Jackson (1996) (8262) and Seidensticker et al. (1999) (5000–7000) for the late 1990s, for the past three generations (21–30 years). Goodrich

Fig. 2.24 Leopard in mixed forest habitat. *Source* Bushland Adventure Travel-Pexels

Fig. 2.25 Leopard prey in India: Chital. *Source* Shravan Khare-Pexels

et al. (2022) present the argument that considering the most recent estimates, the total global populations of tigers may be between 3726 and 5578 individuals in ten countries, and assuming a general matured percentage of 70%, the adult population may be 3140 (2608–3905) individuals, which compared with the estimates from the 1990s above (Nowell & Jackson, 1996; Seidensticker et al., 1999) suggests a decline in population of 22% to 63% for the estimate of Seidensticker et al. (1999) and a decline of 53–68% for the estimate of Nowell and Jackson (1996). For the first

Fig. 2.26 Impala, a common leopard prey species in Africa. *Source* Tkirkgoz-Pexels

Fig. 2.27 Snow leopard. *Source* Pixel Mixer-Pixabay

estimate the tiger population decline would indicate Near Threatened to Endangered status, and for the second estimate Endangered. Based on this information and current factors for the decline of tiger number and habitats, Goodrich et al. (2022) justify the status of the tiger as endangered, and suggest a tiger population reduction over the past three generations.

In the past, tigers inhabited a wide range of Asian nations, extending from Turkey's west coast to Russia‘s east coast and from the southern Indonesian islands of Java

Fig. 2.28 Head of snow leopard head. *Source* Graeme Travers-Pexels

Fig. 2.29 Common snow leopard habitat, Nepal. *Source* Chandan Chaurasia-Unsplash

and Bali to the northern 55th parallel in Russia's far east. Tigers have vanished in the past century from Southeast Asian nations like Vietnam, Laos, and Cambodia in the 2000s, as well as from Singapore (1930s), Bali (1940s), Java and Hong Kong (1960s), Central Asia (1970s), most temperate regions of China (1980s), and tropical mainland regions of China (1990s) (Sanderson et al., 2023). Tiger populations have decreased from 1,049,430 km^2 in 2001 to 978,293 km^2 in 2020, or a 7% decline, leaving them in less than 7% of their historical range. Currently, Bangladesh, Bhutan, China, India,

Fig. 2.30 Bharal, a common snow leopard prey species. *Source* Jan Blanicky-Pixabay

Fig. 2.31 Siberian ibex, a common snow leopard prey species. *Source* Sharkolot-Pixabay

Indonesia, Malaysia, Myanmar, Nepal, Russia, and Thailand are home to tigers that are part of breeding subpopulations. Although the presence of tigers in Myanmar has been verified, populations in its east and south are probably dependent on immigrants from Thailand. Worldwide, over 60% of tiger subpopulations are thought to occur in protected areas (Sanderson et al., 2023). For example, approximately 65% of India's estimated 2967 tigers occur in tiger reserves, underscoring the importance of protected areas. While conservation efforts in these vital source areas remain

Table 2.1 Conservation status of the tiger

Issue	Description
Habitat	Forest, Savanna, Shrubland, Grassland, Wetlands (inland), Marine Coastal/ Supratidal, Artificial/Terrestrial
Status	Endangered
Population Trend	Decreasing, number of mature individuals 2608-3905,3140
Threats	Annual and perennial non-timber crops, wood and pulp plantations, livestock farming and ranching, oil and gas drilling, mining and quarrying, roads and railroads, hunting and trapping terrestrial animals, logging and wood harvesting, war, civil unrest and military exercises, fire and fire suppression, dams and water management/use, storms and flooding

Source IUCN (2006a) and Goodrich et al. (2022)

a priority, there is increasing evidence of the importance of habitats outside the boundaries of protected areas, which not only serve as critical connectivity corridors, but also have the potential to support subpopulations of tigers or provide recovery potential (Chávez et al., 2016; Goodrich et al., 2022; Jhala et al., 2020; Karanth et al., 2020; Sanderson et al., 2023; Thapa & Kelly, 2017).

Threats to tiger numbers and habitats include hunting (as a retaliation for human and livestock predation, and for body parts), habitat losses (to agriculture, urbanization and transportation development, and forestry and deforestation, related to pronounced human population increases) (Boomgaard, 2013; Campbell, 2022). Goodrich et al. (2022) note that hunting of tigers and their prey was the main reason for the decline in the tiger population during the study period. Poaching for the illegal trade in high-value tiger products such as hides, bones, meat, and tonics is a major threat to tigers, which, together with prey scarcity, has led to their recent disappearance from vast areas of otherwise suitable. Poaching of tigers is identified as a greater threat than habitat loss, as there are approximately 1 million square kilometers of uninhabited tiger habitat. Current evidence suggests that poaching has was the main cause of death for tigers in Russia and in the Lao People's Democratic Republic during the assessment period. Hunting is particularly important in Southeast Asia, with intensive trapping and poisoning, and a strong cross-border trade in tiger parts (Goodrich et al., 2022; see also Kerley et al., 2003; Sanderson et al., 2006; Chapron et al., 2008; Goodrich et al., 2008; Lynam et al. 2010; Oswell, 2010; Stoner & Pervushina, 2013; Johnson et al., 2016; Gray, Crouthers, et al., 2017; Gray, Lynam, et al., 2017; Gray et al., 2018; Hossain et al., 2018; Rasphone et al., 2019; Wong & Krishnasamy, 2019).

Habitat destruction is another factor for tiger population declines through tiger inhabited parts of Asia (Boomgaard, 2013). The primary factors contributing to the loss and fragmentation of tiger habitat are the conversion of forest land for forestry (commercial logging) and agriculture, human settlement, and the development of linear infrastructure. Tigers need a large, healthy base of ungulate prey due to their high food requirements, but they also face competition from livestock and intense hunting pressure from humans. Tiger subpopulations may go extinct as a result of habitat fragmentation brought on by urbanization, land use change, and linear infrastructure development. Where there is a disruption or serious compromise to the connectivity of their habitat, tigers may be at risk of functional or local extinction. The genetic variability (heterozygosity) of tigers from central India may decline, greatly raising the risk of extinction for small and isolated subpopulations, according to land use change simulations combined with connectivity analysis (Campbell, 2017, 2022).

Regulating human activity in already developed areas, like along roads, as well as development and land use changes surrounding protected areas, can help to mitigate these impacts to some extent (Campbell, 2022; Goodrich et al., 2022). Recent studies have also demonstrated how isolation and fragmentation can cause atypical phenotypic variation and inbreeding depression in tiger subpopulations. Linear infrastructure and other threats are changing tiger conservation landscapes across tiger range states at unprecedented rates, as evidence of the effects of habitat fragmentation on tiger and wild prey populations and movements grows (Goodrich et al., 2022; see also Gopal et al., 2010; Mondal & Nagendra, 2011; Joshi et al., 2016; Thatte et al., 2018; Harihar, Chanchani, et al., 2018; Harihar, Ghosh-Harihar, & MacMillan, 2018; Harihar et al., 2020; Carter et al., 2020; Jayadevan et al., 2020; Khan et al., 2021, Sagar et al., 2021; Palmeirim & Gibson, 2021; Campbell, 2022).

Tiger attacks on people and livestock can result in high rates of retaliatory tiger killings, create a persistent challenge for conservation efforts to develop local support for tiger conservation, and contribute to increased intolerance toward tigers among surrounding communities (Campbell, 2022). Certain regions have seen a high death toll from tigers; for instance, between 2000 and 2010, at least 40 people died in the Sundarbans mangrove forest in Bangladesh and India, and between 2014 and 2020, 320 people died in India. Retaliation: It is common knowledge that tigers are killed in retaliation for attacks on humans and livestock, which is one of the main reasons for the decline in tiger populations. But this has never been documented in a systematic way (Goodrich et al., 2022; see also Miquelle et al., 2005; Goodrich et al., 2008; Barlow et al., 2013; Manoj, 2021; Campbell, 2022). Additionally, infectious diseases in tigers are not well studied or understood, but several tiger deaths have been attributed to distemper virus infection, with examples in Sikhote-Alin Zapovednik in Russia and African swine fever affecting tiger prey in Russia and Indonesia (Campbell, 2022; Gilbert et al., 2014; Goodrich et al., 2022; Luskin et al., 2021; Quigley et al., 2010; Seimon et al., 2013).

Goodrich et al. (2022) suggest that despite declines in tiger populations and habitats in recent decades, parts of the tiger range have recently been well protected from these threats to stabilize or even increase tiger numbers. These case studies cover some protected areas in India, Nepal, Thailand, and Northeast Asia. However, in other areas, such as most of Southeast Asia, tiger populations continue to decline. It is anticipated that these regional trends may stabilize, with further declines in Southeast Asia and some increases in South and possibly Northeast Asia. Future increases in the global tiger population could be caused by losses in Southeast Asia being more than offset by increases in South Asia, which has a high tiger density (see also Dhakal et al., 2014; Duangchantrasiri et al., 2016; Jhala et al., 2019; Walston et al., 2010). However, poaching and habitat loss, as well as the recent extinctions of tigers from Vietnam, Cambodia, and the Lao People's Democratic Republic, as well as large areas of habitat throughout the species' range where the species is sparsely present, are placing severe pressure on tiger subpopulations that are either distributed or absent entirely. Therefore, caution should be exercised when assessing population stability or increase (Campbell, 2022; Goodrich et al., 2022).

With regard to conservation measures, the 13 tiger range countries, including Russia, China, Laos, Nepal, and Bangladesh, adopted the Global Tiger Recovery Program (Global Tiger Recovery Program, 2010-2022) at a tiger summit in November 2010 in St. Petersburg, Russia. The aim was to effectively double the number of tigers in the wild by 2022 through measures in the areas of habitats, poaching, smuggling and illegal trafficking, transboundary landscape management, engagement with indigenous and local communities, and tiger and habitat management and reintroducing tigers into their former range. It was held that the future of tigers depends on preventing poaching and illegal trafficking of tigers and their prey, as well as the loss and degradation of their habitat within and outside protected areas, while ensuring connectivity within tiger conservation landscapes. In addition, as only 21% of tiger habitat is protected, increased protected area systems are required, including areas outside protected areas where tigers contact resident people, and conflicts emerge. Connectivity was determined to be vital between protected habitat remnants, and the management effectiveness of existing protected areas is crucial to the recovery (Hossain et al., 2018; Stoner & Pervushina, 2013). In many areas, this requires empowering and empowering communities living in and around tiger habitats to actively contribute to conservation. Tigers have been successfully reintroduced to a number of protected areas in India and Russia (Kolipaka et al., 2017; Rozhnov et al., 2021; Sankar et al., 2010; Sarkar et al., 2016), and reintroduction is considered into former range countries including Kazakhstan and Cambodia (Chestin et al., 2017; Gray, Lynam, et al., 2017).

Conservation Biogeography of the Lion

The conservation of the lion (*Panthera leo* Linnaeus, 1758), a dominant apex predator, essential ecological mainstay, and popular wildlife symbol, is one of the most important topics in current conservation biology research and, in fact, a relevant topic for other disciplines such as applied ecology, wildlife studies, biogeography, and political ecology (Angelici et al., 2015; Becker et al., 2022; Campbell, 2018). The current conservation status of the lion is shown in the Table 2.2, as vulnerable based on the most recent assessment of the lion in 2014 (Bauer et al., 2016). Recent classification recognizes two subspecies of lions: *P. l. leo* (Linnaeus, 1758), which is the nominate subspecies, including the Asiatic lion, the extinct Barbary lion, and those in West and north-central Africa, and *P. l. melanochaita* (Smith, 1842) including the Cape lion and populations in East and Southern African regions (Kitchener et al., 2017).

Table 2.2 Conservation status of the lion

Issue	Description
Habitat	Forest, savanna, Shrubland, grassland, desert
Status	Vulnerable
Population trend	Decreasing, 23,000–39,000 mature individuals, a reduction of approximately 43% over the past 21 years (approximately three Lion generations, 1993–2014)
Threats	Housing and urban areas, annual and perennial non-timber crops Livestock farming and ranching, hunting and trapping terrestrial animals Logging and wood harvesting, war, civil unrest and military exercises

Source IUCN (2006a) and Bauer et al. (2016)

Previous classifications recognized the African lion (*P. l. leo*) and the Asiatic lion (*P. l. persica*) (Bauer et al., 2015, 2016).

Lion populations vary regionally, in terms of numbers, density, human relations, and survival prospects (Campbell, 2018). For the African lion, according to Bauer and Van Der Merwe's (2004) research, there are four types of African countries based on the presence of lions: those where lions are extinct (North Africa); those where few records were made at the time (Sudan and Somalia); those where lions roam both inside and outside of protected areas (most of Central, Eastern, and Southern Africa); and those where the few surviving lions are restricted to protected areas (some West African countries). Historically, lions ranged from Northern Africa through to Southwest Asia, Europe, and India. Extinctions occurred over these regions over centuries (Bauer et al., 2015, 2016; see also Nowell & Jackson, 1996, Sunquist & Sunquist, 2002; West & Packer, 2013).

Currently, in South Africa, lions are faring better than elsewhere, as noted by Bauer et al. (2015): "African lion populations are declining everywhere, except in four southern countries (Botswana, Namibia, South Africa, and Zimbabwe)." Population models indicate a 67% chance that lions in West and Central Africa decline by one-half, while estimating a 37% chance that lions in East Africa also decline by one-half over two decades," and "At a regional scale, lion populations in West, Central, and East Africa are likely to suffer a projected 50% decline over the next two decades, whereas lion populations are only increasing in southern Africa." This is a serious issue, as Lindsey et al. (2018) argue that protected areas (PAS) with lions require between US $1 and 2 billion or US $1000 and US $2000 per km^2 per year, but receive only US $381 million or US $200 per km^2 (median) per year. The majority of PAs that have lions are underfunded; their deficits range from $0 to $21 billion. To halt future lion and wildlife declines and realize the economic, social, and environmental benefits of healthy protected areas, governments and donors urgently need to make significant investments in protected areas.

In West Africa, the survival of the lion is at serious risk (Campbell, 2018). The factors for this situation include current issues facing the area such as the growing human population and its impact on urbanization, agriculture, and land cover changes; this also includes the depletion of prey populations and the resulting human unrest due to livestock depletion and human deaths. However, considering relationships between large carnivores and humans in relation to other species on other continents, as well as lion status in other parts of Africa, the problems facing lions in West Africa are comparatively difficult. Lion predation on people, companion animals and livestock, and public perception of these and other disruptive behaviors are major problems that could hinder conservation efforts. Lion conservation faces challenges because of problems associated with protecting carnivores on other continents, less dangerous mammals (such as jaguars and all species of bears), similarly dangerous species (such as leopards and slightly more dangerous tigers), and more adaptable species (such as cougars). Furthermore, predator conservation may be less developed and socially supported than on other continents due to a lack of international funding and support, fewer protected areas, and a lack of research results. More thorough research using parameters derived from studies of other species would be the first step.

In Asia, the only lion population is that of the Asiatic lion, (or Indian or Persian lion), which is currently limited to Gujarat, India (mostly Gir National Park and surrounding areas) (Venkataraman, 2016). The previous range was over most Southwest Asia and northern India, in some places sharing habitats with the tiger (Schnitzler & Hermann, 2019; Singh, 2017a, 2017b; Singh & Gibson, 2011; Venkataraman, 2016). Some studies point to a recovery of the Asiatic lion population. For example, Ram et al. (2022) note that the Asiatic lion is well-known for its tenacity and for being an excellent example of management and conservation. After leaving the Gir forest, lions have been settling in the nearby coastal environments for roughly 25 years. The local community's support and committed conservation management have allowed the Asiatic lion population to grow steadily over the past 60 years. Because of these efforts, the IUCN Red List has down listed the Asiatic lion from the "Critically Endangered" to the "Endangered" category. The dispersal of lions into the surrounding land cover outside Gir Protected Area is another effect of this population growth (Ram et al., 2022).

Bauer et al. (2016) offer justifications for the classification of the global lion populations as vulnerable, pointing out that the total lion population may have undergone a reduction of about 43% over the past 21 years, which approximates three Lion generations, 1993–2014). This estimation is based on time trend analysis of census data for 47 monitored Lion subpopulations, totaling an estimated 7500 lions in 2014. Bauer et al. (2016) argue that although sample Lion subpopulations have increased by 12% in four southern African countries (Botswana, Namibia, South Africa, and Zimbabwe), and in India, an observed decline of 60% in sample subpopulations outside these countries is inferred for the remainder of the African range, which is why the species is classified as vulnerable overall.

It is argued that several lion subpopulations have recently been stable, these including the subpopulation in the Gir Forest area of Gujarat, India, and some subpopulations in southern Africa. However, lion subpopulations in Angola, Central African Republic, and South Sudan are under-researched, but may be declining. Although the subpopulations stability occurs where conservation management is financially supported, many subpopulations occur in areas with little such funding, resulting in numerical decline and sometimes extinction, as in parts of West Africa (Bauer et al., 2016; see also Packer et al., 2013; Henschel et al., 2014). Hence, lions are extinct over the majority of their former range, possibly now occurring over 1,654,375 km^2, or 8% of their historical range (Bauer et al., 2016; see also Riggio et al., 2012).

Therefore, the lion population fulfills the criteria for endangered for most of its range, with an estimated rate of decline exceeding 50% in three generations; however, a small number of subpopulations within a limited geographic range numerically offset this trend (Bauer et al., 2016). This phenomenon is reflected in variable, regional assessments: Least Concern in South Africa, Endangered in India, and Critically Endangered in West Africa. Additionally, "Lions have been recently extirpated in 12 African countries and we suspect possible recent extirpation in another four" (Bauer et al., 2016; see also Breitenmoser et al., 2008; Henschel et al., 2014). Bauer et al. (2016) defend the lion population trend assessment: "Previous assessments were based on a suspected decline of the total estimated number of Lions, which necessarily included low-quality data. In the present assessment, we did not use total Lion estimates, because we had a better alternative. We now have enough good quality data for a representative sub-set of Lion subpopulations to calculate an observed decline, from which we infer a decline for the species as a whole."

Conservation Biogeography of the Jaguar

The conservation of the jaguar (*Panthera onca*), another dominant apex predator, is justified as it is an ecologically important species, and a popular wildlife symbol. The current conservation status of the jaguar is shown in the Table 2.3, as vulnerable based on the most recent assessment of the jaguar (Quigley et al., 2017). Wozencraft (2005a) lists nine subspecies recognized as of 2005. However, this is disputed. Quigley et al. (2017) note that status of jaguar subspecies is uncertain, as "although eight subspecies have been recognized, morphological and genetic analyses do not support the existence of discrete subspecies" (citing Seymour, 1989; Larson, 1997; Eizirik et al., 2001; Ruiz-Garcia et al., 2006). Additionally, "while not elevating the regional differences to the subspecies level" they note that Eizirik et al. (2001) "found evidence for four incompletely isolated phylogeographic groups: Mexico and Guatemala, southern Central America, northern South America, and South America south of the Amazon River" (Quigley et al., 2017). Ruiz-Garcia et al. (2006) are also cited as finding "that the Andes Mountains incompletely isolates Jaguar populations in Colombia" (see Quigley et al., 2017).

Integrated Taxonomic Information System (ITIS) (2022) listed some subspecies, but noted that several subspecies are not recognized (Larson, 1997; Pocock, 1939; Ruiz-Garcia et al., 2006). Pocock (1939) defined eight jaguar subspecies which were recognized by some experts (Larson, 1997, p. 107). Sixteen subspecies of jaguar were defined Nelson and Goldman (1933) before Pocock's study, but Pocock disregarded 8 of these (Larson, 1997, p. 107). The subspecies listed by the Integrated Taxonomic Information System (ITIS) (2022) were the Arizona jaguar (*Panthera*

Table 2.3 Conservation status of the jaguar

Issue	Description
Habitat	Forest, savanna, Shrubland, grassland, Wetlands (inland), artificial/terrestrial
Status	Near threatened
Population trend	Decreasing
Threats	Housing and urban areas, commercial and industrial areas, tourism and recreation areas, annual and perennial non-timber crops, wood and pulp plantations, livestock farming and ranching, oil and gas drilling, mining and quarrying, roads and railroads, utility and service lines, hunting and trapping terrestrial animals, gathering terrestrial plants, logging and wood harvesting, recreational activities, work and other activities, fire and fire suppression, dams and water management/use, other ecosystem modifications, domestic and urban waste water, industrial and military effluents, agricultural and forestry effluents

Source IUCN (2006b) and Quigley et al. (2017)

onca arizonensis Goldman, 1932), Central American jaguar (*Panthera onca centralis* Mearns, 1901), Yucatan Peninsula jaguar (*Panthera onca goldmani* Mearns, 1901), West Mexican jaguar (Panthera onca hernandesii J. E. Gray, 1857), East Brazilian jaguar (*Panthera onca onca* Linnaeus, 1758), Pantanal jaguar (*Panthera onca palustris* Ameghino, 1888), Paraguay jaguar (*Panthera onca paraguensis* Hollister, 1914), Peruvian Jaguar (*Panthera onca peruviana* de Blainville, 1843), and Northeastern jaguar (*Panthera onca veraecrucis* Nelson & Goldman, 1933) (Integrated Taxonomic Information System (ITIS), 2022).

Paviolo et al. (2016) contend that the jaguar, which presently ranges from the southern United States to northern Argentina, is the apex predator in the Neotropical zone. Its disappearance from roughly 54% of its original range can be attributed to habitat loss, reduction in the prey base, and persecution by humans. While it is listed as Near Threatened by the IUCN, in certain places it is critically endangered or locally extinct. Quigley et al. (2017) justify the classification of the jaguar's status as Near Threatened, with a probable 20–25% decline in settlement area, abundance, and habitat quality, and levels of exploitation over the last three generations (21 years). Threats to jaguars have either remained the same or increased grown since the 2008 Red List assessment. Sanderson et al. (2002) conducted the first expert mapping of the jaguar range and created a map that served as the basis for the Red List evaluation in 2008 and estimated that jaguars occupied about 46% of their historic range. Quigley et al. (2017) state that "with our improved knowledge of Jaguar range, this percentage is set at 51% currently." Jaguar populations are increasingly fragmented, especially in eastern and southeast Brazil, northern Venezuela, and the Maya Forest (Selva Maya) in Mexico and Guatemala. The jaguar's range may have decreased by 20% from 8.77 million km^2 to 7.02 million km.^{2} over the course of the last 14 years (2002–2015) (Quigley et al., 2017).

Quigley et al. (2017) argue that "with a generation length of 6.84 years, we suspect at least a 20–25% loss in mature individuals over the past 21 years (three generations) because there are documented population declines together with habitat loss for most of the range countries" citing various sources (Azevedo et al., 2016; Ceballos et al., 2011; Chávez et al., 2016; Costa et al., 2005; Di Bitetti et al., 2016; Díaz-Santos et al., 2016; Espinosa et al., 2016; García-Anleu et al., 2016; González-Maya et al., 2016; Hoogesteijn et al., 2016; Jedrzejewski et al., 2016; Maffei et al., 2016; McCain & Childs, 2008; Mora et al., 2016; Moreno et al., 2016; Olsoy et al., 2016; Paviolo et al., 2016; Payán et al., 2016; Payán, Carbone, et al., 2013; Payán, Castaño-Uribe, et al., 2013; Wallace et al., 2013). Key problems include jaguar-livestock conflict throughout their range, human activities like illegal hunting even in protected areas, and loss of connectivity between jaguar populations at local and regional scales. Isolated populations have fewer individuals and are more susceptible to local extinctions (Quigley et al., 2017, see also Quigley & Crawshaw Jr, 1992; Medellín et al., 2002; Sollmann et al., 2008; Ceballos et al., 2011; Hoogesteijn & Hoogesteijn, 2011; Payán, Carbone, et al., 2013; Payán, Castaño-Uribe, et al., 2013; Petracca, Hernández-Potosme, et al., 2014; Petracca, Ramírez-Bravo, & Hernández-Santín, 2014; Quigley et al., 2015; De la Torre et al., 2016; Olsoy et al., 2016; De la Torre, Gonzalez-Maya, et al., 2017; De la Torre, Nuñez, & Medellín, 2017).

The decline of the jaguar is most pronounced in the arid, open terrain of its northern ranges, including the American states of Arizona and New Mexico, and northern Mexico and even parts of the forests of Central America and some northern parts of South America (Azevedo et al., 2016; Campbell, 2017, 2022; Campbell & Torres Alvarado, 2011; Di Bitetti et al., 2016; Johnson & Van Pelt, 2016; Pereira-Garbero & Sappa, 2016). The Amazon basin rainforest and its surrounding regions of the Pantanal and Gran Chaco, tropical moist lowland forests in Mesoamerica (the Selva Maya of Guatemala, Mexico, and Belize), and a narrow stretch of the Choco-Darien from Panama and Colombia to northern Honduras are among the places where jaguars have a high chance of surviving (Sanderson et al., 2002; Hernández-SaintMartín et al., 2015). The Atlantic Tropical Forest and the Cerrado in Brazil; portions of the Chaco in northern Argentina; the Gran Sabana in northern Brazil, Venezuela, and Guyana; sections of Venezuela's coastal dry forest; and the remainder of their range in Central America and Mexico is among the places where jaguars have the lowest chances of surviving (Campbell, 2017, 2022; Campbell & Torres Alvarado, 2011; Sanderson et al., 2002).

Jaguar populations are assessed regionally (Quigley et al., 2017). Jaguars have been extirpated from Uruguay (Pereira-Garbero & Sappa, 2016), El Salvador, and the south-western United States (Campbell, 2012, 2017, 2022; Johnson & Van Pelt, 2016). For the northern populations in Mexico, population densities are estimated from 0.75 to 6 adults per 100 km^2 (Ceballos et al., 2011; Chávez et al., 2016). For example, Ceballos et al. (2011) note in their study of two census results that the total jaguar population was estimated at about 4000 individuals in the 2010 census and about 4800 individuals in the 2018 census. The Yucatan Peninsula was the region with the largest population in both censuses, with about 2000 jaguars. Quigley et al. (2017) note that the northern and central populations are becoming increasingly isolated and jaguar populations are declining in areas where they were relatively well recorded (see also Grigione et al., 2009; Rosas-Rosas & Bender, 2012).

In Central America, jaguars are probably extinct in El Salvadoe (Campbell & Torres Alvarado, 2011). Jaguar densities in the rest of Mesoamerica were estimated to be 0.74 to 11.2/100 km^2 between 2000 and 2010, with density estimates higher in tropical moist forests than tropical montane or deciduous forests and higher in national parks than multiple-use forest areas (Maffei et al., 2011). The population density of jaguars in the rainforest of Belize is estimated at 7.5–8.8/100 km^2 (Silver et al., 2004). Jaguar density in the Cockscomb Basin Wildlife Sanctuary in Belize is estimated to be 3.5 ($\pm$SE = 0.7) to 11.0 ($\pm$3.1)/100 km^2 (Harmsen et al., 2011). Jaguar densities declined in the human-influenced landscape, which include unprotected forests, savannahs, agricultural landcover and semi-rural and rural settlements at varying distances from the protected forest (Foster, 2008). Jaguar populations in protected areas in Guatemala, Honduras, and Nicaragua are threatened by deforestation and hunting (Petracca, Hernández-Potosme, et al., 2014; Petracca, Ramírez-Bravo, & Hernández-Santín, 2014).

In South America, Quigley et al. (2017) identify multiple populations for evaluation. For example, depending on the technique employed (telemetry or camera traps, respectively), the estimated density of jaguars in the Brazilian Pantanal was 6.6–6.7/100 km^2 or 10.3–11.7/100 km^2 (Soisalo & Cavalcanti, 2006). The estimated density of jaguars in the Bolivian Amazon forest was 2.8/100 km^2 (Silver et al., 2004). Estimates of jaguar densities in the Colombian Amazon range from 4 to 2 points per 100 km^3 (Payán et al., 2016). According to Silveira (2004), jaguar densities are estimated to be 2/100 km^2 in the Brazilian Cerrado savannahs, 3.5/100 km^2 in the semi-arid shrublands of the Caatinga, and 2.2/100 km^2 in the Atlantic Forest. In the Bolivian Gran Chaco, they are 2.2–5 per 100 km^2 (Maffei et al., 2011). It has been estimated that there are 200 $\pm$ 80 adults in Brazil's Atlantic Forest subpopulation (Leite et al., 2002). Jaguar populations are low-density and under severe threat from cattle and persecution in the Chaco region of northern Argentina and Brazil as well as in Caatinga, Brazil (Altrichter et al., 2006). Some populations are assessed as Least Concern (for example, the Amazonia subpopulation, which is estimated to be 89% of the total species population) while others are assessed as Critically Endangered or Endangered (for example, most of 33 identified populations) (see De la Torre, Gonzalez-Maya, et al., 2017; De la Torre, Nuñez, & Medellín, 2017; Quigley et al., 2017).

Threats against jaguar populations include habitat loss and fragmentation, killing of prey species, human population growth and direct exploitative killing and persecution for livestock predation (Bernal-Escobar et al., 2015; Campbell, 2022; Ceballos et al., 2011; Medellín et al., 2002). Deforestation in Latin America ranks with sub-Saharan Africa as the highest in the world (D'Annunzio et al., 2015). Industrial agriculture is a factor for nearly 70% of deforestation in Latin America. The result of such landuse is fragmentation with jaguar populations left in isolated patches, seeking nearby livestock as prey, resulting in jaguar-livestock owner conflict, which occurs throughout the jaguar's range when near human presence (De la Torre et al., 2016; Hoogesteijn & Hoogesteijn, 2011; Quigley et al., 2015). Despite Latin America's comparatively low human population, there is high population growth, with landuse expansion and exploitation of jaguar habitats and prey species (Campbell & Torres

Alvarado, 2011). (Altrichter et al., 2006; De la Torre, Gonzalez-Maya, et al., 2017; De la Torre, Nuñez, & Medellín, 2017; Foster, Harmsen, & Doncaster, 2010; Medellín et al., 2002; Paviolo et al., 2016; Petracca, Hernández-Potosme, et al., 2014; Petracca, Ramírez-Bravo, & Hernández-Santín, 2014). The white-lipped peccary (*Tayassu pecari* Link, 1795) an important jaguar prey species has been extirpated from 21% of its historical range over the past century (Altrichter et al., 2012; Foster, Harmsen, Valdes, et al., 2010; Keuroghlian et al., 2013).

The killing of livestock by jaguars and retaliatory killings of jaguars by people are a significant threat for jaguars (Campbell, 2022). Fear of jaguars in non-cattle contexts is also important. As noted by Knox et al. (2019) knowledge of human-jaguar relations in non-cattle areas is important, as "most studies on human-jaguar relationships have been carried out in cattle ranching areas, originally on large properties in the Pantanal of Brazil...In contrast, relatively little is known about human-jaguar relationships and, in particular, jaguar persecution in forested land within and around protected areas, and in indigenous territories relying primarily on natural resources" (see Hoogesteijn & Chapman, 1997; Polisar et al., 2003; Zimmermann et al., 2005; Azevedo & Murray, 2007; Foster, Harmsen, & Doncaster, 2010; Foster, Harmsen, Valdes, et al., 2010; Rosas-Rosas & Bender, 2012; Mexico: Figel et al., 2011; Costa Rica: Amit et al., 2013; Guatemala: Soto-Shoender & Main, 2013; Belize: Steinberg, 2016; Venezuela: Jędrzejewski et al., 2017). Knox et al. (2019) also contend that although negative perceptions of jaguars are not unusual, they are frequently linked to the economic costs and livestock losses, but negative attitudes can also be reinforced by fears and opportunistic encounters. For example, indigenous communities may have differing opinions about jaguars; Mayan hunters in Belize were found to have negative opinions, while indigenous/community protected areas in Mexico had positive opinions (Knox et al., 2019; see also Zimmermann et al., 2005; Cavalcanti et al., 2010; Rosas-Rosas & Bender, 2012; Figel et al., 2011; Castaño-Uribe et al., 2016; Porfirio et al., 2014; Parker et al., 2014; Steinberg, 2016; Amit & Jacobson, 2017).

Knox et al. (2019) note that "persecution by humans is one of the most pressing threats to jaguars... throughout the Americas, yet few studies have examined the killing of jaguars outside cattle-ranching communities. Although over one-third of the jaguar's range is formally protected, relatively little is known about human-jaguar relationships within protected areas and indigenous territories." The authors also write that "jaguar attacks on humans are remarkably rare compared to other large felids, but in their study of the Bolivian Amazon, people either feared jaguars of had information on people who had been attacked by jaguars, although many did not have first-hand experience. Some people thought that jaguar-killing was good, and "this perception of jaguar killing being common and socially accepted, combined with high rates of past killing and a growing illegal trade of jaguar parts, may create an atmosphere conducive to widespread jaguar persecution in the Bolivian Amazon" (Knox et al., 2019).

In terms of conservation actions, jaguar hunting is banned in Argentina, Brazil, Colombia, Costa Rica, French Guiana, Honduras, Mexico, Nicaragua, Panama, Paraguay, Suriname, the United States, and Venezuela, and restricted in Guatemala and Peru. However, in some protected areas, jaguars may be illegally hunted (Ceballos et al., 2011; Medellín et al., 2002; Payán, Carbone, et al., 2013; Payán, Castaño-Uribe, et al., 2013; Petracca, Ramírez-Bravo, & Hernández-Santín, 2014; Quigley & Crawshaw Jr, 1992; Sollmann et al., 2008). The commercial hunting for jaguar pelts has declined since the 1970s, due to political action in western nations, but there is some local demand for paws and teeth, and there is some replacement of tiger parts with jaguar parts for Asian medicine in Latin America (Nowell & Jackson, 1996). Habitat corridors are also planned to counteract habitat fragmentation. Due to the impact of retaliatory jaguar killing, livestock production is a high priority in conservation efforts. Other measures include prevention of jaguar predation for sporting, commercial and subsistence purposes, sustainable hunting practices and the development of international, regional, and local monitoring programs for jaguars and prey species (Kay et al., 2015; Nowell & Jackson, 1996; Rabinowitz & Zeller, 2010).

Conservation Biogeography and the Cougar

The conservation of the cougar (*Puma concolor*), another dominant apex predator, is an ecologically important species, and popular wildlife symbol. The current conservation status of the cougar is shown in the Table 2.4, as vulnerable based on the most recent assessment of the cougar in 2020 (Nielsen et al., 2015). Wozencraft (2005b) cites 6 subspecies *P. c. concolor* (Linnaeus, 1771) *P. c. puma* (Molina, 1782), *P. c. couguar* (Kerr, 1792), *P. c. costaricensis* (Merriam, 1901), *P. c. anthonyi* (Nelson & Goldman, 1933), and *P. c. cabrerae* (Pocock, 1939). Another list that of Culver et al. (2000): *P. c. cougar* (North America), *P. c. Costaricensis* (Central America), *P. c. capricornensis* (eastern South America), *P. c. Concolor* (northern South America), *P. c. Cabrerae* (central South America), and *P. c. Puma* (southern South America) (see also Nielsen et al., 2015).

Table 2.4 Conservation status of the cougar

Issue	Description
Habitat	Forest, savanna, Shrubland, grassland, desert
Status	Least concern
Population trend	Decreasing
Threats	Housing and urban areas, commercial and industrial areas, annual and perennial non-timber crops, wood and pulp plantations, livestock farming and ranching, roads and railroads, hunting and trapping terrestrial animals, logging and wood harvesting, war, civil unrest and military exercises, fire and fire suppression, dams and water management/use, problematic native species/diseases

Source IUCN (2006b) and Nielsen et al. (2015)

The cougar is listed as Least Concern because it is one of the most widely distributed mammals in the Western Hemisphere. Despite having been eliminated from its previous range in eastern and midwestern North America (Nowell & Jackson, 1996), it is making a comeback in these areas (LaRue et al., 2012; Thompson & Jenks, 2010) and populations are robust enough for Western North America to allow controlled harvesting. The species is thought to be in decline in other parts of its range, though, and because it is a large carnivore that is closely related to other wildlife and habitat associations, managing, and conserving, it presents many social and political difficulties (Nielsen et al., 2015).

In the early 1990s, population estimates for cougars in the western United States were 10,000 and Canada 3500–5000, respectively (Nowell & Jackson, 1996). Although the population of cougars in the Amazon basin's dense rainforest is unknown, the population in Central and South America is possibly higher (Nowell & Jackson, 1996), despite the near or total extinction of this species in El Salvador (Campbell & Torres Alvarado, 2011). A reintroduction of cougars from Texas increased the isolated 100–180-member Florida subpopulation (Florida Fish and Wildlife Conservation Commission, 2014; Sunquist & Sunquist, 2002). In Brazil, the cougar is classified as Near Endangered, but subspecies found outside of the Amazon are vulnerable (Machado et al. (2005). In Argentina (Díaz-Santos et al., 2016), Colombia (Rodriguez-Mahecha et al., 2006), and Peru (INRENA (Instituto Nacional De Recursos Naturales), 2006), the cougar has been classified as Near Threatened, but information is lacking for the population in Chile (CONAMA, 2005). A ranking of cougar density estimates (per 100 km^2) in selected parts of North and South America includes those of Washington (5.03), Idaho (0.77–1.04), and Utah (0.3–0.5) for the United States (see Hemker et al., 1984; Laundre & Clark, 2003; Robinson et al., 2008). For Central America, there is information on Belize (2–5) and El Salvador (>0/=0) (Campbell & Torres Alvarado, 2011; Kelly et al., 2008). For South America and Mexico, there is Bolivia (5–8), Patagonia (6), the Pantanal (4.4), western Mexico (3–5), Peru (2.4), and Argentina (0.5–0.8) (Franklin et al., 1999; Janson & Emmons, 1990; Kelly et al., 2008; Nowell & Jackson, 1996; Nunez et al., 2000).

Threats to cougars include habitat loss, fragmentation, and poaching of prey species (Campbell & Torres Alvarado, 2011). Like most big cats, they are killed by livestock owners in retaliation for predation and may be killed due to real or perceived threats to human life (Campbell, 2017). Cougars are legally hunted in some western US states. Traffic fatalities are the leading cause of death in the endangered Florida panther subpopulation, and busy roads present a major barrier to cougar movements (Sunquist & Sunquist, 2002). The cougar is protected throughout most of its range, with hunting restricted or banned in most of Argentina, Brazil, Bolivia, Chile, Colombia, Costa Rica, French Guiana, Guatemala, Honduras, Nicaragua, Panama, Paraguay, Suriname, Venezuela, and Uruguay, and strong regulations in Canada, Mexico, Peru, and the United States (Nowell & Jackson, 1996). Conservation actions require reducing conflicts caused by livestock predation and more information about the impacts of cougar and jaguar predation on livestock. Cougar killing of humans is rare but has occurred primarily in North America.

Conservation Biogeography of the Leopard

The conservation of the leopard (*Panthera pardus*), another dominant apex predator, is an ecologically important species, and popular wildlife symbol. The current conservation status of the leopard is shown in the Table 2.5, as vulnerable based on the most recent assessment of the leopard in 2020 (Stein et al., 2020, 2023). Recent classifications recognize eight subpspecies of leopards: African leopard (*P. p. pardus*, Linnaeus, 1758), Indian leopard (*P. p. fusca*, Meyer, 1794), Javan leopard (*P. p. melas,* Cuvier, 1809), Arabian leopard (*P. p. nimr*, Hemprich and Ehrenberg, 1830), Persian or Anatolian leopard (*P. p. tulliana*, Valenciennes, 1856), Amur leopard (*P. p. Orientalis*, Schlegel, 1857), Indochinese leopard (*P. p. Delacouri*, Pocock, 1930), and Sri Lankan leopard (*P. p. Kotiya*, Deraniyagala, 1956) (Kitchener et al., 2017). The African leopard is defined as the nominate subspecies (Miththapala et al., 1996; Uphyrkina et al., 2001).

Laguardia et al. (2017, p. 153) state that the leopard "has the widest distribution of any wild felid species, ranging from western and southern Africa to the Russian Far East and Java. This is largely a result of the species' high adaptability, as it can occupy diverse ecosystems ranging from tropical rainforests to boreal forests and arid savannahs and take a large variety of prey species." Leopard populations vary across their wide range (Campbell, 2017; Campbell & Torres Alvarado, 2011). Stein et al. (2023) note that leopards are found throughout Africa and Asia, but they have become extinct in much of their former range due to population decline and isolation. Leopards are challenging to classify as a single species due to their wide geographic range, secretive nature, and ability to adapt to different types of habitat. Leopard populations are affected through the impacts of growing human populations, habitat loss and fragmentation through agricultural and urban development, an increase in the illegal wildlife trade, an abundance of prey for ceremonial use of pelts, a decline in the prey base, and badly managed trophy hunting (Stein et al., 2023). An important factor for leopard population declines is human killings of prey species, mostly

Table 2.5 Conservation status of the leopard

Issue	Description
Habitat	Forest, savanna, shrubland, grassland, rocky areas (e.g., inland cliffs, mountain peaks), desert
Status	Vulnerable
Population trend	Decreasing
Threats	Housing and urban areas, annual and perennial non-timber crops, wood and pulp plantations, livestock farming and ranching, mining and quarrying, roads and railroads, hunting and trapping terrestrial animals, logging and wood harvesting, war, civil unrest and military exercises, fire and fire suppression, dams and water management/use, other ecosystem modifications

Source IUCN (2006a) and Stein et al. (2020)

medium-sized and large herbivores, largely due to bushmeat trades (Hayward et al., 2007; Lindsey, Balme, Becker, et al., 2013; Marker & Dickman, 2005).

Major related impacts of the bushmeat trade include a nearly 60% decline in Leopard prey species populations across in protected areas in West, East, and southern Africa between 1970 and 2005 (Craigie et al., 2010), and despite increased ungulate populations in southern Africa, decreases of prey species numbers by 52% in East Africa and 85% in West Africa (Craigie et al., 2010), and similar declines in Mozambique and Zambia (Becker et al., 2013; Fusari & Carpaneto, 2006; Hatton et al., 2001; Lindsey et al., 2014). Stein et al. (2023) "with such reductions to leopard prey, we infer a >50% loss of Leopard populations across East and West Africa." Trophy hunting also impacts on leopard populations in southern Africa (Balme et al., 2009; Pitman et al., 2015). Similar problems exist in Asia, as poaching has decimated prey species, such as Sambar Deer in Malaysia (Corbett, 2007; Kawanishi et al., 2014).

These factors have contributed to significant declines and regional extinctions of leopards in North, East, and West Africa, the Middle East, and East and Southeast Asia. Stein et al. (2023) further argue that between 1994 and 2014, the human population increased by 2.57% per year, which led to a 57% rise in the conversion of potential leopard habitat to agricultural land between 1975 and 2000 (Stein et al., 2023; see also Hatton et al., 2001; Du Toit et al., 2004; Fusari & Carpaneto, 2006; Datta et al., 2008; Balme et al., 2009; Brink & Eva, 2009; Sodhi et al., 2010; Miettinen et al., 2011; Thorn et al., 2013; Lindsey et al., 2014; Selvan et al., 2014).

Jacobson et al. (2016) claim that the leopard now occupies 25–37% of its historical range, but this obscures important differences between subspecies. Of the nine recognized subspecies, three (P. p. pardus, fusca, and saxicolor) represent 97% of the leopard‘s existing range, while another three (*P. p. orientalis, nimr,* and *japonensis*) have each lost up to 98% of their historical range. Isolation, small areas and few remaining areas pose further threats to the six subspecies, each with a range of less than 100,000 km (2). Jacobson et al. (2016) also note that about 17% of existing leopard range is protected, but some of the more endangered subspecies range over much smaller areas. Another finding was that research on leopard conservation was increasing, but research efforts were primarily focused on the subspecies with the largest remaining ranges, while the rarer subspecies that needed the most urgent attention were under-researched (Jacobson et al., 2016).

In Africa, leopards are recorded from North Africa, West Africa, Central Africa, East Africa, and Southern Africa (Hunter, 2015; Sunquist & Sunquist, 2002). The leopard‘s distribution in North Africa has been reduced to 97% of its former range, leaving only isolated remnant populations (Durant et al., 2014). Stein et al. (2023) argue that recently, remnant populations in Morocco and Algeria "are likely extirpated as recent surveys have failed to confirm their presence" and the "Egyptian population is considered very small with only a few observed signs in Elba National Park in southeastern Egypt, whereas there has been no recorded sign in Sinai since 1995," and the North African leopard population is probably Critically Endangered based on an IUCN assessment (Stein et al., 2023).

Stein et al. (2023) have described a decline in the distribution of leopards in West Africa. In addition to more thorough survey efforts that have confirmed its presence—or likely absence—across the region, habitat fragmentation identified as probably the primary cause of this decline. A few leopard sightings have been noted in Niger, near the southwest border with Benin and Burkina Faso, and there are unverified reports of sightings in the Air and Tenere national reserves (Stein et al., 2023). From Senegal in the west to Nigeria in the east, leopards are typically limited to a small number of protected areas. Leopard sightings have been reported from southern Senegal, which includes the Parc National de Niokolo-Koba. Small remnant populations can be found in Sierra Leone close to the Gola National Forest, Outamba Kilimi National Park, and the country's eastern border with Guinea/Liberia (Stein et al., 2023). In Liberia, leopard populations have been observed in Sapo National Park in the east and Lofa-Mano National Park in the west. Leopards can be found in Ghana in the Mole National Park to the north and west as well as along the Ivory Coast border, and in Benin along the northern border. Few leopard sightings are reported from Nigeria (Stein et al., 2023).

In Central Africa, in the Democratic Republic of the Congo, the leopard's range has been significantly reduced in the Democratic Republic of the Congo, especially in areas where human impact has increased and in areas that are relatively easily accessible, making them targets for illicit hunting and bushmeat trafficking (Stein et al., 2023). Leopards can be found in both the north and south of Cameroon, through most of Gabon, with the exception of a few small, uninhabited areas in the southeast and southwest, and in the southwest, center, and east of the Central African Republic (CAR). Except for the Sudd wetland, leopards can also be found all over South Sudan (Stein et al., 2023).

In East Africa, the distribution of leopards has decreased, especially in Somalia, Kenya, Ethiopia, and central Tanzania. Populations in Djibouti, Eritrea, and northern Sudan may still exist (Stein et al., 2023). In Somalia, they're practically nonexistent. Leopards can be found in parts of Uganda, western, central, and southern Kenya, as well as southern Ethiopia. In Tanzania, leopards inhabit the south and west as well as the Serengeti-Ngorogoro crater system. From the southeast border of Lake Victoria to central Tanzania, it is thought that there are no leopards (Stein et al., 2023).

Southern Africa may have a higher leopard population than the rest of their range. Stein et al. (2023) describe southern Africa as the "so called stronghold of the leopard" where however "there is no evidence to suggest that Leopard populations have remained stable." While the estimate for suitable leopard range globally is over 30% for the last three generations (over 20 years), the estimate from southern Africa is about 21%. Beyond areas dominated by humans, it is generally accepted that leopard populations in Angola, Zambia, Zimbabwe, and Mozambique are healthy but declining (Stein et al., 2023). Leopards are found throughout most of Namibia, except for the heavily populated northern area, the arid farmlands in the southeast, and the desert coast (Stein et al., 2020). In the north and west of Botswana, there is a significant leopard population. With dense populations in the Limpopo region, leopards can be found in South Africa along its borders with Namibia, Botswana,

Zimbabwe, and Mozambique. In South Africa's Cape Provinces, leopards are also recorded (Stein et al., 2023).

In Asia, leopards occur in Southwest Asia and the Caucasus, South and Central Asia, and Southeast and East Asia, and in most leopard populations face serious local threats (Stein et al., 2023). Concerning the Indian Leopard (*P. p. fusca*) Stein et al. (2023) note that the known range has increased in comparison with earlier evaluation, possibly due to assessments of more thorough surveys conducted in suburban and wild areas. Threats to Indian leopards include poaching for the wildlife trade, the lethal control of problematic individuals, decline in prey species, and recovery of tiger populations which may contribute to declines in leopard populations, through dominant competition and/or killing of leopards (Athreya et al., 2013; Datta et al., 2008; Harihar et al., 2011; McDougal, 1988; Mondal & Nagendra, 2011; Odden et al., 2010; Raza et al., 2012; Selvan et al., 2014). Leopards are also reported in nearby Bhutan, Nepal, and Sri Lanka (including the Sri Lankan Leopard *P. p. Kotiya*) (Athreya et al., 2013; Harihar et al., 2009).

In South West Asia (the Middle East), the Arabian Leopard (*P. p. nimr*) is mainly found in southwestern Oman and northeastern Yemen, and parts of Saudi Arabia (Kaboodvandpour et al., 2021; Khorozyan et al., 2014; Spalton, & al Hikmani, H., 2014; Zaffar-ul-Islam et al., 2014). Leopards have been recorded in Israel, but may be extinct in Egypt, Jordan, and the United Arab Emirates (Edmonds et al., 2006; Mallon et al., 2008; Perez et al., 2006; Qarqaz & Abu Baker, 2006; Spalton, & al Hikmani, H., 2014; Spalton et al., 2006). Leopards have been recorded across Iran, including northeastern part of Iran near the border with Turkmenistan (Erfanian et al., 2013; Hamidi et al., 2014; Sanei & Zakaria, 2011). There are also reports (*P. p. saxicolor* or *P. p. ciscaucasica*) from Iraq, Turkey, Armenia, and Azerbaijan (Avgan et al., 2012; Khorozyan, 2008; Sarukhanova, 2014; Voskyanyan, 2014) and North Ossetia, Georgia (Stein et al., 2023; Yarovenko, 2010).

There are also reports from South and Central Asia, including Uzbekistan, Afghanistan (Moheb & Bradfield, 2014). Leopards are reported from Pakistan, in some cases in acacia scrub forest (Roberts, 1997; Shehzad et al., 2014). Stein et al. (2023) state the critically endangered status of the leopards of the Arabian Peninsula (P. p. nimr), Java (*P. p. melas*), and Russian Far-East (*P. p. orientalis*) with low populations and fragmented habitats (Xiao et al., 2014). Stein et al., 2023) note that in the Russian Far-East, Leopards occur over about 7000 km^2 along the eastern slopes of the East Manchurian Mountains on the border with China but are rare (see also Hebblewhite et al., 2011).

In Southeast and East Asia, the leopard populations ("classically described as the subspecies *P. p. Delacouri*") have declined, but have been reported in protected areas of Thailand, Myanmar, Malaysia, Cambodia, southern China, and Java, Indonesia but less so from Lao PDR and Viet Nam (Stein et al., 2023). In this region, poaching is a serious factor for leopard population decimation, with a total estimation of less than 2500 individuals throughout Southeast Asia. As described by Stein et al. (2023) "targeted poaching for wildlife trade is probably the greatest factor that contributed to the recent range collapse of Leopards in South-east Asia. Leopard parts are used as substitutes for Tiger parts for medicinal purposes in China and South-east Asia, and Leopard skins are highly sought after as luxury items." Other factors include the decline or local extirpation of prey species, and destruction of suitable habitats, sometimes also within protected areas, as "the deforestation rate in South-east Asia is the highest of all tropical regions, and the rate is still increasing" (Stein et al., 2023; see also Sodhi et al., 2010; Steinmetz et al., 2010; Miettinen et al., 2011; Wilcove et al., 2013; Kawanishi et al., 2014).

In northern East Asia, the North China Leopard occurs in Central China to eastern Tibet regions (Stein et al., 2023). Vitekere et al. (2020, p. 1) record that four subspecies of leopard occur in China (*P.p. orientalis, P.p. fusca, P.p. japonensis, P.p. delacouri*), among them being the North China leopard (*P. p. japonensis*, Gray, 1862), the ecology and status of which is under-researched. *P.p. delacouri* may be absent from China (Laguardia et al., 2015; Song et al., 2014). The findings of their research indicate that the North China leopard's current distribution "has drastically changed, only 2% of its historical distribution remains occupied by around 174–348 individuals" (Vitekere et al., 2020, p. 1). Factors for leopard population include habitat fragmentation, human retaliation, for predation and decline populations of prey species. However, this species had no "reliable population estimates and information on problematic individuals which are responsible of conflict when an interaction with local people occurred" (Vitekere et al. (2020, p. 2).

Laguardia et al. (2017, p. 153) add that "at least three subspecies have been reported to occur throughout much of the country, and in 1998 the population was estimated to be 1000" (see Ma, 1998, for the possibly outdated figure) with current populations being small and fragmented, mostly in isolated protected areas, with a current population of 174–348 for the endemic North China leopard, and less than 30 individuals each for the remaining species, which also range outside China. Laguardia et al. (2017, pp. 154–156) assessed the conservation status of each of the four leopard subspecies that occur in China. The Amur leopard (*P. p. orientalis*) may possibly still range between Russia and China. The Indochinese leopard (*P. p. delacouri*) has been recorded near the border between China and Myanmar, and "might be on the verge of extirpation in the country" (Laguardia et al., 2017, p. 156). The Indian leopard, *P. pardus fusca*, may have been recorded in Tibet (Xizang Autonomous Region), but the authors note that such records are "uncertain" arguing that the Tibet sightings could be "P. pardus fusca, this subspecies has been considered synonymous with *P. pardus delacouri* in southern China" (Laguardia et al., 2017, p. 154, see also Smith & Xie, 2008) and that "the Tibet population is separated from the main distribution of *P. pardus delacouri* by high mountain ranges, although the Tibet population occupies similar habitat and shares a continuous distribution with *P. pardus japonensis*" and "we consider it to be P. pardus japonensis. Leopards from southcentral Tibet, however, such as those near Mount Everest, are almost certainly P. pardus fusca" (Laguardia et al., 2017, p. 154). Both *P. p. fusca* and *P. pardus delacouri in Tibet and southeastern China respectively may be less than 20 each* (Laguardia et al., 2017).

Problems for leopards in China include attacks on people, killings by poachers, livestock predation, habitat fragmentation through logging, farming, mining urbanization and transport networks, and prey species losses, especially those of wild ungulates, some in protected areas. Laguardia et al. (2017, p. 156) argue that "the reasons for the decline of leopard populations and their disappearance from many areas in China have yet to be investigated. Leopards are known to be successful at adapting to altered habitat and can persist as long as there is an adequate prey base, but they cannot withstand intense persecution" and human actions such as "retaliatory killings as a result of human–wildlife conflicts and poaching for the wildlife trade are reported in several provinces." Stein et al. (2023) argue that such "threats are expected to continue in the future, indicating that Leopard populations may continue to decline."

Conservation Biogeography of the Snow Leopard

The conservation of the snow leopard (*Panthera uncia*), another dominant apex predator and an ecologically important species, and popular wildlife symbol is another important topic. The current conservation status of the snow leopard is shown in the Table 2.6, as vulnerable based on the most recent assessment of the snow leopard by the IUCN (McCay-Peet & Quan-Haase, 2017). The scientific name of the snow leopard, *Uncia uncia,* was stated by used by Reginald Innes Pocock in 1930 from the examination of skins and skulls of *Panthera* species from Asia (Sunquist & Sunquist, 2002). Although initially classified in the monotypic genus *Uncia*, the snow leopard was later classified within the genus *Panthera* as *Panthera uncia* from the results of later studies (Davis et al., 2010; Johnson et al., 2006; Kitchener et al., 2016, 2017; O'Brien & Johnson, 2007; Sunquist & Sunquist, 2002; Wozencraft, 2005c). McCarthy and Mallon (2016) noted that the snow leopard is most closely related to the tiger, and the two species diverged over 2 million years ago (O'Brien & Johnson, 2007), and that two subspecies were defined due to morphological differences, which were not confirmed by genetic analysis (Medvedev, 2000). More recent research has suggested three subspecies (Kitchener et al., 2017). These are the nominate *P. u. uncia* which ranges between Central Asia and Pakistan mostly in the Pamir Mountains, *P. u. irbis* in Mongolia, and *P. u. uncioides* in the Himalayas and Qinghai in northwestern China. Recent research has developed these classifications (Janecka et al., 2017, 2018, 2020; Senn et al., 2018).

McCay-Peet and Quan-Haase (2017) justify the designation of the snow leopard as Vulnerable by the IUCN based on its total population, between 10,000 and 2500 mature individuals, with a projected decline over three generations (22.62 years), based several population assessments of mature adults (Jackson et al., 2010; McCarthy & Chapron, 2003; McCarthy, Mallon, Jackson, et al., 2016; Snow Leopard Working Secretariat, 2013), with the most recent being 7463 to 7980 (McCarthy & Mallon, 2016; Thinley et al., 2016). The vast range of the snow leopard may limit breeding opportunities for the current population (Riordan et al., 2015; McCarthy & Mallon, 2016). The secretive nature of the snow leopard, along with its generally low population density, sparse distribution, and remote terrain habitat, contributes to generally low detection rates and small sample sizes, which can make extrapolation of population survey results or estimates difficult (e.g., Alexander et al., 2015;

Table 2.6 Conservation status of the snow leopard

Issue	Description
Habitat	Forest, Shrubland, grassland, rocky areas (e.g., inland cliffs, mountain peaks)
Status	Vulnerable
Population trend	Decreasing, number of mature individuals 2710-3386
Threats	Housing and urban areas, tourism and recreation areas, livestock farming and ranching, mining and quarrying, roads and railroads, hunting and trapping terrestrial animals, war, civil unrest and military exercises, problematic native species/diseases, other impacts

Source IUCN (2006b) and McCay-Peet and Quan-Haase (2017)

McCarthy, Mallon, Sanderson, & Zahler, 2016; Shrestha et al., 2013; Thinley et al., 2014). Population modeling can be used to estimate the proportion of snow leopards in each age group. For breeding adults, these are estimates based on breeding age, the percentage of each age class that produced offspring.

Both negative and positive population trends have been documented, across the best range of the snow leopard. Negative impacts on populations are frequently estimated, due to increased poaching, retaliatory killings for livestock predation and increased livestock rearing and infrastructural developments in snow leopard range areas, which based on some studies suggest a 10% decline over 3 generations (22.62 years) (Nowell & Pervushina, 2014; Taubmann et al., 2015). McCarthy and Mallon (2016) and McCarthy, Mallon, Sanderson, and Zahler (2016) note that population declines in the Central Asian republics, which comprise 7% of the global occupied range, may have declined by 40–75% after the collapse of the Soviet Union during the 1990s, this decline may have been temporary (McCarthy, Mallon, Jackson, et al., 2016). Other studies suggest local population stability or even possible increases, examples including Bhutan; the Sagarmatha area, Nepal; the Vakhan Range of Tajikistan; Gissar Nature Reserve in Uzbekistan; and South Gobi, Mongolia (see Diment et al., 2012; Ferretti et al., 2014; Lovari et al., 2009; Sharma et al., 2014; Shrestha et al., 2013; Thinley et al., 2014; Wangchuk & Tharchen, 2016). Other supporting publications focus on the Eastern Pamirs between Pakistan and Central Asia; Ladakh-Spiti, India; and Sarychat-Ertash State Reserve, Kyrgyzstan (see Jumabay-Uulu et al., 2014; Kachel, 2014; Koshkarev & Vyrypaev, 2000; Suryawanshi et al., 2013).

There is evidence that the number of snow leopards in China is stable or has selectively increased, even in regions that are not on record or where the animals were previously unknown, like Inner Mongolia (Riordan & Shi, 2016). McCarthy, Mallon, Jackson, et al. (2016) report that since the first assessment in 1995, snow leopards and their ecosystems appear to have improved nationally, according to China's State Forestry Administration's Second National Biodiversity Survey. Examples include populations in the Sanjiangyuan region; Suojia and Sanjiangyuan; the Qilian Shan range (Alexander et al., 2015; Li, 2012; Liu et al., 2016; Schaller, 2014). McCay-Peet and Quan-Haase (2017) add that "the current situation for Snow Leopard in China is a result of new information emerging as survey effort improves, as well as positive effects of the illegalization of firearms in China in 1989, along with hunting bans in key provinces, and greater awareness of the national importance of Snow Leopards" and at the Asian level, these figures indicate "that Snow Leopards are likely more numerous than historic estimates suggest, perhaps substantially so. It should be noted that this represents better knowledge and not necessarily a population increase."

Currently, livestock predation, dwindling prey, disease, illegal trade, climate change, and resource theft are the main threats to snow leopards, although these can vary regionally depending on the species' range (McCay-Peet & Quan-Haase, 2017). Risks include competition with livestock, habitat degradation, and reduction in prey populations. Snow leopards may kill livestock and stakeholders kill the predators in retaliation. There is also the illegal trade in bones, hides, and other body parts. More recent threats include mining, climate change, large infrastructure, and barriers such as fenced railways or roads, as well as a lack of awareness and proactive action by institutions and stakeholders for cross-border cooperation (McCarthy & Mallon, 2016; Snow Leopard Network, 2014; Wingard et al., 2014; Zahler, 2016).

Nowell and Pervushina (2014) noted that five countries account for over 90% of incidents of poaching of snow leopards: China (103–236 animals poached annually), Mongolia (34–53), Pakistan (23–53), India (21–45), and Tajikistan (20–25). Although China, which contains the majority of the snow leopard's range, has the highest rates of both illegal trade and poaching, other countries that make up a smaller portion of the global range have also reported disproportionately high rates of estimated poaching (Nepal and Pakistan) and illegal snow leopard trade (Afghanistan and Russia). In contrast to observations of poaching and illegal trafficking, Afghanistan, Bhutan, and Kazakhstan recorded noticeably lower seizure rates, suggesting the need for more law enforcement in these nations (McCay-Peet & Quan-Haase, 2017; Nowell & Pervushina, 2014).

Conservation actions include new protected areas, antipoaching actions (including actions in Russia, Tajikistan, Kyrgyzstan, and Kazakhstan, see Paltsyn et al., 2016, McCay-Peet & Quan-Haase, 2017), actions to reduce herder-leopard conflicts, confiscation of weapons used for hunting, and community engagement, these actions noted in workshops in China (e.g., the Rangewide Priority Setting workshop in Beijing in 2008, see Sanderson et al., 2006), Kyrgyz Republic (Global Snow Leopard, and Ecosystem Protection Program, see Snow Leopard Working Secretariat, 2013) and the United States (Snow Leopard Survival Strategy, see the Snow Leopard Network, 2014). Protected area development, with varying success, includes projects in Afghanistan, Kyrgyzstan, Mongolia, and Tajikistan (Hotham et al., 2016; Li et al., 2014; Mallon & Kulikov, 2015; Moheb & Paley, 2016). There are also transboundary international actions, Central Asia, including between China, Kyrgyzstan, and/or Kazakhstan, creating links and sharing actions and knowledge (for example, the Khan Tengri nature conservation park Kyrgyzstan linking Naryn Reserve, Sarychat-Ertash Reserve, and others in Kyrgyzstan with Tomur Nature Reserve in China and a project between Kyrgyzstan and Kazakhstan) (see Mallon & Kulikov, 2015; McCay-Peet & Quan-Haase, 2017; Rosen & Zahler, 2016).

There are also infrastructural projects, including protections for livestock, which may involve the building of corrals in Russia, Afghanistan, India, Pakistan, and Tajikistan, which significantly reduced livestock killing by snow leopards (McCarthy & Mallon, 2016; Mohammad et al., 2016; Moheb & Paley, 2016; Paltsyn et al., 2016). Other actions include developments in local government, training of rangers, regulation enforcement, alternate livelihood activity development (to replace snow leopard poaching), livestock insurance and vaccination, participative ecotourism projects, education projects, and participative promotion of links between scientific and traditional knowledge (Bayarjargal et al., 2016; Hillard et al., 2016; Jackson, 2015; Jackson & Wangchuk, 2004; Jacobson et al., 2016; Kunkel et al., 2016; Mishra et al., 2016; Namgail & Dadul, 2016; Nawaz et al., 2016; Wangchuk et al., 2010; Zahler, 2016).

Conclusions

In this case study, the biogeographical and ecological background of conservation biogeography was discussed using the example of big cats. These species were chosen for a case study because, unlike large, carnivorous birds and large herbivorous mammals, big cats are considered a threat to human life due to their attacks and even consumption of humans, pets, and livestock, and therefore form a unique relationship with humans. They are also considered charismatic and powerful in the human psyche, are heavily involved in human-animal conflicts, and require large, pristine habitats, which may be required for agricultural, urban, transport, and infrastructural development. In studying such animals, political ecology and conservation psychology provide an explanatory framework for understanding human interactions and perceptions behind conservation decisions in conservation biogeography. The next chapter examines the relationship between these large mammals and humans in more detail, taking the perspectives of political ecology and conservation psychology. This contributes to knowledge about protected area management and policies related to conservation.

References

Alexander, J. S., Gopalaswamy, A. M., Shi, K., & Riordan, P. (2015). Face value. Towards robust estimates of snow leopard densities. *PLoS One, 10*(8), e0134815. https://doi.org/10.1371/journal.pone.0134815

Altrichter, M., Boaglio, G., & Perovic, P. (2006). The decline of jaguars Panthera onca in the Argentine Chaco. *Oryx, 40*, 302–309.

Altrichter, M., Taber, A., Beck, H., Reyna-Hurtado, R., Lizarraga, L., Keuroghlian, A., & Sanderson, E. W. (2012). Range-wide declines of a key Neotropical ecosystem architect, the near threatened white-lipped peccary Tayassu pecari. *Oryx, 46*(1), 87–98. https://doi.org/10.1017/S0030605311000421

Amit, R., Gordillo-Chávez, E. J., & Bone, R. (2013). Jaguar and puma attacks on livestock in Costa Rica. *Human-Wildlife Interactions, 7*, 77–84. https://doi.org/10.26077/885q-4818

Amit, R., & Jacobson, S. K. (2017). Understanding rancher coexistence with jaguars and pumas: A typology for conservation practice. *Biodiversity and Conservation, 26*, 1353–1374. https://doi.org/10.1007/s10531-017-1304-1

Angelici, F. M., Mahama, A., & Rossi, L. (2015). The lion in Ghana: Its historical and current status. *Animal Biodiversity and Conservation, 38*(2), 151–162.

Athreya, V., Odden, M., Linnell, J. D. C., Krishnaswamy, J., & Karanth, U. (2013). Big cats in our backyards: Persistence of large carnivores in a human dominated landscape in India. *PLoS ONE, 8*(3), 1–8. https://doi.org/10.1371/journal.pone.0057872

Avgan, B., Huseynali, T. T., Ismayilov, A., Fatullayev, P., Askerov, E., & Breitenmoser, U. (2012). First hard evidence of leopard in Nakhochivan. *Cat News, 57*, 33.

Azevedo, F. C. C., Gomes de Oliveira, T., de Paula, C. R., Bueno de Campos, C., Moraes, E. A., Jr., Cavalcanti, S. M. C., et al. (2016). Estatus del jaguar (Panthera onca) en Brasil. In R. A. Medellín, J. A. de la Torre, H. Zarza, C. Chávez, & G. Ceballos (Eds.), *El Jaguar en el Siglo XXI: La Perspectiva Continental, Fondo de Cultura Económica*. Universidad Nacional Autónoma de México.

Azevedo, F. C. C., & Murray, D. L. (2007). Spatial organization and food habits of jaguars (Panthera onca) in a floodplain forest. *Biological Conservation, 137*, 391–402.

Balme, G. A., Slotow, R., & Hunter, L. T. B. (2009). Impact of conservation interventions on the dynamics and persistence of a persecuted leopard (Panthera pardus) population. *Biological Conservation, 142*(11), 2681–2690.

Balme, G. A., Slotow, R., & Hunter, L. T. B. (2010). Edge effects and the impact of nonprotected areas in carnivore conservation: leopards in the Phinda–Mkhuze Complex, South Africa. *Animal Conservation, 13*, 315–323. https://doi.org/10.1111/j.1469-1795.2009.00342.x

Barlow, A. C. D., Ahmad, I., & Smith, J. L. D. (2013). Profiling tigers (Panthera tigris) to formulate management responses to human-killing in the Bangladesh Sundarbans. *Wildlife Biology in Practice, 9*, 30–39.

Bauer, H., & Van Der Merwe, S. (2004). Inventory of free-ranging lions Panthera leo in Africa. *Oryx, 38*(1). https://doi.org/10.1017/s0030605304000055

Bauer, H., Chapron, G., Nowell, K., Henschel, P., Funston, P., Hunter, L. T., MacDonald, D. W., & Packer, C. (2015). Lion (Panthera leo) populations are declining rapidly across Africa, except in intensively managed areas. *Proceedings of the National Academy of Sciences of the United States of America, 112*(48), 14894–14899. https://doi.org/10.1073/pnas.1500664112

Bauer, H., Packer, C., Funston, P. F., Henschel, P., & Nowell, K. (2016). Panthera leo (errata version published in 2017). *The IUCN Red List of Threatened Species*. https://doi.org/10.2305/IUCN.UK.2016-3.RLTS.T15951A107265605.en

Bayarjargal, A., Allen, P., Dashzeveg, U., Midgiddorj, T., & Snell-Rullman, J. (2016). Snow leopard enterprises, Mongolia. In T. McCarthy & D. Mallon (Eds.), *Snow leopards* (pp. 169–172). Elsevier.

Becker, M., McRob, R., Watson, F., Droge, E., Kanyembo, B., Murdoch, J., & Kakumbi, C. (2013). Evaluating wire-snare poaching trends and the impacts of by-catch on elephants and large carnivores. *Conservation Biology, 158*, 26–36. https://doi.org/10.1016/j.biocon.2012.08.017

Becker, M. S., Almeida, J., Begg, C. M., Bertola, L. D., Breitenmoser, C., Breitenmoser, U., Coals, P., Funston, P. J., Gaylard, A., Groom, R. J., Henschel, P., Ikanda, D., Jorge, A. A., Kruger, J., Lindsey, P. A., Maimbo, H., Mandisodza-Chikerema, R., Maude, G., Mbizah, M. M., Miller, S. M., Mudongo, E. I., Mwape, H., Mweetwa, T., Naude, V. N., Nyirenda, V. R., Parker, A., Parker, D. M., Reid, C., Robson, A., Sayer, E., Selier, S. A., Sichande, M., Simukonda, C., Uiseb, K., Williams, V. L., Zimba, D., & Hunter, L. T. (2022). Guidelines for evaluating the conservation value of African lion (Panthera leo) translocations. *Front Conserv Sci, 3*, 963961. https://doi.org/10.3389/fcosc.2022.963961

Bernal-Escobar, A., Payán, E., & Cordovez, J. M. (2015). Sex dependent spatially explicit stochastic dispersal modeling as a framework for the study of jaguar conservation and management in South America. *Ecological Modelling, 299*, 40–50.

Boomgaard, P. (2013). *Frontiers of fear: Tigers and people in the Malay World, 1600–1950*. Yale University Press.

Breitenmoser, U., Mallon, D.P., Ahmad Khan, J., & Driscoll, C. (2008). Panthera leo ssp. persica. IUCN red List of threatened species. Version 2013.2

Brink, A. B., & Eva, H. D. (2009). Monitoring 25 years of land cover change dynamics in Africa: A sample based remote sensing approach. *Applied Geography, 29*, 501–512.

Bruskotter, J. T., Vucetich, J. A., Manfredo, M. J., Karns, G. R., Wolf, C., Ard, K., Carter, N. H., López-Bao, J. V., Chapron, G., Gehrt, S. D., & Ripple, W. (2017). Modernization, risk, and conservation of the world's largest carnivores. *Bioscience, 67*(7), 646–655. https://doi.org/10.1093/biosci/bix049

Campbell, M. (2012). The relevance of age and gender for public attitudes to Brown bears (Ursus arctos), black bears (Ursus americanus), and cougars (Puma concolor) in Kamloops, British Columbia. *Society & Animals, 21*(2013), 341–359. https://doi.org/10.1163/15685306-12341260

Campbell, M. (2017). *Biological conservation in the 21st century: A conservation biology of large wildlife*. Nova Science Publishers.

Campbell, M. (2018). The ecology and conservation of the West African lion. In O. P. Jenkins (Ed.), *Advances in animal science and zoology* (pp. 73–110). Nova Science Publishers.

Campbell, M. (2022). The two tigers: The jaguar ('El Tigre') of the Americas and the Tiger of Asia. In K. K. Jha & M. Campbell (Eds.), *Dynamics and interrelations between nature, science, and society* (pp. 83–123). Nova Science Publishers.

Campbell, M., & Torres Alvarado, M. E. (2011). Public perceptions of jaguars Panthera onca, pumas Puma concolor and coyotes Canis latrans in El Salvador. *Area, 43*(3), 250–256. https://doi.org/10.1111/j.1475-4762.2011.00996.x

Carter, N., Killion, A., Easter, T., Brandt, J., & Ford, A. (2020). Road development in Asia: Assessing the range-wide risks to tigers. *Science Advances, 6*(18), 9619.

Castaño-Uribe, C., Lasso, C., Hoogesteijn, R., Díaz, A., & Payán, E. (2016). *Conflicto Entre Felinos y Humanos en América Latina, Editado por Carlos Lasso. Vol. 2. Serie Fauna Silvestre Neotropical*. Instituto de Investigaciones Biológicas Alexander von Humboldt.

Cavalcanti, S., Marchini, S., Zimmermann, A., Gese, E., & Macdonald, D. (2010). Jaguars, livestock, and people in Brazil: Realities and perceptions behind the conflict. In D. W. Macdonald & A. J. Loveridge (Eds.), *Biology and conservation of wild felids* (pp. 383–402). Oxford University Press.

Ceballos, G., Chávez, C., List, R., Zarza, H., & Medellín, R. (2011). *Jaguar conservation and managment in México: Case studies and perspectives*. Universidad Nacional Autónoma de México.

Chapron, G., Miquelle, D. G., Lambert, A., Goodrich, J. M., Legendre, S., & Colbert, J. (2008). The impact on tigers of poaching versus prey depletion. *Journal of Applied Ecology, 45*(6), 1667–1674.

Chávez, C., Zarza, H., de la Torre, J. A., Medellín, R. A., & Ceballos, G. (2016). Distribución y estado de conservación del Jaguar en México. In R. A. Medellín, J. A. de la Torre, H. Zarza, C. Chávez, & G. Ceballos (Eds.), *El jaguar en el Siglo XXI: La Perspectiva continental, Fondo de Cultura Económica*. Universidad Nacional Autónoma de México, Ciudad de México.

Chestin, I. E., Paltsyn, M. Y., Pereladova, O. B., Legorova, L. V., & Gibbs, J. P. (2017). Tiger re-establishment potential to former Caspian tiger (Panthera tigris virgata) range in Central Asia. *Biological Conservation, 205*, 42–51.

CONAMA. (2005). *Conselho Nacional do Meio Ambiente - Resolução*. CONAMA.

Corbett, R. T. (2007). The impact of hunting on the mammalian fauna of tropical Asian forests. *Biotropica, 39*(3), 292–303.

Costa, L. P., Leite, Y. L. R., Mendes, S. L., & Ditchfield, A. D. (2005). Mammal conservation in Brazil. *Conservation Biology, 19*, 672–679.

Craigie, I. D., Baillie, J. E. M., Balmford, A., Carbone, C., Collen, B., Green, R. E., & Hutton, J. M. (2010). Large mammal population declines in Africa's protected areas. *Biological Conservation, 143*(9), 2221–2228.

Crooks, K. R. (2002). Relative sensitivities of mammalian carnivores to habitat fragmentation. *Conservation Biology, 16*(2), 488–502. https://doi.org/10.1046/j.1523-1739.2002.00386.x

Culver, M., Johnson, W. E., Pecon-Slattery, J., & O'Brien, S. J. (2000). Genomic ancestry of the American puma (Puma concolor). *Journal of Heredity, 91*, 186–197.

D'Annunzio, R., Lindquist, E. J., & MacDicken, K. G. (2015). Global forest land-use change from 1990–2010: An update to the a global remote sensing survey of forests.

Datta, A., Anand, M. O., & Naniwadekar, R. (2008). Empty forests: Large carnivore and prey abundance in Namdapha National Park, north-East India. *Biological Conservation, 141*(5), 1429–1435. https://doi.org/10.1016/j.biocon.2008.02.022

Davis, B. W., Li, G., & Murphy, W. J. (2010). Supermatrix and species tree methods resolve phylogenetic relationships within the big cats, Panthera (Carnivora: Felidae). *Molecular Phylogenetics and Evolution, 56*(1), 64–76. https://doi.org/10.1016/j.ympev.2010.01.036

De la Torre, J. A., Ceballos, G., Chavez, C., Zarza, H., & Medellín, R. A. (2016). Prioridades y recomendaciones. In R. Medellín, J. A. de la Torre, H. Zarza, C. Chavez, & G. Ceballos (Eds.), *El Jaguar en el Siglo XXI: La Perspectiva Continental* (pp. 495–505). Universidad Nacional Autónoma de México.

De la Torre, J. A., Gonzalez-Maya, J. F., Zarza, H., Ceballos, G., & Medellín, R. A. (2017). The jaguar's spots are darker than they appear: Assessing the global conservation status of the jaguar Panthera onca. *Oryx, 51*, 1–16.

De la Torre, J. A., Nuñez, J. M., & Medellín, R. (2017). Habitat availability and connectivity for jaguars in the southern Mayan Forest: Conservation priorities for a fragmented landscape. *Biological Conservation, 206*, 270–282.

Dhakal, M., Karki, M., Jnawali, S. R., Subedi, N., Pradhan, N. M. B., Malla, S., Lamichhane, B. R., Pokheral, P. C., Thapa, G. J., Oglethorpe, J., Subba, S. A., Bajracharya, P. R., & Yadav, H. (2014). *Status of Tigers and Prey in Nepal*. https://doi.org/10.13140/2.1.3290.2407

Di Bitetti, M. S., De Angelo, C., Quiroga, V., Altrichter, M., Paviolo, A., Cuyckens, G. A. E., & Perovic, P. G. (2016). Estado de conservaci6n del Jaguar en Argentina. In R. A. Medellín, J. A. de la Torre, C. Chávez, H. Zarza, & G. Ceballos (Eds.), *El Jaguar en el Siglo XXI: La Perspectiva Continental, Fondo de Cultura Económica*. Universidad Nacional Autónoma de México, Ciudad de Mé.

Díaz-Santos, F., Polisar, J., Maffei, L., & Santos-Díaz, F. G. (2016). Avances en el conocimiento de los Jaguares en Nicaragua. In R. A. Medellín, J. A. de la Torre, C. Chávez, H. Zarza, & G. Ceballos (Eds.), *El Jaguar en el Siglo XXI: La Perspectiva Continenta, Fondo de Cultura Económica*. Universidad Nacional Autónoma de México, Ciudad de México.

Diment, A., Mallon, D. P., & Hotham, P. (2012). First biodiversity survey of Zorkul reserve, Pamir Mountains, Tajikistan. *Oryx, 46*, 13–14.

Du Toit, J. T., Walker, B. H., & Campbell, B. M. (2004). Conserving tropical nature: Current challenges for ecologists. *Trends in Ecology and Evolution, 19*(1), 12–17.

Duangchantrasiri, S., Umponjan, M., Simcharoen, S., Pattanavibool, A., Chaiwattana, S., Maneerat, S., Kumar, N. S., Jathanna, D., Srivathsa, A., & Karanth, K. U. (2016). Dynamics of a low-density tiger population in Southeast Asia in the context of improved law enforcement. *Conservation Biology, 30*(3), 639–648.

Durant, S. M., Wacher, T., Bashir, S., Woodroffe, R., De Ornellas, P., Ransom, C., Newby, J., Abaigar, T., Abdelgadir, M., El Alqamy, H., Baillie, J., Beddiaf, M., Belbachir, F., Belbachir-Bazi, A., Berbash, A. A., Bemadjim, N. E., Beudels-Jamar, R., Boitani, L., Breitenmoser, C., Cano, M., Chardonnet, P., Collen, B., Cornforth, W. A., Cuzin, F., Gerngross, P., Haddane, B., Hadjeloum, M., Jacobson, A., Jebali, A., Lamarque, F., Mallon, D., Minkowski, K., Monfort, S., Ndoassal, B., Niagate, B., Purchase, G., Samaila, S., Samna, A. K., Sillero-Zubiri, C., Soultan, A. E., Price, M. R. S., & Pettorelli, N. (2014). Fiddling in biodiversity hotspots while deserts burn? Collapse of the Sahara's megafauna. *Diversity and Distributions, 20*, 114–122.

Edmonds, J.-A., Budd, K. J., Al Midfa, A., & Gross, C. (2006). Status of the Arabian leopard in United Arab Emirates. *Cat News, 1*, 33–39.

Eizirik, E., Bonatto, S. L., Salzano, F. M., Johnson, F. M., O'Brien, S. J., Crawshaw, P. G., Vie, J.-C., & Brousset, D. M. (2001). Phylogeography, population history and conservation genetics of jaguars (Panthera onca, Mammalia, Felidae). *Molecular Ecology, 10*, 65–79.

Erfanian, B., Mikarimi, S. H., Mahini, A. S., & Rezaei, H. R. (2013). A presence-only habitat suitability model for Persian leopard Panthera pardus saxicolor in Golestan national park, Iran. *Wildlife Biology, 19*(2), 170–178.

Espinosa, S., Albuja, L., Tirira, D. G., Zapata-Río, G., Araguillin, E., Ultreras, V., & Andrew, N. (2016). Análisis del estado de conservación del Jaguar en el Ecuador. In R. A. Medellín, J. A. de la Torre, C. Chávez, H. Zarza, & G. Ceballos (Eds.), *El Jaguar en el Siglo XXI: La Perspectiva Continental, Fondo de Cultura Económica*. Universidad Nacional Autónoma de México, Ciudad de México.

Fernández-Sepúlveda, J., & Martín, C. A. (2022). Conservation status of the world's carnivorous mammals (order Carnivora). *Mammalian Biology, 102*, 1911–1925. https://doi.org/10.1007/s42991-022-00305-8

Ferretti, F., Lovari, S., Minder, I., & Pellizzi, B. (2014). Recovery of snow leopard in Sagarmatha (Mt Everest) National Park: Effects on main prey. *European Journal of Wildlife Research, 60*, 559–562.

Figel, J. J., Durán, E., & Bray, D. B. (2011). Conservation of the jaguar Panthera onca in a community-dominated landscape in montane forests in Oaxaca, Mexico. *Oryx, 45*, 554–560. https://doi.org/10.1017/S0030605310001353

Florida Fish and Wildlife Conservation Commission. (2014). *Annual report on the research and management of Florida panthers: 2013-2014*. Fish and Wildlife Research Institute & Division of Habitat and Species Conservation.

Foster, R. J. (2008). The ecology of jaguars in a human-influenced landscape. PhD thesis, University of Southampton, UK.

Foster, R. J., Harmsen, B. J., & Doncaster, C. P. (2010). Habitat use by sympatric jaguars and pumas across a gradient of human disturbance in Belize. *Biotropica, 42*, 724–731.

Foster, R. J., Harmsen, B. J., Valdes, B., Pomilla, C., & Doncaster, C. P. (2010). Food habits of sympatric jaguars and pumas across a gradient of human disturbance. *Journal of Zoology, 280*, 309–318.

Franklin, W. L., Johnson, W. E., Sarno, R. J., & Iriarte, J. A. (1999). Ecology of the Patagonia puma Felis concolor patagonica in southern Chile. *Biological Conservation, 90*, 33–40.

Fusari, A., & Carpaneto, G. M. (2006). Subsistence hunting and conservation issues in the game reserve of Gile, Mozambique. *Biodiversity and Conservation, 15*(8), 2477–2495.

García-Anleu, R., McNab, R. B., Polisar, J., Ramos, V. H., Moreira, J., Ponce-Santizo, G., et al. (2016). Estatus del Jaguar en Guatemala, informe del año 2013. In R. A. Medellín, J. A. de la Torre, C. Chávez, H. Zarza, & G. Ceballos (Eds.), *El Jaguar en el Siglo XXI: La Perspectiva Continental*. Fondo de Cultura Económica, Universidad Nacional Autónoma de México, Ciudad de México.

Gilbert, M., Miquelle, D. G., Goodrich, J. M., Reeve, R., Cleaveland, S., Matthews, L., & Joly, D. O. (2014). Estimating the potential impact of canine distemper virus on the Amur tiger population (Panthera tigris altaica) in Russia. *PLoS One, 9*, e110811.

Gittleman, J., Funk, S., MacDonald, D., & Wayne, R. (2001). *Carnivore conservation*. Cambridge University Press.

Global Tiger Recovery Program. (2010–2022). (English). World Bank Group. http://documents.worldbank.org/curated/en/874191468331048098/Global-tiger-recovery-program-2010-2022

González-Maya, J. F., Bustamante, A., Moreno, R., Salom-Pérez, R., Tavares, R., & Schipper, J. (2016). Estado de conservación y prioridades para el Jaguar en Costa Rica. In R. A. Medellín, J. A. de la Torre, C. Chávez, H. Zarza, & G. Ceballos (Eds.), *El Jaguar en el Siglo XXI:La Perspectiva Continental*. Ciudad de México. Fondo de Cultura Económica, Universidad Nacional Autónoma de México.

Goodrich, J., Wibisono, H., Miquelle, D., Lynam, A. J., Sanderson, E., Chapman, S., Gray, T. N. E., Chanchani, P., & Harihar, A. (2022). *Panthera tigris. The IUCN red list of threatened species 2022.* https://doi.org/10.2305/IUCN.UK.2022-1.RLTS.T15955A214862019.en

Goodrich, J. M., Kerley, L. L., Smirnov, E. N., Miquelle, D. G., McDonald, L., Quigley, H. B., Hornocker, M. G., & McDonald, T. (2008). Survival rates and causes of mortality of Amur tigers on and near the Sikhote-Alin biosphere Zapovednik. *Journal of Zoology, 276*, 323–329.

Gopal, R., Qureshi, Q., Bhardwaj, M., Singh, R. J., & Jhala, Y. V. (2010). Evaluating the status of the endangered tiger Panthera tigris and its prey in Panna Tiger Reserve, Madhya Pradesh, India. *Oryx, 44*(3), 383–389.

Gray, T. N., Crouthers, R., Ramesh, K., Vattakaven, J., Borah, J., Pasha, M. K. S., Lim, T., Phan, C., Singh, R., Long, B., & Chapman, S. (2017). A framework for assessing readiness for tiger Panthera tigris reintroduction: A case study from eastern Cambodia. *Biodiversity and Conservation, 26*(10), 2383–2399.

Gray, T. N., Hughes, A. C., Laurance, W. F., Long, B., Lynam, A. J., O'Kelly, H., Ripple, W. J., Seng, T., Scotson, L., & Wilkinson, N. M. (2018). The wildlife snaring crisis: An insidious and pervasive threat to biodiversity in Southeast Asia. *Biodiversity and Conservation, 27*(4), 1031–1037.

Gray, T. N., Lynam, A. J., Seng, T., Laurance, W. F., Long, B., Scotson, L., & Ripple, W. J. (2017). Wildlife-snaring crisis in Asian forests. *Science, 355*, 255–256.

Grigione, M., Menke, K., López-González, C., List, R., Banda, A., Carrera, J., Carrera, R., Giordano, A., Morrison, J., & Sternberg, M. (2009). Identifying potential conservation areas for felids in the USA and Mexico: Integrating reliable knowledge across an international border. *Oryx, 43*, 78–86.

Hamidi, A. K., Ghoddousi, A., Soufi, M., Ghadirian, T., Jowkar, H., & Ashayeri, S. (2014). Camera trap study of Persian leopard in Golestan National Park, Iran. *Cat News, 60*, 12–14.

Harihar, A., Chanchani, P., Borah, J., Crouthers, R. J., Darman, Y., Gray, T. N. E., Mohamad, S., Rawson, B. M., Rayan, M. D., & Roberts, J. L. (2018). Recovery planning towards doubling wild tiger Panthera tigris numbers: Detailing 18 recovery sites from across the range. *PLoS One, 13*(11), e0207114.

Harihar, A., Ghosh-Harihar, M., & MacMillan, D. C. (2018). Losing time for the tiger Panthera tigris: Delayed action puts a globally threatened species at risk of local extinction. *Oryx, 52*(1), 78–88.

Harihar, A., Pandav, B., Ghosh-Harihar, M., & Goodrich, J. (2020). Demographic and ecological correlates of a recovering tiger (Panthera tigris) population: Lessons learnt from 13-years of monitoring. *Biological Conservation, 252*(2), 108848.

Harihar, A., Pandav, B., & Goyal, S. P. (2009). Density of leopards (Panthera pardus) in the Chilla range of Rajaji National Park, Uttarakhand, India. *Mammalia, 73*, 68–71.

Harihar, A., Pandav, B., & Goyal, S. P. (2011). Responses of leopard Panthera pardus to the recovery of a tiger Panthera tigris population. *Journal of Applied Ecology, 48*(3), 806–814.

Harmsen, B. J., Foster, R. J., Silver, S. C., Ostro, L. E. T., & Doncaster, C. P. (2011). Jaguar and puma activity patterns in relation to their main prey. *Mammalian Biology, 76*(3), 320–324. https://doi.org/10.1016/j.mambio.2010.08.007

Hatton, J., Couto, M., & Oglethorpe, J. (2001). *Biodiversity and war: A case study of Mozambique.* WWF Biodiversity Support Program.

Hayward, M. W., O'Brien, J., & Kerley, G. I. H. (2007). Carrying capacity of large African predators: Predictions and tests. *Biological Conservation, 139*, 219–229.

Hebblewhite, M., Miquelle, D. G., Murzin, A. A., Aramilev, V. V., & Pikunov, D. G. (2011). Predicting potential habitat and population size for reintroduction of the far eastern leopards in the Russian Far East. *Biological Conservation, 139*, 219–229.

Hemker, T. P., Lindzey, F. G., & Ackerman, B. B. (1984). Population characteristics and movement patterns of cougars in southern Utah. *Journal of Wildlife Management, 48*, 1275–1284.

Henschel, P., Coad, L., Burton, C., Chataigner, B., Dunn, A., MacDonald, D., et al. (2014). The lion in West Africa is critically endangered. *PLoS One, 9*(1), e83500. https://doi.org/10.1371/journal.pone.0083500

Hernández-SaintMartín, A. D., Rosas-Rosas, O. C., Palacio-Núñez, J., Tarango-Arambula, L. A., Clemente-Sánchez, F., & Hoogesteijn, A. L. (2015). Food habits of jaguar and puma in a protected area and adjacent fragmented landscape of Northeastern Mexico. *Natural Areas Journal, 35*(2), 308–317. https://doi.org/10.3375/043.035.0213

Hillard, D., Weddle, M., Padmanabhan, S., Ale, S., Khuukhenduu, T., & Almashev, C. (2016). Environmental education for snow leopard conservation. In T. McCarthy & D. Mallon (Eds.), *Snow leopards* (pp. 245–255). Elsevier.

Hoogesteijn, A., Hoogesteijn, R., Boede, E. O., Gonzalez-Fernandez, A., Isasi-Catala, E., Yerena, E., & Torres, D. (2016). Situacion de las poblaciones del Jaguar en Venezuela, estudio retrospectivo. In R. Medellin, C. Chavez, A. de la Torre, H. Zarza, & G. Ceballos (Eds.), *El Jaguar en el Siglo XXI: La Perspectiva continental*. Mexico.

Hoogesteijn, R., & Chapman, C. A. (1997). Large ranches as conservation tools in the Venezuelan llanos. *Oryx, 31*, 274–284. https://doi.org/10.1046/j.1365-3008.1997.d01-16.x

Hoogesteijn, R., & Hoogesteijn, A. (2011). *Estrategias anti-depredación para fincas ganaderas en Latinoamérica: una guía*. Panthera.

Hossain, A. N. M., Lynam, A. J., Ngoprasert, D., Barlow, A., Barlow, C. G., & Savini, T. (2018). Identifying landscape factors affecting tiger decline in the Bangladesh Sundarbans. *Global Ecology and Conservation, 13*, 382.

Hotham, P., Howard, P., Nyul, H., & Whitten, T. (2016). Corporate business and the conservation of the snow leopard: Worlds that need not collide. In T. McCarthy & D. Mallon (Eds.), *Snow leopards* (pp. 277–289). Elsevier.

Hunter, L. (2015). *Wild cats of the world*. Bloomsbury Publishing.

INRENA (Instituto Nacional De Recursos Naturales). (2006). *Plan Nacional de Reforestacion Perù 2005–2024*. INRENA.

Inskip, C., Carter, N., Riley, S., Roberts, T., & MacMillan, D. (2016). Toward human-carnivore coexistence: Understanding tolerance for tigers in Bangladesh. *PLoS One, 11*(1), e0145913. https://doi.org/10.1371/journal.pone.0145913

Integrated Taxonomic Information System (ITIS). (2022). Results of: Search in every Kingdom for Common Name containing 'jaguar'.

IUCN. (2006a). *Conservation strategy for the lion in west and Central Africa*. IUCN SSC Cat Specialist Group.

IUCN. (2006b). *Conservation strategy for the lion in eastern and southern Africa*. IUCN SSC Cat Specialist Group.

Jackson, R., Mishra, C., McCarthy, T. M., & Ale, S. B. (2010). Snow leopards, conflict and conservation. In D. W. MacDonald & A. Loveridge (Eds.), *Biology and conservation of wild felids* (pp. 417–430). Oxford University Press.

Jackson, R. M. (2015). HWC ten years later: Successes and shortcomings of approaches to global snow leopard conservation. *Human Dimensions of Wildlife, 20*, 310–316.

Jackson, R. M., & Wangchuk, R. (2004). A community-based approach to mitigating livestock depredation by snow leopards. *Human Dimensions to Wildlife, 9*, 307–315.

Jacobson, A. P., Gerngross, P., Lemeris, J. R., Jr., Schoonoer, R. F., Anco, C., Breitenmoser-Würsten, C., Durant, S. M., Farhadinia, M. S., Henschel, P., Kamler, J. F., Laguardia, A., Rostro-García, S., Stein, A. B., & Dollar, L. (2016). Leopard (Panthera pardus) status, distribution, and the research efforts across its range. *PeerJ, 4*, e1974. https://doi.org/10.7717/peerj.1974

Janecka, J. E., Hacker, C., Broderick, J., Pulugulla, S., Auron, P., Ringling, M., Nelson, B., Munkhtsog, B., Hussain, S., Davis, B., & Jackson, R. (2020). Noninvasive genetics and genomics shed light on the status, phylogeography, and evolution of the elusive snow leopard. In J. Ortega & J. E. Maldonado (Eds.), *Conservation genetics in mammals. Integrative research using novel approaches* (pp. 83–120). Springer International Publishing. https://doi.org/10.1007/978-3-030-33334-8_5

Janecka, J. E., Janecka, M. J., Helgen, K. M., & Murphy, W. J. (2018). The validity of three snow leopard subspecies: Response to Senn et al. *Heredity, 120*(6), 586–590. https://doi.org/10.1038/s41437-018-0052-7

Janecka, J. E., Zhang, Y., Li, D., Munkhtsog, B., Bayaraa, M., Galsandorj, N., Wangchuk, T. R., Karmacharya, D., Li, J., Lu, Z., & Uulu, K. Z. (2017). Range-wide snow leopard Phylogeography supports three subspecies. *Journal of Heredity, 108*(6), 597–607. https://doi.org/10.1093/jhered/esx044

Janson, C. H., & Emmons, L. H. (1990). Ecological structure of the nonflying mammal community at Cocha Cashu Biological Station, Manu National Park, Peru. In A. H. Gentry (Ed.), *Four neotropical forests* (pp. 314–338). Yale University Press.

Jayadevan, A., Nayak, R., Karanth, K. K., Krishnaswamy, J., DeFries, R., Karanth, K. U., & Vaidyanathan, S. (2020). Navigating paved paradise: Evaluating landscape permeability to movement for large mammals in two conservation priority landscapes in India. *Biological Conservation, 247*, 108613.

Jedrzejewski, W., Boede, E. O., Abarca, M., Sanchez-Mercado, A., Ferrer-Peris, J. R., Lampo, M., Velasquez, G., Carreno, R., Viloria, A. L., Hoogesteijn, R., Robinson, H. S., Stachowicz, I., Cerda, H., del Mar, W., Barros, M., Rivas, T. R., Borges, G. A., Molinari, G., Lew, J., Takiff, D., & Schmidt, K. (2016). Predicting carnivore distribution and extirpation rate base on human impacts and productivity factors; assessment of the state of jaguar in Venezuela. *Biological Conservation, 206*, 132–142.

Jędrzejewski, W., Carreño, R., Sánchez-Mercado, A., Schmidt, K., Abarca, M., Robinson, H. S., et al. (2017). Jaguar conflicts and the relative importance of retaliatory killing and hunting for jaguar (Panthera onca) populations in Venezuela. *Biological Conservation, 209*, 524–532. https://doi.org/10.1016/j.biocon.2017.03.025

Jhala, Y. V., Banerjee, K., Chakrabarti, S., Basu, P., Singh, K., Dave, C., & Gogoi, K. (2019). Asiatic lion: Ecology, economics, and politics of conservation. *Frontiers in Ecology and Evolution, 7*, 312. https://doi.org/10.3389/fevo.2019.00312

Jhala, Y. V., Qureshi, Q., & Nayak, A. K. (2020). *Status of tigers, copredators and prey in India, 2018*. National Tiger Conservation Authority, Government of India.

Johnson, A., Goodrich, J., Hansel, T., Rasphone, A., Saypanya, S., Vongkhamheng, C., Venevongphet, & Strindberg, S. (2016). To protect or neglect? Design, monitoring, and evaluation of a law enforcement strategy to recover small populations of wild tigers and their prey. *Biological Conservation, 202*, 99–109.

Johnson, T., & Van Pelt, B. (2016). Jaguares en el Borde: Evaluación y Perspectivas Conservació del Jaguar Continental. In R. A. Medellín, J. A. de la Torre, H. Zarza, C. Chávez, & G. Ceballos (Eds.), *El Jaguar en el Siglo XXI: La Perspectiva Continental, Fondo de Cultura Económica*. Universidad Nacional Autónoma de México.

Johnson, W. E., Eizirik, E., Pecon-Slattery, J., Murphy, W. J., Antunes, A., Teeling, E., & O'Brien, S. J. (2006). The late Miocene radiation of modern Felidae: A genetic assessment. *Science, 311*, 73–77.

Joshi, A. R., Dinerstein, E., Wikramanayake, E., Anderson, M. L., Olson, D., Jones, B. S., Seidensticker, J., Lumpkin, S., Hansen, M. C., Sizer, N. C., Davis, C. L., Palminteri, S., & Hahn, N. R. (2016). Tracking changes and preventing loss in critical tiger habitat. *Science Advances, 2*(4), e1501675.

Jumabay-Uulu, K., Wegge, P., Mishra, C., & Sharma, K. (2014). Large carnivores and low diversity of optimal prey: A comparison of the diets of snow leopards Panthera uncia and wolves Canis lupus in Sarychat-Ertash Reserve in Kyrgyzstan. *Oryx, 48*, 529–535.

Kaboodvandpour, S., Almasieh, K., & Zamani, N. (2021). Habitat suitability and connectivity implications for the conservation of the Persian leopard along the Iran-Iraq border. *Ecology and Evolution, 11*(19), 13464–13474. https://doi.org/10.1002/ece3.8069

Kachel, S. M. (2014). *Evaluating the efficacy of wild ungulate trophy hunting as a tool for snow leopard conservation in the Pamir mountains of Tajikistan*. University of Delaware.

Karanth, K. U., Kumar, N. S., & Karanth, K. K. (2020). Tigers against the odds: Applying macro-ecology to species recovery in India. *Biological Conservation, 252*, 108846.

Kawanishi, K., Rayan, D. M., Gumal, M. T., & Shepherd, C. R. (2014). Extinction process of the sambar deer in peninsular Malaysia. *IUCN Deer Specialist Group Newsletter, 26*, 48–59.

Kay, E., Dickerson, A., Urbina, Y., Lizama, D., Correa, E., Cruz, F., Garcia, R., Thompson, R., Williams, L., Quintana, R., Young, J., Andrewin-Bohn, J., Cawich, V., Joseph, R., Humes, S., & Mai, A. (2015). *Central belize corridor: Conservation action plan 2015-2018.*

Kelly, M. J., Noss, A. J., Di Bitetti, M., Maffei, L., Arispe, L. R., Paviolo, A., De Angelo, C. D., & Di Blanco, Y. E. (2008). Estimating puma densities from camera trapping across three study sites: Bolivia, Argentina, and Belize. *Journal of Mammalogy, 89*(2), 408–418.

Kerley, L. L., Goodrich, J. M., Miquelle, D. G., Quigley, H. B., Hornocker, M. G., & Smirnov, E. N. (2003). Reproductive parameters of wild female Amur (Siberian) tigers (Panthera tigris altaica). *Journal of Mammalogy, 84*, 288–298.

Keuroghlian, A., Desbiez, A., Reyna-Hurtado, R., Altrichter, M., Beck, H., Taber, A., & Fragoso, J. M. V. (2013). *Tayassu pecari. The IUCN Red List of Threatened Species 2013.* https://doi.org/10.2305/IUCN.UK.2013-1.RLTS.T41778A44051115.en

Khan, A., Patel, K., Shukla, H., Viswanathan, A., van der Valk, T., Borthakur, U., Nigam, P., Zachariah, A., Jhala, Y. V., Kardos, M., & Ramakrishnan, U. (2021). Genomic evidence for inbreeding depression and purging of deleterious genetic variation in Indian tigers. *Proceedings of the National Academy of Sciences USA, 118*(49), e2023018118. https://doi.org/10.1073/pnas.2023018118

Khorozyan, I. (2008). Panthera pardus ssp. saxicolor. The IUCN Red List of Threatened Species 2008. https://doi.org/10.2305/IUCN.UK.2008.RLTS.T15961A5334217.en

Khorozyan, I., Stanton, D., Mohammed, M., Al-Ra'il, W., & Pittet, M. (2014). Patterns of co-existence between humans and mammals in Yemen: Some species thrive while others are nearly extinct. *Biodiversity and Conservation, 23*, 1995–2013.

Kitchener, A. (1999). Tiger distribution, phenotypic variation and conservation issues. In J. Seidensticker, S. Christie, & P. Jackson (Eds.), *Riding the Tiger: Tiger conservation in human-dominated landscapes* (pp. 19–39). Cambridge University Press.

Kitchener, A. C., Breitenmoser-Würsten, C., Eizirik, E., Gentry, A., Werdelin, L., Wilting, A., Yamaguchi, N., Abramov, A. V., Christiansen, P., Driscoll, C., Duckworth, J. W., Johnson, W., Luo, S.-J., Meijaard, E., O'Donoghue, P., Sanderson, J., Seymour, K., Bruford, M., Groves, C., Hoffmann, M., Nowell, K., Timmons, Z., & Tobe, S. (2017). A revised taxonomy of the Felidae: The final report of the cat classification task force of the IUCN cat specialist group. *Cat News, 11*, 71–73.

Kitchener, A. C., Driscoll, C. A., & Yamaguchi, N. (2016). What is a snow leopard? Taxonomy, morphology, and phylogeny. In T. McCarthy & D. Mallon (Eds.), *Snow leopards*. Academic Press.

Knox, J., Negrões, N., Marchini, S., Barboza, K., Guanacoma, G., Balhau, P., Tobler, M. W., & Glikman, J. A. (2019). Jaguar persecution without "Cowflict": Insights from protected territories in the Bolivian Amazon. *Frontiers in Ecology and Evolution, 7*, 494. https://doi.org/10.3389/fevo.2019.00494

Kolipaka, S. S., Tamis, W. L. M., van Zelfde, M., Persoon, G. A., & de Iongh, H. H. (2017). New insights into the factors influencing movements and spatial distribution of reintroduced Bengal tigers (Panthera tigris tigris) in the human-dominated buffer zone of Panna Tiger Reserve, India. *Mammalia, 83*(3), 207–217.

Koshkarev, E. P., & Vyrypaev, V. (2000). The snow leopard after the break-up of the Soviet Union. *Cat News, 32*, 9–11.

Kunkel, K., Hussain, S., & Khatiwada, A. (2016). A review of lessons, successes, and pitfalls of livestock insurance schemes. In T. M. McCarthy & D. P. Mallon (Eds.), *Snow leopards*. Elsevier.

Laguardia, A., Kamler, J., Li, S., Zhang, C., Zhou, Z., & Shi, K. (2017). The current distribution and status of leopards Panthera pardus in China. *Oryx, 51*(1), 153–159. https://doi.org/10.1017/S0030605315000988

Laguardia, A., Wang, J., Shi, F. L., Shi, K., & Riordan, P. (2015). Species identification refined by molecular scatology in a community of sympatric carnivores in Xinjiang, China. *Zoological Research, 36*, 72–78.

Larson, S. E. (1997). Taxonomic re-evaluation of the jaguar. *Zoo Biology, 16*, 107–120. https://doi.org/10.1002/(SICI)1098-2361(1997)16:2<107::AID-ZOO2>3.0.CO;2-E

LaRue, M. L., Nielsen, C. K., Dowling, M., Miller, K., Wilson, B., Shaw, H., & Anderson, C. R., Jr. (2012). Cougars are recolonizing the Midwest: Analysis of cougar confirmations during 1990-2008. *Journal of Wildlife Management, 76*, 1364–1369.

Laundre, J., & Clark, T. W. (2003). Managing puma hunting in the western United States: Through a metapopulation approach. *Animal Conservation, 6*, 159–170.

Leite, M. R. P., Boulhosa, R. L. P., Galvao, F., & Cullen, L., Jr. (2002). Conservacion del jaguar en las areas protegidas del Bosque Atlantico de la costa de Brasil. In R. A. Medellin, C. B. Chetkiewicz, A. R. Rabinowitz, K. H. Redford, J. G. Robinson, E. W. Sanderson, & A. B. Taber (Eds.), *El jaugar en el nuevo milenio* (pp. 25–42). Universidad Nacional Autonoma de Mexico/ Wildlife Conservation Society.

Li, J. (2012). *Ecology and conservation strategy of snow leopard (Panthera uncia) in Sanjiangyuan area on the Tibetan plateau*. University of Peking.

Li, J., Wang, D., Yin, H., Zhaxi, D., Jiagong, Z., Schaller, G. B., Mishra, C., McCarthy, T. M., Wang, H., Wu, L., Xiao, L., Basang, L., Zhang, Y., Zhou, Y., & Lu, Z. (2014). Role of Tibetan Buddhist monasteries in snow leopard conservation. *Conservation Biology, 28*(1), 87–94. https://doi.org/10.1111/cobi.12135

Lindsey, P. A., Balme, G., Becker, M., Begg, C., Bento, C., Bocchino, C., Dickman, A., Diggle, R. W., Eves, H., Henschel, P., Lewis, D., Marnewick, K., Mattheus, J., McNutt, J. W., McRobb, R., Midlane, N., Milanzi, J., Morley, R., Murphree, M., Opyene, V., Phadima, J., Purchase, G., Rentsch, D., Roche, C., Shaw, J., Van der Westhuizen, H., Van Vliet, N., & Zisadza-Gandiwa, P. (2013). The bushmeat trade in African savannas: Impacts, drivers, and possible solutions. *Biological Conservation, 160*, 80–96. https://doi.org/10.1016/j.biocon.2012.12.020

Lindsey, P. A., Miller, J. R. B., Petracca, L. S., Coad, L., Dickman, A. J., Fitzgerald, K. H., Flyman, M. V., Funston, P. J., Henschel, P., Kasiki, S., Knights, K., Loveridge, A. J., Macdonald, D. W., Mandisodza-Chikerema, R. L., Nazerali, S., Plumptre, A. J., Stevens, R., Van Zyl, H. W., & Hunter, L. T. B. (2018). More than $1 billion needed annually to secure Africa's protected areas with lions. *Proceedings of the National Academy of Sciences of the United States of America, 115*(45), E10788–E10796. https://doi.org/10.1073/pnas.1805048115

Lindsey, P. A., Nyirenda, V. R., Barnes, J. I., Becker, M. S., McRobb, R., Tambling, C. J., Taylor, W. A., Watson, F. G., & t'Sas-Rolfes, M. (2014). Underperformance of African protected area networks and the case for new conservation models: Insights from Zambia. *PLoS One, 9*(5), e94109. https://doi.org/10.1371/journal.pone.0094109

Linnell, J. D. C., Swenson, J. E., & Andersen, R. (2001). Predators and people: Conservation of large carnivores is possible at high human densities if management policy is favourable. *Animal Conservation, 4*(4), 345–349. https://doi.org/10.1017/S1367943001001408

Liu, Y., Weckworth, B., Li, J., Xiao, L., & Zhao, X. (2016). Conservation status and opportunities of snow leopard in the Sanjiangyuan region of the Tibetan plateau. In T. McCarthy & D. Mallon (Eds.), *Snow leopards* (pp. 514–521). Elsevier.

Lovari, S., Boesi, R., Minder, I., Mucci, N., Randi, E., Dematteis, A., & Ale, S. B. (2009). Restoring a keystone predator may endanger a prey species in a human-altered ecossystem: The return of the snow leopard to Sagarmatha National Park. *Animal Conservation, 12*, 559–570.

Lozano, J., Olszańska, A., Morales-Reyes, Z., Castro, A. A., Malo, A. F., Moleón, M., & Martín-López, B. (2019). Human-carnivore relations: A systematic review. *Biological Conservation, 237*, 480–492. https://doi.org/10.1016/j.biocon.2019.07.002

Luskin, M. S., Meijaard, E., Surya, S., Walzer, C., & Linkie, M. (2021). African swine fever threatens Southeast Asia's 11 endemic wild pig species. *Conservation Letter, 14*(3), 12784.

Ma, Y. Q. (1998). Panthera pardus. In S. Wang (Ed.), *China red data book of endangered animals: Mammalia* (pp. 116–118). Science Press.

Machado, A. B. M., Drummond, G. M., & Martins, C. S. (2005). *Lista da Fauna Brasileira Ameaçada de Extinção: Incluindo as Espécies Quase Ameaçadas e Deficientes em Dados*. Fundação Biodiversitas.

Maffei, L., Polisar, J., Garcia, R., Moreira, J., & Noss, A. J. (2011). Perspectives from ten years of jaguar (Panthera onca) camera trapping in Mesoamerica. *Mesoamericana, 15*, 49–59.

Maffei, L., Rumiz, D., Arispe, R., Cuéllar, E., & Noss, A. (2016). Situación del Jaguar en Bolivia. In R. A. Medellín, J. A. de la Torre, H. Zarza, C. Chávez, & G. Ceballos (Eds.), *El Jaguar en el Siglo XXI: La Perspectiva Continental, Fondo de Cultura Económica*. Universidad Nacional Autónoma de México.

Mallon, D., & Kulikov, M. (2015). *Transboundary snow leopard conservation in Central Asia*. Fauna & Flora International and Convention on Migratory Species.

Mallon, D. P., Breitenmoser, U., & Ahmad Khan, J. (2008). *Panthera pardus ssp. NIMR. The IUCN red list of threatened species 2008*. https://doi.org/10.2305/IUCN.UK.2008.RLTS.T15958A5333919.en

Manoj, E. M. (2021). *40 killed in tiger attacks in the country last year. The Hindu 13 Sept 2021*. Retrieved from https://www.thehindu.com/news/national/kerala/40-killed-in-tiger-attacks-in-the-country-last-year/article36442241.ece

Marker, L., & Dickman, A. J. (2005). Factors affecting leopard spatial ecology, with particular reference to Namibian farmlands. *South African Journal of Wildlife Research, 35*(2), 105–115. https://hdl.handle.net/10520/EJC117223

McCain, E. B., & Childs, J. L. (2008). Evidence of resident jaguars (Panthera onca) in the southwestern United States and the implications for conservation. *Journal of Mammalogy, 89*, 1–10.

McCarthy, T., Mallon, D., Jackson, R., Zahler, P., & McCarthy, K. 2016. Panthera uncia. The IUCN red list of threatened species 2017. https://doi.org/10.2305/IUCN.UK.2017-2.RLTS.T22732A50664030.en

McCarthy, T., Mallon, D., Sanderson, E., & Zahler, P. (2016). Biogeography and status overview. In T. McCarthy & D. Mallon (Eds.), *Snow leopards* (pp. 23–41). Elsevier.

McCarthy, T. M., & Chapron, G. (2003). *Snow leopard survival strategy*. Snow Leopard Trust and Snow Leopard Network.

McCarthy, T. M., & Mallon, D. (2016). *Snow leopards: Biodiversity of the world: Conservation from genes to landscapes*. Elsevier. https://doi.org/10.1016/C2014-0-01301-9

McCay-Peet, L., & Quan-Haase, A. (2017). What is social media and what questions can social media research help us answer. In L. Sloan & A. Quan-Haase (Eds.), *The SAGE handbook of social media research methods* (pp. 13–26). SAGE.

McDougal, C. (1988). Leopard and tiger intereactions at Royal Chitwan National Park, Nepal. *Journal of the Bombay Natural History Society, 85*, 609–610.

Medellín, R. A., Chetkiewicz, C., Rabinowitz, A., Redford, K. H., Robinson, J. G., Sanderson, E., & Taber, A. (Eds.). (2002). *Jaguares en el nuevo milenio: una evaluación de su estado, detección de prioridades y recomendaciones para la conservación de los Jaguares en América*. Universidad Nacional Autónoma de México / Wildlife Conservation Society.

Medvedev, D. G. (2000). Morfologichskiye otlichiye isbisa iz yuzhnogo Zabaikalia. *Vestnik Irkutskoi Gosudarstvennoi Selskokhozyaistvennoi Akademii, 20*, 20–30.

Miettinen, J., Shi, C., & Liew, S. C. (2011). Deforestation rates in insular Southeast Asia between 2000 and 2010. *Global Change Biology, 17*(7), 2261–2270.

Miquelle, D. G., Nikolaev, I. J., Goodrich, G., Litvinov, B., Smirnov, E. N., & Suvorov, E. (2005). Searching for the co-existence recipe: A case study of conflicts between people and tigers in the Russian Far East. In R. Woodruffe & S. Thirgood (Eds.), *People and wildlife: Conflict or co-existence?* (pp. 305–322). Cambridge University Press.

Mishra, C., Bhatnagar, Y. V., Trivedi, P., Timbadia, R., Bijoor, A., Sonam, K., Thinley, S., Namgail, T., & Prins, H. T. (2016). The role of village reserves in revitalizing the natural prey base of the snow leopard. In T. McCarthy & D. Mallon (Eds.), *Snow leopards* (pp. 184–187). Elsevier.

Miththapala, S., Seidensticker, J., & O'Brien, S. J. (1996). Phylogeographic subspecies recognition in leopards (Panthera pardus): Molecular genetic variation. *Conservation Biology, 10*(4), 1115–1132. https://doi.org/10.1046/j.1523-1739.1996.10041115.x

Mohammad, G., Mostafawi, S. N., Dadul, J., & Rosen, T. (2016). Corral improvements. In T. McCarthy & D. Mallon (Eds.), *Snow leopards* (pp. 180–183). Elsevier.

Moheb, Z., & Bradfield, D. (2014). Status of the common leopard in Afghanistan. *Cat News, 61*, 15–16.

Moheb, Z., & Paley, R. (2016). Snow leopard status and conservation in Afghanistan. In T. McCarthy & D. Mallon (Eds.), *Snow leopards* (pp. 409–417). Elsevier.

Mondal, P., & Nagendra, H. (2011). Trends of forest dynamics in tiger landscapes across Asia. *Environmental Management, 48*(4), 781.

Mora, J. M., Polisar, J., Portillo, H., & Franklin, C. (2016). Estado del Jaguar en Honduras. In R. A. Medellín, J. A. de la Torre, C. Chávez, H. Zarza, & G. Ceballos (Eds.), *El Jaguar en el Siglo XXI:La Perspectica Continental, Fondo de Cultura Económica*. Universidad Nacional Autónoma de México, Ciudad de México.

Moreno, R., Bustamante, A., Méndez-Carvajal, P., & Moreno, J. (2016). Jaguares (Panthera onca) en Panamá, Estado Actual y Conservación. In R. A. Medellín, J. A. de la Torre, C. Chávez, H. Zarza, & G. Ceballos (Eds.), *El Jaguar en el Siglo XXI: La Perspectiva Continental, Fondo de Cultura Económica*. Universidad Nacional Autónoma de México, Ciudad de México.

Namgail, T., & Dadul, J. (2016). Himalayan homestays; fostering human-snow leopard conservation coexistence. In T. M. McCarthy & D. P. Mallon (Eds.), *Snow leopards*. Elsevier.

Nawaz, M. A., Ud Din, J., & Buzdar, H. (2016). The ecosystem health program: A tool to promote the co-existence of livestock owners and snow leopards. In T. McCarthy & D. Mallon (Eds.), *Snow leopards* (pp. 188–195). Elsevier.

Nelson, E. W., & Goldman, E. A. (1933). Revision of the jaguars. *Journal of Mammalogy, 14*(3), 221–240. https://doi.org/10.2307/1373821

Nielsen, C., Thompson, D., Kelly, M., & Lopez-Gonzalez, C. A. (2015). Puma concolor (errata version published in 2016). The IUCN Red List of Threatened Species 2015. https://doi.org/10.2305/IUCN.UK.2015-4.RLTS.T18868A50663436.en

Nowell, K., & Jackson, P. (Eds.). (1996). *Wild cats: Status survey and conservation action plan*. IUCN/SSC Cat Specialist Group.

Nowell, K., & Pervushina, N. (2014). *Review of implementation of resolution Conf. 12.5 (rev. CoP16) on conservation and trade in tigers and other appendix-I Asian big cats*. IUCN and TRAFFIC report prepared for the CITES Secretariat, 65th meeting of the CITES standing committee, Geneva, 7-11 July

Nunez, R., Miller, B., & Lindzey, F. (2000). Food habits of jaguars and pumas in Jalisco, Mexico. *Journal of Zoology, 252*(3), 373–379. https://doi.org/10.1111/j.1469-7998.2000.tb00632.x

O'Brien, S. J., & Johnson, W. E. (2007). The evolution of cats. *Scientific American, 7*, 68–75.

Odden, M., Wegge, P., & Fredriksen, T. (2010). Do tigers displace leopards? If so, why? *Ecological Research, 25*, 875–881.

Olsoy, P. J., Zeller, K. A., Hicke, J. A., Quigley, H. B., Rabinowitz, A. R., & Thornton, D. H. (2016). Quantifying the effects of deforestation and fragmentation on a range-wide conservation plan for jaguars. *Biological Conservation, 203*, 8–16.

Oswell, A. (2010). *The big cat trade in Myanmar and Thailand*. TRAFFIC Southeast Asia.

Packer, C., Canney, S., Loveridge, A., Garnett, S. T., Zander, K. K., Balme, G., Bauer, H., Begg, C., Begg, K., Bhalla, S., Bonham, R., Brink, H., Burton, C., Caro, T. M., Clegg, B., Dloniak, S., Frank, L., Funston, P., Groom, R., Heath, B., Hill, T., Hunter, L., DeIongh, H. H., Joubert, D., Kissui, B., Knocker, W., Leatham, B., Lindsey, P. A., Maclennan, S. D., MacNutt, T., Nicholls, K., Patterson, B., Plumptre, A., Salerno, J., Slotow, R., Sogbohossou, E., Stratford, K., Winterbach, C., Winterbach, H., & Polasky, S. (2013). Conserving large carnivores: Dollars and fence. *Ecology Letters, 16*(5), 635–641. https://doi.org/10.1111/ele.12091

Palmeirim, A. F., & Gibson, L. (2021). Impacts of hydropower on the habitat of jaguars and tigers. *Community Biology, 4*, 1358.

Paltsyn, M., Poyarkov, A., Spitsyn, S., Kuksin, A., Istomov, S., Gibbs, J. P., ... & Rozhnov, V. (2016). *Northern range: Russia* (pp. 501–511). Elsevier EBooks. https://doi.org/10.1016/b978-0-12-802213-9.00040-7

Parker, D. M., Whittington-Jones, B. M., Bernard, R. T. F., & Davies-Mostert, H. T. (2014). Attitudes of rural communities toward dispersing African wild dogs in South Africa. *Human Dimensions of Wildlife Wildlife, 19*, 512–522. https://doi.org/10.1080/10871209.2014.926575

Paviolo, A., De Angelo, C., Ferraz, K., Morato, R. G., Martinez Pardo, J., Srbek-Araujo, A. C., de Mello Beisiegel, B., Lima, F., Sana, D., de Silva, M. X., Valazquez, M. C., Cullen, L., Crawshaw, P., Jr., Jorge, M. L., Galetti, P. M., Di Bitetti, M. S., Cunha de Paula, R., Eizirik, E., Mitchell Aida, T., Cruz, P., Perilli, M. L., Souza, A. S., Quiroga, V., Nakano, E., Ramirez Pinto, F., Fernandez, S., Costa, S., Moraes, E., Jr., & Azevedo, F. (2016). A biodiversity hotspot losing its top predator: The challenge of jaguar conservation in the Atlantic Forest of South America. *Scientific Reports, 6*, 37147. https://doi.org/10.1038/srep37147/

Payán, E., Carbone, C., Homewood, K., Paemelaere, E., Quigley, H. B., & Durant, S. (2013). Where will jaguars roam? The importance of survival in unprotected lands. In M. Ruiz-Garcia & J. Shostell (Eds.), *Molecular population genetics, phylogenetics, evolutionary biology and conservation of the Neotropical carnivores* (pp. 603–628). Nova Science.

Payán, E., Castaño-Uribe, C., Gonzalez-Maya, J. F., Valderrama, C., Ruiz-Garcia, M., & Soto, C. (2013). Distribución y estado de conservación del Jaguar en Colombia. In E. Payán & C. Castaño-Uribe (Eds.), *Grandes Felinos de Colombia* (pp. 23–36). Panthera Colombia, Fundación Herencia Ambiental Caribe, Conservación Internacional Colombia y Cat Specialist Group UICN/SSC.

Payán, E., Soto, C., Ruiz-García, M., Nijhawan, S., González-Maya, J. F., Valderrama, C., & Castaño-Uribe, C. (2016). Unidades de conservación, conectividad y calidad del hábitat de Jaguar en Colombia. In R. A. Medellín, J. A. de la Torre, H. Zarza, C. Chávez, & G. Ceballos (Eds.), *In El Jaguar en el Siglo XXI: La Perspectiva Continental, Fondo de Cultura Económica.* Universidad Nacional Autónoma de México.

Pereira-Garbero, R., & Sappa, A. (2016). Historia del Jaguar en Uruguay y la Banda Oriental. In R. A. Medellín, J. A. de la Torre, H. Zarza, C. Chávez, & G. Ceballos (Eds.), *El Jaguar en el Siglo XXI: La Perspectiva Continental, Fondo de Cultura Económica.* Universidad Nacional Autónoma de México.

Perez, I., Geffen, E., & Mokady, O. (2006). Critically endangered Arabian leopard Panthera pardus nimr in Israel: Estimating population parameters using molecular scatology. *Oryx, 40*, 295–301.

Petracca, L. S., Hernández-Potosme, S., Obando-Sampson, L., Salom-Pérez, R., Quigley, H., & Robinson, H. S. (2014). Agricultural encroachment and lack of enforcement threaten connectivity of range-wide jaguar (Panthera onca) corridor. *Journal for Nature Conservation, 22*(5), 436–444.

Petracca, L. S., Ramírez-Bravo, O. E., & Hernández-Santín, L. (2014). Occupancy estimation of jaguar Panthera onca to assess the value of east-Central Mexico as a jaguar corridor. *Oryx, 48*, 133–140.

Pitman, R. T., Swanepoel, L. H., Hunter, L., Slotow, R., & Balme, G. A. (2015). The importance of refugia, ecological traps and scale for large carnivore management. *Biodiversity and Conservation, 24*, 1975–1987.

Pocock, R. I. (1939). The races of jaguar (Panthera onca). *Novitates Zoologicae, 41*, 406–422.

Polisar, J., Maxit, I., Scognamillo, D., Farrell, L., Sunquist, M. E., & Eisenberg, J. F. (2003). Jaguars, pumas, their prey base, and cattle ranching: Ecological interpretations of a management problem. *Biological Conservation, 109*(2), 297–310.

Porfirio, G., Sarmento, P., Leal, S., & Fonseca, C. (2014). How is the jaguar Panthera onca perceived by local communities along the Paraguai River in the Brazilian Pantanal? *Oryx, 1*, 1–6. https://doi.org/10.1017/S0030605314000349

Qarqaz, M., & Abu Baker, M. (2006). The leopard in Jordan. *Cat News, 1*, 9–10.

Quigley, H., Foster, R., Petracca, L., Payan, E., Salom, R., & Harmsen, B. (2017). Panthera onca (errata version published in 2018). The IUCN Red List of Threatened Species 2017. https://doi.org/10.2305/IUCN.UK.2017-3.RLTS.T15953A50658693.en

Quigley, H., Hoogesteijn, R., Hoogesteijn, A., Foster, R., Payan, E., Corrales, D., Salom-Perez, R., & Urbina, Y. (2015). Observations and preliminary testing of jaguar depredation reduction techniques in and between core jaguar populations. *PARKS, 21*(1), 63–72.

Quigley, H. B., & Crawshaw, P. G., Jr. (1992). A conservation plan for the jaguar Panthera onca in the Pantanal region of Brazil. *Biological Conservation, 61*, 149–157.

Quigley, K. S., Evermann, J. F., Leathers, C. W., Armstrong, D. L., Goodrich, J. M., Duncan, N. M., & Miquelle, D. G. (2010). Morbillivirus infection confirmed in a wild Siberian tiger in the Russian Far East. *Journal of Wildlife Diseases, 42*, 1252–1256.

Rabinowitz, A., & Zeller, K. A. (2010). A range-wide model of landscape connectivity and conservation for the jaguar, Panthera onca. *Biological Conservation, 143*, 939–945.

Ram, M., Sahu, A., Tikadar, S., Jayawant, H., Jhala, L., Zala, Y., & Venkataraman, M. (2022). Living on the seacoast: Ranging and habitat distribution of Asiatic lions. *Scientific Reports, 12*, 19235. https://doi.org/10.1038/s41598-022-23761-1

Rasphone, A., Kéry, M., Kamler, J. F., & Macdonald, D. W. (2019). Documenting the demise of tiger and leopard, and the status of other carnivores and prey, in Lao PDR's most prized protected area: Nam et-Phou Louey. *Global Ecology and Conservation, 20*, 1–15.

Raza, R. H., Chauhan, D. S., Pasha, M. K. S., & Sinha, S. (2012). *Illuminating the blind spot: A study on illegal trade in leopard parts in India (2001–2010)*. Traffic International.

Riggio, J., Jacobson, A., Dollar, L., Bauer, H., Becker, M., Dickman, A., et al. (2012). The size of savannah Africa: A lion's (Panthera leo) view. *Biodiversity and Conservation, 22*, 17–35. https://doi.org/10.1007/s10531-012-0381-4

Riordan, P., & Shi, K. (2016). *China: Current state of snow leopard conservation in China* (pp. 523–531). https://doi.org/10.1016/b978-0-12-802213-9.00042-0

Riordan, P., Cushman, S. A., Mallon, D., Shi, K., & Hughes, J. (2015). Predicting global population connectivity and targeting conservation action for snow leopard across its range. *Ecography, 39*(5), 419–426. https://doi.org/10.1111/ecog.01691

Ripple, W. J., Estes, J. A., Beschta, R. L., Wilmers, C. C., Ritchie, E. G., Hebblewhite, M., Berger, J., Elmhagen, B., Letnic, M., Nelson, M. P., Schmitz, O., Smith, D. W., Wallach, A. D., & Wirsing, A. J. (2014). Status and ecological effects of the world's largest carnivores. *Science, 343*, 1241484. https://doi.org/10.1126/science.1241484

Roberts, T. J. (1997). *The mammals of Pakistan*. Oxford University Press.

Robinson, H. S., Wielgus, R. B., Cooley, H. S., & Cooley, S. W. (2008). Sink populations in carnivore management: Cougar demography and immigration in a hunted population. *Ecological Applications, 18*, 1028–1037.

Rodriguez-Mahecha, J. V., Alberico, M., Trujillo, F., & Jorgenson, J. (2006). Libro Rojo de los Mamíferos de Colombia. Serie Libros Rojos de Especies Amenazadas de Colombia. Conservación Internacional Colombia & Ministerio de Ambiente, vivienda y Desarrollo Territorial

Rosas-Rosas, O. C., & Bender, L. C. (2012). Population status of jaguars (Panthera onca) and pumas (Puma concolor) in northeastern Sonora, Mexico. *Acta Zoológica Mexicana, 28*, 86–101.

Rosen, T., & Zahler, P. (2016). Transboundary initiatives and snow leopard conservation. In T. McCarthy & D. Mallon (Eds.), *Snow leopards* (pp. 267–276). Elsevier.

Rozhnov, V. V., Naidenko, S. V., Hernandez-Blanco, J. A., Chistopolova, M. D., Sorokin, P. A., Yachmennikova, A., Blidchenko, E. Y., Kalinin, A. Y., & Kastrikin, V. A. (2021). Restoration of the Amur Tiger (Panthera tigris altaica) population in the northwest of its distribution area. *Biology Bulletin of the Russian Academy of Sciences, 48*, 1401–1423.

Ruiz-Garcia, M., Payán, E., Murillo, A., & Alvarez, D. (2006). DNA microsatellite characterisation of the jaguar (Panthera onca) in Colombia. *Genes and Genetic Systems, 81*, 115–127.

Sagar, V., Kaelin, C. B., Natesh, M., Reddy, P. A., Mohapatra, R. K., Chhattani, H., Thatte, P., Vaidyanathan, S., Biswas, S., Bhatt, S., & Paul, S. (2021). High frequency of an otherwise rare phenotype in a small and isolated tiger population. *Proceedings of the National Academy of Sciences, 118*(39), e2025273118.

Sanderson, E., Forrest, J., Loucks, C., Ginsberg, J., Dinerstein, E., Seidensticker, J., Leimgruber, P., Songer, M., Heydlauff, A., O'Brien, T., Bryja, G., Klenzendorf, S., & Wikramanayake, E. (2006). *Setting priorities for the conservation and recovery of wild tigers: 2005–2015. The technical assessment*. WCS, WWF, Smithsonian, and NFWF-STF.

Sanderson, E. W., Miquelle, D. G., Fisher, K., Harihar, A., Clark, C. G., Moy, J., Potapov, P., Robinson, N. P., Royte, L., Sampson, D., Sanderlin, J. S., Yackulic, C. B., Belecky, M., Breitenmoser, U., Breitenmoser-Würsten, C., Chanchani, P., Chapman, S., Deomurari, A., Duangchantrasiri, S., & Facchini, E. (2023). Range-wide trends in tiger conservation landscapes, 2001–2020. *Frontiers in Conservation Science, 4*, 1191280. https://doi.org/10.3389/fcosc.2023.1191280

Sanderson, E. W., Redford, K. H., Chetkiewicz, C. B., Medellin, R. A., Rabinowitz, A. R., Robinson, J. G., & Taber, A. B. (2002). Planning to save a species: The jaguar as a model. *Conservation Biology, 16*(1), 58.

Sanei, A., & Zakaria, M. (2011). Distribution pattern of the Persian leopard (Panthera pardus saxicolor) in Iran. *Asia Life Sciences, 7*(Supplement), 7–18.

Sankar, K., Qureshi, Q., Nigam, P., Malik, P. K., Sinha, P. R., Mehrotra, R. N., Gopal, R., Bhattacharjee, S., Mondal, K., & Gupta, S. (2010). Monitoring of reintroduced tigers in Sariska Tiger Reserve, Western India: Preliminary findings on home range, prey selection and food habits. *Tropical Conservation Science, 3*(3), 301–318.

Sarkar, M. S., Ramesh, K., Johnson, J. A., Sen, S., Nigam, P., Gupta, S. K., Murthy, R. S., & Saha, G. K. (2016). Movement and home range characteristics of reintroduced tiger (Panthera tigris) population in Panna Tiger Reserve, Central India. *European Journal of Wildlife Research, 62*(5), 537–547.

Sarukhanova, S. (2014). *Leopards Boooom in Azerbaijan. Caucasus ecoregion newsletter* (p. 6). WWF Caucasus Program.

Schaller, G. B. (2014). *Trip report: Qinghai Province, China, June 5–July 12, 2014*. Panthera.

Schnitzler, A., & Hermann, L. (2019). Chronological distribution of the tiger Panthera tigris and the Asiatic lion Panthera leo persica in their common range in Asia. *Mammal Review, 49*(4), 340–353. https://doi.org/10.1111/mam.12166

Seidensticker, J., Christie, S., & Jackson, P. (1999). *Preface. Riding the tiger. Tiger conservation in human-dominated landscapes*. Cambridge University Press.

Seimon, T. A., Miquelle, D. G., & Chang, T. Y. (2013). Canine distemper virus: An emerging disease in wild endangered Amur tigers (Panthera tigris altaica). *Molecular Biology, 4*, 00410.

Selvan, K. M., Lyngdoh, S., Habib, B., & Gopi, G. V. (2014). Population density and abundance of sympatric large carnivores in the lowland tropical evergreen forest of Indian eastern Himalayas. *Mammalian Biology, 79*(4), 254–258.

Senn, H., Murray-Dickson, G., Kitchener, A. C., Riordan, P., & Mallon, D. (2018). Response to Janecka et al. 2017. *Heredity, 120*(6), 581–585. https://doi.org/10.1038/s41437-017-0015-4

Seymour, K. L. (1989). Panthera onca. *Mammalian Species, 340*, 1–9.

Sharma, K., Bayrakcismith, R., Tumursukh, L., Johansson, O., Sevger, P., McCarthy, T., & Mishra, C. (2014). Vigorous dynamics underlie a stable population of the endangered snow leopard Panthera uncia in Tost Mountains, south Gobi, Mongolia. *Plos One, 9*(7), e1011319.

Shehzad, W., Nawaz, M. A., Pompanon, F., Coissac, E., Riaz, T., Shah, S. A., & Taberlet, P. (2014). Forest without prey: Livestock sustain a leopard Panthera pardus population in Pakistan. *Oryx, 49*(2), 248–253.

Shrestha, R., Tenzing, D., Tashi, N., & Wangdi, G. (2013). *A report on snow leopard (Panthera uncia) population survey in the central range of Wangchuk Centennial Park, Bhutan*. WWF-US, Eastern Himalayas Program.

Silveira, L. (2004). Ecologia comparada e conservação da onça-pintada (Panthera onca) e onça-parda (Puma concolor), no Cerrado e Pantanal. Ph.D. thesis, Univ. de Brasília, Brasília, Brazil.

Silver, S. C., Ostro, L. E. T., Marsh, L. K., Maffei, L., Noss, A. J., Kelly, M. J., Wallace, R. B., Gomez, H., & Ayala, G. (2004). The use of camera traps for estimating jaguar Panthera onca abundance and density using capture/recapture analysis. *Oryx, 38*, 148–154.

Singh, A. P. (2017a). The Asiatic lion (Panthera leo persica): 50 years journey for conservation of an endangered carnivore and its habitat in Gir protected area, Gujarat, India. *Indian Forester, 143*(10), 993–1003.

Singh, H. S. (2017b). Dispersion of the Asiatic lion Panthera leo persica and its survival in human-dominated landscape outside the Gir forest, Gujarat, India. *Current Science, 112*(5), 933–940. https://doi.org/10.18520/cs/v112/i05/933-940

Singh, H. S., & Gibson, L. (2011). A conservation success story in the otherwise dire megafauna extinction crisis: The Asiatic lion (Panthera leo persica) of Gir forest. *Biological Conservation, 144*(5), 1753–1757. https://doi.org/10.1016/j.biocon.2011.02.009

Smith, A. T., & Xie, Y. (2008). *A guide to the mammals of China*. Princeton University Press.

Snow Leopard Network. (2014). *Snow leopard survival strategy. Version 2014.1.* Snow Leopard Network. Retrieved from www.snowleopardnetwork.org

Snow Leopard Working Secretariat. (2013). *Global snow leopard and ecosystem protection program*. Snow Leopard Working Secretariat.

Sodhi, N. S., Koh, L. P., Clements, R., Wanger, T. C., Hill, J. K., Hamer, K. C., Clough, Y., Tscharntke, T., Posa, M. R. C., & Ming Lee, T. (2010). Conserving southeast Asian forest biodiversity in human-modified landscapes. *Biological Conservation*. https://doi.org/10.1016/j.biocon.2009.12.029

Soisalo, M. K., & Cavalcanti, S. M. C. (2006). Estimating the density of a jaguar population in the Brazilian Pantanal using camera-traps and capture-recapture sampling in combination with GPS radio-telemetry. *Biological Conservation, 129*, 487–496.

Sollmann, R., Torres, N. M., & Silveira, L. (2008). Jaguar conservation in Brazil: The role of protected areas. *Cat News, 2*, 15–20.

Song, D., Wang, B., Jiang, J., Wan, S., Cui, S., Wang, T., & Feng, L. (2014). Using camera trap to monitor a north Chinese leopard (Panthera pardus japonensis) population and their main ungulate prey. *Biodiversity Science, 22*, 733–736.

Soto-Shoender, J. R., & Main, M. B. (2013). Differences in stakeholder perceptions of the jaguar Panthera onca and puma Puma concolor in the tropical lowlands of Guatemala. *Oryx, 47*, 109–112. https://doi.org/10.1017/S003060531200107X

Spalton, J. A., & al Hikmani, H. (2014). *The Arabian leopards of Oman*. Stacey International Press.

Spalton, J. A., Al Hikmani, H. M., Willis, D., & Said, A. S. B. (2006). Critically endangered Arabian leopards Panthera pardus nimr persist in the Jabal Samhan nature reserve, Oman. *Oryx, 40*(3), 287–294. https://doi.org/10.1017/S0030605306000743

Stein, A. B., Athreya, V., Gerngross, P., Balme, G., Henschel, P., Karanth, U., Miquelle, D., Rostro-Garcia, S., Kamler, J. F., Laguardia, A., Khorozyan, I. & Ghoddousi, A. (2020). Panthera pardus (amended version of 2019 assessment). The IUCN red list of threatened species 2020. https://doi.org/10.2305/IUCN.UK.2020-1.RLTS.T15954A163991139.en

Stein, A. B., Athreya, V., Gerngross, P., Balme, G., Henschel, P., Karanth, U., Miquelle, D., Rostro-García, S., Kamler, J. F., Laguardia, A., Khorozyan, I., & Ghoddousi, A. (2023). Panthera pardus. The IUCN Red List of Threatened Species 2023. https://doi.org/10.2305/IUCN.UK.2023-1.RLTS.T15954A215195554.en

Steinberg, M. K. (2016). Jaguar conservation in southern Belize: Conflicts, perceptions, and prospects among Mayan hunters. *Conservation and Society, 14*(1), 13–20. https://doi.org/10.4103/0972-4923.182801

Steinmetz, R., Chutipong, W., & Seuaturien, N. (2010). Population recovery patterns of tropical ungulates after poaching in Thailand. *Biological Conservation, 143*, 42–51.

Stoner, S. S., & Pervushina, N. (2013). *Reduced to skin and bones revisited: An updated analysis of Tiger seizures from 12 Tiger range countries (2000–2012)*. TRAFFIC.

Sunquist, M., & Sunquist, F. (2002). *Wild cats of the world*. University of Chicago Press.

Suryawanshi, K. R., Bhatnagar, Y. V., Redpath, S., & Mishra, C. (2013). People, predators and perceptions: Patterns of livestock depredation by snow leopards and wolves. *Journal of Applied Ecology, 50*, 550–560. https://doi.org/10.1111/1365-2664.12061

Taubmann, J., Sharma, K., Uulu, K. Z., Hines, J. E., & Mishra, C. (2015). Status assessment of the endangered snow leopard panthera uncia and other large mammals in the Kyrgyz Alay, using community knowledge corrected for imperfect detection. *Oryx, 50*(2), 220–230. https://doi.org/10.1017/s0030605315000502

Thapa, K., & Kelly, M. J. (2017). Density and carrying capacity in the forgotten tigerland: Tigers in the understudied Nepalese Churia. *Integrative Zoology, 12*, 211–227.

Thatte, P., Joshi, A., Vaidyanathan, S., Landguth, E., & Ramakrishnan, U. (2018). Maintaining tiger connectivity and minimizing extinction into the next century: Insights from landscape genetics and spatially-explicit simulations. *Biological Conservation, 218*, 181–191.

Thinley, P., Dagay, L., Dorji, C., Namgyel, S., Yoenten, P., & Dorji, T. (2014). *Estimating snow leopard (Panthera uncia) abundance and distribution in Jigme Dorji National Park using camera traps: A technical report*. KUENSEL Corporation Ltd.

Thinley, P., Lham, D., Wangchuk, S., & Wangchuk, N. (2016). *National snow leopard survey of Bhutan 2014–2016*. Department of Forests and Park Services.

Thompson, D. J., & Jenks, J. A. (2010). Dispersal movements of subadult cougars from the Black Hills: The notion of range expansion and recolonization. *Ecosphere, 1*, 1–11.

Thorn, M., Green, M., Scott, D., & Marnewick, K. (2013). Characteristics and determinants of human-carnivore conflict in South African farmland. *Biodiversity and Conservation, 22*, 1715–1730.

Uphyrkina, O., Johnson, E. W., Quigley, H., Miquelle, D., Marker, L., Bush, M., & O'Brien, S. J. (2001). Phylogenetics, genome diversity and origin of modern leopard, Panthera pardus. *Molecular Ecology, 10*(11), 2617–2633. https://doi.org/10.1046/j.0962-1083.2001.01350.x

Venkataraman, M. (2016). Wildlife and human impacts in the Gir landscape. In P. K. Agrawal, A. Verghese, S. R. Krishna, & K. Subaharan (Eds.), *Human animal conflict in agro-pastoral context: Issues & policies* (pp. 32–40). Indian Council of Agricultural Research.

Vitekere, K., Consolee, K., Zaman, M., Karanja, H., Hua, Y., & Jiang, G. (2020). Insights on the North China leopard (Panthera Pardus Japonensis Gray, 1862): Challenges in distribution, population status, threats, and implications for conservation. *Journal of Animal and Plant Sciences, 31*(1), 1–18. https://doi.org/10.36899/JAPS.2021.1.0187

Voskyanyan, G. (2014). *One male and two females of Caucasian leopard live in southern Armenia. Caucasus ecoregion newsletter* (p. 6). WWF Caucasus Program.

Wallace, R., Lopez-Strauss, H., Mercado, N., & Porcel, Z. (2013). Diversity, distribution and conservation of Bolivian carnivores. In M. Ruiz-Garcia & J. Shostell (Eds.), *Molecular population genetics, evolutionary biology and biological conservation of the Neotropical carnivores* (pp. 659–709). Nova Science.

Walston, J., Robinson, J. G., Bennett, E. L., Breitenmoser, U., da Fonseca, G. A. B., Goodrich, J., Gumal, M., Hunter, L., Johnson, A., Karanth, K. U., Leader-Williams, N., MacKinnon, K., Miquelle, D., Pattanavibool, A., Poole, C., Rabinowitz, A., Smith, J. L. D., Stokes, E. J., Stuart, S. N., Vongkhamheng, C., & Wibisono, H. (2010). Bringing the Tiger Back from the Brink—The six percent solution. *PLoS Biology, 8*(9), e1000485.

Wangchuk, R., Jackson, R., & Brewer, W. (2010). Snow leopards and Himalayan homestays: Catalysts for community-based conservation in mountain areas. *Mountain Forum Bulletin, 2010*, 25–28.

Wangchuk, T., & Tharchen, L. (2016). Snow leopard conservation in Bhutan. In T. McCarthy & D. Mallon (Eds.), *Snow leopards* (pp. 449–456). Elsevier.

West, P. M., & Packer, C. (2013). Panthera leo. In J. Kingdon & M. Hoffmann (Eds.), *The mammals of Africa. Volume V: Carnivores, pangolins, equids and rhinoceroses*. Bloomsbury Publishing.

Wilcove, D. S., Giam, X., Edwards, D. P., Fisher, B., & Koh, L. P. (2013). Navjot's nightmare revisited: Logging, agriculture, and biodiversity in Southeast Asia. *Trends in Ecology and Evolution, 28*(9), 531–540.

Wilting, A., Courtiol, A., Christiansen, P., Niedballa, J., Scharf, A. K., Orlando, L., Balkenhol, N., Hofer, H., Kramer-Schadt, S., Fickel, J., & Kitchener, A. C. (2015). Planning tiger recovery: Understanding intraspecific variation for effective conservation. *Science Advances, 1*(5), e1400175. https://doi.org/10.1126/sciadv.1400175

Wingard, J., Zahler, P., Victurine, R., Bayasgalan, O., & Buuveibaatar, B. (2014). *Guidelines for addressing the impact of linear infrastructure on migratory large mammals in Central Asia.* Convention on Migratory Species Technical Report.

Wong, R., & Krishnasamy, K. (2019). *Skin and bones: Unresolved*. Retrieved from https://www.traffic.org/publications/reports/skin-and-bones-unresolved/

Wozencraft, W. C. (2005a). Species Panthera tigris. In D. E. Wilson & D. M. Reeder (Eds.), *Mammal species of the world: A taxonomic and geographic reference* (3rd ed.). Johns Hopkins University Press.

Wozencraft, W. C. (2005b). Species Panthera onca. In D. E. Wilson & D. M. Reeder (Eds.), *Mammal species of the world: A taxonomic and geographic reference* (3rd ed., pp. 546–547). Johns Hopkins University Press.

Wozencraft, W. C. (2005c). Species Uncia uncia. In D. E. Wilson & D. M. Reeder (Eds.), *Mammal species of the world: A taxonomic and geographic reference* (3rd ed.). Johns Hopkins University Press.

Xiao, W. H., Feng, L. M., Zhao, X. D., Yang, H. T., Dou, H. L., Cheng, Y. C., Mou, P., Wang, T. M., & Ge, J. P. (2014). Distribution and abundance of Amur tiger, Amur leopard and their ungulate prey in Hunchun National Nature Reserve, Jilin. *Biodiversity Science, 22*(6), 717–724. https://doi.org/10.3724/sp.j.1003.2014.14184

Yarovenko, Y.A. 2010. Status and distribution of leopards (Panthera pardus) in the mountains of Dagestan, Russia. Final report..

Zaffar-ul-Islam, M., Boug, A., As-Sheheri, A., & Al Jaid, M. (2014). Poisoning of endangered Arabian leopard in Saudi Arabia and its conservation efforts. *Cat News, 60*, 16–17.

Zahler, P. (2016). Linear infrastructure and snow leopard conservation. In T. McCarthy & D. Mallon (Eds.), *Snow leopards* (pp. 123–126). Elsevier.

Zimmermann, A., Walpole, M. J., & Leader-Williams, N. (2005). Cattle ranchers 'attitudes to conflicts with jaguar Panthera onca in the Pantanal of Brazil. *Oryx, 39*, 406–412. https://doi.org/10.1017/S0030605305000992

Chapter 3
Voluntary Village Relocation and Sustainability: A Case of Ecological Component in Kanha Tiger Reserve, Central India

Sanjay Kumar Pathak and Kaushalendra Kumar Jha

Abstract This chapter is based on a research study in one of the most populous Tiger Reserves of Central India where people and tigers coexisted before human activities began to affect tiger presence. While tiger populations were reduced almost to extinction in Tiger Reserves like Sariska and Panna and with tiger extinction possibly imminent in some other Tiger Reserves of the country, steps were taken to save the apex species of the forest ecosystem to keep the ecological balance. Measures taken first focussed on population increase. Later, tiger habitat became the main focus at least the core area or the critical tiger habitat, which required the exclusion of any anthropogenic activities, which in some cases required the relocation of the human population from the core areas and resettlement outside these areas without compromising human's economic and social requirements. To have the "win-win" situation of sustainable development, human habitation from the core area was relocated in consonance with the extant legal provisions. This chapter examines the pre- and post-relocation situation in terms of the three constituents of sustainability i.e. environmental/ecological, social, and economic parameters. The chapter presents the results of environmental or ecological components in detail. The results suggest, in most of the responses (77.61%), that humans expressed full satisfaction towards choice of relocation. The response of 91% of respondents showed that affected people were assisted by the Forest Department on all occasions. The M-STrIPES-generated maps showed that wild animals, both carnivores and herbivores frequently used the vacated areas. There was significant change in the vegetation after relocation and introduction of grassland species. The number of the grazing cattle decreased significantly post-relocation. Relocated people began to use fuel other than

S. K. Pathak
National Tiger Conservation Authority, New Delhi, India

Indian Institute of Public Administration, New Delhi, India

Uttar Pradesh Forest Corporation, Lucknow, India

K. K. Jha (✉)
Department of Environment, Forests and Climate Change, Lucknow, Uttar Pradesh, India
e-mail: jhakk1959@gmail.com

K. K. Jha and M. O. Campbell (eds.), *Intersections of Conservation Biogeography and Wildlife Management*, https://doi.org/10.1007/978-981-95-0121-2_3

firewood from forest. They experienced almost negligible human-wildlife conflicts, crop, and cattle depredation post-relocation, which reduced wildlife conflicts, and fostered the survival of both wildlife as well as human beings. The findings of this study provide a background for policymakers and conservationists involved in tiger conservation and broader issues of wildlife conservation elsewhere.

Keywords Village relocation · Ecosystem sustainability · Voluntariness · Ecological parameters · Hand holding and satisfaction · Madhya Pradesh

Introduction

The study on "Voluntary village relocation and sustainability of ecosystem in Kanha Tiger Reserve" was conducted considering issues and challenges of voluntary village relocation in the wider framework of ecosystem sustainability. Millions of people were displaced globally during the establishment of Protected Area (PA) Networks to create wildlife habitats (West et al., 2006). As many as 1 million to 16 million people were resettled in Africa (Geisler & De Sousa, 2001) while 0.6 million tribal people were resettled in India (IITR, 2011 according to reference note No. No.30/RN/Ref./December/2013 of Lok Sabha secretariat; Fanari, 2019). Displacements were part of the PA management policies in the past, however, there was little concrete documentation on such displacements. As a result, there was extensive criticism which could lead to improvements in resettlement policies (Agrawal & Redford, 2009; Lasgorceix & Kothari, 2009; West et al., 2006). Moreover, few landscape-level studies assessed relocation impacts on conservation outside the PA boundary, and the well-being of people after resettlement (Cernea & Schmidt-Soltau, 2006; Milgroom et al., 2014).

Relocation or displacement of people has been forced due to natural or man-made causes. Loss of habitation due to natural calamities or incidences such as floods, earthquakes, and landslide resulted in the displacement of human populations. Some unavoidable man-made causes also contributed to large-scale displacements. Recently, developmental projects have led to forced displacement of people who claimed residence rights and hindered such projects. The affected people were compensated as per the extant rules and regulations, but their concurrence or agreement was not fully accepted. The examples of such development-induced displacements in India are large irrigation projects like Sardar Sarovar Project in Gujarat, Tehri Dam Project in Uttarakhand; infrastructural projects such as Highways, Railway lines, Thermal & Nuclear power stations, large-scale mining, etc. (Aiyar & Kaushal, 2018; Asthana, 2012; Das & Das, 2014; Parasuraman, 1999).

In India, Project Tiger[1] came into being on 1 April 1973 and eventually became one of the most successful ventures for species conservation in the world. The endangered

[1] Project Tiger is a wildlife conservation movement initiated in India to protect the endangered tiger. It was initiated in 1973 by the Ministry of Environment, Forest, and Climate Change of the

tiger populations reached an effective path of recovery. Jhala et al. (2021) noted that "India has managed to increase its adult tiger population from around 1700 in 2010 to about 3000 by 2019 with tigers currently occupying almost 90,000 km^2". Creating inviolate space has always been considered as one of the efficacious steps to foster the cause of tiger conservation and for that, people were relocated from some of the tiger reserves, but the track record of such relocation exercises has been neither consistent or praise-worthy everywhere and in many cases, people who have been relocated have either returned or are today contributing to the pressure on the park from the fringes (Karanth et al., 2018; RTTF, 2005; Shahabuddin & Bhamidipati, 2014). Prior to 2008, even conservation-induced displacements or relocation often used to be involuntary in effect, where people were hardly able to do much negotiation about their choices (Torri, 2011). This situation was anything but conservation friendly which provided directives for the policymakers and scientists and suggested revision of guidelines pertaining to voluntary village relocation from the core area or critical tiger habitat.[2]

As per the legal/statutory provisions laid down in Section 38V (4) (i) of thc Wildlife (Protection) Act, 1972, as amended in 2006 (WPA, 1972) as well as the Section 4 (2) of Scheduled Tribes and Other Traditional Forest Dwellers (Recognition of Forest Rights) Act, 2006 (FRA, 2006), rights of forest dwelling Scheduled Tribes (FDST) and other traditional forest dwellers (OTFD) must be recognized in forest areas within core or critical tiger habitat of tiger reserves or critical wildlife habitat of protected areas and may be modified and resettled for providing inviolate spaces[3] for tigers or other wild animals. Compensation (rights settlement in addition to the relocation package offered under the Centrally Sponsored Scheme of Project Tiger) needed to be paid in such cases as per the stipulated Project Tiger guidelines. There is provision laid down for the acquisition of rights in or over land declared by the State Government under section 18 (for constituting a Sanctuary) or section 35 (for constituting a National Park) in Chapter IV of WPA, 1972 (section 24). The Collector is authorized under Sub-section (2) of section 24 of WPA to acquire such land or rights. Therefore, rights are modified or settled statutorily through payment of compensation for the immovable property of people. Jhala et al. (2021) reviewed that the establishment of Tiger Conservation Foundations, sharing of tiger-profits with buffer zone communities, and a handsome package for incentivized voluntary relocation have gone a long way in winning over local communities.

Research on tiger ecology has indicated that a minimum number of 20 tigresses of breeding age are required to maintain a viable population of 80–100 tigers (in and around the core) required an inviolate space of 800–1200 km^2 (Gopal et al., 2007).

Government of India. Currently, there are 55 Protected Areas designated as tiger reserves holding around 3682 wild tigers in India—almost 75% of the world's wild tiger population.

[2] Critical Tiger Habitats are wildlife habitats built in National Parks and Sanctuaries that must remain unchanged to save species. Critical 'tiger'habitats also known as core areas of tiger reserves—are identified under the Wildlife Protection Act (WLPA), 1972.

[3] Spaces devoid of human presence, serving as a cornerstone for species conservation. This separates wildlife populations from human-induced threats.

Tiger conservation, which covers an umbrella-species,[4] will also ensure viable populations of other wild animals (co-predators, prey) and forest, thereby ensuring the ecological viability of the entire area and habitat. Therefore, it became an ecological imperative to keep the core areas of tiger reserves inviolate for the survival of source populations of tiger and other wild animals. To create inviolate space within the tiger reserve, relocation of villages was considered mandatory. The major paradigm shift in the policy was that relocation was made voluntary from tiger reserves unlike development induced relocation where still choice of people hardly mattered. This was a unique feature and very conducive for conservation per se. As per data available with the National Tiger Conservation Authority (NTCA), New Delhi, around 10 thousand families, out of a total of about 54 thousand families residing in the core or critical tiger habitat of tiger reserves surveyed till the year 2018–2019 had been voluntarily relocated from the tiger reserves of the country under the extant package.

Voluntary village relocation had been exercised in many tiger reserves of the country inter alia, including those from the states of Maharashtra, Madhya Pradesh, Rajasthan, Karnataka, Odisha, etc. (Dash & Behera, 2018; Indira, 2024; Karanth et al., 2018; Sarma & Barpujari, 2023; Torri, 2011). In Maharashtra, major relocation had taken place in Melghat, Tadoba, and Pench Tiger Reserves. Kanha Tiger Reserve in Madhya Pradesh is one of the examples where entire core or critical tiger habitat has become inviolate (office record of Kanha Tiger Reserve). However, there is little research on the impact of voluntary village relocation on people's well-being and overall health of forest and wildlife which includes all the three components of sustainability (Fig. 3.1; Adams, 2006).

Problem Statement

A few studies have assessed the impact of relocation from PAs but the findings are not elaborate on account of sustainability. Although many people are displaced worldwide by "fortress" conservation strategies and many more are current or future displacement, displacement from PAs is generally under-researched in India and South Asia (Kabra, 2009). In India and elsewhere, examples of successful resettlement and rehabilitation of households affected by conservation-induced displacement are rare, and unfulfilled promises and reduced livelihood security after relocation continue to be the norm (Rossi, 2000; Chatty & Colchester, 2002; Geisler & De Sousa, 2001; Schmidt-Soltau, 2005). The case of Kanha tiger reserve has been examined against the backdrop of this inference to see how far this statement may be generalized. The studies as referred above are based on experiences in different

[4] An umbrella species is defined as a species whose conservation is expected to confer protection to many naturally co-occurring species. This concept has been proposed as a tool for determining the minimum size for conservation areas, selecting sites to be included in reserve networks, and setting minimum standards for the composition, structure, and processes of ecosystems (Roberge & Angelstam, 2004).

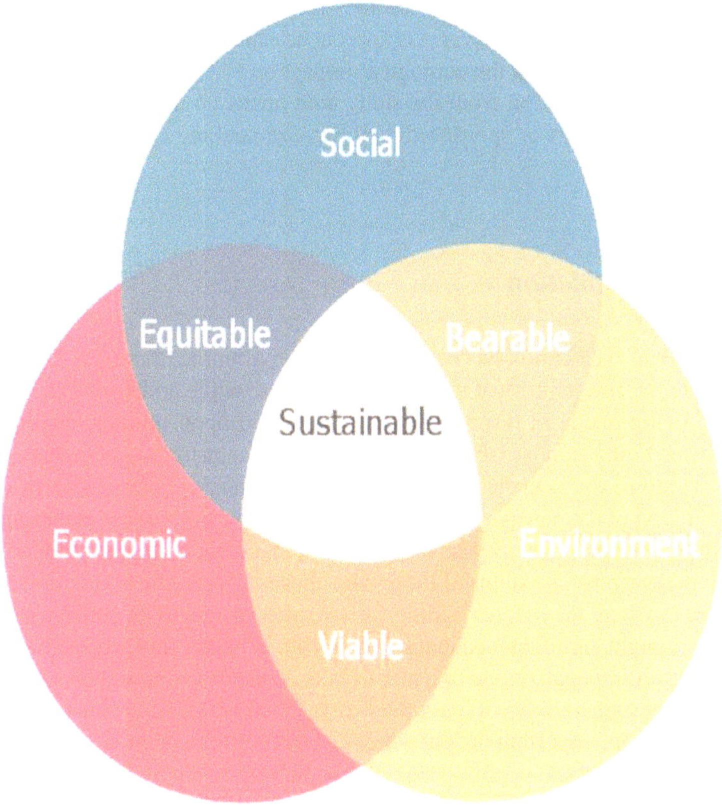

Fig. 3.1 Venn diagram of Sustainable Development: at the confluence of three constituent parts. *Source* Adams (2006)

parts of the world. How far such observations hold good for Kanha tiger reserve has also been assessed through the study.

The shortcomings in the studies done so far in this regard include comparison amongst different sets of households (one those which have been relocated and another which are there in the core area as of now) has been done; the same set of people have not been assessed for their pre- and post-relocation status; almost no deliberation on biodiversity conservation aspect except for the prescription that people's perception matters for sustainable use and management of biodiversity and constant engagements and dialogue between local authorities and indigenous forest communities can improve their perception, thereby resulting in conflict mitigation amongst themselves.

The objective of this study is to assess the ecological gains to the habitat besides the socio-economic conditions of the village dwellers after the voluntary relocation from the Kanha Tiger Reserve. Relocations may ignore the need for desirable settings. The new settings offering new habitats to people need acclimatization to ensure a conducive place to live in, besides making space for wildlife. The study was conducted considering the various approaches and dimensions of the relocation of human habitat as well as the ecological impact on forest and wildlife. Moreover, voluntary village relocation from the study area opens up a new window of scholarship where certain set of sponsored facilities can heal the tyranny and malady of development.

Rationale/Justification

Voluntary village relocation has been a flagship programme, the main aim of which was to make the core or critical tiger habitat inviolate while addressing the concerns of people living inside in respect of better livelihood opportunities, quality of life and enhanced living standard, bringing them into the mainstream, preventing human-wildlife negative interface, etc. by taking them out in the developmental world. After implementation of the new package of compensation, whether these envisaged goals were fulfilled or not, have not been studied systematically and is a matter of setting up different research. The status of those who chose to go for the option of relocation from the area where they were residing for a very long time to an entirely new place must have brought significant changes in their life. Whether this process has yielded a positive result or negative one or a mix of both, it needs a comprehensive analysis. The findings of such a study will pave the way for future relocation plans. The study is likely to provide an important insight which can foster future planning by considering the boon and bane of the process. This will be helpful in designing the process in a way in which it can provide maximum benefit to the field of conservation without antagonising people.

Kanha Tiger Reserve was taken as a case because several families were relocated from there and the entire core area became inviolate. As per the status in the year 2018–2019, there were around 12 tiger reserves, where relocation was ongoing, but their core area was yet to become inviolate as villages were still present inside. Therefore, such tiger reserves might not have provided a true picture of what were the visible changes once the area became inviolate.

Objectives

The objective of the research was to investigate the implementation of statutory provisions as laid down in WPA and the Project Tiger Guidelines, besides studying the overall contribution of this step towards achieving sustainability. The main research objectives are listed below:

(i) To find out whether relocation of families had been voluntary in a formal or in an effective manner in Kanha Tiger Reserve, whether this is a utopian concept, or it is practically achievable, whether guidelines laid down has adequately been adhered to or not,
(ii) To study the ecological contribution of voluntary village relocation towards achieving sustainability of overall ecosystem (ecological aspect includes the impact on vegetation and wildlife, biotic pressure status, in situ conservation of tigers, and other wildlife besides the change in their habitat, climate change and adaptation),
(iii) To find out the perceived satisfaction level of the target population towards the choice and whole process of voluntary village relocation, and
(iv) To examine whether a relationship may be established between voluntary village relocation and sustainability of the ecosystem.

Voluntary Village Relocation from Tiger Reserves

Successful relocation initiatives often require a long-term commitment, adaptive management, and a willingness to address challenges as they arise. Balancing the needs of both wildlife and people is a delicate task that requires careful planning and continuous evaluation to achieve mutually beneficial outcomes (Ali et al., 2023). According to Brockington and Igoe (2006), as quoted by Kabra (2009), a review of the existing literature on conservation-induced displacement has found that there are limited in-depth studies of resettlement of indigenous people from PAs worldwide. Moreover, very scarce and noncomprehensive study exists from India, while only one such study has been conducted (for Nepal) in the entire South Asia.

Maximum studies have been conducted on development induced involuntary/forced displacement. There have been studies on biological or socio-economic aspects but none of them have integrated the concept of sustainability with respect to the overall ecosystem. Voluntary village relocation is ongoing in some tiger reserves, but its role in achieving sustainability is yet to be studied comprehensively. Mostly, the downside of relocation has been highlighted (Jain & Sajjad, 2015; Meena, 2015). One can notice a shortfall in the primary information. Either the data collected is from different sets of villages, one relocated while another present within the core area or the buffer area of the tiger reserve, or based on grey literature, newspaper reports or secondary information which the authors considered reliable.

To have knowledge about the relocation of villages from the tiger reserves in the past and genesis of voluntary village relocation besides other prescriptions related to conservation of tigers after the Sariska debacle where the said park became devoid of any tiger presence because of poaching, the report of the Tiger Task Force, Joining the Dots (2005) was reviewed. More than 40 thousand families were residing within the Core areas of Tiger Reserves of India (www.projecttiger.nic.in) in the year 2018–2019.

As per the report of the Tiger Task Force,[5] a total of 80 villages and 2904 families have been relocated from different tiger reserves since the inception of Project Tiger in the early 1970s and till submission of the report in 2005 under the previous packages, out of which maximum was from the Kanha tiger reserve in Madhya Pradesh, from where 27 villages and 656 families were relocated. The report observed—"the track record of that relocation exercise has also not been uniform or praise-worthy. In fact, in many cases, people who had been relocated had either returned or are today contributing to the pressure on the park from the fringes". Many recommendations were made by the said Task Force on various aspects, inter alia, including the issue of voluntary village relocation.

The package for voluntary village relocation got revised as per the report of the agency entrusted to carry out formulation of the action plan, keeping in view recommendations of the Tiger Task Force. Since 2008, two options are available under relocation package which are:

(a) Option-I[6]—Payment of the entire package amount (INR 10 lakhs per family) to the family in case the family opts so, without involving any rehabilitation or relocation process by the Forest Department.
(b) Option-II—Carrying out relocation or rehabilitation of villages from a protected area or tiger reserve by the Forest Department.

Kanha Tiger Reserve: The Study Area

To investigate the history of the case study, a review was conducted of the management practices, the flagship species and other aspects including voluntary village relocation done in the past of the Kanha Tiger Reserve, the prevalent Tiger Conservation Plan (2011–2012 to 2021–2022) and a mandatory statutory document required for management of tiger reserves as per section 38V (3) of the WPA. Kanha Tiger Reserve (Fig. 3.2; 80°26′ 10″ and 81°4′44″E longitudes and 22°15″ and 22°27′48″ N latitude; Chandra et al., 2006) one of the premier tiger reserves of India is situated in the State of Madhya Pradesh, even considered one of the best habitats for tigers in the world. Being nestled essentially on the northern slopes of the Maikal hills of the Satpura range in Central Indian highlands, it is one of the first nine tiger reserves launched in 1973–1974 at the onset of Project Tiger, the flagship programme of Government of India for conservation of tigers in the wild. This proved to be one of the most successful ventures of species conservation of all time in the entire world. Being a classical habitat for the tigers, it is also famous as the only reserve where hard ground Barasingha (*Cervus duvauceli branderi*) has been thriving since long. A wide spectrum of flora and fauna and especially the grass meadows add considerable significance to the entire landscape.

---

[5] The Ministry of Environment and Forests (Project Tiger) set up the task force vide notification no. 6(4)/ 2005-PT dated April 19, 2005 to review the management of tiger reserves.

[6] The amount has been revised to INR 15 lakh in 2020–21 (Indira, 2024)

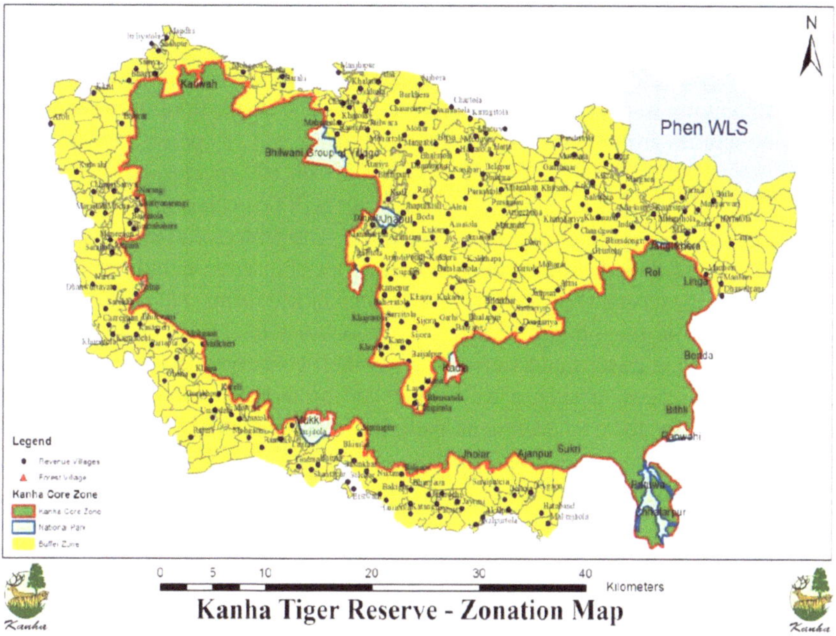

Fig. 3.2 A zonation map of Kanha Tiger reserve (*Source* Tiger Conservation Plan of Kanha Tiger Reserve, 2011–2012 to 2021–2022)

The Tiger Reserve consists of the core zone (Critical Tiger Habitat; green washed area in Fig. 3.2), having a total area of 917.43 km^2 which is part of the erstwhile Kanha National Park and notified as such by the MP State Government. The buffer zone (multiple use area), having the area of the buffer zone Forest Divisions (1134.31 km^2), consists of forest land, revenue land, and private holdings. Except for some of the eastern part, the buffer almost surrounds the core zone. As per the International Union for the Conservation of Nature (IUCN), a lead organization of wildlife conservationists and scientists and instrumental in the selection, establishment, and management of protected areas (a protected area is defined as an area of land and/or sea especially dedicated to the protection and maintenance of biological diversity), natural and associated cultural resources which are managed through legal or other effective means. The Kanha National Park falls in the category-II (Ecosystem conservation and recreation—National Park) of the IUCN protected area management classification system.

The protected area also harbours a wide range of faunal species, some of which figure prominently in the IUCN Red List of Threatened Species. Some of these species are *Cuon alpinus* (Dhole or wild dog), *Melursus ursinus* (Sloth Bear), *Lutra perspicillata* (Smooth-Coated Otter), *Panthera pardus* (Leopard), *Panthera tigris tigris* (Tiger), *Bos gaurus* (Gaur), and *Python molurus* (Python). There are 43 species of mammals and 26 species of reptiles recorded in this reserve. The floral diversity comprises around 850 species of 10 varieties of angiosperms belonging to 506 genera and 134 families, and 22 species of Pteridophyte belonging to 14 genera and 14 families (Pandey & Namdeo, 2009). The floral diversity also includes two Gymnosperm species belonging to two genera and two families. The above flora of the protected area also includes around 50 species of aquatic plants and 18 species of rare plants. The dominant family in the core zone is Poaceae with 109 species, representing 66 genera. The core zone is also rich in avifauna. Over 300 species of birds have been reported by various workers (Chandra et al., 2006). The protected area also supports a variety of lesser fauna, including numerous varieties of insects, including butterflies, moths, and beetles; reptiles, fishes, and other smaller life forms. All these contribute significantly to the functioning of this wildlife ecosystem. The unprecedented rate of extinction of the lesser fauna outside makes their conservation extremely important. The significance of such a well preserved protected area also concerns the improvement of the local environment, including stabilization of hydrological cycle, conservation of soil and moisture; impartation of conservation education, providing recreational facilities, and protection of rich and impressive heritage for the posterity.

History of Village Relocation

The village relocation from the forests was initiated in 1976. Further, after the introduction of the revised package as per the 2008 guidelines of the Centrally Sponsored Scheme of Project Tiger, as notified by the NTCA, several villages got shifted from the core/ critical tiger habitat of the tiger reserve. Now, the whole core area is inviolate or free from human presence. Altogether, 38 villages (including 27 villages and 656 families, which were relocated as per earlier package) comprising about 2300 families have been voluntarily relocated.

Methodology Adopted for Impact Study

Research Design

The methodological framework adopted for the study was treated with due diligence to avoid any illegitimate threat of tautological and teleological errors. The research design strictly followed the explanatory approach. Both quantitative and qualitative methods of research were used. A balanced mix of data from primary as well as secondary sources was collected. One of the most important stakeholders, i.e. the villagers acted as primary source of information while reports, guidelines, provisions of acts, studies, records, and data available with the NTCA and concerned tiger reserve administration, publications and other relevant research works on various thematic areas served as secondary source which fostered in building the grounds to prepare tools for the study.

Secondary Data Collection

The books, journals and published articles, reports released by Government of India and various other organizations, besides documentation done by Kanha Tiger Reserve Administration, served as the secondary source of information. A total of 11 villages had been relocated under the revised package. Details of villages and households relocated, present place of establishment of those households, and their locational details were procured from the office of Chief Conservator of Forests & Field Director, Kanha Tiger Reserve. Geographical Information System-based maps of the area of influence, besides maps generated by the Tiger Reserve Administration as part of M-STrIPES[7] exercise, were also referred to as having an idea about ecological impact.

Sampling of Relocated Villages

As per information provided by the tiger reserve office, 1623 households and 4203 individuals were voluntarily relocated from the core or critical tiger habitat of Kanha Tiger Reserve between 2009–2010 and 2014–2015. The number of families so relocated varied from 53 to 273. Based on the number of families relocated, these villages were categorized: Small—Category-I (up to 100 households), Medium—Category-II (more than 100 and up to 200), and Big—Category-III (more than 200). Two random villages from each category were selected to have a mix of the smallest and the largest villages in terms of number of families so relocated in each category. To ensure the confidence level of more than 90% and margin of error less than 10%, a total of 67 households/families needed to be selected. Several respondents from each village were selected on a pro-rata basis. Out of eleven villages, a total of six villages were selected, two villages each in all the categories. Villages selected under the Category-I were Kariwah (53 families) and Rol (80 families); under Category-II were Bithli (102 families) and Jaami (153 families); and under Category-III were Sukdi (210 families) and Ajanpur (273 families). To decide the number of respondents from each of these villages, the number of families in each village was divided by the sum of the total number of families in these six villages. Based on the percentage

[7] M-STrIPES, short for Monitoring System for Tigers—Intensive Protection and Ecological Status, is a software-based monitoring system launched across Indian tiger reserves by the Indian government's National Tiger Conservation Authority in 2010. The system's objective is to strengthen patrolling and surveillance of the Endangered Bengal tiger.

share of each village, their representation was decided amongst 67 respondents: 21 in Ajanpur, 16 in Sukdi, 12 in Jaami, 8 in Bithli, 6 in Rol and 4 in Kariwah.

Sampling of Respondents

To select the respondent individuals randomly from each village, the total number of families in that village was divided by the number of villages which constituted to total sample, i.e. 67. For example, to select who all are to be the respondents from village Sukdi, the total number of families in village Sukdi, i.e. 210 was divided by 16 (which was the representation derived for the said village), and consequently the person(s), whose name found mention at serial number 13 and its multiples were selected. A similar technique was adopted for other villages too. In some cases, minor adjustments had to be made and there were certain departures from the select list owing to the death of the person so selected, or due to non-availability of the respondent concerned with very fewer chances of finding him/her on another occasion. Two–three randomly selected respondents went far off to places like Durg and Hyderabad. Around 5–6 respondents so selected had to be substituted owing to their death with next of their kin available. Where the beneficiary's wife was available, her opinion was taken. In case of her absence, the viewpoint of one of their heirs was taken. In one case, where a 90-year-old lady respondent who was infirm, the opinion of her daughter was considered. In another case, the respondent was contacted telephonically. Though the opinion of 67 respondents had been formally taken, there was fair involvement of other members of the family too in providing some information. It needs to be clarified that the ones who were substituted for other respondents for being their heir (wife not included) were themselves a beneficiary of the said voluntary relocation scheme.

Primary Source Data Collection

The primary source of information was the villagers who got relocated voluntarily. Initially, the questionnaire survey was thought of, but while validating the questions, it became apparently clear that the said method would not be helpful in extracting the desired information as the target population was mostly illiterate. Hence, a structured interview schedule was found to be a more prudent approach to seek the opinion of stakeholders as an elaborate explanation would have been required to give the proper understanding of what information is required. Similar methodology shifting from questionnaire to semi-structured interview was adopted in Jha et al. (2023). The main purpose of the interview schedule was to extract information on ecological aspects in addition to social, and economic issues. Questions were framed in a manner to know broadly about every bits and pieces of information on each aspect. The questionnaire relating to the structured interview schedule is in Appendix. Altogether, 26 parameters were incorporated to assess the general composition, demographic details and total composition of the sample, information relating to age-class, gender, location now and then, land holding and occupational details before and after relocation.

Voluntariness Assessment

Opinion of respondents was sought through three questions to find out whether relocation of families had been voluntary in a formal or in an effective manner in Kanha tiger reserve and whether guidelines laid down had adequately been adhered to or not. These questions were on the choice how and why they opted for relocation besides their awareness regarding the voluntary village relocation (Questions 1–3 may be seen in Appendix).

Ecological Contribution/Impact Assessment

Besides secondary source of information, endeavour was put to come out with certain primary information from the respondents on parameters like number of cattle they were having while residing inside the core or critical tiger habitat and current number after relocation, distance of their present residence from any kind of forest and type of fuel used for cooking before and after relocation, in order to assess the biotic pressure faced by the forest and wildlife habitat (see general information besides Questions 5 and 6 Appendix).

Independent sample t-test for the number of cattle (before and after relocation) was applied to see if there was a significant difference in the number of cattle before and after relocation. The hypothesis tested was whether any significant difference existed in owning animals before and after relocation. Chi-square test was applied to find out whether there was any significant change in fuel using behaviour of the relocated populace post-relocation in comparison with the past.

Satisfaction Level Assessment

Opinion was sought from the respondents to assess the perceived satisfaction level of the target population regarding the choice of getting relocated and the process of relocation (Questions 15 and 16 may be seen in the Questionnaire at Appendix).

Hand-Holding Assessment

Whether proper hand holding was done by the Forest Department or the concerned Tiger Reserve Administration was also assessed through the opinion of the respondents (Question 17 of the Questionnaire at Appendix may be seen). Documentary records were also screened to assess the level of interaction the department was having with the voluntarily relocated populace. The researcher's objective observation was another tool used to make out the kind of involvement between the two.

Lastly, respondents were requested to come out with their suggestions, if any, regarding steps to be taken for making voluntary village relocation more effective besides further betterment of wildlife in future.

Relocation and Sustainability of Ecosystem Assessment

The findings on each of the three aspects of sustainability, namely, ecological, social, and economic, besides perceived satisfaction level of the target population and their further hand holding by the Tiger Reserve Administration, were used in a qualitative way to assess the kind of relationship between the two, i.e. Voluntary Village Relocation and Sustainability of Ecosystem.

Limitations

The target households had been relocated under Option-I of the relocation package which is a cash package, without the involvement of the forest department in the rehabilitation process. The relocated households had settled on their own at different places far and wide as per their own convenience. Some had gone far up to and beyond 300 km. At the most 10–15 families had settled at one place and somewhere even one or two families had shifted. Mostly, they had purchased land for home and agriculture at a place where they had family relatives. Thus, some of them had gone to places in the neighbouring State of Chhattisgarh. The researcher was to contact at least 65 participants to achieve a 90% confidence interval. Due to the wide dispersal of participants, it was not possible to contact more than 10–15 participants in a day even after spending 15–16 hours a day. The researcher visited the new residing places of participants from 2 to 9 February, 2019. It could become possible to contact around 67 participants in the said span of time with assistance from the Tiger reserve staff. Journey by road and on foot was performed to contact the participants. Primarily, the target population was tribal people who were mostly illiterate and were not so well conversant. So, interaction with the target population was not easy because of their own perception about things. Relocation-related issues were to be elaborately explained to facilitate them to come out with their view point on different aspects.

Hence, a simple questionnaire survey was not possible, and consequently, a structured interview schedule was resorted to. Some of the randomly selected participants were not available due to their death or migration/shifting to other places for work. In their absence, their relatives were to be interviewed. For confidence building one must spend reasonable time to extract the desired information. Frequent visits to their area and constant interaction with them would have been helpful but was not possible in given time. However, with available resource base and access thereto, information from the field had been collected as proposed and documented in the framework of study.

Observed Ground Realities and Results

The data collected from respondents through structured interview schedules and observation made through perusal of secondary sources, besides the researcher's objective observation, was processed and analysed keeping in view the research propositions. Responses on each question were analysed to find out the opinion of voluntarily relocated sample populations on ecological parameters besides social, and economic. However, we focussed on ecological understanding of relocation activities directed towards wildlife conservation.

Respondents' Profile

Number of respondents in the sampled villages varied from a minimum of 4 to a maximum of 21, depending upon the overall population of the village. Altogether 44 male and 23 female respondents were interviewed. The average age of female and male respondents respectively came out to be 44 years and 43 years with the overall average age being 43 and a half years. The average family size was 5.73 i.e. about 6 members per family. The overall educational profile of the respondents illustrated that most of them were illiterate (59.70%). Only about 9% of the respondents had the education of high school and above level. Some of the respondents/beneficiaries' status and researcher of the presented study are given in Figs. 3.3, 3.4, 3.5, and 3.6.

Responses on Voluntariness

Majority of the (61%) responses were for the option "As per own wish", followed by "Due to pursuance by fellow villagers" (34%), while those considering pursuance by Forest Department were 3% only, leaving a miniscule percentage (1.5%) of those who considers pursuance by family members as their option for relocation. The reason for

Fig. 3.3 Beneficiary Shamu (name changed), along with his sister in front of his newly constructed house in the relocated area, very proudly informed that he was able to consume pigeon pea (*Arhar Dal*, generally consumed by effluents in the society) in every meal and he had been harvesting a quintal of the same every year, while during pre-relocation times, he was hardly able to consume it, even once a week

Fig. 3.4 Jaggu (name changed) claimed that quality of life has changed enormously and it was a hundred thousand times better than that during the pre-location period

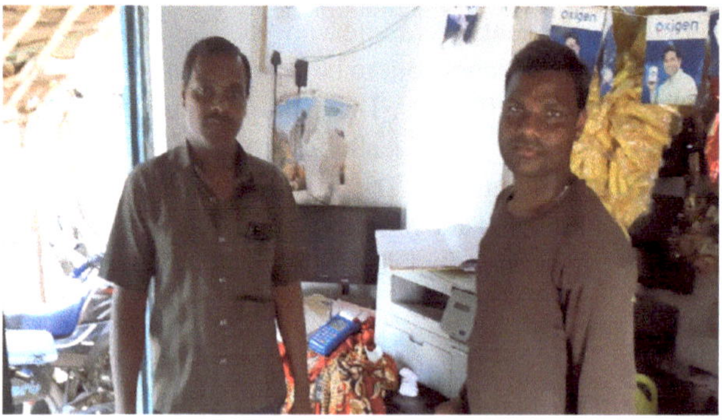

Fig. 3.5 Mani (name changed) started the business of general goods and was facilitating cashless transactions through POS Machine at his shop (Since April 2018, more than 2100 cashless transactions were made for which he had been maintaining a separate register and showed it to the researcher with a sense of pride)

Fig. 3.6 Raghu (name changed) responded that earlier he needed at least 2 days for getting a petty work done owing to non-availability or restricted availability of means of transport, but at his new home the same work was done within a couple of hours

opting relocation was mainly the lack of amenities and most pertinent conflict with wildlife. Developmental concerns and lucrative packages were the other reasons.

Regarding the awareness on relocation, about 30% of respondents admitted that they had complete knowledge of voluntary village guidelines whereas about 60% observed that they had some knowledge. About 10% of respondents had no knowledge at all.

Responses on Ecological Parameters

Changes in wildlife sighting pattern including camera trap photographs of wild animals taken from the areas which have become inviolate through voluntary relocation, Google maps of pre- and post-relocation period, change in vegetation and status of establishment of palatable grass species in the vacated area obtained after voluntary relocation as well as comparison of status of biotic interference through records maintained by the Tiger Reserve Administration were referred and compared to. Besides this, records available at the NTCA on Tiger Monitoring Project of the WII-NTCA including Phase-IV Exercise report, were also perused. Maps generated as part of M-STrIPES exercise by the tiger reserve were also considered.

As primary source, the information on cattle owned by the villagers while they were residing in the core or critical tiger habitat and when voluntarily relocated was sought from the respondent villagers. The fuelwood use pattern before and after voluntary relocation, based on the respondents' perception was also examined. Information regarding distance of place of their habitation from any kind of forest before and after was also sought from the respondents. A brief on six relocated villages, location (Fig. 3.7), pre- and post-facto signature of land use in Google images (Figs. 3.8, 3.9, 3.10, 3.11, 3.12, and 3.13), extracted from secondary sources, and some primary information is presented below:

1. Jami village: This was surrounded by dense Sal Forest in the tri-junction of Lodhabarra, Kamkodadar, and Jamsahera beats of Supkhar Range in Kanha Tiger Reserve. The village was relocated in the year 2010 under Option-I Scheme of the NTCA. This village had an area of 105.21 ha. The Google imageries of the time series changes in this village are Fig. 3.8a, b.
2. Rol village: This was situated in the Rol Beat of Supkhar Range with an area of 85.66 ha and was relocated in 2013 under Option-I Scheme of the NTCA. The Google imageries of the time series changes in this village are Fig. 3.9a, b.
3. Bithli village: This was situated in the Bithli Beat of Supkhar range with an area of 128.94 ha. It was relocated in the year 2012 under Option-I scheme of NTCA relocation policy. The Google imageries of the time series changes in this village are Fig. 3.10a, b.
4. Ajanpur village: This village was one of largest villages with an area of 574.65 ha. It was relocated under Option-I Scheme of NTCA voluntary village relocation policy. Situated in the Ajanpur Forest beat of Bhaisanghat Range, this was fully relocated in 2014. The Google imageries of the time series changes in this village are Fig. 3.11a, b.
5. Kariwah village: This village was situated in Sarahi Range of Kanha Tiger Reserve and was occupying an area of 75.27 ha. It got relocated in the year 2015. The Google imageries of the time series changes in this village are Fig. 3.12a, b.
6. Sukdi Village: This village was situated in Sukdi Beat of Bhainsaghat range and was relocated in 2016. Area occupied by this village was 568.17 ha. The Google imagery of time series changes in this village are Fig. 3.13a, b).

Besides these six villages, there were five more namely—Jholar (220 families), Linga (139 families), Benda (68 families), Ranwahi (227 families), and Sajaalgaan (123 families), which got relocated but not included as sample in the Study.

Change in Vegetation Composition

The relocated villages were predominantly occupied by agricultural crops as cultivation was the main occupation in these villages. There was significant change in the vegetation after relocation and introduction of grassland species could be noticed. Following Table 3.1 provides the pre- and post-relocation comparison of crops and vegetation.

Wildlife Sighting

As per the records of Bhainsaghat Range, Kanha Tiger Reserve, the number of herbivores increased in the vacated area of Ajanpur and Sukdi. While there was hardly any presence of hard ground Barasingha earlier in this area, the presence of

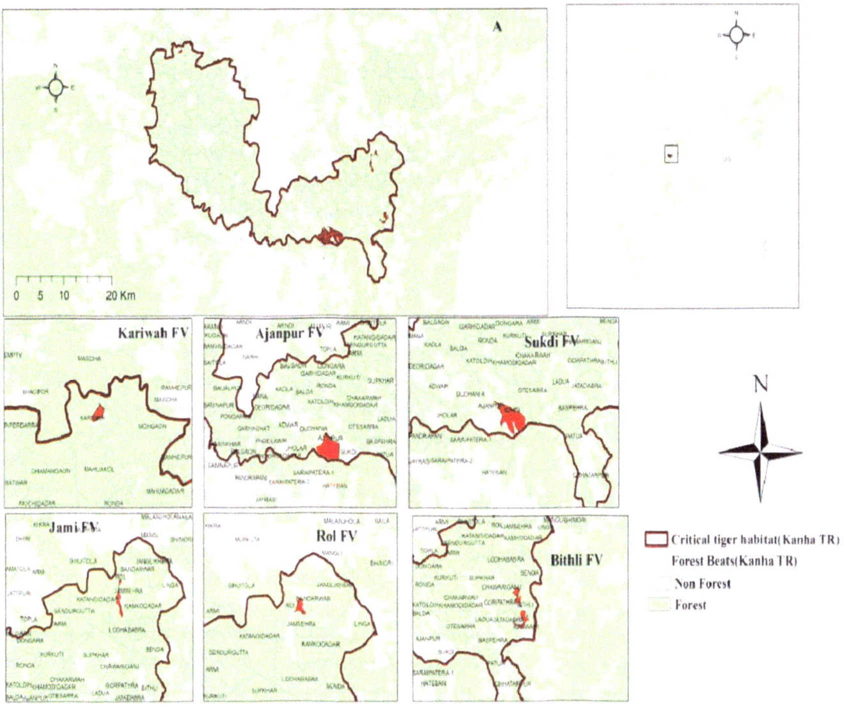

Fig. 3.7 The Kanha Tiger reserve (dark red polygon) and the relocated villages. *Source* NTCA-WII Tiger cell

9 individuals in Ajanpur was noticed in the year 2018. In Sukdi, their number was found to be 18 in the year 2017 and increased to 34 in 2018. Number of Gaur using Ajanpur area was increased from 15 in 2016 to 23 in 2018, whereas in Sukdi, their number increased from 17 in 2016 to 24 in 2018. Number of Spotted Deer in 2016 was only 40 in Ajanpur and 91 in Sukdi in 2016, which increased to 270 in Ajanpur and 360 in Sukdi in 2018. Fifteen Sambars each were reported from both the vacated area. Four tigers (3 Male and 1 Female) were using the area since 2016.

The Tiger Reserve has been using the state of art patrolling and ecological module, namely "M-STrIPES" since 2012. The patrolling data, inter alia, included sighting of wild animals and got plotted on the map through GIS tool. It was observed in such maps that the wild animals, both carnivores as well as herbivores, had been using these areas frequently. The indicative sighting map(s) for the month of December 2018 pertaining to carnivore sighting (both direct and indirect) and herbivore sighting is placed below on the following pages (Figs. 3.14, 3.15, 3.16, 3.17, and 3.18). The photographs of wild animals captured in camera traps are also further testimony of their presence in the vacated areas (Figs. 3.19 and 3.20).

a: Household and agriculture land in Jami village in 2009. Source: Digital Globe, 2019, Google Earth

b: Grassland in relocated Jami village in 2018. Source: Digital Globe, 2019, Google Earth

Fig. 3.8 (**a**) Household and agriculture land in Jami village in 2009. *Source* Digital Globe, 2019, Google Earth. (**b**) Grassland in relocated Jami village in 2018. *Source* Digital Globe, 2019, Google Earth

a: Land-use in Rol village in 2008 (with agricultural land & households). Source: Digital Globe, 2019, Google Earth

b: Rol village After relocation, converted as grassland in 2017. Source: Digital Globe, 2019, Google Earth

Fig. 3.9 (**a**) Land-use in Rol village in 2008 (with agricultural land & households). *Source* Digital Globe, 2019, Google Earth. (**b**) Rol village after relocation, converted as grassland in 2017. *Source* Digital Globe, 2019, Google Earth

a: Bithli village before relocation in 2012. Source: Digital Globe, 2019, Google Earth

b: Bithli village after relocation in 2018. Source: Digital Globe, 2019, Google Earth

Fig. 3.10 (**a**) Bithli village before relocation in 2012. *Source* Digital Globe, 2019, Google Earth. (**b**) Bithli village after relocation in 2018. *Source* Digital Globe, 2019, Google Earth

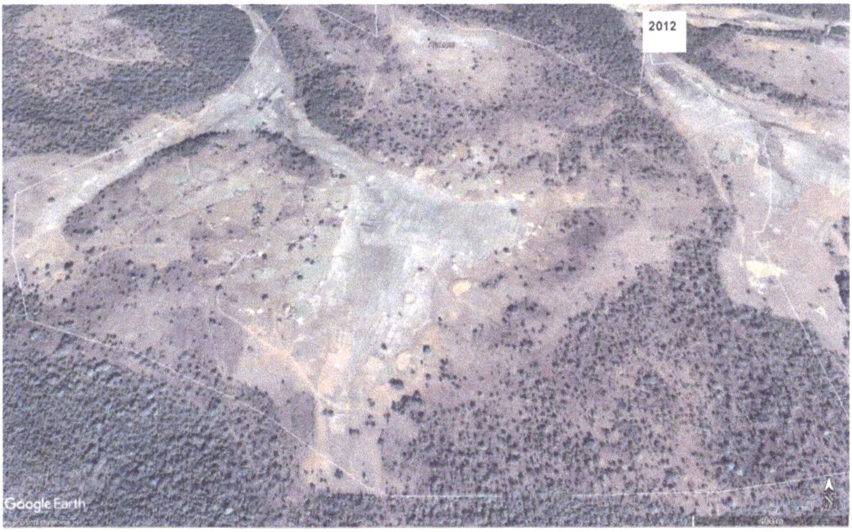

a: Ajanpur village in Pre-relocation stage in 2012 (several built up areas are visible along with agriculture). Source: Digital Globe, 2019, Google Earth

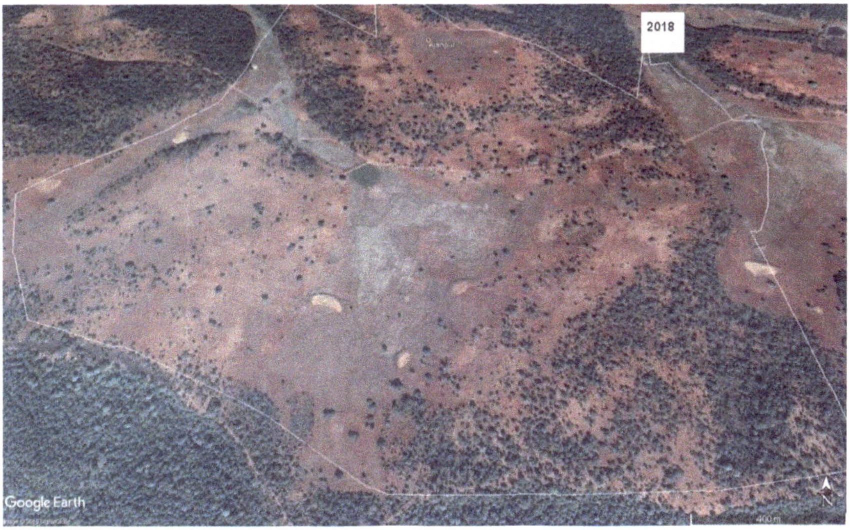

b: Post relocation imagery of Ajanpur, 2018 (Agriculture area got converted into grassland). Source: Digital Globe, 2019, Google Earth.

Fig. 3.11 (**a**) Ajanpur village in pre-relocation stage in 2012 (several built-up areas are visible along with agriculture). *Source* Digital Globe, 2019, Google Earth. (**b**) Post-relocation imagery of Ajanpur, 2018 (agriculture area got converted into grassland). *Source* Digital Globe, 2019, Google Earth

a: Pre relocation satellite image of Sukdi village in 2012. Source: Digital Globe, 2019, Google Earth

b: Post-relocation satellite image of Sukdi village in 2018 showing grassland patches. Source: Digital Globe, 2019, Google Earth

Fig. 3.12 (**a**) Pre-relocation satellite image of Sukdi village in 2012. *Source* Digital Globe, 2019, Google Earth. (**b**) Post-relocation satellite image of Sukdi village in 2018 showing grassland patches. *Source* Digital Globe, 2019, Google Earth

a: Pre-relocation satellite image of Kariwah village in 2015. Source: Digital Globe, 2019, Google Earth

b: Post-relocation satellite image of Kariwah village in 2018. Source: Digital Globe, 2019, Google Earth

Fig. 3.13 (**a**) Pre-relocation satellite image of Kariwah village in 2015. *Source* Digital Globe, 2019, Google Earth. (**b**) Post-relocation satellite image of Kariwah village in 2018. *Source* Digital Globe, 2019, Google Earth

Table 3.1 Changes in vegetation pattern from occupied to vacated stages of the land in studied villages

Village	Pre-relocation vegetation	Post-relocation vegetation
Jami	Paddy	*Saccharum spontaneum*, *Themeda* spp. grassland community
Rol	Paddy	*Imperata cylindrica*, *Heteropogon* spp. community
Bithli	Mostly paddy	*Imperata cylindrica*, *Heteropogon* spp. community
Sukdi	Paddy and Leguminous vegetables	*Saccharum spontaneum*, *Themeda quadrivalvis*, *Bothriochloa odorata*, *Chloris virgata*, *Heteropogon contortus*, *Cynodon dactylon*
Ajanpur	Paddy and Leguminous vegetables	*Saccharum spontaneum*, *Themeda* spp. (annual), *Dichanthium annulatum* (perennial), *Cynodon dactylon* (perennial), *Heteropogon contortus* (perennial)
Kariwah	Paddy	*Cynodon dactylon*

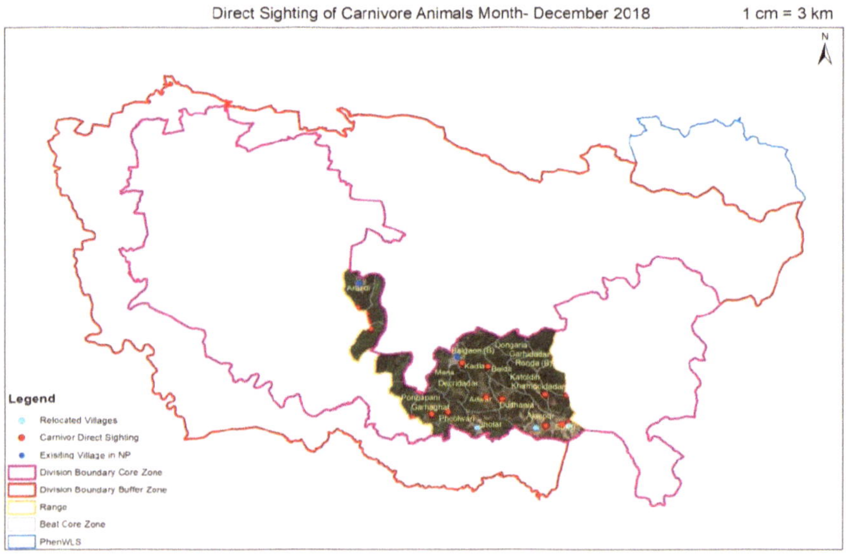

Fig. 3.14 Direct sighting of carnivores in Ajanpur and Sukdi (December, 2018). *Source* M-STrIPES Database, Kanha Tiger Reserve. Detail below

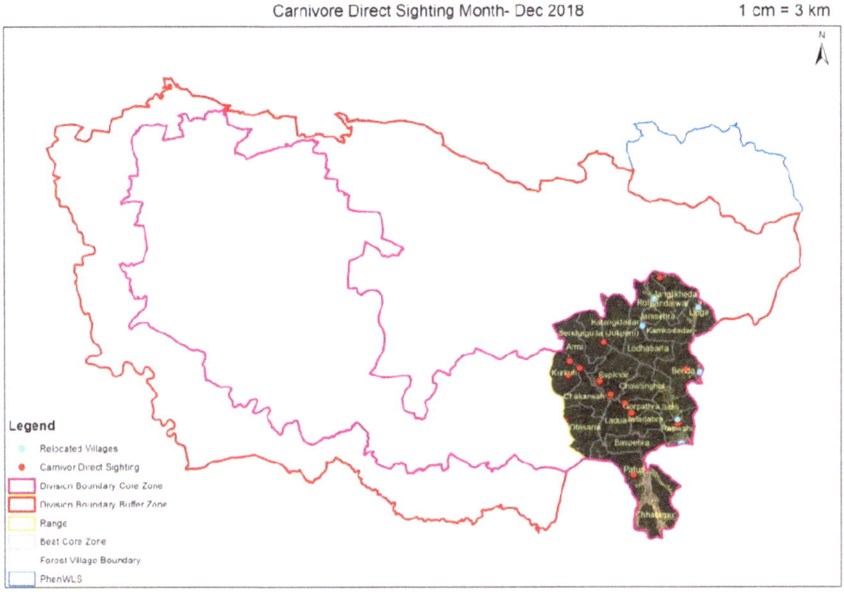

Fig. 3.15 Direct sighting of carnivores in Rol and Bithli area (December, 2018). *Source* M-STrIPES Database, Kanha Tiger Reserve. Detail below

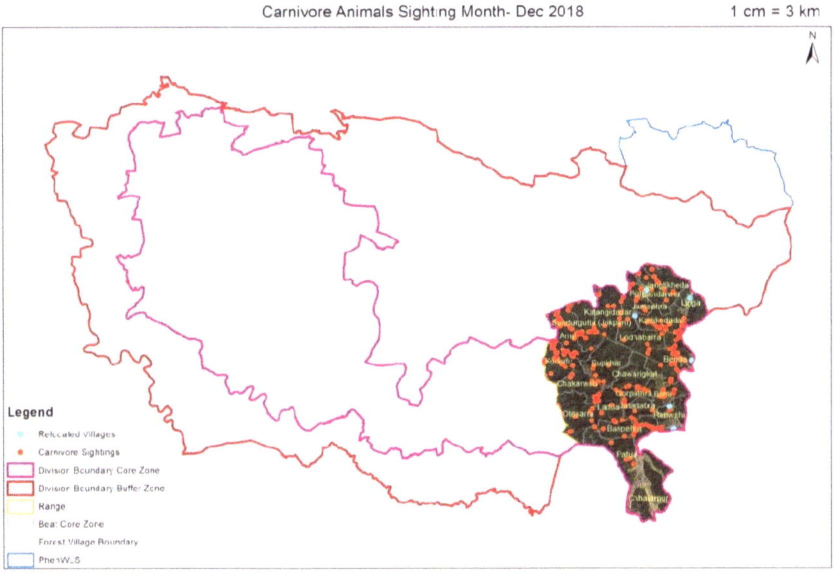

Fig. 3.16 Carnivore Sightings in Rol and Bithli area (December, 2018). *Source* M-STrIPES Database, Kanha Tiger Reserve. Detail below

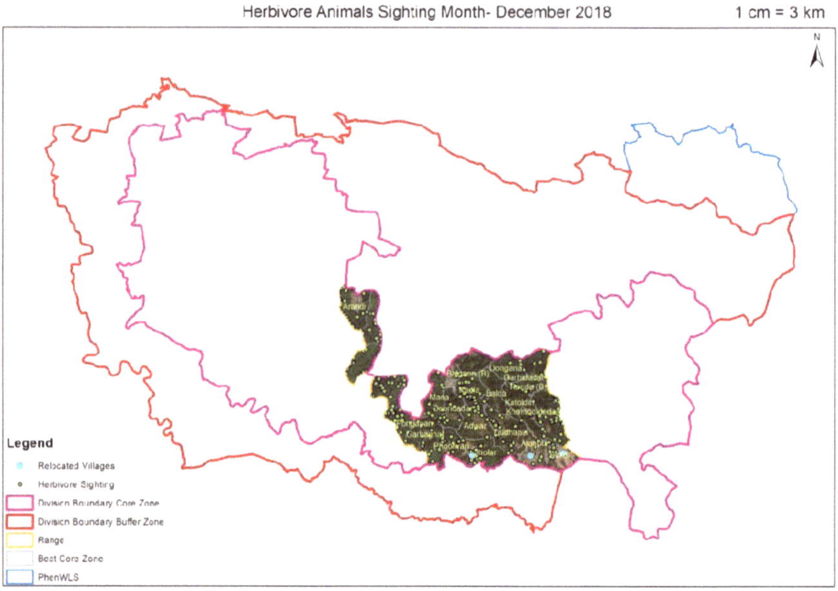

Fig. 3.17 Herbivore sighting in Ajanpur and Sukdi area (December, 2018). *Source* M-STrIPES Database, Kanha Tiger Reserve. Detail below

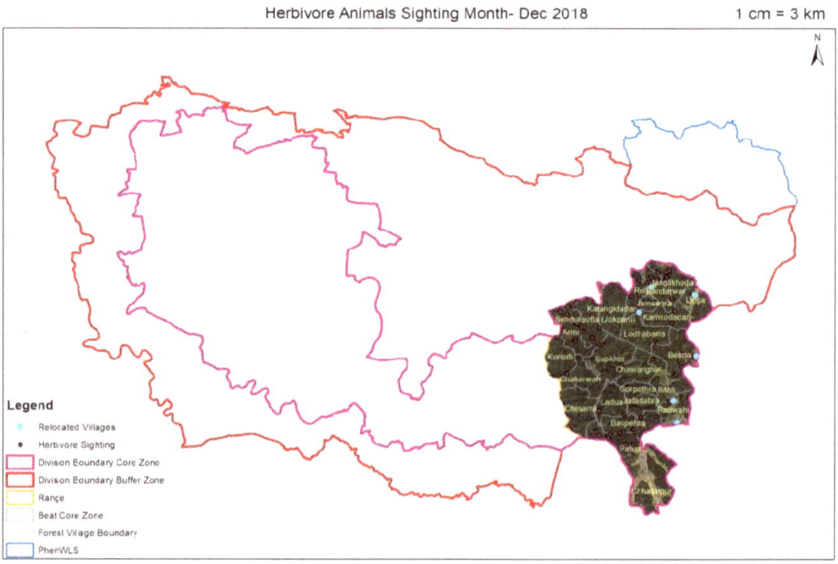

Fig. 3.18 Herbivore sighting in Rol and Bithli area (December, 2018). *Source* M-STrIPES Database, Kanha Tiger Reserve. Detail below

Fig. 3.19 Photographs captured in Camera Trap in Bithli area Post-relocation (*Source* NTCA-WII). Top row: left A sloth bear (*Melursus ursinus*), Right a leopard (*Panthera pardus*), bottom row: left a sambar Deer (*Rusa unicolor*), right a spotted deer (*Axis axis*)

Biotic Pressure

Biotic pressure included the negative influence of human and other living forms except wildlife on the forest. Mainly, it is in the form of grazing by cattle, fuelwood collection and frequent movement of people which compromised the tranquillity and inviolate nature of forest due to which wild animals are negatively impacted.

The comparison of cattle number and its impact before and after relocation was analysed through an independent sample t-test to see whether there was a significant difference in the number of cattle before and after relocation. It was discovered that the number of the animals had significantly decreased post-relocation ($T = 3.579$ and P value $= 0.000$. Since, P Value is 0.000).

Significant behavioural change vis-à-vis fuel use may be noticed, post-relocation. While everyone was using fuelwood brought from forest with about 13% using cow-dung cake occasionally as another alternative during the pre-location period, people started using other alternative fuels also, post-relocation. Statistical analysis using chi-square test showed significant behavioural change in use, post-relocation ("Fuel brought from forest" (Chi Square $= 11.984$, P value 0.001), "Fuel wood brought from elsewhere" (Chi Square 16.891, P value 0.000), "Cow dung cake and others"

Fig. 3.20 Photographs captured in Camera Trap in Jami area Post-relocation (*Source* NTCA-WII). Top row: left a tiger (*Panthera tigris tigris*), right a leopard (*Panthera pardus*), bottom row: left a sambar (*Rusa unicolor*), right a sloth bear (*Melursus ursinus*)

(Chi Square 48.639, *P* value 0.000) and "LPG" (Chi Square 27.766, *P* value 0.000)). The villagers were living deep inside the core area of the forest. Such status changed post-relocation to the average distance from any kind of forest to ~3 km.

Satisfaction and Hand Holding

Most of the responses (77.6%) were indicative of full satisfaction towards choice. Around 13% respondents found themselves satisfied while 4.5% fell in the dissatisfied category. As far as satisfaction with the process was concerned, more than 68% of respondents felt very satisfied. While more than 28% felt satisfied, just 1.5% were dissatisfied. The response of 91% of participants revealed that they had been assisted by the Forest Department on all occasions, whenever it was required, post-relocation. Six per cent of respondents never required assistance after relocation. One and a half per cent each responded that they got required assistance, sometimes or so many times.

Voluntariness: A Reality or Fallacy

The essence behind making the process of village relocation voluntary is to avoid any antagonism towards the concept amongst the villagers who opt for this. It was considered that the situation would be beneficial for both human beings and wildlife through enabling people to avail such facilities which their counterparts residing in areas outside are enjoying and ensure safe and undisturbed inviolate space for the wildlife as envisaged under statutory provisions. One of the biggest challenges was to ensure transparency in the whole process of disbursement of funds to the beneficiaries besides making it voluntary in letter and spirit. It means that no coercive measure had to be taken to evict the resident villager forcefully from the core or critical tiger habitat against his or her wishes.

Official records showed due diligence at each stage. An effective mechanism of associating the District Administration, the Panchayati Raj Institutions, and the Bank as stipulated was thus adopted. The entire due amount was transferred to the District Administration. During the initial phase, a third-party NGO (WWF-India in this case) was also inducted to assist/ facilitate the villagers as part of a trust-building exercise. INR 1 lakh (= a hundred thousand) was given for meeting immediate expenses which got deposited under a saving bank account opened as a joint account of both wife and husband of the beneficiary household. Remaining INR 9 lakhs got deposited under Fixed Deposit Receipt (it's a joint account of Collector and the beneficiary), withdrawal from which needed the concurrence of District Magistrate and recommendation of Assistant Director of tiger reserve for purposes like purchase of land and construction of house, etc. (power delegated to Sub-Divisional Magistrate of the concerned sub-division). The disbursal of funds used to be made in instalments of INR 3 lakhs or more depending upon the requirement, to the beneficiaries subject to fulfilment of requisite steps like the presentation of agreement for the purchase of land for house construction and for agriculture, roping in the Revenue and Forest administration. Fixed interest accrued during the month got automatically transferred to the saving bank account of the beneficiary. A lock-in period of 3 years for withdrawal of the remaining amount after completion of all purchases was in vogue, to ensure judicious use of the amount and to avoid misuse.

The responses given by the respondents concerning the question on their choice for opting relocation clearly showed that no coercive measures or pressure tactics had been used by the Forest Department or any Government Department to evict the villagers. While more than 61% of respondents observed that they had relocated as per their own wish, more than 34% responded that they had agreed owing to pursuance by fellow villagers. Therefore, it may be inferred that relocation was not done forcibly.

The most prominent reason suggested for opting relocation was lack of amenities, mainly transportation, electricity, and employment opportunities as most of these habitations were deep inside the forest and to find out any type of facility, the villagers were constrained to go on foot for minimum 7–15 km. during pre-relocation times.

The second most pertinent reason was conflict with wildlife followed by developmental concerns like health and education issues and the last reason given was the offer of a lucrative package. Karanth et al. (2018) also recorded human-wild animal conflict as the top reason for relocation choice. However, there was no taker for the option of "Frequent Disputes with Forest Department" during the interaction. Here also, it becomes apparently clear that there was a higher degree of discontentment amongst the villagers with the kind of life they were leading while residing inside the forest. The analysis revealed that they would have gone for an amicable settlement as they were fed up with the prevailing conditions, pre-relocation.

Regarding awareness about voluntary village relocation guidelines, only around 10% of respondents confessed that they had no knowledge of such guidelines. Around 60% admitted that they have some knowledge and around 30% admitted that they had complete knowledge. Moreover, those admitting ignorance relied on the fellow villagers and 90% amongst them were female respondents.

Ecological Contribution/Impact of Voluntary Village Relocation

Biotic pressure is the most negative interaction that can happen with any forest or wildlife. People living inside the forest or in the fringe area of forest are mostly dependent on forest for their day-to-day needs. They extract small timber for fuelwood; let their cattle do grazing inside the forest, harvest minor forest produce for their personal use and trade it for income. Consequently, all these affect the entire ecosystem.

The perusal of records, Google maps of the area in question, both pre- and post-relocation, the change in vegetation, the monthly maps of wild carnivore and herbivore sightings generated under M-STrIPES Patrolling exercise, revealed a significant change on the ground on ecological front. A drastic change in vegetation cover was not possible within a short span of 4–8 years, as there are different stages of succession and accordingly there are some ecological requirements of each stage. The areas belonging to relocated villages were earlier occupied by the agricultural crops and the nearby areas were facing biotic pressure in the form of cattle grazing and fuelwood collection. Such areas had very little evidence of grassland-dependent ungulates, like Spotted Deer or Chital (*Axis axis*). The *Cervus duavuceli branderi* (hard ground Barasingha) was hardly present in the adjoining beats. Later the relocation of villages provided a good grassland patch after conversion of abandoned agricultural land. The evidence of all endangered species was recorded through camera traps by the Wildlife Institute of India, Dehradun (WII) in collaboration with the NTCA.

As per the classical ecological theory, succession is a process involving several phases (Clements, 1916). The vacated areas passed through nudation, migration, and ecesis phase and were in the competition phase while reaction and stabilization were yet to take place. The time scale for these changes could be decades or even millions of years (Sahney & Benton, 2008). Despite such fact, the Google imageries of pre- and post-relocation period gave a fair insight towards the change achieved in vegetation. However, Kassa et al. (2023) suggested that the degraded forests could recover to their original condition more effectively through combination of assisted and enriched planting in relatively quicker time. Zhang et al. (2018) emphasized converting abandoned area into more forest to blend in with the surrounding land-use categories. Much earlier, Majid (2000) recorded that ex-cultivation areas in tropical forests could be planted with light-demanding species. But these efforts were aimed at enhancing productivity of the area by raising timber species. Therefore, it would be wise of the managers to develop the abandoned agriculture and built-up land in favour of wild animals of the area, preferably grassland with low height browsing plants in the early phase to attract local herbivores/ ungulates. A caveat regarding stoppage of invasion of alien species in abandoned/vacated area is important since Indian forests or tiger reserves are infested with several invasive species (*Lantana camara*, *Pontederia crassipes*, *Parthenium hysterophorus*, *Chromolaena odorata*, *Mimosa pudica*, *Ageratum conyzoides*, *Argemone mexicana*, *Leucaena leucocephala*, *Mikania micrantha*, *Ricinus communis*) which could enter such area easily if pre-emptive vegetation management is not done (Bhatta et al., 2023; Hiremath & Sundaram, 2014; Kishore et al., 2024). Introduction of grasses and leguminous species suggested by Love et al. (2009) may be adopted for quick vegetating of barren area.

The vegetation mapping of those areas showed the emergence of several palatable annual and perennial grass species. Besides this, wild animals like Gaur (*Bos gaurus*), Spotted Deer (*Axis axis*), Sambhar (*Rusa unicolor*), and the hard ground Barasingha (*Cervus duavuceli branderi*) were not only sighted in the area, but also showed increasing trend as far as their numbers are concerned. The area was used by four tigers, which is a positive sign of success of the venture. It is well known that human presence and infrastructure impact wildlife and low human pressure changes the movement pattern of animals (Lewis et al., 2021; Procko et al., 2023; Smith et al., 2024).

Due to the voluntary relocation of 1652 families and more than 4000 people along with around 2000 cattle as per the admission of respondents, biotic pressure has become negligible on the forest of Kanha Tiger Reserve. People moved from Kanha but some were near adjoining buffer areas, where the mean distance from the forest was about 3 km. People were still using fuelwood from the forest and from other areas, but significant change had been noticed in fuel using behaviour. People began to use other forms of fuel rather than fuelwood from the forest. In totality, the carbon loss was minimized which was a mitigating factor for climate change and fostered adaptation to climate change. The forest and wildlife benefitted on the ecological front.

Status of Satisfaction Level and Hand Holding after Voluntary Relocation

The people who opted for voluntary relocation appeared to be satisfied, both with the choice as well as the process. When asked about the cause of dissatisfaction, one of the respondents with a pucca house, cattle shed, water facilities and adjoining farm land, observed that earlier she was working as an Anganwadi worker and was getting INR 1000 as an honorarium. Due to relocation, she had to leave that work resulting in loss of income which was later raised to INR 5000. Another respondent observed that she had lost her male family members after relocation which she thought could have been avoided had she not shifted to a new place. Some respondents were critical about the bank's working, particularly citing technical glitches like server error, etc. by the bank staff for not performing their work while a few raised the issue of absence of the Sub-Divisional Magistrate on stipulated dates in his office.

The departmental influence was evident through the response of beneficiaries. It was observed that the staff had a very good rapport with the relocated villagers. The way in which staff were greeted besides the body language of villagers, revealed a good bonding. The staff were aware of the villagers' new place of stay and non-hesitantly took the researcher to each of the respondent villager's house, even though some of them were living as far as 110–120 km away from the Range headquarters. While trying to make out the reason behind, it came to notice that one of the Deputy Rangers was made officer in charge of one village and he was responsible for all liasioning and facilitation jobs. Constant interaction had brought them closer and there was a feeling of empathy for the villagers in the hearts of forest staff. At formal level also, the relocated villagers must be the prime beneficiaries for jobs relating to park safari and tourism in tiger reserves.

Relationship Between Voluntary Village Relocation and Sustainability of Ecosystem

Voluntary village relocation had a positive bearing on all the three pillars of sustainability, i.e. environment apart from society, and economy. The process had been able to fulfil all the commitments required for the sustainability of an ecosystem. The Kanha relocation experience may be replicated to other reserves ensuring sustainable development on the one hand and mainstreaming the vulnerable communities, on the other.

General Discussion

The study on "Voluntary Village Relocation and Sustainability of Ecosystem" was conducted in the backdrop of problems being faced by people while they were residing deep inside the forest and got relocated from the core or critical tiger habitat of the Kanha Tiger Reserve during the period of 2010–2016 as part of the statutory requirement of making the core area inviolate as mandated under the Wildlife (Protection) Act, 1972 as amended in 2006. The study begun by setting up the objectives, viz. to investigate the level of adherence to the concept of voluntariness as well as studying the ecological (and socio-cultural, and economic: not forming the part of this text) contribution of voluntary village relocation towards achieving sustainability besides assessing the satisfaction level of the relocated people and their further hand holding. To get focussed responses on the objectives, a balanced mix of primary and secondary information was garnered.

This study observed that the whole process of relocation showed the voluntary nature of the scheme. The population made it clear that no coercive measures were adopted. People had reasons to opt for relocation, foremost amongst which was lack of amenities and conflict with wild animals as well as other developmental concerns (Indira, 2024). The affirmative response towards assistance by the Forest Department was a further testimony of the empathetic attitude of the department even after completion of the exercise. Therefore, it is clear that the purpose of voluntariness had been completely realized. The success achieved by the department proved that it is achievable. Awareness regarding voluntary village relocation guidelines was found to be respectable as around 90% of the population was having complete or some knowledge of the guidelines in this regard.

The process of voluntary village relocation had been instrumental in bringing the people in the mainstream, who otherwise were subsisting on the meagre resources, with few opportunities to improve their quality of life. The incidences of human-wildlife negative interactions post-relocation had become negligible (Indira, 2024). More than 70% of the populace found no conflict while around 24% found a decline in such incidences. It was a favourable situation which could eliminate the antagonistic relationship and would foster the survival of both wildlife and people. It's a win-win situation for both the wildlife as well as the villagers. The villagers used to face problems like crop depredation and cattle depredation during pre-relocation times, due to which they were constrained to face heavy economic losses thereby worsening their state of poverty. The wild animals on the other hand were facing tough competition with the domestic cattle, and the threat of several communicable diseases from them. Loss of habitat and other disturbances were creating an adverse situation for their well-being. Therefore, it can be concluded that the very process of voluntary village relocation had an efficacious impact on the social parameters while fostering the social pillar of sustainability in a constructive way.

On the ecological front, there was a positive outcome in the emergence of annual and perennial grass species in the vacated area, which was slowly converted into grasslands. It is pertinent to note that while the forest of Kanha was bearing the brunt of fuelwood collection (every year, every month, and every day by more than 1600 families, not only for their bona fide use but also for sale) it was free of that pressure. This is a major contribution for carbon sequestration and is fostering the mitigation towards climate change by locking enormous amounts of carbon inside the forest. Indira (2024) has also recorded some positive impacts of re-translocation like, natural forest restoration, free movement of wildlife, no land encroachment, reduction in cattle grazing pressure on the forest, reduction in human-animal conflict, etc. Therefore, it can be concluded that the voluntary village relocation had immensely contributed to the environmental pillar of sustainability.

Voluntary village relocation, thus, has been a better proposition for human beings as well as forest and wildlife. Both have gained from this exercise and the sustainability of the ecosystem is achieved through this step. Not only, it has been proved to be improving the social and economic well-being of people so relocated, it has also enhanced the ecological stability, resulting in substantive area being available for the wildlife, not only in physical terms but also in a qualitative way.

The way of execution of a programme decides the satisfaction with it and if not executed efficiently, it may backfire and prove to be a defeatist step. The satisfaction of stakeholders is an essential pre-requisite and, on that front, Kanha has passed the test with the majority admitting satisfaction with both the choice and the process. Proper hand holding is a must for sustainability and on that front also, the approach has been appreciable as the majority opinion says that people are getting assistance even after the process has seen its culmination.

There is a relationship between the three pillars of sustainability and voluntary village relocation. The weakness of any pillar could have jeopardized the whole process as has been observed in the relocations from tiger reserves in the past (RTTF, 2005). Examples include the case of the Sariska Tiger Reserve where the relocated people came back to their old place of habitation and could be evicted only through legal recourse (office records of the NTCA). Therefore, it would not be appropriate to say that a relationship cannot be established between the two, as any shortcoming on any of the social, economic, or ecological front could have a deleterious effect on the overall health of both villagers as well as the forest. However, the current study showed the perfect balance amongst the environment, economic and social components (latter two not the part of this study) in the sustainability of the relocation project.

Conclusion

This study focused on impact of voluntary relocation of human population from critical tiger habitat in terms of sustainability keeping in view all the three pillars of sustainable development. However, only the environmental component of sustainability is discussed in detail with passing reference of the two other components: social and economic. The results suggested that the re-translocated indigenous community form the core or inviolate area were reasonably satisfied on the choice of option and process of relocation. The department provided supports during and after migration. The objective of relocation was completely achieved as the conservation strategy started working in the vacated area. The earlier inhabited area was free for unrestricted movement of the wild animals: both herbivores and carnivores were recorded by camera traps and direct observation. In a few years, the abandoned areas began to revegetate automatically in addition to successful efforts of artificial vegetation. The cattle grazing, fuelwood collection, human-animal conflict and other pressures were drastically reduced to support future tiger conservation. Voluntary village relocation is an efficient mechanism for achieving desired results of well-being for both the human beings as well as the wild denizens including tigers. This experiment has shown that when statutory provisions are implemented in true letter and spirit with due involvement of all the stakeholders, it yields positive results. The tiger and other wild animals and the habitat, all benefitted from the relocation of villagers, who were deprived of all developmental amenities which their counterparts living outside the forest were enjoying. This case of Kanha Tiger Reserve's voluntary re-translocation proved to be successful, hence, deserved to be replicated elsewhere in the country.

Appendix: Questionnaire

Name of Respondent

Age…………… (in years) Sex- Male/ Female/ Trans-gender

Address

Annual Income (Before relocation)

(i) Below 25,000/- (ii) 25,000 to 50,000 (iii) 50,000 to 75,000 (iv) 75,000 to 1,00,000 (V) Above 1,00,000/-

Annual Income (After relocation)

(i) Below 25,000/- (ii) 25,000 to 50,000 (iii) 50,000 to 75,000 (iv) 75,000 to 1,00,000 (V) Above 1,00,000/-

Occupation (Before relocation) ……………………………………………………………………..

Occupation (After relocation) ………………………………………………………………………….

Educational Qualification……………………………………………………………………………

No. of Family members…………………………………………………………………………….

No. of Cattle (before relocation) ……………………………………………………………………

No. of Cattle (after relocation) ………………………………………………………………………

Distance of Forest from habitation (before relocation) ………………………………………………………..

Distance of Forest from habitation (after relocation) ……………………………………………………………

Questionnaire

1. You opted for relocation
(a) As per your own wish
(b) Due to pursuance by forest department
(c) Due to pursuance by fellow villagers
(d) Due to pursuance by family
(e) Any other reason (Please state)…………………………………………………

2. Reason for opting relocation
(a) Lucrative package
(b) Conflict with wildlife
(c) Lack of amenities
(d) Developmental concerns
(e) Frequent disputes with forest department
3. Awareness about voluntary village relocation guidelines
(a) Have complete knowledge
(b) Have some knowledge
(c) No knowledge
(d) Trying to know
(e) Can't say

4. Your present living standard vis-à-vis past (Food, Platter, clean drinking water, cloth, residence, toilet, clean fuel for cooking, lighting arrangement/ electricity, days of illness, other amenities)

(a) Much better
(b) Better
(c) No change
(d) Worse
(e) Worst

5. Type of Fuel used for cooking before relocation
(a) Fuelwood brought from forest
(b) Fuelwood brought from elsewhere
(c) Mix of both a & b
(d) Cow dung cake and others
(e) LPG

6. Type of Fuel used for cooking after relocation
(f) Fuelwood brought from forest
(g) Fuelwood brought from elsewhere
(h) Mix of both a & b
(i) Cow dung cake and others
(j) LPG

7. Association with the main stream now vis-à-vis in the past (Ration card, Aadhar, Bank A/C, Voting opportunities, proximity to amenities like access to market, local self govt. institutions and Govt. offices, means of transport, means of communication, access to modern gazettes viz. TV, Computer, Mobile/ Telephone)
(a) Very much associated
(b) Associated
(c) Associated to an extent
(d) No change
(e) Completely dissociated

8. Access to Educational facilities now as compared to past (Availability of elementary, primary, secondary, intermediate schools and college, distance then and now, availability of teachers and study materials, education of girl child)
   (a) Much better
   (b) Slightly better
   (c) Same
   (d) Worse
   (e) Worst

9. Access to Health facilities now vis-à-vis that in the past (Distance of PHC, Big Hospital, Availability of Doctor and other Facilities, medicine availability, maternity facility, mother and child care, vaccination, preventive health care and awareness, Infant Mortality Rate, veterinary care unit)
(a) Much better
(b) Slightly better
(c) same
(d) Worse
(e) Worst

10. Status of conflict with wildlife now as compared to past
(a) No conflict at all
(b) Decreased to an extent
(c) Almost similar
(d) Increased
(e) Increased significantly

11. Impact on community life or cultural aspects
(a) Very adversely affected
(b) Adversely affected
(c) Affected to an extent
(d) No affect
(e) Not at all affected

12. Employment opportunities now compared to past and earning (type of employment, range, distance travelled to get work, beneficiary schemes, minimum wages, availed man-days, employment to women folk)?
(a) Very much increased
(b) Increased
(c) Increased to an extent
(d) No increase
(e) No increase at all

13. Availability of number of working days and work opportunities for members of family
(a) Very much increased
(b) Increased
(c) Decreased
(d) No increase, no decrease
(e) Can't say

14. Field of employment availability after relocation
(a) As Daily wagers in Govt. Schemes only
(b) Private Sector
(c) Govt. Service
(d) Status quo
(e) Can't say

15. Satisfaction level with the choice of getting relocated
(a) Very much satisfied
(b) Satisfied
(c) No change
(d) Dissatisfied
(e) Very much dissatisfied

16. Satisfaction level with the process, the way in which it has been executed
(a) Very much satisfied
(b) Satisfied
(c) Satisfied to an extent
(d) Dissatisfied
(e) Very much dissatisfied

17. Assistance from Forest Department after relocation
(a) Every time
(b) So many times
(c) Sometimes
(d) Never
(e) Didn't require

18. Your suggestions, if any, regarding effective steps to be taken for betterment of wildlife as well as relocated people
(a)…………………………………………
(b)…………………………………………..
(c)………………………………………………
(d)………………………………………………
(e)…………………………………………………

References

Adams, W. M. (2006). The future of sustainability: Re-thinking environment and development in the twenty-first century. In *Report of the IUCN renowned thinkers meeting* (Vol. 29, p. 31). IUCN.

Agrawal, A., & Redford, K. (2009). Place, conservation, and displacement. *Conservation and Society, 7*(1), 56.

Aiyar, S. S. A., & Kaushal, N. (2018). *Are resettled oustees from the sardar sarovar dam project better off today than their former neighbors who were not ousted? Working paper 24423.* National Bureau of Economic Research. Retrieved from http://www.nber.org/papers/w24423

Ali, S., Vasudev, L., Salma, & Khan, M. (2023). Village relocation from tiger reserves: Needs and challenges. *Vigyan Varta, 4*(11), 158–161.

Asthana, V. (2012). Forced displacement: A gendered analysis of the Tehri dam project. *Economic and Political Weekly, 47*(47), 96–102. Retrieved from http://www.jstor.org/stable/41720414

Bhatta, S., Shrestha, B. B., & Pyšek, P. (2023). Invasive alien plants in South Asia: Impacts and management. *NeoBiota, 88*, 137–167. https://doi.org/10.3897/neobiota.88.104118

Brockington, D., & Igoe, J. (2006). Eviction for conservation: A global overview. *Conservation and Society, 4*(3), 424.

Cernea, M. M., & Schmidt-Soltau, K. (2006). Poverty risks and national parks: Policy issues in conservation and resettlement. *World Development, 34*(10), 1808–1830.

Chandra, K., Mahabal, A., Gupta, R. P., & Nema, D. K. (2006). *The avifauna of Kanha National Park*. Zoological Survey of India.

Chatty, D., & Colchester, M. (2002). *Conservation and mobile indigenous People: Displacement, Forced Settlement and Sustainable Development*. Berghahn Books.

Clements, F. E. (1916). *Plant succession: An analysis of the development of vegetation (No. 242).* Carnegie Institution of Washington.

Das, M. K., & Das, L. (2014). Development-induced displacement and resettlement in India. In *Gender dimensions of displacement and resettlement*. SSDN Publisher.

Dash, M., & Behera, B. (2018). Biodiversity conservation, relocation and socio-economic consequences: A case study of Similipal Tiger Reserve, India. *Land Use Policy, 78*, 327–337.

Fanari, E. (2019). Relocation from protected areas as a violent process in the recent history of biodiversity conservation in India. *Ecology, Economy and Society - The INSEE Journal, Indian Society of Ecological Economics, 2*(1), 304023. https://doi.org/10.22004/ag.econ.304023

FRA. (2006). *The scheduled tribes and other traditional forest dwellers (recognition of forest rights) act, 2006*. Government of India.

Geisler, C., & De Sousa, R. (2001). From refuge to refugee: The African case. *Public Administration and Development, 21*(2), 159–170.

Gopal, R., Sinha, P. R., Mathur, V. B., Jhala, Y. V., & Qureshi, Q. (2007). Guidelines for preparation of Tiger conservation plan. A technical document of the National Tiger Conservation Authority. *Ministry of Environment and Forests, Government of India.*

Hiremath, A. J., & Sundaram, B. (2014). Invasive plant species in Indian protected areas: Conserving biodiversity in cultural landscapes. In L. C. Foxcroft, P. Pysek, D. M. Richardson, & P. Genovesi (Eds.), *Plant invasions in protected areas: Patterns, problems and challenges, invading nature. Springer series in invasion ecology*. Springer. https://doi.org/10.1007/978-94-007-7750-7_12

IITR. (2011). *Development projects vs. internally displaced populations in India: A literature based appraisal* (p. 6). Indian Institute of Technology. Retrieved August 13, 2024, from https://loksabhadocs.nic.in/Refinput/New_Reference_Notes/English/DisplacementandRehabilitation.pdf

Indira, R. (2024). Is relocation an answer to conservation? Experiences from the Kali Tiger Reserve, Uttara Kannada District, Karnataka. *Journal of Social and Economic Development*. https://doi.org/10.1007/s40847-023-00318-0

Jain, P., & Sajjad, H. (2015). Analysis of willingness for relocation of the local communities living in the critical Tiger habitat of the Sariska Tiger Reserve, India. *Local Environment*. https://doi.org/10.1080/13549839.2015.1129605

Jha, R., Jha, K. K., & Kanaujia, A. (2023). Humans and vultures: Sociocultural and conservation perspective in northern India. *Human Ecology, 51*, 107–118. https://doi.org/10.1007/s10745-022-00377-7

Jhala, Y., Gopal, R., Mathur, V., et al. (2021). Recovery of tigers in India: Critical introspection and potential lessons. *People and Nature, 3*, 281–293. https://doi.org/10.1002/pan3.10177

Kabra, A. (2009). Conservation-induced displacement: A comparative study of two Indian protected areas. *Conservation and Society, 7*(4), 249–267.

Karanth, K. K., Kudalkar, S., & Jain, S. (2018). Re-building communities: Voluntary resettlement from protected areas in India. *Frontiers in Ecology and Evolution, 6*, 183. https://doi.org/10.3389/fevo.2018.00183

Kassa, K., Gebeyehu, G., Bekele, G., & Abebe, G. (2023). Rehabilitation options for degraded secondary forests in Adiyo district of Kaffa zone, Southwest Ethiopia. *Trees, Forests and People, 12*, 100394. https://doi.org/10.1016/j.tfp.2023.100394

Kishore, B. S. P. C., Kumar, A., Saikia, P., Lele, N., Srivastava, P., Pulla, S., Suresh, H., Bhattarcharya, B. K., Khan, M. L., & Sukumar, R. (2024). Mapping of understorey invasive plant species clusters of Lantana camara and Chromolaena odorata using airborne hyperspectral remote sensing. *Advances in Space Research, 73*(2), 1379–1396. https://doi.org/10.1016/j.asr.2022.12.026

Lasgorceix, A., & Kothari, A. (2009). Displacement and relocation of protected areas: A synthesis and analysis of case studies. *Economic and Political Weekly, 44*, 37–47.

Lewis, J. S., Spaulding, S., Swanson, H., Keeley, W., Gramza, A. R., VandeWoude, S., & Crooks, K. R. (2021). Human activity influences wildlife populations and activity patterns: Implications for spatial and temporal refuges. *Ecosphere, 12*(5), e03487. https://doi.org/10.1002/ecs2.3487

Love, A., Babu, S., & Babu, C. R. (2009). Management of Lantana, an invasive alien weed, in forest ecosystems of India. *Current Science, 97*(10), 1411–1429.

Majid, N. M. N. A. (2000). Rehabilitation of degraded and abandoned forestland, underutilized sandy soils and landfill sites. In *UPM research report 1997–2000, vol II, section 2-extended abstracts* (pp. 246–247). UPM.

Meena, M. L. (2015). Relocation and livelihood concerns of Sariska Tiger project, Rajasthan: A pride or plight? *Space and Culture, India, 3*(2), 72–80.

Milgroom, J., Giller, K. E., & Leeuwis, C. (2014). Three interwoven dimensions of natural resource use: Quantity, quality and access in the great Limpopo Transfrontier conservation area. *Human Ecology, 42*(2), 199–215.

Pandey, R. K., & Namdeo, P. (2009). *Floral diversity of Kanha Tiger Reserve*. State Forest Research Institute.

Parasuraman, S. (1999). Displacement due to the Sardar Sarovar project. In *The development dilemma*. Palgrave Macmillan. https://doi.org/10.1007/978-1-349-27248-8_10

Procko, M., Naidoo, R., LeMay, V., & Burton, A. C. (2023). Human presence and infrastructure impact wildlife nocturnality differently across an assemblage of mammalian species. *PLoS One, 18*(5), e0286131. https://doi.org/10.1371/journal.pone.0286131

Roberge, J.-M., & Angelstam, P. (2004). Usefulness of the umbrella species concept as a conservation tool. *Conservation Biology, 18*, 76–85. https://doi.org/10.1111/j.1523-1739.2004.00450.x

Rossi, G. (2000). L'Ingérence écologique. Environnement et développement rural du Nord au Sud. Paris: CNRS

RTTF. (2005). *Report of tiger task force: Joining the dots*. NTCA.

Sahney, S. & Benton, M. J. (2008). Recovery from the most profound mass extinction of all time. *Proceedings of the Royal Society B: Biological Sciences, 275*, 759–765. https://doi.org/10.1098/rspb.2007.1370

Sarma, U. K., & Barpujari, I. (2023). Realizing a rights-based approach to resettlement from protected areas: Lessons from Satpura Tiger Reserve, Madhya Pradesh (India). *Land Use Policy, 125*(X), 106494. https://doi.org/10.1016/j.landusepol.2022.106494

Schmidt-Soltau, K. (2005). The environmental risks of conservation related displacements in central Africa. In Ohta, & Y. D. Gebre (Eds.), *Displacement Risks in Africa* (pp. 282–311). Trans Pacific Press.

Shahabuddin, G., & Bhamidipati, P. L. (2014). Conservation-induced displacement: Recent perspectives from India. *Environmental Justice, 7*(5), 122–129. https://doi.org/10.1089/env.2014.0012

Smith, A. F., Kasper, K., Lazzeri, L., Schulte, M., Kudrenko, S., Say-Sallaz, E., Churski, M., Shamovich, D., Obrizan, S., Domashevsky, S., Korepanova, K., Bashta, A.-T., Zhuravchak, R., Gahbauer, M., Pirga, B., Fenchuk, V., Kusak, J., Ferretti, F., Kuijper, D. P. J., Schmidt, K., & Heurich, M. (2024). Reduced human disturbance increases diurnal activity in wolves, but not Eurasian lynx. *Global Ecology and Conservation, 53*, e02985. https://doi.org/10.1016/j.gecco.2024.e02985

Torri, M. C. (2011). Conservation, relocation and the social consequences of conservation policies in protected areas: Case study of the Sariska Tiger Reserve, India. *Conservation and Society, 9*(1), 54–64.

West, P., Igoe, J., & Brockington, D. (2006). Parks and peoples: The social impact of protected areas. *Annual Review of Anthropology, 35*, 251–277.

WPA. (1972). *Wildlife (protection) act 1972, as amended in 2006*. Government of India.

Zhang, L., Zhang, S., Huang, Y., Xing, A., Zhuo, Z., Sun, Z., Li, Z., Cao, M., & Huang, Y. (2018). Prioritizing abandoned mine lands rehabilitation: Combining landscape connectivity and pattern indices with scenario analysis using land-use modeling. *ISPRS International Journal of Geo-Information, 7*(8), 305. https://doi.org/10.3390/ijgi7080305

Chapter 4
Rise of Tigers from a Steep Fall: Panna Tiger Reserve in Case

Kaushalendra Kumar Jha and Sanjay Kumar Pathak

Abstract This is a case study of the tigers' fall from a high and then rise from a considerable low, in general, but in Panna Tiger Reserve (PTR) in central-north-east India, the population recovered from a nadir. The chapter is based on our field work, informal interviews of reserve managers and review of literature: relevant books, peer-reviewed journal articles, committee reports, and some media reporting. It is apparent that in the beginning, tigers were numerous in their range countries, but there was a sharp decline in their populations, after which there was a marginal recovery due to conservation efforts. Indian tiger populations also declined but recovered after serious conservation interventions at the federal government level: forming Project Tiger institution and promulgating Wildlife (Protection) Act, around 50 years ago. In the case of PTR, the situation was different and unique, as the tiger population arguably declined to total extirpation and then recovered rapidly due to tiger relocation-based natural breeding management and strict control on external factors like poaching by 24 × 7 monitoring. The phenomenal success of this intervention could be replicated elsewhere under similar conditions. However, a long-term management plan would be required for the management of the rapidly increasing tiger populations in PTR.

Keywords Panna's tiger saga · Tiger relocation · New challenges · Post reintroduction research · Reintroduction and breeding · Human-tiger conflict

K. K. Jha (✉)
Indian Institute of Forest Management, Bhopal, India
e-mail: jhakk1959@gmail.com; jhakkifs@gmail.com

S. K. Pathak
National Tiger Conservation Authority, New Delhi, India

Uttar Pradesh Forest Corporation, Lucknow, India

K. K. Jha and M. O. Campbell (eds.), *Intersections of Conservation Biogeography and Wildlife Management*, https://doi.org/10.1007/978-981-95-0121-2_4

Introduction

Tiger populations in some Eurasian countries are currently declining due to habitat destruction, prey-base depletion, killing, and illegal trade of body parts. Tiger conservation efforts in these countries intensified after a meeting of world leaders and conservation practitioners in St. Petersburg, Russia, in 2010 to discuss strategies for tiger recovery (Jhala et al., 2021; Joshi et al., 2016). India, being the highest contributor to the global tiger population, played a crucial role in these efforts. The populations of tigers in various tiger reserves in India were declining on similar pattern but in two of them, the Panna Tiger Reserve (PTR) and Sariska Tiger Reserve (STR), tigers had reached extinction in the recent past. This was partially due to cultural factors. Therefore, the historical background of the Indian context of the PTR is important.

Historical Background

In ancient India, during the Chanakya's period (375–283 BCE), as recorded the chronicle the *Arthashastra*, the dense and richly biodiverse forests were classified as per use. For example, there were: (1) recreational forests for hunting; (2) economically useful forests for collection of forest products such as timber; and (3) elephant forests near the frontier to capture elephants. In addition, special areas were allocated for ascetics and brahmins (Rangrajan, 1990). Dangerous wild animals (such as tigers, crocodiles, animal herds, bird flocks, snakes) were assessed and in some cases excluded. The abundance of these animals was not recorded, but they were numerous. Velmurugan (2017) noted that the great Indian Emperor Ashoka established several sanctuaries for the protection of wild animals, and that "Wildlife experts quoted Ashoka's edicts on guardianship of animals in their attempts to stop the killing of endangered species. Emperor Ashoka brought game laws under the category of *Dharmaniyama* or the Law of Piety. He prohibited the killing of certain species of birds, fishes, and wild animals throughout his empire and all classes of living creatures on certain specified days; there was also prohibition for killing primates and carnivorous animals for a total period of 72 days in a lunar year. All hunters, trappers, bird catchers and fishermen were strictly ordered to obey the royal injunctions prohibiting slaughter." For medieval India, records on wild animals are few but the kings were still using elephants in war and some were engaged in occasional hunting. Velmurugan (2017) further recorded that "the Mughal emperors from the thirteenth to the sixteenth centuries were famous as hunters." Akbar the Great is said to have kept 1000 trained cheetahs with which he hunted antelopes, and his successor shot, among other animals, a total of 889 nilgai (blue bulls). Though the Mughal period saw declines in wildlife, during the British colonization of India, there was active killing of tigers. "Fewer tigers meant more cultivation and more revenue, their elimination a blessing of imperium after the elimination of an oriental despot. Larger rewards were given out for killing tigresses and special prizes for finishing off cubs" (Velmurugan, 2017). The royal classes were infamous for tiger destruction, though they tried to protect them in their forests but mostly for their hunting pleasure and cultural perceptions of valor. As Ranjitsinh (2017) recorded "the status of wildlife and its conservation varied from state to state, according to personal interest and, often the idiosyncrasies of the ruler." He also noted that out of 617 national parks and sanctuaries currently notified, half of them were hunting reserves—87 of British and 277 of Indian princes. At that time tigers were mostly hunted to prove sporting skill and masculine bravery. Such activities were numerous, for example, a record states that "Fateh Singhji shot more than 500 tigers in his reign of 45 years. Unfortunately, by the 1960s, Udaipur lost all its tigers" (Ranjitsinh, 2017). Sankhla (1977) gives more information:

> "The traditional Mughal type of hunting was replaced by the Western style during the time of British Raj, and the wealthy nawabs, rajas and maharajas joined with the British in the slaughter. Each one of them wanted the status symbol of scoring a century, and then to repeat it as many times as possible. Maharaja of Surguja killed 1707 tigers during his life time (up till 1958) and held the record for the highest score. Maharaja Ganga Singh shot 104 tigers, of which 17 were killed in ten days. Others who scored century included Maharana of Udaipur, Maharaja of Kotah, Jaipur, Vijanagaram and nawab of Tonk. Tiger hunting in Rewa deserve special mention: in three years 163 tigers were shot and over the period of 1923–1969 the total was 364. Maharaja Scindia of Gwalior bagged more than 700 tigers with his own rifle and his guests accounted for 200".

Other records exist from other parts of the tiger's range in India. Examples are from the Panna Tiger Reserve, pre- and post-independence, except for the period of extremely low tiger populations due to various factors before 2009 (Ahmad, 2009; Chundawat, 2023) and then the increased populations up to 2022 (official records of PTR). Chundawat et al. (2016) also stated that the Tropical Dry Forest is the largest tiger habitat in India and that tiger populations are declining at a much faster rate here than from any other habitat. They (Chundawat et al., 2016) further recorded that sharp decline in PTR was not unique to this site but part of a larger pan-India problem that was seen in historic trends.

Tiger Havens

"The tiger (*Panthera tigris* L. 1758) occupies diverse habitats throughout its range, from the cold temperate forests of the Russian Far East to the hot and humid rainforests and tidal forests of the South and Southeast Asia (Wikramanayake et al., 2004)." Tigers range across 13 countries of Eurasia: the Indian subcontinent (India, Bangladesh, Bhutan, Nepal), Southeast Asia (Cambodia, Lao PDR, Malaysia, Myanmar, Thailand, Viet Nam, Indonesia), and parts of China and Russia (Dutta & Krishnamurthy, 2024; Luo et al., 2004; Sunquist & Sunquist, 2002). In the Indian subcontinent, Tropical Dry Forest is the largest habitat type that supports tiger populations (Chundawat, 2023; Smith et al., 2011; Wikramanayake et al., 1998). However,

India's tiger population declined to approximately 1800 at the beginning of Project Tiger from a pre-colonial estimate of about 40,000 tigers (Panwar, 1982). Karanth et al. (2020) reviewed "Hunting and habitat degradation are driving steep population decline of large mammal species (Ceballos et al., 2005; Ripple et al., 2017). The tiger has suffered a historical range collapse of 93% and breeding populations now occupy <1% of the remaining potential habitat of ~1.6 million km^2 in Asia (Walston et al., 2010)."

Ranjitsinh (2017), in the chapter 'Indira's India' in his book 'A Life with Wildlife,' discussed a landmark decision of the then Prime Minister of India who supported the Wildlife (Protection) Act, 1972 and established the Project Tiger in 1973 to conserve tigers and wildlife in general. Subsequently, India has managed to improve the habitats and increase its adult tiger population from around 1700 in 2010 to about 3000. By 2019, tigers were ranging across almost 90,000 km^2 in various landscapes. This is no minor feat by global standards, since many tiger range countries that are economically more advanced have recorded declines in their tiger populations and occupied range (Jhala et al., 2021). Karanth et al. (2020) also reported that "Enactment of the Wildlife (Protection) Act of 1972, which created wildlife reserves, barred hunting and commercial forestry, a constitutional amendment that enhanced the Central Government's role in tiger conservation in 1976, the Forest Conservation Act of 1980 that reduced forest diversion, and the Environmental Protection Act of 1984, which regulated land-use adjacent to forests, were key legal actions that underpinned India's tiger recovery."

The earlier census of tiger populations was questioned for its accuracy but later it was improved and scientifically well accepted. As pointed out by Karanth et al. (2020), "From 1970 onward India's forestry officials, however, did try to assess tiger populations country-wide, using the 'pugmark census method'." Based on censuses conducted at four-year intervals, officials claimed that the numbers of tigers rose from 1872 to 4005 between 1972 and 2004. Karanth (1987) indicated flaws in the pugmark census method and later collaborated with other scientists to offer a comprehensive critique. Responding to these critiques, the Tiger Task Force (2005) finally abolished the pugmark census method (Karanth et al., 2020). However, this method was replaced by "the monitoring protocol [which] evaluates the status of co-predators, prey and habitat. The protocol involves ground surveys by […] personnel who collect information using the mobile phone GPS enabled application MSTrIPES (Ecological module) in an occupancy (MacKenzie et al., 2017) and distance sampling framework (Buckland et al., 2005). Spatial sampling units consist of forest beats […] grids (Jha et al., 2020). Spatially explicit data are digitally recorded with photo-evidence of wildlife signs stamped with date, time and geographic coordinates, that are sampled as occupancy surveys; ungulate sightings along line transects; vegetation, pellet density and human disturbance are sampled on plots. Subsequently, camera traps are used to obtain data on photo-captures that are analyzed in a spatially explicit capture–recapture framework (Borchers & Efford, 2008) to estimate densities of tigers and leopards (Jhala et al., 2021)."

As per the latest census 2022 and preceding ones from 2006, the number of tigers in India is shown in Table 4.1. Pictorial representation of tiger presence in India is

Table 4.1 A time series tiger population in India in different tiger landscapes

Landscape	2006	2010	2014	2018	2022
Shivalik hills & Gangetic plains landscape	297	353	485	646	804
Central Indian Highlands & Eastern Ghats Landscape	601	601	688	1033	1161
Western Ghats landscape	402	534	776	981	824
North Eastern Hills & Brahmaputra Plains Landscape	100	148	201	219	194
Sundarbans landscape	Not sampled	70	76	88	100
India	1411	1706	2226	2967	3080

given in Fig. 4.1 (Qureshi et al., 2023). However, PTR story in numbers is no different from India, except a unique reporting of crisis between 2006 and 2010 and then a grand success of reintroduction and rewilding which is discussed in the following sections.

Panna's Legacy

Destruction Phase

Before independence Indian territories were governed through Presidencies and Princely states. Both had gun power and abundant animals of various categories in the vast expanses of biodiverse and dense forests. The princes and other *hukmaran* (rulers) wielded their gun skills against game animals in the name of sports. They promoted scenes of involvement in hunting, where royal members stood next to the dangerous animals they had killed, sometimes with their foot on the carcasses of dangerous animals. Most of the animals were tracked, driven into areas of easy targeting, trapped, and killed. These actions contributed to wildlife population reduction.

Panna, besides Ajaygarh, Chhatarpur, Dongargarh, etc., was one of the Princely States of north eastern Central India covered by highly biodiverse forests. Ahmad (2009) recorded that Panna Forest was known as Kehrvan (kehr = tiger, van = forest) or forest of tigers: "[there was] a rigid ban on shooting by common public [but] the exclusive privilege of shooting was vested with only the ruling princes and their kith and kin. There would not have been so much concern for the tigers if rulers had gone just on occasional shootings. But killing a tiger had become a symbol of valor. Kings and princes had become proud of the number of tigers they had shot. As if that was not enough, the rulers of other princely states were invited to shoot the tigers [in Panna Forest]. This hunting spree continued [for long, at least hundred years of recorded incidents]." Without such actions, tiger populations in that area would have been much higher. Even during independent India tiger killing, legal or illegal continued, as recorded in the section 'Tigers fall: the PTR controversy.'

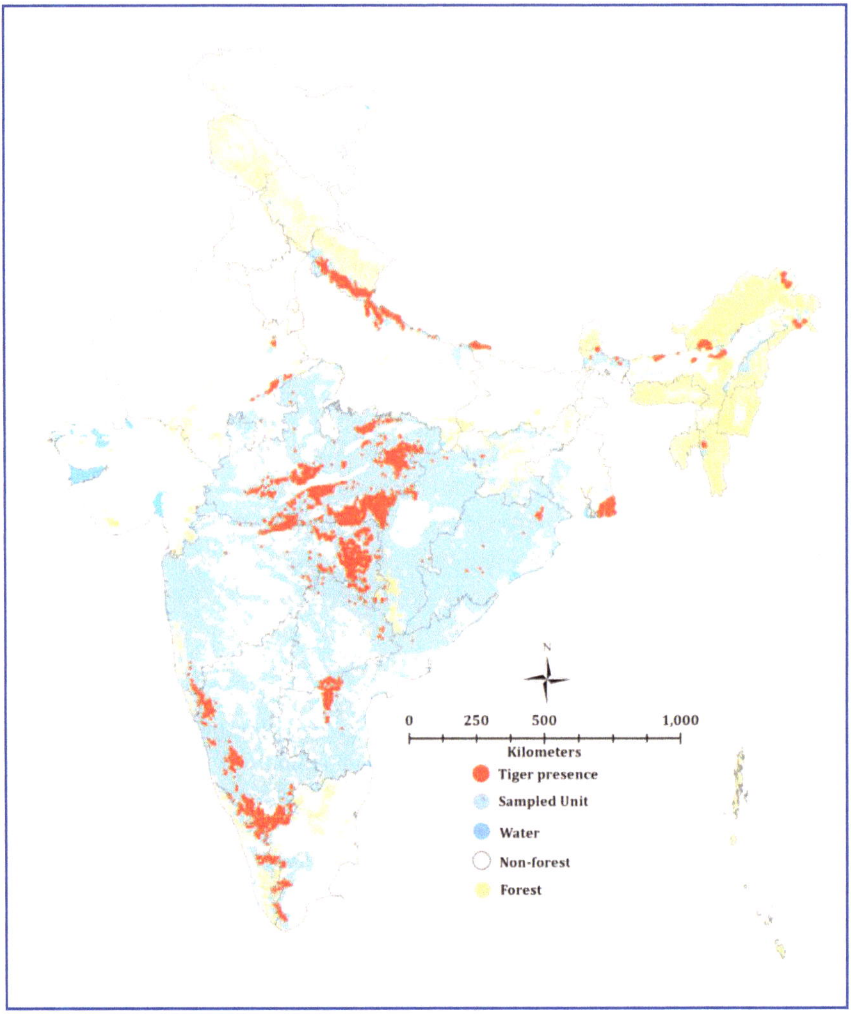

Fig. 4.1 Tiger presence locations based on 2022 census. Curtsey: Qureshi et al. (2023)

Initial Turning Point

Shri Lokendra Singh (Ex MLA and MP), one of the royal members of the lineages of the Princely State of Panna in independent India, who was a hunter changed to conservationist. One record states that after he killed a tiger at the age of 15, his heart changed. While he had a long career as a representative in the State Assembly and the Central Parliament, Lokendra Singh advocated for tiger conservation and succeeded in creating National Parks (Ahmad, 2009). Gangau Sanctuary, a part of Panna Forest, was established in 1975 and then extended and renamed to Panna

Sanctuary in 1979. In 1981, the sanctuary notification was superseded and the area was elevated to the higher protection of a National Park. In 1982, 22 tigers were reported in Panna and by 2000 the number had gone up to 32; though this suffered a drastic fall later (Ranjitsinh, 2017). Being an important tiger habitat, the national park was converted to a Tiger Reserve (PTR) for focused conservation of the flagship and apex predator species *Panthera tigris* L. 1758.

Natural History

The area of PTR has historically supported high species biodiversity. The area is one of outstanding natural beauty, with escarpments, seasonal waterfalls, and emerald canopies, which may entice tourists. Teak forest, mixed forest, *Boswellia* forest, bamboo brakes, grasslands, etc. are thriving, as are many wild animal species. In scientific terms, the forests are categorized (Champion & Seth, 1968) as Southern tropical dry deciduous teak forest (5A/C1b), Northern tropical dry deciduous mixed forest (5B/C2), Dry deciduous scrub forest (5/S1), *Boswellia* forest (5/E2), Dry bamboo brakes (5/E9), and *Anogeissus pendula* forest (5/E1).

The biodiversity of this area was recorded by Chundawat (2023). The species list included tigers (*Panthera tigris tigris*), common leopards (*Panthera pardus*), wild dog (*Cuon alpinus*). Sloth bear (*Melursus ursinus*), hyena (*Hyaena hyaena*), wolf (*Canis lupus laniger*), desert cat (*Felis sylvestris*), jungle cat (*Felis chaus*), caracal (*Caracal caracal schmitzi*), jackals (*Canis aureus*), Indian fox (*Vulpes bengalensis*), etc. The extensive open thorny forests with shorter grasslands are the ideal habitat for antelopes namely, *Chausingha* or four-horned antelope (*Tetracerus quadricornis*), *Nilgai* or blue bull (*Boselaphus tragocamelus*) and *Chinkara* (*Gazella benetti*). Ungulate species like Sambar (*Cervus unicolor*) and *Chital* or spotted deer (*Axis axis*) also occur, and may be tiger preys. Scavengers, such as the Indian Vulture (*Gyps indicus*), are common in the cliffs near waterbodies. The Red-headed Vulture (*Sarcogyps calvus*) occasionally perches on trees. Some of the photographs of the landscape, mammals, and scavengers from this Reserve are presented in Figs. 4.2, 4.3, 4.4, 4.5, and 4.6.

The Tigers' Fall: The PTR Controversy

India had the largest wild tiger population in the world and the Tropical Dry Forest areas constituted the largest habitat for them. Recent extinctions, however, from two high-profile tiger reserves [STR and PTR], highlighted the vulnerability of tiger in this habitat (Chundawat et al., 2016). Chundawat (2023), a whistle blower (Sehgal 2023 in Chundawat, 2023), based on long-term research claimed that "the increase in tiger population [of PTR] from 1996 to 2002 portrayed a steep graph but the fall in 2002–2005 was even steeper. The population could not recover from these losses

Fig. 4.2 This is one of the tiger cubs born after the crisis period in Panna Tiger Reserve. Curtsey: K. Ramesh, WII

Fig. 4.3 Sambhar, a common tiger prey species in Panna Tiger Reserve. Curtsey: K. Ramesh, WII

and went extinct in 2009." He (Chundawat, 2023) did not quote the number but the ambience was reverberating with zero in 2005 (Mazharullah., 2005). Runacres (2023) gave similar view and recorded that "The Panna Tiger Reserve is probably best known for the loss of its local tiger population due to poaching and mismanagement in 2009. Despite the warnings of researchers and conservationists, the Panna and Madhya Pradesh forest departments failed to prevent the disappearance of the area's tigers, the second such crisis in India after Sariska Tiger Reserve in Rajasthan and lost its tigers in 2004 (Shahabuddin et al., 2007)." Runacres (2023) further alleged, in tune with Chundawat (2023), that "The decline of the tiger population started in 2002, and repeated denials by forest officials, falsified wildlife surveys, and smear

Fig. 4.4 Dry deciduous forest in Panna Tiger Reserve during the monsoon. Curtsey: K. Ramesh, WII

Fig. 4.5 Red-headed and Indian Vultures perched on *Semal* tree (*Bombax ceiba*) in Panna Tiger Reserve. Curtsey: K. Ramesh, WII

campaigns against those raising alarm bells prevented any successful intervention to stop Panna's tigers either leaving the landscape or being poached from it." An interesting point in this statement is the tiger leaving the landscape in addition to poaching.

Gopal et al. (2010) note that Panna was occupied by dacoits in late 2006 and anti-insurgent activities caused further disturbances to tigers. As Adam Runacres from London could not be contacted for triangulation, the authors hope that he would have credible research supporting his statement. However, though PTR and neighboring forests were sensitive to tiger poaching (SIT, 2009), they (Chundawat, 2023; Runacres, 2023) overlooked the prevailing adverse conditions and hard work

Fig. 4.6 Painted Francolin in the grass of Panna Tiger Reserve. Curtsey: K. Ramesh, WII

of the park management in removing bad elements who were possibly involved in poaching and park area encroachment directly or indirectly. This fact resurfaced from the informal interactions with erstwhile park managers (most of them not in service now due to superannuation but available online or in person). One of the managers, on the condition of anonymity, reported that the whistleblowing acts, though proved useful later, were due to conflict of interest between the management and other parties. However, media reports periodically cited, such difficulties faced by the park staff during protection. In one such report, Saxena (2017) printed that Panna was infested with dacoits and operating from their hides in the forests. An article "Rehabilitated to the Panna Tiger Reserve in 2005, the Dala Community Faces Eviction" published in the "Science the Wire" informed the presence of encroachment in PTR (https://science.thewire.in/politics/rights/rehabilitated-to-the-panna-tiger-reserve-in-2005-the-dala-community-faces-eviction/). The dacoit infestation and their retaliatory action harming forest staff and wildlife is well recorded in an expert committee report titled 'Lessons from Panna' (ECR, 2010). The online interview respondents, though were skeptical about retaliation in the form of poaching due to tough stands by the management, admitted that eviction efforts of dacoits and encroachers helped in improving the park management in favor of biodiversity conservation. One of the PTR Directors who was in charge long term, admitted on camera around a couple of weeks ago that once upon a time tiger population was very low along with prey base and it could be revived after relocation and wilding of non-Panna tigers in PTR (https://www.youtube.com/watch?v=gX6tAk0EFzU accessed 26 August 2024).

Gulf News had reported in 2005 that the Madhya Pradesh Government had ordered a fresh census of tigers in the Panna National Park which was moved after an allegation that not a single tiger had been sighted in the park for the last 2 or 3 years (Mazharullah., 2005). The allegation was initially contested by the PTR management based on presence of 34 tigers which was the tiger census results conducted by traditional 'pug mark' method. But this method is treated with malice, showing that it has flaws in assessing the population with accuracy. In recent research, Karanth et al. (2020) recorded that "Karanth [...] pointed out flaws in the pugmark census method, and later collaborated with other scientists to offer a comprehensive critique [...]. Responding to these critiques, the Tiger Task Force (2005) finally abolished the pugmark census method." The pugmark census method was also replaced in Madhya Pradesh by a new and holistic protocol jointly conceptualized and tested by NTCA, WII, and Madhya Pradesh Forest Department (Shukla, 2018). Ahmad (2009) recorded that a pioneer and well-established institute in wildlife research, the Wildlife Institute of India, Dehradun, conducted population estimation exercise with a new technology of 'camera trap' showing photographic evidence of seven tigers in 2006 (Fig. 4.7) which silenced the critics. The final tiger estimation figure of this, specially organized, census in PTR was minimum 15 to maximum 32 (average 24) which vindicated the allegation of decimation in 2005. However, WII had also suggested in 2006 that skewed male-female ratio was causing imbalance (SIT, 2009). SIT (2009) also recorded that maximum decline took place from 2003 to 2005 which continued till 2007–2008 and PTR lost its last tiger in 2009. The later research of 2012–2013 by Ramesh et al. (2013), 11–13 tigers were estimated in PTR based on the camera trap method. The camera trap sampling was in line with Phase IV protocol of All India Tiger Population Estimation (Jhala et al., 2011). This number was highly encouraging keeping in view the decimation or local extinction in 2009 (Dutta & Krishnamurthy, 2024; Ramesh et al., 2013), though this result was due to highly successful relocation plan. This number kept increasing in the following years: credit to natural breeding in base population.

Fig. 4.7 Camera trap images of seven tigers recorded during 2006 survey by Wildlife Institute of India. *Source* Shahbaz Ahmad, Former Chief Wildlife Warden

An interesting response to media coverage regarding tiger controversy in PTR by the Director, Project Tiger (Later, Member Secretary, NTCA, New Delhi), a highest authority for tiger affairs in India, is worth quoting here:

> … …. The researchers, like many others who have done breast-beating in the media over tigers, also appears to have a hidden agenda. It is learnt he has set up an NGO near the Tiger Reserve recently to further his cause, which perhaps warrants an antisystem posture to gain credibility. … … (Chundawat, 2023).

For the above clarification, a responsible technocrat was denied his integrity by the man who created the controversy. In his book "Rise and Fall of Emerald Tigers," Chundawat (2023) recorded praise of some Directors and Deputy Directors of PTR (for example, P. K. Chaudhary and Shashi Malik) but some succeeding managers got his flak. He also mentioned in the book that he was denied park entry even as a tourist along with Joanna though he maintained a good relationship with the staff during his long-time research in the park. He also mentioned that the Field Director cautioned him for his interference in park management. During an informal interaction a Deputy Director of the park shared that many persons were denied entry to park during the crisis period due to strict vigilance on poisoning and suspected killing of tigers. Nevertheless, Ahmad (2009) reported that the park was adjudged "Best managed tourist friendly national park of the country" in 2006 by the Government of India.

While the committee believes that "the tigers were lost to poaching, it maintains that the population stress was the primary reason for tiger mothers ranging out and leaving the breeding area with cubs. Once outside, the females and their cubs became exposed to poaching in absence of a viable dispersal habitat that would have been available if a functional buffer zone existed" was the observation of Expert Committee constituted for studying Panna debacle. One more point worth mentioning here recorded the committee is that the researchers ignored the stress factor resulting into oddities. The timely detection by researchers could have helped the management in resolving stress issue (ECR, 2010).

However, a former Chief Wildlife Warden of Madhya Pradesh admitted (August 15, 2024) that "At this point of time we are in a position to take the crisis, contested steep fall, as a lesson, while rejoicing the success we take pledge not to succumb to any complacency and move forward to replicate the learning of rejuvenation elsewhere."

Reintroduction

Reintroductions are an important and powerful tool in the conservation and recovery of endangered species (Zecherle et al., 2021). This is an attempt to reestablish a species in an area which was once part of its historical range. A taxon is extinct when there is no reasonable doubt that the last individual has died (IUCN, 1995). This is a last resort to establish a species in a habitat and is an effective tool for the recovery of locally extinct species (Hilbers et al., 2020). Generally, in reintroduction, such animals are translocated which have been living in similar habitat and have been enjoying similar sort of prey-base. Where the species is extinct from the country, related sub-species may be selected. But there are certain pre-requisites for the reintroduction of any species. Reintroduction provides an opportunity to study the area, to determine the reasons for the extinction of the species from the area and provides an opportunity for experimentation through proper monitoring of the whole operation (Sutherland et al., 2004).

India is experienced in the translocation of carnivores, such as tigers and leopards but there are few post-monitoring findings. A few incidents of translocation of tigers have been reported from the Indian subcontinent and the Russian Far East (Goodrich & Miquelle, 2005; Seidensticker, 1976). These were the conflict cases where tigers were captured and released into the wild and the post-release monitoring was reported. Historically, the central Indian zone comprising Sarguja and Jaipur witnessed several plans of reintroducing large carnivores (Rangrajan, 2001). Sariska Tiger Reserve, infamous for the disappearance of its tigers in the year 2004 (Doubleday & Rubino, 2022; Narain et al., 2005), was the first venue where India could write the chapter of tiger reintroduction followed by another chapter in Panna.

Turning Around

The alleged controversy became an opportunity for the PTR managers who perfected the tiger translocation tool with tiger-centric intensive management. Supposedly no tigers in 2009 (Chundawat, 2023) changed through 13 in 2012–2013 to 59 individuals in 2021 (Ahmad, 2009; Dutta & Krishnamurthy, 2024). This must be considered difficult and committed work of employees of PTR with the able leadership of the Directors and Deputy Directors under proper guidance from headquarters supported by National Tiger Conservation Authority of Government of India. Efforts of hundreds of days and thousands of hours are summarized below, extracted from literature and official records of PTR, NTCA (National Tiger Conservation Authority), and WII (Wildlife Institute of India).

A comprehensive strategy was developed with the help of WII, Dehradun, and the Madhya Pradesh Forest Department at the behest of the NTCA. The biggest paradox for PTR was that the ecological determinants of tiger population viz. prey abundance and undisturbed habitat were still intact, when the decline and final local extinction took place (WII, 2022). This indicated that the tigers were either poached or became the victim of retaliatory killing or dispersed completely from the area. The home range of tigers of this area in the past was large and included areas of territorial divisions in vicinity (Chundawat et al., 1999).

Kolipaka et al. (2017) reviewed that "Madhya Pradesh Forest Department initiated a tiger reintroduction programme in 2009 after local tigers became extinct due to poaching (Gopal et al., 2010; Sarkar et al., 2016). Following successful breeding, tiger numbers in PTR increased from six founder animals to over 30 individuals between 2009 and 2014 (Sarkar et al., 2016)." By mid-2022 tiger number reached 57–60 (official records PTR). The success story is recorded in subsequent paragraphs.

Fig. 4.8 Translocated (base) population in Panna Tiger Reserve. Upper row: T 1 (left) and T 2 (right); Lower row: T 3 (left) and T 6 (right). These are the camera trap photographs taken in 2020. *Source* PTR and WII (2020)

Skewed Population Offsetting

As per statutory requirement of tiger translocation under Sec. 12(bb) of the Wildlife (Protection) Act, 1972, due permission from the Government of India was taken for moving tigers from neighboring populations. Under the report/assumption of presence of only male tigers after the crisis, two females, namely T1 and T2 were translocated from the Bandhavgarh Tiger Reserve (by road) and from Kisli Range of Kanha Tiger Reserve (air-lifted), respectively in March 2009 as a part of tiger recovery plan (Ahmad, 2009). They got themselves established very soon. However, no sooner it was discovered that the male population of PTR also disappeared. Then, a male tiger T-3 was translocated in November 2009 from Pench Tiger Reserve, Madhya Pradesh (Fig. 4.8).

Founder Population

An orphaned female cub T-4 from Kanha was rewilded in March 2011 and her sister T-5 was released in PTR in November 2011. Later T-6 was also reintroduced in PTR (Fig. 4.9). These translocated tigresses and tiger formed the base for PTR's new progeny tigers as they adopted the new home well in a couple of years, though with some initial difficulties in accepting the new environment, and started breeding satisfactorily.

Fig. 4.9 Panna (progeny) population from translocated parents in Panna Tiger Reserve. Upper row: P 123 (left) and P 433 (right); Lower row: P 653 (left) and P 234–23 (right). These are the camera trap photographs taken in 2020. *Source* PTR and WII (2020)

Dawn of Revival

In fact, within 14 months of translocation, T-1 delivered first litter of four cubs in April 2010 and second litter of another four cubs in February 2012. The second female, T-2 also delivered four cubs each in first and second litter in October 2010 and April 2012. Though some infant mortality was seen the remaining cubs grew well and formed the part of larger PTR population. Successful breeding and survival of cubs from translocated base population threw a bright ray of hopes of revival of tigers in PTR.

Successive Multiplication

The orphaned tigress reared in captivity in Kanha and rewilded in PTR, started hunting, and became a mother of two cubs. The reintroduction and rewilding success of PTR wrote a new chapter in tiger history. The reintroduced tigers and their progeny started and continued breeding successfully. Official records of PTR (https://www.pannatigerreserve.in/2022) detailing the breeding status of Panna tigresses are shown in Table 4.2.

Table 4.2 Status of litter production in June 2022

S. No.	Tigress ID	Age (year)	Remarks
1.	T 6	>11	6th litter
2.	P 222	>9	4th litter
3.	P 234	>8	2nd litter
4.	P 433	>8	2nd litter
5.	P 141	>7	3rd litter
6.	P 142	>7	3rd litter
7.	P 151	>5	2nd litter
8.	P 643	>4	1st litter
9.	P 234-22	>4	1st litter
10.	P 234-23	>4	1st litter
11.	P 141-12	>4	1st litter
12.	P 641	>4	1st litter
13.	P 642	>4	1st litter
14.	P 433 22	>3	1st litter
15.	P 213 62	>3	1st litter
16.	P 234 21	>4	1st litter

Established Population

The above list of breeding tigresses (Table 4.2) does not include tigresses T1 and T2, since they might not further breed due to their old age. T1 did not produce any litter in last 6 years while T2 did not produce litter in last 4 years. So, in addition to 16 breeding tigresses, following 7 tigresses are also likely to be added in this list of breeding tigresses by the end of year 2022: (1) P 152 (2) P 652 (3) P 653 (4) P 433 (5) P 433–21 (6) P 433–23 (7) P 222–32 (Official record of PTR https://www.pannatigerreserve.in/2022).

No Looking Back

The reintroduced tiger population in PTR had a higher average litter size, low cub mortality, short inter-birth interval, and an annual growth rate of ~31%, which perhaps led to the rapid population recovery. Population Viability Analysis indicated a high chance of survival probability of the tiger population in PTR, with different scenarios (WII, 2022). Having 23 breeding tigresses in PTR by 2022 indicated that the population of tigers in PTR has attained a viable stage, unless there is any catastrophe, as reviewed by Jhala et al. (2021) that a demographically viable tiger population requires a minimum of ~20 breeding females which translates to a population between 75 and 100 tigers (Bisht et al., 2019; Chapron et al., 2008; Gopal et al., 2007).

Panna Tiger Reserve can boast its laurel after achieving around 57–60 tigers in 2022 from its no tiger status within a span of less than 15 years (2009–2022). The official records (https://www.pannatigerreserve.in/2022-%20MID%20YEAR%20REVIEW%20tiger%20Popln%20PTR.pdf accessed 15 September 2024) of PTR claim that "there are enough indications based on statistical evidences that the tiger number in PTR will touch 100 much sooner than earlier predicted." Personal communication of the achieving director of PTR (R. Srinivas Murthy) suggested that PTR already had around 90 tigers in mid-2024. However, there is absolutely no doubt about it that in regaining the tigers in Panna haven, PTR establishment gets full credit. A small credit is deserved by Raghu Chundawat of "Fall of Emerald Tiger" fame (for helping PTR management in establishing radio-collaring and monitoring technique) who had gone hammer and tongs mentioning the establishment failure, etc. in his book.

However, recovery entails increasing population abundance, ensuring viability, and eliminating or minimizing threats to improve the likelihood of the species' persistence in their landscapes (Goble, 2009; Harihar et al., 2018; Hutchings et al., 2012) and till date PTR has been able to fulfill the desire of tiger conservation community.

Post-Reintroduction Research

The tiger reintroduction in past decade years in PTR has contributed to practical conservation experience, in terms of species management and organizational, political, and administrative issues (Zamboni et al., 2017). In addition, Sharma and Jerande (2021) envisaged that increasing population of tigers in PTR would throw management challenges in future which needed to be anticipated and dealt with proper prior planning. Malviya et al. (2022) cautioned that restoring a carnivore population may create unforeseen problems for communities by fueling human-wildlife conflict. Therefore, long-term persistence of tigers can only be ensured when the support of these local communities is garnered for conservation efforts, especially in release sites from where they were initially eliminated due to anthropogenic causes. The first step to gaining support for tigers and their reintroduction programs is to understand how local communities perceive these large carnivores. In this light we present some of the significant research carried out recently in PTR.

Working on Conservation translocation and Population dynamics of tiger recovery (Dutta & Krishnamurthy, 2024) observed that "Since reintroduction in 2009 and until the reporting period, 18 females […] recruited 120 cubs from 45 litters. This led to 59 individuals in 2021 with a growth rate of ~26%. The mean litter size was 2.66 (SE 0.1), and the inter-birth interval (IBI) was 19.16 months (SE 0.5). The high survival rate of the reintroduced population (0.82 ± 0.2) helped to achieve the success of reintroduction. We observed non-constant mortality trajectories for both sexes (higher survival probabilities for females) with a moderately higher risk of death in younger (<1 year) and older (>10 years) individuals." It was concluded that positive population growth rate, shorter IBI, and a high cub survival rate helped the reintroduced population to recover quickly (Dutta & Krishnamurthy, 2024).

Research conducted by WII during 2019–2020 revealed that tiger density in core and buffer plus sanctuary area of PTR was 6.5 ± 0.9 and 1.8 ± 0.5 per 100 km^2. The prey density in the core area was 28.16 ± 3.90 km^{-2}, 10.4 ± 4.02 km^{-2}, 8.07 ± 1.74 km^{-2}, 6.00 ± 1.09 km^{-2}, and 5.11 ± 0.79 km^{-2} for wild ungulate, primate, chital, sambar, and nilgai, respectively. The overall occupancy (Ψ) value for tiger was 0.47 ± 0.043, but in core and buffer it was 0.76 ± 0.05 and 0.16 ± 0.05, respectively (PTR & WII, 2020).

Kolipaka et al. (2017) working on factors influencing movement and spatial distribution of tigers reported that "tiger-specific factors—age group (generation) and sex—and environmental factors—seasons and day/night—significantly explain the observed variations in tiger use of the human-dominated buffer zone. For instance, second-generation tigers (sub-adults) spent 40% of their time in the human-use areas, compared to 10% spent by first-generation tigers (adult). When in human-use areas, sub-adult tigers approach areas near villages and spent 30% less time in areas close to water than adult tigers."

The home range size of a species is a fundamental behavioral response that results from an individual's interaction with the environment (Benson et al., 2006; Borger et al., 2008). Working on home ranges of reintroduced tigers and their progenies, Sharma and Jerande (2021) analyzed the movement monitored territories of tigers aimed at dispersal or territory shift (Fig. 4.10a, b) and population stress (Fig. 4.11a, b). They concluded that "There was a shift in territory of all 3 translocated tigers from their introduction year to the present (around 2021). This shift in territory was from core area of the reserve to peripheral areas. Shift in territory had happened for both male and female translocated tigers. The size of the territory had shrunk from introduction year to 2019. Panna progeny being younger were occupying the core area." Similar reporting of home range contraction, shift, and expansion was observed in male and female tigers in PTR by (Chundawat et al., 2016). For example, range of Male-91 contracted from year 1 (1997) to year 4 (1999–2000) and Male-125 shifted its range in year 1 (2000–2001) to year 3 (2003–2004). Female-123 expanded her range from year 3 (2001–2002) to year 4 (2003–2004). Sharma and Jerande (2021) further recorded that the distribution of male and female tiger in PTR was widespread and did not give any indication of population stress. The situation needed to be assessed continuously every 6 months and any stress visible must be tackled.

Malviya (2021) studied human-tiger conflict issues in PTR and found that there are three villages inside the critical tiger habitat (CTH), 49 villages within buffer and seven villages within Gangau Wildlife Sanctuary. Later two groups of villages also exploited the CTH. On the issue of dependency on forest it was found that 84.8% of households were dependent on the forest reserve for fuelwood, fodder, and NTFPs. Although most of the communities were agro-pastoral, some of the communities relied on husbandry of large livestock holdings. The number of households owning livestock was 66.1% and average livestock holding was 4.34 ± 6.8. Only 67% of households that owned livestock, grazed them in the reserve forest. They traveled

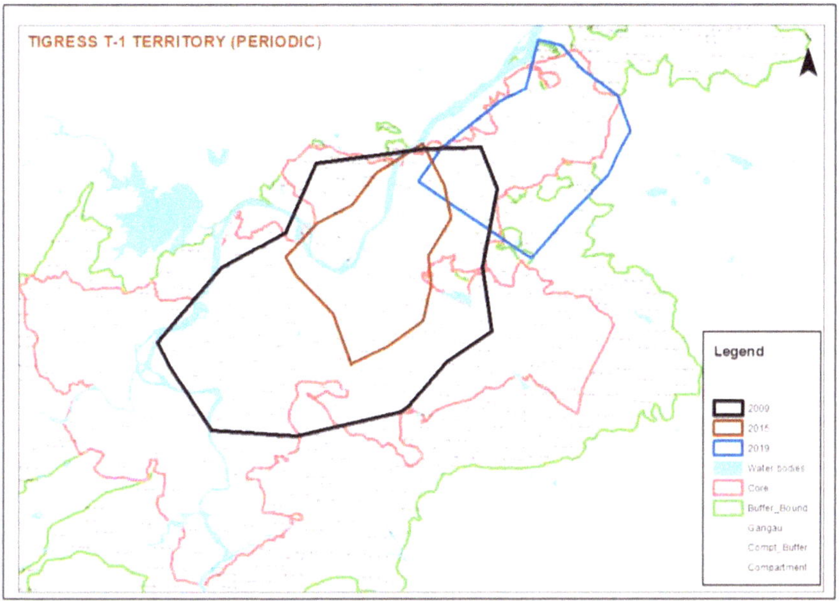

Fig. 4.10 (**a**) Territory distribution of T1 (translocated female) in Panna tiger Reserve. *Source* Sharma and Jerande (2021). (**b**) Territory distribution of T3 (translocated male) in Panna tiger Reserve. *Source* Sharma and Jerande (2021)

longer distance from their villages (on an average 3.8 ± 2.1 km) to collect these resources. They also traveled almost similar distance (average 3.4 ± 1.7 km) to graze livestock. The livestock also ventured on average 3.7 ± 1.7 km from their village, to graze when unaccompanied. These conditions increase possible encounters with tigers and other animals resulting in to conflict.

Tigers and People

Hunting, prey depletion, and habitat destruction (Wikramanayake et al., 2010) along with the continued illegal trade of tiger body parts and products driven by high demand in China and Southeast Asia have threatened the species with local extinction (Dinerstein et al., 2007) and a general global decline (Jhala et al., 2021). These issues are discussed below.

Tiger population declines have been attributed to human-tiger conflict across its range (Norris, 2005). As per limited available information, killing of tigers, as elsewhere in India, existed in Panna. Ahmad (2009) stated that a local prince killed tigers at the age of nine and fifteen in 1955 and in 1962 at Patori, part of PTR (Fig. 4.12). His ancestors or previous rulers took pride in the number of tigers they killed. Princes from neighboring estates were invited for hunting (Fig. 4.13). To make

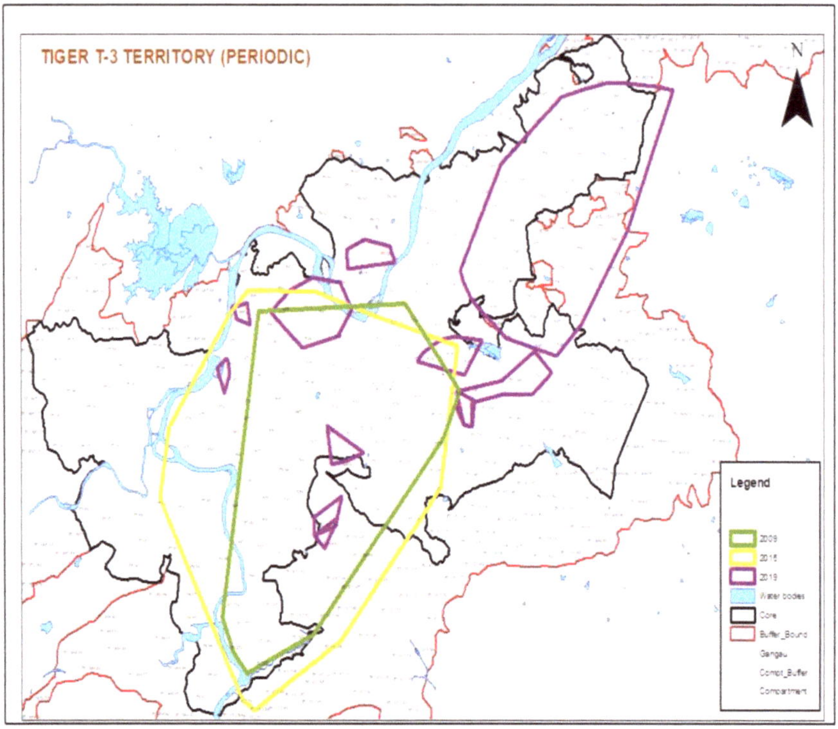

Fig. 4.10 (continued)

hunting easier, one of the kings invented *Phansia* (a trap with a stone-platform and a bait, generally buffalo, at the top, and a surrounding ditch covered with a false roof of vegetative materials. While attacking the bait, the tiger would slip from the tapering wall of platform and land in the ditch which was the only opening for escape in the cage meant to capture the tiger) to trap the tigers. The king had developed a zoo and a garden to keep the captures.

Malviya (2021) noted that "In some of its range states like Russia, about 50% of tiger deaths have reportedly occurred due to retaliatory killing by humans in response to domestic cattle depredation (Inskip & Zimmermann, 2009; Miquelle et al., 2005). Experts attribute the extinction of Bali (*Panthera tigris balica*) and Javan (*Panthera tigris sondaica*) tigers to HTC [Human Tiger Conflict] (Seidensticker, 1987)." Though official records of retaliatory killing are not available, such killings have also occurred in PTR during population decline.

Malviya (2021) also noted that 13 villages had been relocated outside the Panna National Park, still there were three villages within the Park, 49 within the buffer zone, and seven within Gangau Wildlife Sanctuary. These villages depended upon the reserve forest for construction wood, fuelwood, fodder, and NTFPs like Amla, Achar, Tendu, and Mahua (Chowdhry, 1996). Many of the communities are agro-pastoral

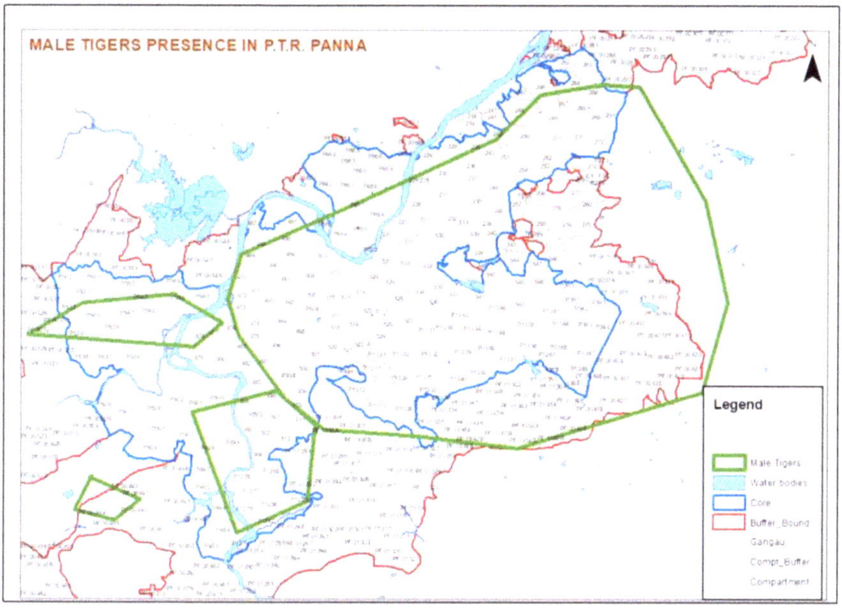

Fig. 4.11 (**a**) Distribution of male tigers in 2019 in Panna tiger Reserve. *Source* Sharma and Jerande (2021). (**b**) Distribution of female tigers in 2019 in Panna tiger Reserve. *Source* Sharma and Jerande (2021)

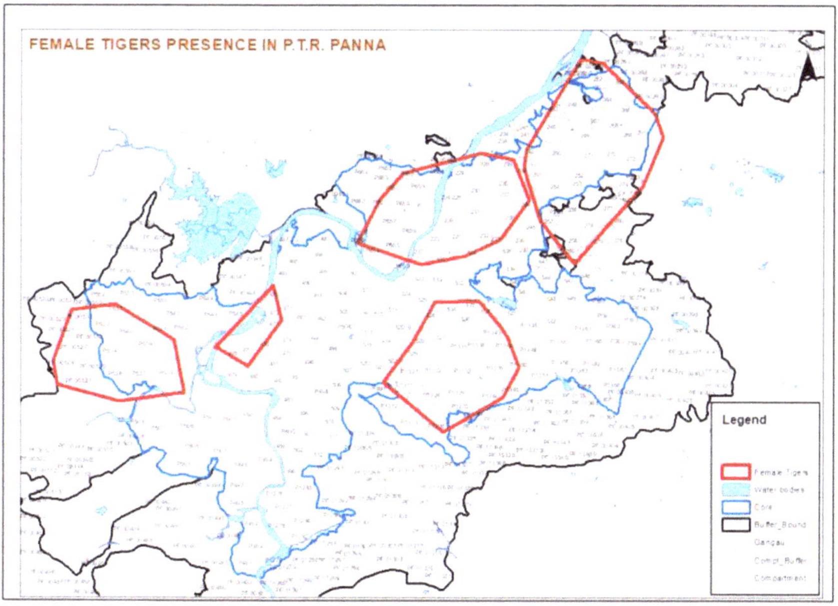

Fig. 4.11 (continued)

Fig. 4.12 Princely killing of tigers. Young Lokendra Singh, killer turned conservationist, posing with a bag of three. *Source* Shahbaz Ahmad, Former Chief Wildlife Warden

Fig. 4.13 A kill of Laxman Singh of Dongargarh, a neighboring state. *Source* Shahbaz Ahmad, Former Chief Wildlife Warden

and grazed their livestock within the reserve forest, including the core area. Illegal timber felling also happened in the peripheral areas. Occasional poaching by forest-dwelling tribes is still reported (Murthy, 2018). Illegal fishing by some communities was frequently reported in parts of reserve forest falling within Chattarpur district (Chowdhry, 1996).

The communities living in and around the Protected Areas (PAs) interact with carnivores at different spatial and temporal scales, but not every human-tiger contact/ interaction (HTI) can be termed as 'conflict.' It is, therefore, important to understand which are the factors that contribute to HTI becoming conflict (Malviya, 2021).

Sharma (2021) reported that humans and tigers both use the same water sources, be it streams or ditches, during the stress period in Panna region. Several conflicts between tiger or bear and female adult or child are reported on way to distant streams in the gorges. Malviya (2021) reported that 27.3% of the households suffered livestock loss by reintroduced tigers: a total of 224 livestock were lost to tiger, between mid-2011 and December 2014.

For conflict management (Malviya, 2021) suggested that "Managers need to see the whole problem and all its facets, before they can manage it. Applying Conflict Determinant Model (CDM) to the problem of HTC in [...] Panna Tiger Reserve, helped in gaining comprehensive understanding of the factors that were operating together resulting in livestock loss due to tigers. The same approach can be applied by managers to mitigate the issue (Fig. 4.14). It can help answer three questions, understanding which are critical for a manager to effectively resolve the problem at hand: When and where to focus? On whom to focus? And, how to focus?"

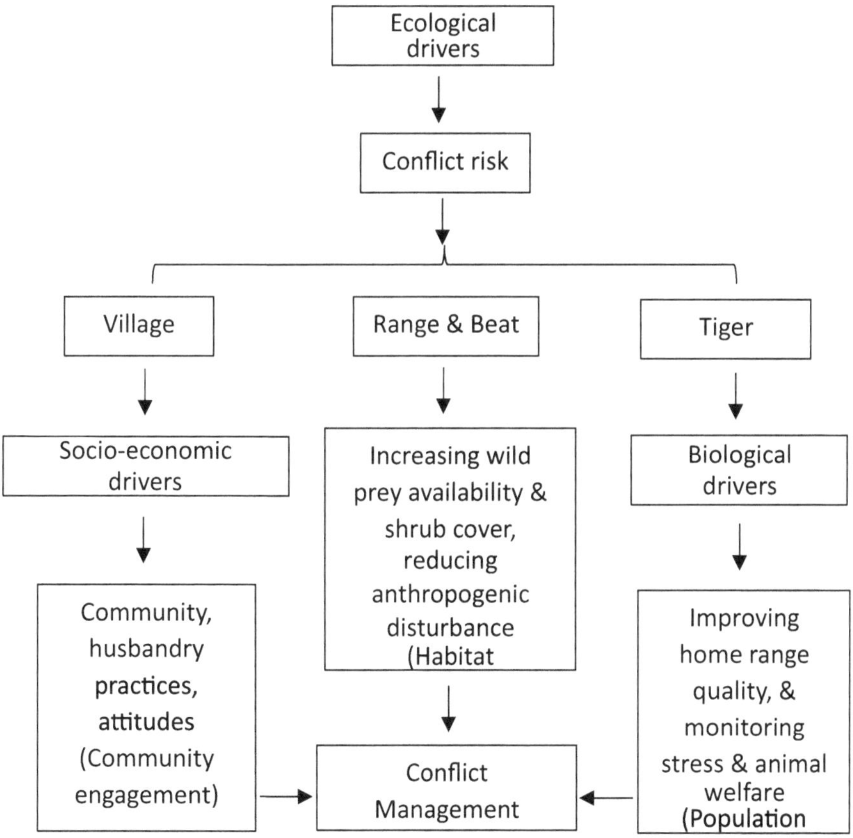

Fig. 4.14 Diagram depicting application of CDM for HTC management in Panna Tiger Reserves. *Source* Malviya (2021)

Currently, the tiger crisis is over beyond recognition and has no major problems, at least externally introduced in the form of poaching. However, a long-term tiger-centric management plan should be implemented in the area adjacent to PTR to conserve and secure the habitat of the entire landscape for the long-term survival of the reintroduced population in a metapopulation framework (Dutta & Krishnamurthy, 2024). It is important to note that large carnivores, like tigers, need a comparatively large area to fulfill their life requirements (Majumder et al., 2017; Tucker et al., 2014). Habitat fragmentation, agricultural field expansion, and human activities are known to constrain habitat size; moreover, smaller protected areas are too small to sustain a long-term viable population of long-ranging carnivores in India (Ranganathan et al., 2008; Wikramanayake et al., 1998). Therefore, landscape-level management (discussed later in the chapter) is obligatory for continuously increasing tiger population in PTR.

New Challenges

The beautiful Ken valley and surrounding area are threatened by a proposed dam (Figs. 4.15 and 4.16). A planned 100 km^2 reservoir would fill the gorge and inundate not only the tiger habitat but also the unique cliffs for vulture (Chundawat, 2023). More than 55 km^2 area will be affected if Ken-Betwa project is implemented, noted (Nayak, 2020). However, a well-researched Integrated Landscape Management Plan of Greater Panna Landscape (Fig. 4.17) for 2022–2023 to 2032–2033 is in hand for management of any contingency (WII, 2022). It envisages that "an integrated landscape management approach is imperative to understand the importance of conservation, sustainable use of biodiversity, and fair and inclusive benefit sharing. The Greater Panna Landscape (GPL, 47,621 km^2 area), with Panna Tiger Reserve (PTR) as the central focus, offers significant biodiversity, historic, cultural, and ecosystem service values. With the local extinction and successful reintroduction of tigers and the presence of other wildlife, PTR has earned global recognition. However, this region is also known for water scarcity, poverty, and sound health deficit, which affect livelihoods and development. The proposed Ken-Betwa Link Project (KBLP), therefore, aims to alleviate the recurring water stress and contribute to the overall development of the region. However, the livelihood and development aspirations will affect wildlife habitats and species populations, requiring options that will enable win-win outcomes (WII, 2022)." This plan includes the following major aspects.

Tiger Conservation

There are over 60 tigers in PTR, with an annual growth rate of 31%. Over 60% of PTR is suitable for tigers, with several unoccupied areas. The carrying capacity of tigers

Fig. 4.15 The Ken River is a lifeline for wild animals as a permanent waterhole. It is proposed to be dammed, and this may submerge the habitat. *Source* Jha et al. (2020)

Fig. 4.16 Cliff with monsoonal waterfall. This is the roosting place for endangered scavengers and birds of prey. *Source* K Ramesh, WII

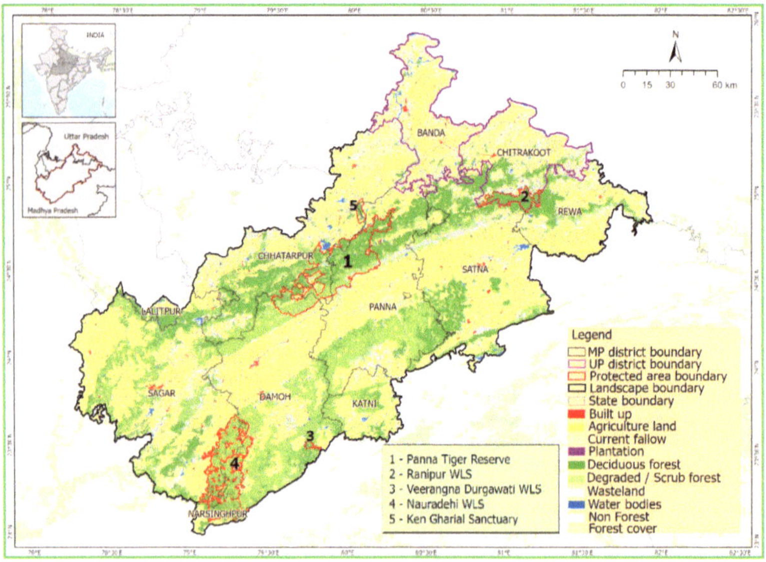

Fig. 4.17 Location of Greater Panna Landscape (81°53′25.706″E to 78°17′12.562″E longitude and 24°53′13.772″N to 24°14′4.787″N latitude) comprising eight districts of Madhya Pradesh (Sagar, part Narsinghpur, Damoh, Rewa, Katni, Panna, Satna, Chhatarpur) and three districts of Uttar Pradesh (Chitrakoot, Banda, part Lalitpur). *Source* WII (2022)

in PTR could be around 130 individuals. Thirty-seven adults/sub-adults have already dispersed in the landscape between 2013 and 2021. Twenty-eight habitat cores and 67 potential corridor links have been identified. Prescriptions for tiger conservation are provided in this plan as below (Box 4.1):

Box 4.1 Tiger Conservation Prescription

	Interventions
1	Special Performance Unit (SPU)
2	Continuous Camera-trap Monitoring System (CCMS)
3	Crime Deduction and Alert System (CDAS)
4	Intelligence Network and Incentives
5	Un-manned Aerial System (UAV)
6	Strengthening Staff Infrastructure and Logistics
7	Virtual Fence Network
8	Social Fencing Network (SFN)
9	Fine-scale Distribution Mapping and Monitoring
10	Spatially Explicit Population Estimation and Monitoring
11	Demographic Management
12	Genetic Management
13	Establishment of Controlled Population Management Facility
14	Landownership-transfer of Compensatory Village Area
15	Landownership-transfer of Revenue Open Areas and Community Reserve
16	Restoration of Consolidated Village Areas
17	Construction of Structural Connectivity across Ken River
18	Diversion of Road Traffic Volume, Strategic Fencing and Alert System
19	Retrofitting of Under-passes and Over-passes
20	Recognition and Protection of Important Tiger Habitats (satellite cores, steppingstones and corridors)
21	Research, Monitoring and Capacity Building

Source WII (2022)

Vulture Conservation

Seven vulture species have been recorded in PTR and GPL. Of the ~3000 vultures in GPL, PTR alone accounts for 34%. Panna Tiger Reserve also accounts for 14% of highly suitable cliff-nesting habitats in GPL. Elevated areas and ravines offer tree-nesting habitats in PTR. Geographical Positioning System-tagged ($n = 25$) vultures offer new insights for conservation. Vultures use the entire landscape for feeding. Prescriptions for vulture conservation are provided in this plan as below (Box 4.2):

Box 4.2 Vulture Conservation Prescription

1	Establishment of Electronic Surveillance System
2	Fencing of Carcass Dump Sites
3	Deployment of Alert System
4	Aerial Surveillance of Fire and Prevention
5	Establishment of Network of Vulture Champions
6	Gaushala Coordination Committee
7	Robust Population Estimation and Monitoring
8	Demographic and Genetic Structure Profiling
9	Roosting Site and Nest-site Monitoring
10	Biochemical Analyses and Monitoring
11	Establishment of Vulture Rehabilitation Centres (VRC)
12	Awareness and Engagement of People and Veterinary Professionals
13	Implementation of Flyway Action Plan
14	Disease Surveillance and Monitoring
15	Augmentation of Riparian Tree Species
16	Management of Dumping Site/ Feeding Point
17	Developing Community/ Conservation Reserve
18	Vulture-based Tourism
19	Research, Monitoring and Capacity Building

Source WII (2022)

Crocodilians Conservation

Due to the low population size of the critically endangered gharials (*Gavialis gangeticus*) in Ken River, restocking efforts have been ongoing since 1983. More than 160 individuals have been released but with low retention of released individuals. Suitable habitat for gharials in PTR is ~13 km^2. Muggers (*Crocodylus palustris*) are relatively abundant in PTR and Ken Ghariyal Sanctuary, with estimates of 38–62 individuals. Prescriptions for crocodilian conservation are provided in this plan as below (Box 4.3):

Box 4.3 Crocodilian Conservation Prescription

1	Protection of Crocodile Priority and Restricted Zones
2	Enforcement of Ban or Strategic Regulation on Fishing
3	Protection and Co-Management of River Habitats
4	Deployment of Unmanned Aerial System
5	Population Estimation and Monitoring
6	Demographic and Genetic Management
7	Coordinated Gharial Recovery Program
8	Crocodile Rehabilitation Centre
9	Management of Environment Flow and Sediments
10	Prey Population Management
11	Management of Artificial Sandbanks
12	Alternative Livelihood Opportunities
13	Awareness, Training and Stakeholder Workshop
14	Research, Monitoring and Capacity Building

Source WII (2022)

Biodiversity Conservation

The Greater Panna landscape has wide habitat diversity and species richness, along with distinct biodiversity-rich riparian habitats. Panna Tiger Reserve alone has over 360 species of plants, 30 species of mammals, 210 species of birds, 50 species of herpetofauna, 50 species of fishes. Several other taxa are to be explored in PTR and GPL. Prescriptions for biodiversity conservation are provided in this plan (Box 4.4):

Box 4.4 Biodiversity Conservation Prescription

1	Management of Lantana and Van Tulsi in PTR
2	Management of Lantana and Van Tulsi in GPL
3	Grassland Management in PTR in Core and Buffer
4	Management and Augmentation of Existing Ponds/Water Bodies
5	Augmentation/Restoration of Riparian Habitats inside PTR
6	Establishment of Riparian Conservation Zone / Community Reserve
7	Conservation Measures for Vegetation Species
8	Conservation Measures for Mammal Species
9	Conservation Measures for Bird Species
10	Conservation Measures for Herpetofauna Species
11	Spatial Delineation of Freshwater Fish Safe Zone
12	Conservation Awareness and Species Recovery Plan for Mahseer
13	Research, Monitoring and Capacity Building

Source WII (2022)

Community Management

People in the landscape are mostly engaged in farming and manual labor. They have a high level of forest dependency and face low level of socio-economic development. Issues concerning water, health, NTFP collection, and human-wildlife conflict are prevalent. On top of that the community institutions are non-functional. Prescriptions for community welfare are provided in this plan as below (Chart 4.1):

Integrated Management

Forested landscapes are generally intact, with a mosaic of large patches of dense and degraded forests. There are more forests area in PAs now than before, while substantial restoration options are available outside PAs and human-dominated areas. The landscape harbors a wide variety of wildlife species, with tiger, vulture, and gharial being umbrella species, representing terrestrial, arboreal, and aquatic systems, respectively. Among the least developed regions in the country (with high poverty, malnutrition, and infant mortalities), livelihood comes from multiple sources, the major being agriculture-based and forest-based economies posing significant human-wildlife interface issues. Prescriptions suggested for integrated management are as below (Box 4.5):

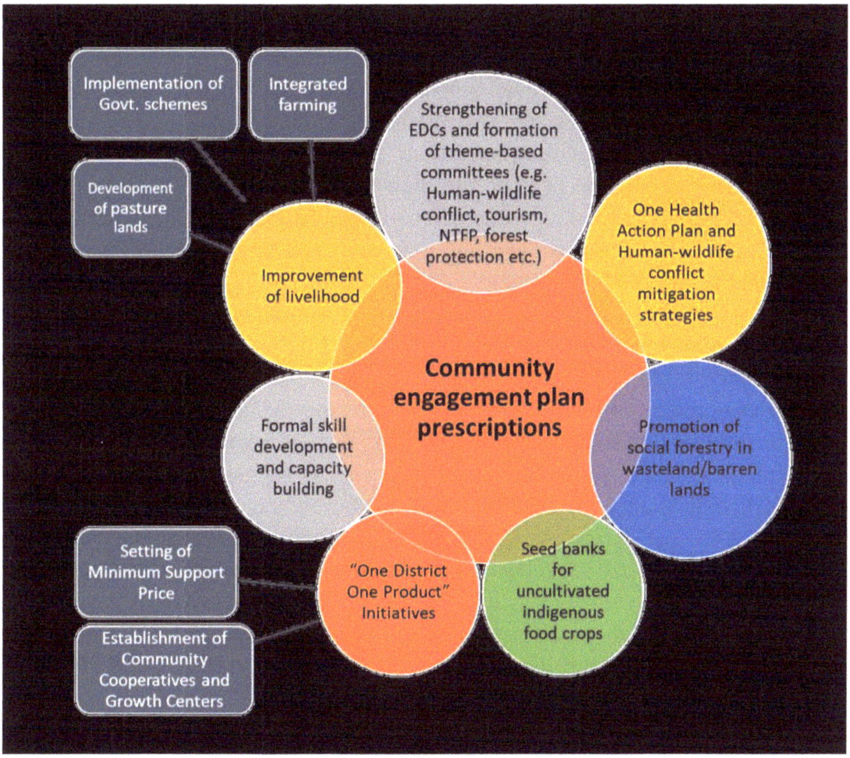

Chart 4.1 Community management prescription. *Source* WII (2022)

Box 4.5 Integrated Conservation Prescription

1	Establishment and Declaration of Interstate Tiger Reserve, Conservation Reserves, Community Reserves, River-zone and Important Conservation Units including Satellite Cores, Steppingstones and Corridors
2	Structural Interventions
3	Restoration Programs based on One Village and One Hectare Concept
4	Engagement of Biodiversity Monitoring Committee, Joint Forest Management Committee and Gaushala Coordination Committee
5	Establishment of Green Economy Cooperatives and Growth Centre
6	Establishment and Functioning of Integrated Research and Learning Centre
7	Notification and Functioning of Greater Panna Landscape Management Council under the Special Purpose Vehicle
8	Integrated Research, Monitoring, Capacity Building and Feedback for Adaptive Management

Source WII (2022)

Way Forward

The tiger population existed, even when at a minimum. Then the priority was to reintroduce the females first and next the males in the following years to balance the gender requirement of the population. The breeding posed the challenge of saving the new population of tiger and their habitat from external factors, mainly poaching. Based on prevailing adverse condition, Ahmad (2009) suggested some concrete solutions. These included (1) long-term and large-scale landscape planning should be adopted, (2) principle of zero waste economics should be applied, (3) organic farming and herbal healthcare system should be adopted, (4) adequate compensation for damage by wild animals should be provided, (5) in no circumstances should killing or torturing of wild animals be allowed for sport or revenue, and (6) development of villages within the landscape should be eco-compatible.

While Ahmad (2009) envisaged the development of whole ecosystem including tiger under the bigger umbrella, Chundawat (2023) suggested some tiger-centric actions. He emphasized the need to create an environment where the population of tigers can grow and occupy the larger landscape. Some possible steps could be (1) removal of ad hoc decisions related to annual plan of operation, (2) since poaching threat to tiger will never die away it should be combated effectively by having long-term management plan for this, (3) genetic profiling of entire tiger and leopard population be done to curb illegal trading, and (4) conservation oriented wildlife research should be taken on larger scale, etc.

For impending crisis due to river linkages and dam construction affecting PTR biodiversity and local populace, an effective integrated management (WII, 2022), as suggested in 'new challenges' section must be initiated at the earliest and be followed for at least a decadal time span. This apparently looks like a hybrid of the suggestions of Ahmad (2009) and Chundawat (2023) for future management. However, such subsequent decennium plan should be revised based on previous experiences and research inputs.

Conclusion

This article shows that PTR followed a similar pattern of tiger population swing in India. Panna forest had a speculatively high number of tigers which faced indiscriminate killing for sports by princes and their guests before legal arrangement for wildlife protection in the Union. The controversial fall from a good number of populations to zero was revealed to be between 2000 and 2005. But it was accepted to be between 2006 and 2009, since a neutral agency, WII of India counted good tiger presence in 2006 by a proven census method. This crisis of tiger disappearance forced the management to take the serious and concrete step of reintroduction of founder population which were translocated from nearby Tiger Reserves like Kanha and Bandhavgarh. The translocated population adapted well to their new niche and subsequently bred well. New generations were known as Panna progeny which further provided more generations. The number increased and dispersed in the buffer and core area of PTR. The last census reported 59 tigers which are bound to increase further. Research suggested that current populations have no serious problems but their future would need a readjustment in current management, considering the PTR's limited carrying capacity. However, a new landscape management plan, the Greater Panna Landscape, is in place to conserve the biodiversity of PTR which focusses on three major species such as tigers, vultures, and crocodilians, including others. A successful model of relocation and reintroduction of tigers in PTR may be replicated in new areas with similar crises.

References

Ahmad, S. (2009). *The crisis of the Tiger: Is there away out?* Publish America.

Benson, J. F., Chamberlain, M. J., & Leopold, B. D. (2006). Regulation of space use in a solitary felid: Population density or prey availability? *Animal Behaviour, 71*, 685–693. https://doi.org/10.1016/j.anbehav.2005.08.005

Bisht, S., Banerjee, S., Qureshi, Q., & Jhala, Y. (2019). Demography of a high-density tiger population and its implications for tiger recovery. *Journal of Applied Ecology, 56*(7), 1725–1740. https://doi.org/10.1111/1365-2664.13410

Borchers, D., & Efford, M. (2008). Spatially explicit maximum likelihood methods for capture-recapture studies. *Biometrics, 64*, 377–385. https://doi.org/10.1111/j.1541-0420.2007.00927.x

Borger, L., Dalziel, B. D., & Fryxell, J. M. (2008). Are there general mechanisms of animal home range behaviour ? A review and prospects for future. *Ecological Letters, 11*, 637–650. https://doi.org/10.1111/j.1461-0248.2008.01182.x

Buckland, S. T., Anderson, D. R., Burnham, K. P., & Laake, J. L. (2005). Distance sampling. *Encyclopedia of Biostatistics, 2*, 1–6. https://doi.org/10.1002/0470011815.b2a16019

Ceballos, G., Ehrlich, P. R., Soberon, J., Salazar, I., & Fay, J. P. (2005). Global mammal conservation: What must we manage? *Science, 309*, 603–607. https://doi.org/10.1126/science.1114015

Champion, H. G., & Seth, S. K. (1968). *The revised survey of Forest types of India*. Manager of Publications, Government of India.

Chapron, G., Miquelle, D. G., Lambert, A., Goodrich, J. M., Legendre, S., & Clobert, J. (2008). The impact on tigers of poaching versus prey depletion. *Journal of Applied Ecology, 45*(6), 1667–1674. https://doi.org/10.1111/j.1365-2664.2008.01538.x

Chowdhry, P. K. (1996). *Management plan for Panna National Park*. Madhya Pradesh Forest Department.

Chundawat, R. (2023). *The rise and fall of the emerald tiger*. Speaking Tigers Book LLP.

Chundawat, R. S., Gogate, N., & Johnsingh, A. J. (1999). Tigers in Panna: Preliminary results from an Indian tropical dry forest. In J. Seidensticker, S. Christie, & P. Jackson (Eds.), *Riding the tiger: Tiger conservation in human dominated landscapes* (pp. 123–129). Cambridge University Press.

Chundawat, R. S., Sharma, K., Gogate, N., Malik, P. K., & Vanak, A. T. (2016). Size matters: Scale mismatch between space use patterns of tigers and protected area size in a tropical dry Forest. *Biological Conservation, 197*, 146–153. https://doi.org/10.1016/j.biocon.2016.03.004

Dinerstein, E., Loucks, C., Wikramanayake, E., Ginsberg, J., Sanderson, E., Seidensticker, J., & Songer, M. (2007). The fate of wild tigers. *Bioscience, 57*(6), 508–514.

Doubleday, K. F., & Rubino, E. C. (2022). Tigers bringing risk and security: Gendered perceptions of tiger reintroduction in Rajasthan, India. *Ambio, 51*(5), 1343–1351. https://doi.org/10.1007/s13280-021-01649-0

Dutta, S., & Krishnamurthy, R. (2024). Successful conservation translocation: Population dynamics of tiger recovery in Panna Tiger Reserve, Central India. *Ecological solutions and Evidence, 5*, e12337. https://doi.org/10.1002/2688-8319.12337
ECR. (2010). *Lessons from Panna: Report of the expert committee on deimation of tigers in the Panna Tiger Reserve*. Expert Committee.
Goble, D. D. (2009). The endangered species act: What we talk about when we talk about recovery. *Natural Resource Journal, 49*, 1–44.
Goodrich, J. M., & Miquelle, D. G. (2005). Translocation of problem Amur tigers Panthera tigris altaica to alleviate tiger-human conflicts. *Oryx, 39*, 454–457. https://doi.org/10.1017/S0030605305001146
Gopal, R., Qureshi, Q., Bhardwaj, M., Singh, R. K., & Jhala, Y. V. (2010). Evaluating the status of the endangered tiger Panthera tigris and its prey in Panna Tiger Reserve, Madhya Pradesh, India. *Oryx, 44*(3), 383–389. https://doi.org/10.1017/S0030605310000529
Gopal, R., Sinha, P. R., Mathur, V. B., Jhala, Y. V., & Qureshi, Q. (2007). *Guidelines for preparation of tiger conservation plan*. National Tiger Conservation Authority, Ministry of Environment and Forests.
Harihar, A., Chanchani, P., Borah, J., Crouthers, R. J., Darman, Y., & Gray, T. N. (2018). Recovery planning towards doubling wild tiger Panthera tigris numbers: Detailing 18 recovery sites from across the range. *PLoS One, 13*(11), e0207114. https://doi.org/10.1371/journal.pone.0207114
Hilbers, J. P., Huijbregts, M. A., & Schipper, M. A. (2020). Predicting reintroduction costs for wildlife populations under anthropogenic stress. *Journal of Applied Ecology, 57*, 192–201. https://doi.org/10.1111/1365-2664.13523
Hutchings, J. A., Butchart, S. H., Collen, B., Schwartz, M. K., & Waples, R. S. (2012). Red flags: Correlates of impaired species recovery. *Trends in Ecology and Evolution, 27*, 542–546. https://doi.org/10.1016/j.tree.2012.06.005
Inskip, C., & Zimmermann, A. (2009). Human-felid conflict: A review of patterns and priorities worldwide. *Oryx, 43*(1), 18. https://doi.org/10.1017/S003060530899030X
IUCN. (1995). *Guidelines for re-introductions: SSC re-introduction specialist group*. IUCN Council. Retrieved from http://www.iucnsscrsg.org/
Jha, K. K., Campbell, M. O., & Jha, R. (2020). Vultures, their population status and some ecological aspects in an Indian stronghold. *Notulae Scientia Biologicae, 12*(1), 124–142. https://doi.org/10.15835/nsb12110547
Jhala, Y., Gopal, R., Mathur, V., Ghosh, P., Negi, H. S., Narain, S., & Qureshi, Q. (2021). Recovery of tigers in India: Critical introspection and potential lesson. *People and Nature, 3*, 281–293. https://doi.org/10.1002/pan3.10177
Jhala, Y. V., Qureshi, Q., Gopal, R., & Sinha, P. R. (2011). *Status of the tigers, copredators, and prey in India, 2010*. National Tiger Conservation Authority, Govt. of India, and Wildlife Institute of India.
Joshi, A. R., Dinerstein, E., Wikramanayake, E., Anderson, M. L., Olson, D., Jones, B. S., & Hahn, N. R. (2016). Tracking changes and preventing loss in critical tiger habitat. *Science Advances, 2*(4), e1501675. https://doi.org/10.1126/sciadv.1501675
Karanth, K. U. (1987). Tigers in India: A critical review of field censuses. In R. L. Tilson & U. Seal (Eds.), *Tigers of the world: The biology, biopolitics, management, and conservation of an endangered species* (pp. 118–132). Noyes Publications.
Karanth, U. K., Kumar, S., & Karanth, K. K. (2020). Tigers against the odds: Applying maroecology to species recovery in India. *Biological Conservtion, 252*, 108846. https://doi.org/10.1016/j.biocon.2020.108846
Kolipaka, S. S., Tamis, W. L., Zelfde, M., Persoon, G. A., & de Iongh, H. H. (2017). New insights into the factors influencing movements and spatial distribution of reintroduced Bengal tigers (Panthera tigris tigris) in the human-dominated buffer zone of Panna Tiger Reserve, India. *Mammalia, 82*(3), 207–217. https://doi.org/10.1515/mammalia-2016-0126
Luo, S. J., Kim, J. H., Johnson, W. E., Walt, J. V., Martenson, J., Yuhki, N., & Tilson, R. (2004). Phylogeography and genetic ancestry of tigers (Panthera tigris). *PLoS Biology, 2*(12), e442.

MacKenzie, D. I., Nichols, J. D., Royle, J. A., Pollock, K. H., Bailey, L., & Hines, J. E. (2017). *Occupancy estimation and modelling: Inferring patterns and dynamics of species occurrence*. Elsevier.

Majumder, A., Qureshi, Q., Sankar, K., & Kumar, A. (2017). Long-term monitoring of a Bengal tiger (Panthera tigris tigris) population in a human-dominated landscape of Central India. *European Journal of Wildlife Research, 63*(1), 1–11.

Malviya, M. (2021). *Proximate drivers of human-tiger interface and conflict in Sariska and Panna Tiger reserves, India*. Saurashtra University.

Malviya, M., Kalyanasundaram, S., & Krishnamurthy, R. (2022). Paradox of success-mediated conflicts: Analysing attitudes of local communities towards successfully reintroduced tigers in India. *Frontiers in Conservation Science, 2*, 783467. https://doi.org/10.3389/fcosc.2021.783467

Mazharullah. (2005). *Fresh census of big cats ordered in Panna reserve*. Gulf News Asia India. Retrieved from https://gulfnews.com/world/asia/india/fresh-census-of-big-cats-ordered-in-panna-reserve-1.281538

Miquelle, D., Nikolaev, I., Goodrich, J., Litvinov, B., Smirnov, E., & Suvorov, E. (2005). Searching for the coexistence recipe: A case study of conflicts between people and tigers in the Russian Far East. In R. Woodroffe & S. Thirgood (Eds.), *People and wildlife: Conflict or coexistence?* (pp. 305–322). Cambridge University Press.

Murthy, S. (2018). *Panna tiger conservation plan—2019-20 to 2028–29*. Madhya Pradesh Forest Department.

Narain, S., Panwar, H. S., Gadgil, M., Thapar, V., & Singh, S. (2005). *Joining the dots: The report of the Tiger task force*. Project Tiger Directorate, Union Ministry of Environment, Government of India.

Nayak, A. K. (2020). *Management mantra for Tiger conservation in India*. Victorious Publishers (India).

Norris, K. M. (2005). *Born to be wild: Tiger persecution and advocacy from 1800 to the present*. Virginia Polytechnic Institute and State University.

Panwar, H. S. (1982). What to do when you've succeeded: Project Tiger ten years later. *Ambio, 11*(6), 330–337.

PTR & WII. (2020). *Status of tiger and prey species in Panna Tiger Reserve, Madhya Pradesh. Phase IV monitoring 2019-2020*. National Tiger Conservation Authority.

Qureshi, Q., Jhala, Y. V., Yadav, S. P., & Mallick, A. (2023). *Status of tigers in India - 2022: Photo-captured tigers, summary report*. National Tiger Conservation Authority and Wildlife Institute of India.

Ramesh, K., Johnson, J. A., Sen, S., Murthy, R. S., Sarkar, M. S., Malviya, M., & Roamin, S. (2013). *Status of tiger and prey species in Panna Tiger Reserve, Madhya Pradesh: Capture-recapture and distance sampling estimates*. Wildlife Institute of India, Dehradun and Panna Tiger Reserve, Madhya Pradesh Forest Department.

Ranganathan, J., Chan, K. M., Karanth, K. U., & Smith, J. L. (2008). Where can tigers persist in the future? A landscape-scale, density-based population model for the Indian subcontinent. *Biological Conservation, 141*(1), 67–77.

Rangrajan, L. N. (1990). *The Arthshatra (Kautilya)*. Penguin Random House.

Rangrajan, M. (2001). *India's wildlife history: An introduction*. Permanent Black.

Ranjitsinh, M. K. (2017). *A life with wildlife: From princely India to the present*. Harper Collins Publishers.

Ripple, W. J., Chapron, G., Lopez-Bao, J. V., Durant, S. M., Macdonald, D. W., Lindsey, P., & Zhang, L. (2017). Conserving the world's megafauna and biodiversity: The fierce urgency of now. *Bioscience, 67*, 197–200. https://doi.org/10.1093/biosci/biw168

Runacres, A. (2023). Relocated tigers and relocated villagers: Ferality and human–animal entanglement in Indian conservation. *Modern Asian Studies, 57*, 222–251. https://doi.org/10.1017/S0026749X21000688

Sankhla, K. (1977). *Tiger! The story of the Indian Tiger*. Simon and Schuster.

Sarkar, M. S., Ramesh, K., Johnson, J. A., Sen, S., Nigam, P., Gupta, S. K., & Saha, G. K. (2016). Movement and home range characteristics of reintroduced tiger (Panthera tigris) population in Panna Tiger Reserve, Central India. *European Journal of Wildlife Research, 62*, 537–547.

Saxena, D. (2017). *Third generation of Panna tigers roars in Chitrakoot*. Times of India. Retrieved from https://timesofindia.indiatimes.com/city/bhopal/third-generation-of-panna-tigers-roars-in-chitrakoot/articleshow/58663791.cms

Seidensticker, J. (1976). On the ecological separation between tigers and leopards. *Biotropica, 8*(4), 225–234.

Seidensticker, J. (1987). Bearing witness: Observations on the extinction of Panthera tigris balica and Panthera tigris sondaica. In R. L. Tilson & U. S. Seal (Eds.), *Tigers of the world: The biology, biopolitics, management, and conservation of an endangered species* (pp. 1–8). Noyes Publications.

Shahabuddin, G., Kumar, R., & Shrivastava, M. K. (2007). Creation of "inviolate space": Lives, livelihoods and conflict in Sariska Tiger Reserve. *Economic and Political Weekly, 42*(20), 1855–1862.

Sharma, M. L. (2021). *Human-tiger conflict in Panna. Retrieved from India Water Portal.* https://www.indiawaterportal.org/articles/human-tiger-conflict-panna

Sharma, U. K., & Jerande, I. R. (2021). *Sustaining tiger population and managing human-animal interaction: Future challenges in Panna Tiger Reserve, MP*. Panna Tiger Reserve.

Shukla, S. K. (2018). *Tigers of Kanha*. Kanha Worker's Coperative Society.

SIT. (2009). *Report on disappearance of tigers from panna tiger reserve*. Special Investigation Team. Retrieved from https://www.environmentportal.in/files/Panna.pdf

Smith, D. J., Simchareon, S., Simchareon, A., Cutter, P., Gurung, B., Chundawat, R. S., & Seidensticker, J. (2011). Seasonally dry tropical forest is essential Tiger habitat. In W. J. McShea, S. J. Davies, & N. Bhumpakphan (Eds.), *The ecology and conservation of seasonally dry forests in Asia*. Smithsonian Institution Scholarly Press.

Sunquist, M. E., & Sunquist, F. C. (2002). Tiger. In *Wild cats of the world* (pp. 343–376). University of Chicago Press.

Sutherland, W. J., Pullin, A. S., Dolman, P. M., & Knight, T. M. (2004). The need for evidence-based conservation. *Trends in Ecology & Evolution, 19*(6), 305–308. https://doi.org/10.1016/j.tree.2004.03.018

Tiger Task Force (2005). Joining the dots. Union Ministry of Environment and Forests (Project Tiger). New Delhi.

Tucker, M. A., Ord, T. J., & Rogers, T. L. (2014). Evolutionary predictors of mammalian home range size: Body mass, diet and the environment. *Global Ecology and Biogeography, 23*(10), 1105–1114.

Velmurugan, M. (2017). Historical development of wildlife protection in India. *International Journal of Current Research and Modern Education, 2*(2), 386–390.

Walston, J., Robinson, J. G., Bennett, E. L., Breitenmoser, U. D., Goodrich, J., & Wibisono, H. (2010). Bringing the tiger back from the brink—The six percent solution. *PLoS Biology, 8*, e1000485. https://doi.org/10.1371/journal.pbio.1000485

WII. (2022). *Integrated landscape management plan for greater Panna landscape, Madhya Pradesh and Uttar Pradesh*. Wildlife Institue of India.

Wikramanayake, E., Dinerstein, E., Forrest, J., Loucks, C., Seidensticker, J., Klenzendorf, S., & O'Brien, T. (2010). Roads to recovery or catastrophic loss: How will the next decade end for wild tigers? In R. Tilson & P. Nyhus (Eds.), *Tigers of the world* (pp. 493–506). William Andrew Publishing.

Wikramanayake, E., McKnight, M., Dinerstein, E., Joshi, A., Gurung, B., & Smith, D. (2004). Designing a conservation landscape for tigers in human-dominated environments. *Conservation Biology, 18*, 839–844.

Wikramanayake, E. D., Dinerstein, E., Robinson, J. G., Karanth, U., Rabinowitz, A., Olson, D., & Bolze, D. (1998). An ecology-based method for defining priorities for large mammal conservation: The tiger as case study. *Conservation Biology, 12*, 865–878.

Zamboni, T., Di Martino, S., & Jiménez-Pérez, I. (2017). A review of a multispecies reintroduction to restore a large ecosystem: The Iberá rewilding program (Argentina). *Perspectives in Ecology and Conservation, 15*(4), 248–256. https://doi.org/10.1016/j.pecon.2017.10.001

Zecherle, J. L., Nichols, H. J., Bar-David, S., Brown, R. P., Hipperson, H., & Templeton, A. R. (2021). Subspecies hybridization as a potential conservation tool in species reintroductions. *Evolutionary Applications, 14*(5), 1216–1224. https://doi.org/10.1111/eva.13191

Part II
Human-Wildlife Conflicts, Methods and Developments

This part includes four chapters:

Chapter 5: Human Wildlife Conflict: Will Loco Parentis Be the Answer?

Chapter 6: Ecosystem Management and Trophy Hunting

Chapter 7: Human Relations with Mesopredators

Chapter 8: Composition and plant species diversity in the diet of some wild ungulates in India: A systematic review

Chapter 9: Geomatics Applications to Conservation and Social Developments

This part examines human-animal conflicts and methods developed for analysis, evaluation and further, more comprehensive studies. The managerial and philosophical aspects of conservation biogeography, including the possible human obligation to protect wildlife and the exercise of equal rights of existence of humans and wildlife are also examined. Direct killing of wildlife by humans is a crucial aspect of conservation biogeography, and trophy hunting is an important part of it because of the financial benefits to those who provide the activity. These human actions can be assessed in the context of more visible, charismatic wildlife, with smaller animals neglected in some studies. The effects of habitat loss on herbivores can also be neglected. In the study of these extremely complex issues, computerization, particularly in the field of geomatics, has emerged as crucial to the assessment, measurement and conservation management of environmental contexts. Chapter 5 explores the concept of loco parentis or guardianship of the environment, based on a case study of resource management in India in the context of high population growth and unemployment, limited land resources, illegal wildlife trade and political power issues. Chapter 6 examines trophy hunting from a conservation biogeography approach, examining the idea of ecosystem services and the evidence that trophy hunting as a management strategy has not helped to reduce human-wildlife conflict and is rather an outdated aspect of the colonial mentality. Chapter 7 examines human relationships with mesopredators (smaller or medium-sized carnivores), which are generally neglected in published research, compared to human relationships with larger, more charismatic and possibly dangerous apex predators. Chapter 8 examines the case that wild ungulates can influence the structure and function of their resident and transient ecosystems through their feeding habits, and argues that plant species diversity is declining as herbivores specialize in areas of habitat destruction. Chapter 9 examines the computational geomatics systems that are currently critical for the detailed measurement, analysis, value-added assessment, and predictive capability of the spatial aspects of ecology and social issues. These mostly geomatics-based systems enable more reliable and precise calculations for environmental research with more applicable results.

Chapter 5
Human-Wildlife Conflict: Can *Loco Parentis* Help?

Arvind Kumar Jha

Abstract India's diversity of ecosystems with forest-dependent populations therein makes human-wildlife conflicts varied and complex. In addition to various sociocultural, socioeconomic, environmental, and socio-environmental problems, there is a demand for land, unemployment among the former hunting communities, ongoing illegal wildlife trade, and impact of devolution of authority in local institutions with a concomitant dilution in the Forest Department's mandate to deal with conservation and enforcement. These contribute to the persistence and escalation of conflicts, which can be addressed through a paradigm shift in problem-solving. The concept of 'in loco parentis' or guardian recognizes the need to balance human and non-human interests and rights, challenges the notion of human dominance, and emphasizes a symmetrical view of the natural world in relation to the rights of human and non-human species. This chapter traces the history of the transition of human-wildlife interactions, addresses the development of important perspectives, examines the causes and effects, and attempts to provide some solutions and action points for managing conflicts. Against this background, the reasons for adopting the paradigm of institutionalization of 'in loco parentis' are presented to suggest that in this way conflicts can be minimized, coexistence promoted, and a more holistic approach to sustainable nature conservation can be achieved. This shift from the anthropocentric approach currently implemented requires a concerted effort to change mindsets, develop institutions, and conservation practices supported by policies and statutory instruments, and good governance that will ultimately create a more balanced and sustainable relationship between humans and the natural world. It is suggested that this approach would contribute to a better understanding of human-environment relations; and at the same time to effective, efficient, and sustainable management of environmental resources.

Keywords *In loco parentis* · Deforestation · Forest fragmentation · Habitat degradation · Tiger · Ecosystem

---

A. K. Jha (✉)
Forest Department, Social Forestry, Pune, Maharashtra, India

K. K. Jha and M. O. Campbell (eds.), *Intersections of Conservation Biogeography and Wildlife Management*, https://doi.org/10.1007/978-981-95-0121-2_5

Introduction

Human-environment interactions also termed as nature-society relations, and covered by numerous disciplines such as environmental sociology, cultural biogeography, ecology, and conservation biology, are becoming a dominant issue in the twenty-first century; with international and globally oriented case studies (Bertuol-Garcia et al., 2018; Campbell, 2018; Durant et al., 2019; Yap, 2019). These disciplines have evolved with perceptions of human control and relationships to environmental dynamics, which in some cases have changed environmental knowledge and disciplinary relations (Campbell, 2018, 2020; Campbell & Jha, 2021; Jha & Campbell, 2021). In relation to environmental sociology, Bowden (2017, p. 48) argues that "attention to the relationship between nature and society has been a defining feature of environmental sociology since its inception". Early research, incorporating insights from ecology, argued for the need to (1) theorize the causal connections between nature and society and (2) contextualize these connections in terms of biophysical limits that arise from resource scarcity. Regarding cultural biogeography, Odonne et al. (2021) noted that this subdiscipline can be defined as "the dynamics through space and time of biocultural complexes", and "cultural biogeography has proven an interesting concept to reconstruct the dynamics of biocultural interactions through space and time…", and that "tracing the origin of a knowledge is nevertheless a complex adventure that deserves time and interdisciplinary studies". Concerning conservation biology and ecology, Hintzen et al. (2020, p. 721) argue that "conservation biology was founded on the idea that efforts to save nature depend on a scientific understanding of how it works", hence applying ecological principles to conservation issues. However, "as conservation biology matured, its focus shifted from ecology to social and political aspects of conservation" and possibly due to "increasing recognition that social, economic, and political factors are critical for successful conservation and possibly from rising scepticism about the relevance of contemporary ecological theory to practical conservation".

Several strands of research are currently emerging from these topics, sometimes from multidisciplinary, interdisciplinary, and cross-disciplinary approaches, and are primarily focused on the effects of various human activities on environmental structures, and how humans control and perceive wildlife and the environment (Campbell, 2018; Cox & Gaston, 2018; Saccò et al., 2019). These research priorities can provide support for environmental policy actions. However, as noted by Bertuol-Garcia et al. (2018, p. 1032), the effective use of research findings in policy actions may be limited: "Applying scientific knowledge to confront societal challenges is a difficult task, an issue known as the science-practice gap. In ecology and conservation, scientific evidence has been seldom used directly to support decision-making, despite calls for an increasing role of ecological science in developing solutions for a sustainable future". Commonly researched human activities include general social perceptions and control over the environment, which can manifest themselves within urban development and associated ecological footprints; agriculture, and ecological changes, including deforestation, habitat change, and wildlife extirpation; and industrial development and its links to habitat pollution, including rural and urban biogeochemical changes (Stan et al., 2015).

Internationally, the numerous impacts of these human activities on the environment cannot be overemphasized, particularly the dominant paradigm of control over the environment, and in some cases, the failure of ethical environmental protections and the exercise of human environmental responsibility (Campbell & Jha, 2021; Jha & Campbell, 2021). Deforestation is significant worldwide, with results recorded for all continents except Antarctica (Curtis et al., 2018; Hoang & Kanemoto, 2021; Kinnebrew et al., 2022; Marques, 2021; Pendrill & Persson, 2017; Sharma et al., 2017; Siegel et al., 2022; Taubert et al., 2018). Wildlife loss is also critical, with numerous species declining and even becoming extinct, resulting in costs to human well-being (Edelblutte et al., 2023). Salerno et al. (2020, p. 891) which point out that "interactions between humans and wildlife resulting in negative impacts are among the most pressing conservation challenges globally". In areas with small-scale livestock and crop production, wildlife interactions with people can threaten human well-being and contribute to negative feelings and retaliatory actions against wildlife, which may limit or eliminate conservation objectives. (Salerno et al. (2020).

These major studies highlight the vitality of human relations with the environment, such as the impacts of hunting and reactive wildlife policies at the local level and throw up questions for new relations that may be more effective in conservation management. The quest for such solutions complements and contributes to the broader studies of environmental change and human impacts. In this context, people are increasingly taking responsibility for environmental changes, recognizing the need for a protective and even a 'parental' role in the preserving of local wildlife and their habitats. This highlights the notion of in loco parentis, or acting in a parent role, which was originally developed initially for the educational role of teachers towards students and is applied here as the educational and protective role of people towards the environment, which is perceived as having human rights and status.

The remainder of this chapter examines a case study from India, examining the feasibility of effective environmental management in a scenario of intensive land use demands, competing resource-based activities, illegal wildlife trade, and local policy interventions. As a background, the history of human-environment relations and their application in India are examined, and then finally the concept of *in loco parentis* or guardian, which recognizes the need for a balance between human and non-human interests and rights, and rejects the idea of human dominance of nature, and advocates for a balanced relationship between humans and the environment in terms of the rights of human and non-human species.

The Background of Human-Wildlife Relations

The understanding of human-animal relationships and human dominance over nature must consider and begin with a historical perspective, as the study of prehistoric human and environmental history is a crucial background for any study of the environment, sociocultural developments, and coevolutionary relationships between nature and society (Jha & Campbell, 2021). For example, Damodaran (2015) poses the question: "What is environmental history and why should we do it?" It is then elaborated that environmental history is variably defined as "the interdisciplinary study of the relations of culture, technology, and nature over time by historians such as Donald Worster and as the historically documented part of the story of the life and death, not of human individuals but of societies and species, in terms of their relationship with the world around them by Richard Grove and Mark Elvin". Smyntyna (2003, p. 44) describes prehistoric environmental studies as linked to the theories, concepts, and approaches "used in contemporary prehistory, archaeology, cultural and social anthropology, ecology, sociology, psychology, and demography". A summary of the development of relationships between humans and wildlife, from 'primitive' hunter-gatherer societies, through the more culturally advanced agricultural societies, to the modernist polities that practice seemingly more advanced conservation strategies becomes imperative.

Transition in Human-Wildlife Interactions

The relationships between humans (*Homo sapiens* Linn. 1758) and other animals in their shared environments have changed since prehistoric times, from a state of coexistence of multiple species to a state of human domination over other species. For millennia, humans were positioned somewhere in the middle of the food chain, as they foraged and gathered according to seasonality and hunted and were hunted by other mammals (Harari, 2014). A study by the Department of Archaeology at Tel Aviv University shows that humans were apex predators for about 2 million years and that due to the extinction of larger animals (megafauna) and animal food sources, the plant content in their diet and nutrition increased towards the end of the Stone Age (Ben-Dor et al., 2021). Humans' efforts to reach the top of the food chain as super predators caused significant ecological consequences, imbalances in population dynamics, and species and biodiversity losses. However, it is interesting to note that a paper published by the Proceedings of the National Academy of Sciences assessing human trophic levels places humans between pigs and omnivorous fish such as anchoveta and herring (Olivia, 2020). It remains an undeniable fact that as cultural developments, in addition to sedentary agriculture and settlement development combined with organized resource extraction, choices evolved depending on the options available in different ecosystems and human relationships with the natural world changed.

Sociocultural changes from foraging to agricultural practices and animal husbandry linked political and economic complexes to specific areas. The cognitive revolution increased human adaptability, led to advances in behavioural traits and cultural structures and increased the gap between humans and non-human animals (Harari, 2014). However, cultural structure also included animism, which did not define a barrier between humans, non-human animals, plants, and inanimate environmental structures. The hunters whose activities and livelihoods depended on hunting could therefore ask for forgiveness before or even after killing. "Happily the animistic heart of that ancient worldview survives still today among a thoughtful subset of traditional-values hunters, as exemplified by Aldo Leopold in *A Sand County Almanac*" (Peterson, 2014).

In the Indian context too, worship of nature, including that of predators like the tiger by forest dwellers, is widespread. For example, in a short article about a tiger hunt entitled "The Death of Red Goddess" the description regarding a tiger hunt says "Its growl the fury of the Goddess Devi, its coat, the colour of flame in the tall grass, the tiger is both feared and revered in India. Due to the scarcity of wild game, many individuals have become cattle killers, and perhaps three or four in a thousand are maneaters" and in some cases, the killing of such tigers would be "under the aegis of a prince" (Ripley, 1964). In the Sundarbans mangrove forests of the state of West Bengal in India, villagers worship '*Bonbibi*' (forest goddess) regardless of their personal religious leanings and that regulates their belief system related to the forest through symbolism (Chakrabarty, 2021). In many forest areas of central India sacred places exist inside forests and forest villages dedicated to tiger worship pointing to a long history of societies engaging with human-wildlife interactions related issues while sharing space with potentially dangerous predators (Nair et al., 2021). Hindu religion as well as traditions of indigenous people (called tribals in India) are replete with examples of nature worship and association of wild animals and plants with deities (Pande & Satish, 2016). In Jainism and Buddhism also the emphasis on the respect for and coexistence with flora and fauna innately relates to reverence-based controls for conservation (Pande, 2022; Pande & Satish, 2015).

With the increasing human population and prioritization of economic development, human-wildlife relations continued to be redefined, from a scenario characterized by customs and cultural traditions to one governed by systems of governance, including formal institutions and ever-evolving statutory instruments (Jha, 2022). The fundamental idea was to engineer and maintain the relationships between humans and animals and the associated ecosystems through formal protective structures. However, the ground reality was that resource extraction continued outside formal conservation paradigms and management practices, and over-exploitation in the ever-changing sociocultural, socioeconomic, ecological, and socioecological situations continued to lead to the impoverishment of the biodiversity in forests and wildlife areas (Jha, 2021). Modern economic perspectives have mostly failed to recognize that human economic activities are deeply rooted in and depend for their sustainability on Nature. Instead, they tended to view humanity as merely consumers of natural resources. The pursuit of trade liberalization without considering the impacts on local ecosystems also led to a depletion of area-specific ecological wealth. Various regulatory and maintenance services provided by Nature working in tandem, interfered with others when disturbed (Dasgupta & Simon, 2023). The resulting fragmentation of forests, loss of biodiversity, and damage to wildlife habitats resulting in conflicts between human and wildlife, now, are also influenced by climate change, which increases environmental complexity (Nuntikorn & Takeuchi, 2022).

From Hunting to Sport of Kings and Elites

Hunting emerged as a valid occupation of local populations from prehistoric times, depending on the resource base and the local culture. Hunting was a key to survival and involved community-wide participation (Milks, 2020). Local communities also consolidated their social order by using unique signs and symbols as identification within their group and wore animal parts and produce as ornaments, and embellishments to indicate their position and authority (Harari, 2014). Hunting also evolved from individual or small group activities to more advanced methods involving larger groups. Hence, customary laws were established for hunting, within the larger scenario of social change which developed into unwritten rules that spread across broader social structures, and either constrained or enabled the local actors in the wild spaces (Jha, 2021).

Eventually, hunting transformed from a local survival activity to a sport of elite classes and kings, with the elites sometimes supported by local guides, beaters, trackers, and servants. Hunting methods became more sophisticated, and the pursuit of large game animals became associated with pride, bravery, and social status (Lodh, 2020). The depletion of wild animal populations and species' extinction went often unnoticed amid the glorification of hunting stories and exuberant photography. During British India the control and domination of natural resources was equated with control over India (Lodh, 2020).

Emergence of Statutory and Administrative Framework

Although there were already rules to protect certain wild animals in Ashoka's[1] time for protecting specific wild animals, there were also areas designated for hunting. The first recorded game rules were laid down by Kautilya[2] in the third century B.C.

[1] The third Mauryan Emperor of Magadha in the Indian subcontinent during c. 268 to 232 BCE. His empire covered a large part of the Indian subcontinent, stretching from present-day Afghanistan in the west to present-day Bangladesh in the east, with its capital at Pataliputra (https://en.wikipedia.org/)

[2] an ancient Indian polymath, also known as Chanakya, was active as a teacher, author, strategist, philosopher, economist, jurist, and royal advisor during Mauryan period. He authored the ancient Indian political treatise, the Arthashastra.

The Mughals established hunting reserves called *shikargah* to protect game animals, the killing of predatory or herbivorous animals in human spaces was considered an act of religious merit and since there were few hunting rules in place, large-scale hunting and the resulting decline in wildlife were characteristics of the Mughal period (Velmurugan, 2017).

The British Government accelerated the extermination of wildlife in colonial India (Rangarajan, 1998). The British Government's incentive programmes to kill "dangerous wildlife"through their mass killing by engaging local participation contributed to an imbalance in the food-web and consequently a dangerous decline in viable populations of various species (Jha, 2021). As per a government order issued in 1912, the killing of selected animals termed as beasts was rewarded by District collectors (Jha, 2021). In accordance with a simultaneous move, forests were encouraged to be cleared for cultivation until independent India declared its Forest Policy in 1952. This new legislation emphasized the need to protect animals in general and rare and endangered forest species in particular. It was recognized that the demands of agriculture were stronger than those of forestry. Destruction of habitats due to increasing human population pressure and encroachment in forest areas continued, leading to further deterioration of the already damaged wildlife ecosystems.

Interestingly the first wildlife-related Acts in India adopted the customary practices developed and adopted by hunting communities like *Pardhi* to ensure the preservation of the resource base on which they depended for their livelihood (Jha, 2021). After independence, the Wildlife (Protection) Act[3] 1972 and rules made thereunder were applied to regulate hunting by introducing the concepts of National Parks and Sanctuaries, species-specific hunting regulations, as well as season-dependent open and closed hunting areas. In order to limit hunting, a licensing system was imposed, but this was not strictly implemented, and with poor supervision, produced unsatisfactory results. An amendment to this Act in 1991 completely banned hunting of all wild animals in India, but illegal hunting continued clandestinely, particularly in the non-protected areas, i.e. areas outside National Parks and Sanctuaries. The combination of degraded habitats and the non-availability of prey populations led to increased crop raids and cattle predation. The increasing trend of killing of people by large predators can be observed worldwide (Bombieri et al., 2023).

The dynamics of the relationship between humans and wildlife is also influenced by trade. The illegal wildlife trade is considered as the fourth largest illegal trade in the world in terms of value and exploits various local conditions (Rana & Kumar, 2023). The factors for this include easy money with relatively lesser risk, unemployment among the traditional hunting tribe and exploitation of their skills by the gangs involved in the wildlife trade, proliferation of mythological beliefs such as sorcery and totem, use of wild animal parts in traditional medicine, and poorly staffed forest department with considerable vacancies at the field level, all leading to the decline of wildlife populations (Nasi et al., 2008). The increasing porosity at more and

[3] An Act to provide for the protection of wild animals, birds, and plants and for matters connected therewith or ancillary or incidental thereto with a view to ensuring the ecological and environmental security of the country.

more export points and the decreasing number of wildlife surveillance personnel at international borders are leading to increased smuggling of wildlife as well as their parts and products thereby promoting trade at the local level. The World Wildlife crime report released in 2024 by the United Nations Office on Drugs and Crime (UNODC), reveals that wildlife crime has shown an increasing trend from 2015 onwards.

In India, the conversion of forest land to non-forestry use was 0.14 million ha per year before the Forest Conservation Act[4] 1980 was enacted by the Union Government. After 2014, an estimated 0.14 million ha area has been lost to projects (Gautham, 2023). Among other factors, linear and infrastructure projects in forest areas contributed to habitat fragmentation and increased considerably the threat due to increased exposure of wildlife to human populations. The Scheduled Tribes and Other Traditional Forest Dwellers (Recognition of Forest Rights) Act[5] (Forest Rights Act 2006, in short) (Anon, 2006) was enacted in India in 2006 to recognize and vest 13 types of forest rights including habitation and self-cultivation of forest land in the forest-dwelling Scheduled Tribes and other traditional forest dwellers. Hunting is not one of the forest rights listed in the Act. However, the Act, resulted in assignment, until February 2024, of 5.17-million-acre forest land (including PAs) for self-cultivation and habitation in forest land. Further, the area under community forest rights grew to 13.83 million acres by February 2024 (Ministry of Tribal Affairs, Government of India, 2024). The non-fixation of last date of claim submission is encouraging claims even after 17 years of the notification of Forest Rights Rules and has triggered encroachments also in anticipation of their regularization in future creating large-scale pockmarks and opening up of areas inside the forest lands in the country. The immediate obvious effect is a considerably increased movement of larger number of people within the forest area.

The decriminalization of illegal cutting of bamboo in forests of India occurred in 2017 by excluding bamboo from the definition of 'tree' in the Indian Forest Act[6] 1927. The Biological Diversity Act[7] 2002 was also amended in 2023, decriminalizing biodiversity-related offences that were earlier punishable with imprisonment. These relaxations intended to facilitate business actually resulted in emboldening of violators of the concerned provisions as they could get away simply by paying a penalty amount. It is worth noting that Section 7 of the original Biological Diversity Act already exempted the local communities, including growers and cultivators of

[4] An Act to provide for the conservation of forests and for matters connected therewith or ancillary or incidental thereto. It has the provision of restriction on the de-reservation of forests or use of forest land for non-forestry purpose.

[5] An Act to recognize the rights of the forest dwelling tribal communities and other traditional forest dwellers to forest resources, on which these communities were dependent for a variety of needs, including livelihood, habitation and other sociocultural needs (https://tribal.nic.in/FRA)

[6] An Act to consolidate the law relating to forests, the transit of forest-produce and the duty liveable on timber and other forest-produce.

[7] An Act to provide for conservation of biological diversity, sustainable use of its components, and fair and equitable sharing of the benefits arising out of the use of biological resources, knowledge, and for matters connected therewith or incidental thereto.

biodiversity, and also practitioners of indigenous medicine from penal provisions. Again, the proposed Forest Conservation Bill (2023) provides for relaxations and exclusion of considerable extent of areas earlier covered by the Forest Conservation Act 1980 which regulates the diversion of forest areas for non-forestry purposes (VIDHI Centre for Legal Policy, New Delhi, 2023). The opening up of forest areas and making them more readily available for non-forestry uses, as well as relaxing legal instruments, are likely to increase the potential, area coverage and impact of human-wildlife conflicts (Sinha & Jha, 2023).

Challenges and Complexity Related to Human-Wildlife Conflicts

Forests in India are increasingly occupied by people in search of land for cultivation and resources for subsistence. To quote an example, in the Report No. 6 of 2017 on Karnataka Province, the Comptroller and Auditor General (CAG) noted that forest encroachments increased almost five-fold between 1995 and 2011. The Government failed to evict encroachers, and all provincial Governments regularized encroachments in forest areas until 1980. The minister of Environment and Forests, Government of India in his answer to a parliamentary question on 15 December 2023 informed the Lok Sabha (the House of the People) that encroachments prior to the promulgation of Forest Conservation Act 1980 had been regularized on an area of about 0.366 million ha. Such action sends wrong signals and invites larger numbers of encroachers, who may anticipate further regularizations (CAG Report No.8, Andhra Pradesh State, India. Reddy, 2021).

The encroachments by forest-dwelling scheduled tribes and other traditional forest dwellers, get regularized under the Forest Rights Act, 2006 in the name of recognition of forest rights. The pockmarked areas with individual forest rights recognized under the Forest Rights Act, 2006 may be expanded and even multiplied in cases of their non-demarcation on the ground. The Supreme Court of India while hearing a public interest litigation 109/2008 (Wildlife First vs. Ministry of Forests and Environment and Others) has, on an application by the ministry of tribal department made in 2019, stayed its own order that was passed to get vacated around 1.17 million encroachments as they were rejected claims under the Forest Rights Act 2006 in various states of the country. As there is no effective deadline for filing applications, new applications are still pouring in and the monthly report uploaded to the Ministry of Tribal Affairs shows that, for example, more than 33,000 cases were cleared in March 2023 in Maharashtra state (Website of Ministry of Tribal Affairs, Government of India, 2023). Further, in Telangana state the number of titles distributed just in the month of February 2024 is 0.134 million cases (Website of Ministry of Tribal Affairs, Government of India, 2024).

Interactions between humans and wildlife result from unrestricted access to forest areas because of non-mention of ‘extent’ in the ‘community right’ titles issued by

District Level Committees (DLCs) under the Forest Rights Act 2006. It is also due to cultivation areas scattered within the forests as a result of recognized cultivation and habitat rights. In addition, the grant of management rights even inside tiger reserve areas to Gram Sabhas (village assemblies of all adult members of a village), in violation of sections 3(1)(i) and 4(3) of the Forest Rights Act 2006, coupled with the free access and removal of bamboos from reserved forests has a negative impact on wildlife habitats. The India State of Forest Report 2021 by the Forest Survey of India reveals that the country has lost 10,594 km^2 of bamboo-bearing area since 2019. The fragmentation and depletion of wildlife habitats' resources, and continuous intrusion of humans into the already damaged habitat, may be devastating (The Economic Times News, India, 2023).

Evolving Perceptions and Conceptual Constructs

The human-caused extinctions of wildlife species are going on in the current Anthropocene era (Turvey & Crees, 2019). The literature on human-wildlife conflicts, interactions, and coexistence has experienced exponential growth in recent years. It is considered that an overemphasis on conflict narratives could limit our understanding of the complex nature of human-wildlife interactions. It is important to recognize that these interactions may include sociocultural, socioeconomic, ecological, and socioecological dimensions (König et al., 2020).

A continuum perspective considers the conflict-coexistence continuum as a framework for understanding peoples' different attitudes and behaviours towards wildlife. Rather than dichotomizing situations into conflict or coexistence, recognizing the fluidity of positions and actions allows for more nuanced approaches to conservation. It highlights the fact that policies and practices should aim to address the concerns and priorities of all stakeholders (Frank, 2016). The coexistence perspective is defined as a dynamic and sustainable state in which humans and wildlife adapt to living in shared landscapes. There is a need for effective institutions that ensure the continued existence of wildlife populations, social legitimacy, and tolerable levels of risk. Shifting the focus from conflict reduction to coexistence enables research into the positive relationships between humans and wildlife (König et al., 2020).

Another approach by Shuyao Wu, Binbin V. Li and Shuang Cheng Li considers a social-ecological systems (SES) framework that captures the interactions between wildlife species and social actors through ecosystem services (ES) and ecosystem disservices (ESD). Stakeholders perceive positive outcomes (e.g. food, recreation) and negative outcomes (e.g. livestock loss, crop damage, threat to human life) as a result of these interactions. Understanding these dynamics can inform strategies to mitigate conflicts and enhance coexistence (Lischka et al., 2018; Mehring et al., 2017). An interesting aspect is seen in situations where stakeholder conflict is observed to lead to human-wildlife conflict. Conflicts among stakeholders can arise within the broader dynamics of human-wildlife interaction. As different stakeholders hold diverse moral values and interests, the management of ecosystem services and disservices can create tensions. Balancing these impacts within a social-ecological systems framework is essential for addressing conflicts effectively (Hill, 2021).

A study of the Taita community in Kenya suggests that successful use of indigenous knowledge-based solutions is possible when used in synergy with scientific knowledge in wildlife management (Mwamidi et al., 2012). The community used traditional methods such as releasing a primate that they had captured and painted with red clay and charcoal. This primate's attempt to blend in with the group led the group to move away from him and, accordingly, from the village, which also reduced the likelihood of conflict with the villagers. Similarly, they traditionally used burning wild animal dung and spreading a mixture of leaves of some native plants at the farm boundary, digging trenches and beating drums to scare away elephants and wild boars to avoid conflict.

In the Indian context, while relocation of forest rights holders currently living inside PAs does exist as an available option, the Forest Rights Act 2006 mentions exploration of coexistence option as an essential exercise before deciding on such a measure. Well-defined parameters need to be developed to assess the feasibility of coexistence and its long-term stainability in Critical Wildlife Habitat areas conceptualized under the Forest Rights Act 2006 (Jha, 2022). Restoring a sense of being part of the entire ecosystem rather than viewing wildlife as the property of the managers of the wildlife and its habitat can change the way people approach wildlife. Furthermore, the ability of a democratic institution to balance the value systems of stakeholders and mediate e.g. between nomadic pastoralists and local graziers, can reduce conflicts and improve the resource base where grassland-based wildlife thrives (Jha, 2022).

Compensation is often seen as a solution to the conflict and larger amounts are considered by Governments as most desirable for both sides. In Maharashtra state itself, the level of compensation has been increased many times over within 10 years. The enforcement organizations view this as a tool to pacify the agitated people while those on the receiving end perceive it as a support to the family. While compensation for damage due to wildlife has its advantages, the subjective nature of the damage calculation and a delayed compensation process can create difficulties for the affected people. This model alone fails to promote positive relations and coexistence between humans and wildlife. Portraying compensation as support from a government agency gradually creates the impression that wildlife is solely the government's responsibility, contributing to a sense of alienation so far as the local populations are concerned. However, timely compensation is expected to be an important support for the family of the deceased and is therefore highly desirable (Sinha & Jha, 2023). Studies have shown that compassion for the victim as well as the sensitivity towards the plight of wildlife in shared landscapes is critical to securing the future of coexistence, rather than just a statistical analysis or instrumental or economic responses.

Many scientists do believe that focusing on conflict narratives creates biases that result in human-wildlife interactions being viewed as merely conflictual in nature. However, eliminating the concept of "conflict" from narratives may not be appropriate, as conflict can trigger a change and development of long-term co-occupancy and coexistence between humans and wildlife. Conflicts may be an essential part of how humans and animals live together (Hill, 2021).

Dealing with human-wildlife conflicts initially focused primarily on managing negative wildlife impacts. However, this perspective has evolved over time, with an increasing recognition of the need to move from reducing conflict to promoting positive relationships and coexistence between people and wildlife. It is important to take note of the changing perceptions and conceptual constructs surrounding human-wildlife conflict in different areas while emphasizing the importance of understanding the complexity of human-wildlife interactions and exploring alternative and innovative approaches to addressing such conflicts.

Major Causes and Implications of Human-Wildlife Conflict

Broader Learnings

Historical and current research on human-wildlife conflict has identified agricultural expansion, human settlement, livestock overgrazing, deforestation, illegal grass collection, and poaching as common causal factors (Conover & Decker, 1991). Certain human behaviours have also been cited as proximate causes of such conflicts, such as the reduced ability of livestock to escape, poor animal husbandry practices, and the decline of the natural prey base of carnivores (Conover & Decker, 1991). The resulting consequence is the development of negative attitudes in communities affected by wildlife and increase in hostility towards these animals. In addition, cases where large carnivores attack humans can lead to reactionary attacks on these animals. This has a long-term negative impact on their population demographics. Interestingly, political changes in Namibia towards private ownership and sustainable use of wildlife led to an increase in conflicts between different interest groups due to increased resource competition (Luetkemeier et al., 2023). Stress levels in elephants, indicative of conflict, have been observed to be elevated outside of protected areas (Barnes & De Jager, 1996). Interestingly, a study applying the 'landscapes of fear' theory discovered that changes in female elk behaviour due to the presence of wolves had impacts on ecology of elk and bison that went beyond that due to direct loss of individuals to wolves (Johnsingh & Negi, 2003).

Urbanization Effect

Rapid urbanization, habitat change, and climate change are putting enormous pressure on wildlife diversity and endangered species everywhere. Despite these challenges, some animals have managed to adapt to urban stressors, although this has led to various levels of conflict with humans as they compete for space and resources.

Species that thrive in urban conditions adapt to human-provided resources such as water and food, as well as enjoy the 'human shield' effect creating safety from predators. As species numbers increase, animals start entering human spaces, such as residential accommodation for nesting and protection. The frequency and intensity of such conflicts increases as populations grow and the normal requirements of the species are not met and the demand for space and resources increases. In this context, the disappearance of the erstwhile commonly found sparrow in human habitations has been a sad story in India. On the other hand, in several Indian cities, rock pigeons and monkey species have adjusted and are thriving (Ghai et al., 2023).

Indian Context

Forest fragmentation, habitat degradation and destruction, prey depletion, expansion of human activities into wildlife habitats, intensive forest use by humans, and decreasing tolerance of local communities towards wildlife have been identified as major contributors to human-wildlife conflict in the Indian context (Johnsingh & Negi, 2003; Panwar, 1979; Rishi, 2005; Shaurabh & Radhakrishna, 2017). The focus on wildlife in these conflicts often overshadows the issues related to land use and reform. In India, the recognition of cultivation rights in forests has been and still is a persistent cause of human-wildlife conflict. Local communities show high levels of intolerance towards leopards when conflicts remain unresolved (Kala & Kothari, 2013). However, with ineffective enforcement authorities and agencies, the consequences of such conflicts could be disastrous for ecosystem services that are vital to all stakeholders (Hongbo et al., 2020).

The ineffectiveness of the administration in managing and controlling the changes in the designated land uses prescribed for non-forest areas has also been a major reason for increasing pressure on wildlife habitats in India. One factor behind the increasing encroachment of cattle herds into forest areas, leading to an increase in human-wildlife conflicts, is the diversion of grazing land called *Gairan* or *Gochar* land in villages for various other unauthorized uses. In a hearing on sixth October 2022, in a case heard *suo motu* by the Bombay High Court, it was stated that human settlements, public buildings, industries etc. in Maharashtra state constituted 220,000 illegal structures spread over an area of 0.45 million ha of such lands. The resulting pressure on forest resources from cattle populations pushes wild ungulates to move into non-forest and agricultural areas, leading to conflict situations on both sides of the forest boundary.

Area and crop preference, including seasonality, have also been observed as interesting features in crop raiding with cheetal, sambar, wild pig, and nilgai raiding fields closer to the forest edge when the crops were at the seedling stage. During the same time, nilgai and wild pig raided fields far from the forest edge also (Kumar et al., 2017). A study of sloth bear attacks in central India found that about half of the attacks occurred during the monsoon season, highlighting the role of seasonality in human-wildlife conflict (Singh et al., 2008).

Infrastructure projects in forests as a part of 'development' contribute significantly to wild animal deaths apart from degrading their traditional range and habitats (Nishank, 2022). Changes in animal behaviour and inadequate remedial action by those involved can worsen conflict situations and their associated impacts (Sahana, 2023). The construction and operation of wind turbines have been found to be a significant contributor to wildlife mortality. Wind turbines pose a substantial threat to bats, in addition to the significant damage their construction causes to forests and biodiversity (Artelle et al., 2021). Projects to build high-voltage power lines across grasslands, which are home to many animal species including that of the Great Indian Bustard (Uddin et al., 2021), are often liberally approved because they do not require felling many trees for this purpose. About 66,337 km^2 of the area of occurrence of this critically endangered species is recently cleared for solar project (Sinha & Jha, 2024).

Local Dynamics

Wildlife density and mobility could be seen as creating dispute, but they should be regarded as externalities since poorly controlled land use and badly implemented land reform process are the real reasons for the core conflict. Forest management rights recognized within the interior areas of forests and habitats may remain an ongoing area of human-wildlife conflict and would harm the interests of all stakeholders, including wildlife in the context of overall natural resources (Pinjarkar, 2014). Irregularities in the process are often ignored by people either out of ignorance or as long as rights are generously recognized. Later, however, inevitable relationship conflicts arise as also disputes over the ability of those involved to use shared resources. Struggles over missed opportunities to participate in decisions making also may come up in the later stages when competition for use of shared resources become intense. In the case of regularized encroachments, history has in fact shown local people the benefits of disobeying laws. As the enforcement authority is weakened and sidelined in both its protective and administrative functions in the name of democratization and decentralization of powers, the impact could be multifaceted with catastrophic consequences for the survival of wildlife.

The Numbers

The nature and extent of conflict between humans and wildlife impacts both humans and natural ecosystems (Eva et al., 2021). India has a wide diversity of ecosystems and forest life as well as local cultural and traditional systems. This uniqueness raises site-specific challenges in terms of both the diversity and variety of the impacts of human-wildlife conflict. According to data presented in Parliament on July 2022, between 2018 and 2021, 29 elephants were killed, 222 were electrocuted, 11 died of poisoning, and there were 30 deaths due to train accidents. In addition, 1700 people were killed by tigers and elephants between 2019 and 2021 (Roy, 2022). The Union Government informed Parliament in March 2023 that a total of 551 tigers had died in the previous 5 years. Of these, 114 died from poaching, 128 from natural causes, 19 due to unnatural causes, and 290 death cases were still under observation (Das, 2023). Data from the Wildlife Protection Society of India shows that in Maharashtra state 22 leopards were killed in road accidents in 2018, 19 in road accidents and three in train accidents in 2019, and 19 in road accidents and one in train accident in 2020. Leopards killed 23 people from January to November 2020, the highest number in the decade (Badri, 2020). As for the great bustard, there were three deaths due to contact with high-voltage wires between April 2022 and March 2023, out of a population of fewer than 150 surviving animals in India, in the Jaisalmer district of Rajasthan, where about 95 per cent of the population lives (The Economic Times News, India, 2023). In Tamil Nadu, there were 7562 cases of crop raids by wild animals and in Andhra Pradesh there were 7589 incidents in the 3 years prior to 2020. In Andhra Pradesh, crops on 5543 acres of land were damaged (Chaitanya & Krishna, 2020). In the urban context, the number of bird species was observed to have declined from 39 to 12 between 1974 and 2014 while the number of buildings increased from 2 to 314 in the study area of Domlur second stage in Bengaluru of Karnataka state (Verghese & Mendonce, 2016). The data sets available indicate that the severity of human-animal conflict is significant, more so because it is showing an upward trend.

Strategies and Action Points

Several methods have been attempted to address the problem of human-wildlife conflict in different parts of the world. Some strategies may be universally applicable, while others may be site or species-specific. It is desirable to determine the most effective strategy based on the prevailing situation and existing dynamics in each local context. Given the diversity of the conflict situations and socioeconomic dynamics that could be present in different areas, the general aspects that could characterize the solutions and related action points are summarised below.

Improving Forest and Wildlife Management and Management in Adjacent Areas

Forests should be managed effectively to achieve the required levels of abundance and density of wild prey species which requires effective landscape management with sustainable, local and practical solutions (Fedorca et al., 2021). Wildlife management systems should be integrated into silviculture systems in areas outside protected areas. Approaches such as fire control, improved livestock management, and regulated grazing can help reduce livestock losses and improve public attitudes towards carnivores (Conover & Decker, 1991). Restoration of degraded forest landscapes, reforestation, and afforestation can be effective approaches to address climate challenges and rejuvenate wildlife habitats (Tampakis et al., 2023).

Crops preferred by wild animals should ideally be avoided near forest edges or protected by appropriate measures such as fences. On the social front, improved livestock management coupled with anti-predator strategies are required (Alemayehu & Temesgen, 2015) including the organization of awareness and capacity-building programmes on prevention and mitigation of such conflicts, use/optimization of artificial and natural barriers, and effective waste management systems (Kumar et al., 2017). Consolidating and implementing regulated interventions outside forest boundaries and improving habitats while eliminating unauthorized interventions can be prioritized and addressed with the support of local authorities. People in low productivity areas where crop raids by wildlife regularly occur should be encouraged and supported to plant trees and medicinal plants on their plots. Appropriate programmes must be designed and implemented for the forest-dwelling populations to meet their diverse needs while improving their economic conditions. Communities can benefit from the provisions related to carbon credits and those of The Energy Conservation (Amendment) Bill, 2022[8] through systems that reach out to the masses.

Managing Conflicts in Urban Landscapes

Successful urban wildlife conservation and reduction of human-wildlife conflict require consideration of both sociological and environmental challenges. Measures such as managing anthropogenic noise and light, promoting gardening, and providing food, water, and shelter for urban wildlife need to be prioritized. Human-made features will progressively shape biodiversity. Integrated pest management (IPM) can also be beneficial for the coexistence of humans and animals in urban areas as an ecosystem-friendly alternative to pesticides. Raising people's awareness of conflict mitigation through volunteer work, comprehensive educational experiences, and providing opportunities to bring about beneficial behaviour changes should become part of the discourse with city residents.

Infrastructure Development: Addressing Coordination Issues

Infrastructure-related conflict situations require the cooperation of various authorities. Project developers and proponents should be made aware of the possible conflicts

[8] The Energy Conservation (Amendment) Bill 2022 enables the Union Government to set up a carbon credit trading scheme in the country.

and their ecological implications. Reducing vehicle speeds in high-risk areas, developing meadows, and water sources to restrict wildlife movement, laying underground cables in grassed areas, aligning new roads and tracks to avoid habitats, preparing corridors for wildlife movement etc. can reduce conflicts (Sinha & Jha, 2023).

Designing Forest-Related Statutory Instruments, Mitigation Systems, and Communication Strategies

Carefully drafted wildlife, forest, and land use laws are critical to addressing socio-environmental concerns. Conflict mitigation systems and communication strategies as well as compensation and insurance schemes must be developed and implemented.

Many conflicts during humans and wildlife interactions can be traced back to disputes between societal interest groups based on interests, moral values, social relationships, hierarchies, and political leanings that influence their treatment of wildlife. Ideally, these systems should involve the village to evolve community-owned solutions followed by a rapid and transparent implementation by authorities.

Human-wildlife conflict is a dynamic issue that requires ongoing research to develop site-specific solutions. Indigenous knowledge-based mitigation tools available in the community should also be explored and incorporated into the solutions matrix where appropriate.

Regular recording and analysis of temporal and spatial data on conflict events can help develop appropriate strategies to reduce the frequency of conflicts and mitigate their adverse effects. In addition, environmentally sound strategies, and social measures such as social awareness and capacity-building programmes to prevent and mitigate human-wildlife conflicts are also critical.

Field Staff Training and Community Involvement

Field staff should be regularly trained to comprehensively mitigate conflict (Karanth & Anubhav, 2020). Rapid Rescue Units should be established to manage conflict situations. Local people should be involved in policy design concerning human-wildlife conflicts to prevent the perception of alienation. Compassion for the victim as well as the plight of wildlife in shared landscapes should be an important theme and trained professionals should be deployed to communicate with victims, their families, and villagers.

Loco Parentis: The Alternative

A Case for a Shift in Paradigm

The findings based on the above literature and experiences suggest a reframing of our approach and develop a policy that goes beyond the currently prevalent imperatives. An objective approach to the issue of human-wildlife conflict requires an assessment of our past actions as a species. Humans have been hunting for 2.5 million years, leading to the extinction of thousands of species and damaging ecosystems (Harari, 2014). This behaviour, coupled with constant encroachments on wildlife habitats, diversion of forest land and resources for cultivation and 'development' including relentless urbanization and industrial activities has further exacerbated wildlife extinction in increasingly destructive conflict scenarios. Illegal trade in biodiversity elements and wildlife articles has over-exploited natural resources and disturbed the food web in the wild. In addition, climate change is leading to increase in conflicts between humans and wildlife (Abrahms et al., 2023).

Repairing ecosystems requires curbing the over-exploitation that has become a prevalent practice and perhaps a fait accompli due to the overwhelming human population and its demands fuelling the exploitative and unsustainable economy to which humans have become accustomed. Experience shows that voluntary political and social change is difficult, and it is necessary to innovate institutions and processes that can support wildlife conservation and sustainable ecosystems.

The strategies and action points mentioned in this article have been used singularly or in combination to address human-wildlife conflicts, with the aim of establishing successful coexistence between human communities and wildlife. In many areas, the selection and application of many methods adapted to suit site-specific requirements have been attempted with the expectation that the endeavour would bring about a state of coexistence and ensure peace and balance between human development and wildlife conservation on a sustainable basis. However, the fact remains that the number of victims/damages on both sides is increasing over time, and the quality of human-animal relationships continues to deteriorate. The mindset of people who once viewed wildlife as part of the same ecosystem that supports their life has changed so much that in the event of a livestock death, they blame law enforcement agencies by saying "Your wildlife has killed my domestic animal". In such an atmosphere, the effectiveness as also the efficacy of existing instruments and strategies is not guaranteed.

Humans continue to act as environmental exploiters and are generally unwilling to shed the notion of dominance over all other species. The erroneous belief that all resources in the world are there for humans to exploit dominates their psyche. This mindset which is ego-centric rather that eco-centric needs to be completely changed and a new approach towards forests and wildlife conservation needs to be adopted.

The trend of relegating the protection of ecosystems and resources to human intervention using a variety of political and administrative formats has often added to conflict also because they are, in most cases, disconnected from indigenous knowledge systems. This approach is inherently alienating and presents several challenges. All constitutions and models of Government are designed primarily to prioritize the 'citizens' of the country and to improve human society using institutions, economies, and administrative systems. In India, Article 48A[9] introduced by the 42nd Amendment in 1976 incorporated 'Directive Principles of State Policy' into the Constitution. Although it says that the State shall endeavour to protect and improve the environment and also protect the forests and wildlife of the country, the provision is not justiciable. Currently, technical justifications and financial solutions are used to appease human victims, with little consideration for wildlife, which is forced into artificial behaviour, and suffer stressful situations due to location changes consequent to fragmentation and destruction of their habitats. Systems of control and redressal, operationalized through legal instruments, often become ineffective due to the prevailing mindsets of individuals or because of group-centred decisions of human stakeholders driven by narrow interests and values. A mechanism must be found that balances the interests of both wildlife and humans.

Despite significant resource investments in conflict mitigation, there is a need for reflection and a revision of the approach so that policies and practices are more responsive to the concerns and priorities of all stakeholders. The new model should be based on proactive engagement with civil society and urban people too so that the gravity of the otherwise simple looking isolated incidents of crop raiding by wildlife, carnivores straying into human habitation, cattle killing by a carnivore, killing of human being by wildlife, and poaching cases are understood by the policy makers also in terms of their true dynamics and long-term implications.

The examples of solutions and action points, as well as their levels of success, show that the goal of developing coexistence is itself dynamic, as the expected state of coexistence, if achieved, is subjective and possibly inherently in disequilibrium, requiring consistent and continual management. Competing perspectives and moral values are important to be considered, because in multiple-use areas, conflicts often arise due to diverging moral values and interests among stakeholders. A holistic perspective is therefore required which suitably recognizes the overarching viewpoint of the unique contribution of forests and wildlife to the survival and sustenance of humanity. This can only be achieved by bringing the issue of human-wildlife conflict to a platform that essentially offers the perspective of inclusivity creating a level field for the voiceless stakeholders while embracing a symmetrical view of rights of human and non-human species.

[9] The 42nd Amendment 1976 places an obligation on the State to protect the environment and wildlife.

Legal Recognitions and Loco Parentis

An emerging concept is that of in *Loco Parentis*. Originally, this term referred to "in place of the parent," which "refers to the doctrine describing the legal authority given to public or private institutions that take the place of parents" in terms of child discipline and care" (Levesque, 2011). Applied to environmental protection, this indicates that human stakeholders must protect the environment (which is like a dependent child) and that protecting the environment is a responsibility given to them. This idea of responsibility for a dependent nature has gradually gained importance in global and national environmental policies (Adams et al., 2023; Artelle et al., 2021; Best et al., 2021).

Legal instruments and policies are increasingly being applied to both human and non-human species, with a tendency in many countries to recognize rights in favour of ecosystems and their elements in environmental legislation. For example, while Ecuador assigned rights to Nature under its constitution, it granted Nature the right to exist, persist, and maintain and regenerate its vital cycles, structures, functions and processes. In 2010, Bolivia passed the "Law of the Rights of Mother Earth". This legislation addressed concerns about protection from commercial exploitation and environmental degradation. It shows an "eco-centric orientation", as it has established that in the event of conflicts of interest, "the protection of Mother Earth should prevail" (Munoz, 2023). New Zealand enacted in 2014 the Te Urewara Act which gave the Te Urewara National Park the status of a legal entity with "all the rights, powers, duties and liabilities of a legal person". In March 2017, Whanganui River in New Zealand was recognized as a legal person.

In India, the Supreme Court of India in *Animal Welfare Board of India* v. *A. Nagaraja & Ors* in 2014 recognized that animal freedoms included the freedom from hunger, fear and distress, physical and thermal discomfort, freedom from pain and injury, and freedom to express normal behavioural patterns.

The High Court of Uttarakhand hearing the Public Interest Litigation no. 43 of 2014 stated in their March 2017 judgement "The entire animal kingdom, including avian and aquatic ones, are declared as legal entities having a distinct persona with corresponding rights, duties, and liabilities of a living person. All the citizens throughout the state of Uttarakhand are hereby declared persons in loco parentis as the human face for the welfare/protection of animals". It declared the Ganga and Yamuna rivers as legal persons and ordered that the Director of *Namami Gange* Programme,[10] the Chief Secretary of Uttarakhand, and the Advocate General of the State of Uttarakhand will be persons in *loco parentis*, i.e. as the 'human face' to protect, conserve and preserve these rivers and their tributaries. The court further observed "Rivers Ganga and Yamuna are central to the existence of half of Indian population, their health, and well-being. The rivers have provided both physical and spiritual sustenance to all of us from time immemorial. Rivers Ganga and Yamuna

[10] An Integrated Conservation Mission, approved as 'Flagship Programme' by the Union Government of India in June 2014 to accomplish the twin objectives of effective abatement of pollution, conservation, and rejuvenation of River Ganga.

have spiritual and physical sustenance. They support and assist both the life and natural resources and health and well-being of the entire community. Rivers Ganga and Yamuna are breathing, living, and sustaining the communities from mountains to sea". The verdict obviously reflects the holistic view inherent in Indian culture and an "inclusive approach" that extends to natural resources that have no voice. However, the Supreme Court of India stayed the order in July 2017 following a request from the Government of India which cited administrative problems in implementing the order as the rivers flows through several states. The underlying theme behind the High Court judgement can be furthered and complemented by the continuation of multiple decision-making bodies and institutions such as the National Green Tribunal (Kelly & Mehta, 2019).

A Case for Adoption of in **Loco Parentis**

The legal recognitions open the scope for adopting the principle of *in loco parentis* in shaping a comprehensive approach to managing human-wildlife conflict. The framework is shaped through judicial interventions, strengthened by Indian cultural ethos, and provides for considering the rights of non-human species in conflict mitigation efforts. It bestows an essential responsibility on humanity to act as guardians of wildlife and ecosystems. Implementing this approach in managing human-wildlife conflicts encourages the adoption of a holistic view of integrating the contribution of forests and wildlife to the survival and sustenance of humanity, transcending local issues, interests, and even political leanings.

The adoption of in *loco parentis* provides a voice to the voiceless and encourages humanity to return to the fundamental philosophy of Indian culture that views humans as just one species among many. This approach also promotes a balanced and peaceful coexistence with nature by minimizing exploitation and conflict, and restoring harmony. It reflects the sentiments expressed by the '*Shanti Mantra*' (Chant for Peace) contained in the *Yajurveda*[11] which says, "May there be peace in sky and ethereal space; on earth; in water; in all herbs, trees, and creepers; in the whole universe; in supreme being; in all beings and always". Peace fundamentally signifies a balance in Nature which underscores the absence of conflict.

Declaring wildlife, its habitat, or a natural resource a legal entity and appointing a guardian can create a more symmetrical view of human and non-human rights and justice afforded to natural systems. Furthermore, in a country like India, the public trust doctrine can support this approach by prioritizing long-term public rights over short-term private gain. The precautionary principle, which makes it incumbent on government to anticipate, prevent, and address environmental degradation, can also provide strength to the operationalization of the concept of in *loco parentis*. The

[11] Yajurveda is one of the four canonical texts of Hinduism, the Vedas. It is basically Karma Veda, providing details of the application of knowledge in practical living at the individual as well as collective level.

perspective of inclusivity embedded in the approach can support the principles of circular economy while attaching appropriate value to the natural capital components and even contribute to addressing climate change issues.

The *loco parentis* institution, acting on the directions of the judiciary, must ensure that the orders are carried out through the administrative and enforcement mechanisms and provide feedback on the compliances to the court or it's nominated body, which is in principle responsible for monitoring. The site-specific solutions can be developed to address human-wildlife conflicts at the ground level. With transparency and accountability assured, a responsive field formation can effectively address the issues of human-wildlife conflict.

Initially, developing a functional interface of this institution with the existing administration may pose some challenges, especially in politically sensitive areas and difficult cases. However, given the visible and historically experienced consequences of the dominant anthropocentric paradigm, it will be prudent to move to a more balanced approach. Granting independence to the guardian and developing a shared mindset among resource-dependent stakeholders and the guardian will essentially create better conditions for reduction of conflicts and transition towards coexistence and a just world order.

Conclusion

Throughout history, the increasing human dominance and exclusivity of humans has led to a shift from the coexistence of multiple species to the domination of humans over other species. The destructive consequences of the current anthropocentric paradigm require urgent consideration and a change in approach to addressing human-wildlife conflicts and related problems. This chapter has argued in favour of a paradigm that advocates the institutionalization of the *in loco parentis* approach while offering an alternative perspective by recognizing the rights of non-human species and taking a more symmetrical view of human and non-human interests. This paradigm shift, supported by field-level mitigation measures, can promote coexistence and minimize conflicts. It can also lead to fundamental changes in our relationship with Nature while recognizing that we are part of a larger ecosystem and have a responsibility to protect and preserve it. The global introduction of *in loco parentis* will be a significant milestone in human history and promote harmonious and sustainable coexistence of all species on earth.

References

Abrahms, B., Carter, N. H., Clark-Wolf, T. J., et al. (2023). Climate change as a global amplifier of human–wildlife conflict. *Nature Climate Change, 13*, 224–234. https://doi.org/10.1038/s41558-023-01608-5

Adams, M. S., Tulloch, V. J. D., Hemphill, J., Penn, B., Anderson, L. T., Davis, K., Avery-Gomm, S., Harris, A., & Martin, T. G. (2023). Inclusive approaches for cumulative effects assessments. *People and Nature, 5*(2), 431–445. https://doi.org/10.1002/pan3.10447

Alemayehu, A., & Temesgen, M. (2015). Approaches to human-wildlife conflict management in and around Chebera-Churchura National Park, Southern Ethiopia. *Asian Journal of Conservation Biology, 4*(2), 136–142.

Anon. (2006). *The scheduled tribes and other traditional forest dwellers (recognition of forest rights) act, 2006*. Government of India.

Artelle, K. A., Adams, M. S., Bryan, H. M., Darimont, C. T., Housty, J., Housty, W. G., Moody, J. E., Moody, M. F., Neasloss, D., & Walkus, J. (2021). Decolonial model of environmental management and conservation: Insights from indigenous-led grizzly bear stewardship in the great bear rainforest. *Ethics, Policy & Environment, 24*(3), 283–323. https://doi.org/10.1080/21550085.2021.2002624

Badri, C. (2020). *Maharashtra saw highest leopard deaths in 5 years in 2020*. Hindustan Times.

Barnes, J. I., & De Jager, J. (1996). Economic and financial incentives for wildlife use on private land in Namibia and the implications for policy. *South African Journal of Wildlife Research, 26*, 37–46.

Ben-Dor, M., Barkai, R., & Sirtoli, R. (2021). The evolution of the human trophic level during the Pleistocene. *American Journal of Physical Anthropology*. https://doi.org/10.1002/ajpa.24247

Bertuol-Garcia, D., Morsello, C., El-Hani, C., & Pardini, R. (2018). A conceptual framework for understanding the perspectives on the causes of the science-practice gap in ecology and conservation. *Biological Reviews of the Cambridge Philosophical Society, 93*(2), 1032–1055. https://doi.org/10.1111/brv.12385

Best, L., Fung-Loy, K., Ilahibaks, N., Ramirez-Gomez, S. O. I., & Speelman, E. N. (2021). Toward inclusive landscape governance in contested landscapes: Exploring the contribution of participatory tools in the upper Suriname River basin. *Environmental Management, 68*(5), 683–700. https://doi.org/10.1007/s00267-021-01504-8

Bombieri, G., Penteriani, V., Almasieh, K., Ambarlı, H., Ashrafzadeh, M. R., et al. (2023). A worldwide perspective on large carnivore attacks on humans. *PLoS Biology, 21*(1), e3001946. https://doi.org/10.1371/journal.pbio.3001946

Bowden, G. (2017). An environmental sociology for the anthropocene. *Canadian Review of Sociology, 54*(1), 48–68. https://doi.org/10.1111/cars.12138

Campbell, M. (2018). *Geomatics and conservation biology*. Nova Science Publishers.

Campbell, M. (2020). *The physical geography of El Salvador: A geophysical and ecological approach*. Nova Science Publishers.

Campbell, M., & Jha, K. K. (2021). *Critical research techniques in animal and habitat ecology: International examples*. Nova Science Publishers.

Chaitanya, S. V., & Krishna, A. (2020). *7562 cases of crop-raiding by wild animals reported in last three years across Tamil Nadu*. The New Indian Express.

Chakrabarty, K. (2021). The cult of Bonbibi: A folk paradigm in deltaic Sundarbans. *Journal of History, Art and Archaeology, 1*, 49–61.

Conover, M. R., & Decker, D. J. (1991). Wildlife damage to crops: Perceptions of agricultural and wildlife professionals in 1957 and 1987. *Wildlife Society Bulletin, 19*, 46–52.

Cox, D. T. C., & Gaston, K. J. (2018). Human-nature interactions and the consequences and drivers of provisioning wildlife. *Philosophical Transactions of the Royal Society of London. Series B, Biological Sciences, 373*(1745), 20170092. https://doi.org/10.1098/rstb.2017.0092

Curtis, P. G., Slay, C. M., Harris, N. L., Tyukavina, A., & Hansen, M. C. (2018). Classifying drivers of global forest loss. *Science, 361*(6407), 1108–1111. https://doi.org/10.1126/science.aau3445

Damodaran, V. (2015). Environmental history. In J. D. Wright (Ed.), *International encyclopedia of the social & behavioral sciences* (2nd ed., pp. 747–755). Elsevier. https://doi.org/10.1016/B978-0-08-097086-8.62112-8

Das, S. (2023). *551 tigers died in 5 years, including 114 due to poaching: Govt. tells parliament*. Hindustan Times.

Dasgupta, P., & Simon, L. (2023). Economic factors underlying biodiversity loss. *Philosophical Transactions of the Royal Society B, 378*, 20220197.

Durant, S. M., Groom, R., Kuloba, B., Samna, A., Muzuma, U., Gadimang, P., Mandisodza-Chikerema, R., Ipavec, A., Mitchell, N., Ikanda, D., & Msuha, M. (2019). Bridging the divide between scientists and decision-makers: How behavioural ecologists can increase the conservation impact of their research? *Philosophical Transactions of the Royal Society of London. Series B, Biological Sciences, 374*(1781), 20190011. https://doi.org/10.1098/rstb.2019.0011

Edelblutte, É., Krithivasan, R., & Hayek, M. N. (2023). Animal agency in wildlife conservation and management. *Conservation Biology, 37*(1), 13853. https://doi.org/10.1111/cobi.13853

Eva, M. G., Jayasinghe, N., Polet, G., Wadhwa, R., & Hilderink-Koopmans, F. (2021). *A future for all: The need for human-wildlife coexistence.* WWF.

Fedorca, A., Fedorca, M., Ionescu, O., Jurj, R., Ionescu, G., & Popa, M. (2021). Sustainable landscape planning to mitigate wildlife–vehicle collisions. *Land, 10*, 737.

Frank, B. (2016). Human wildlife conflicts and the need to include tolerance and coexistence: An introductory comment. *Society and Natural Resources, 29*, 738–743. https://doi.org/10.1080/08941920.2015.1103388

Gautham, K. (2023). *Activists seek more time for feedback on forest act changes.* Times of India, TNN.

Ghai, R., Himanshu, N., & Deol, T. (2023). *Conflict of interest. Urban menace.* CSE India.

Harari, Y. N. (2014). *Sapiens: A brief history of humankind.* Harvill Secker.

Hill, C. M. (2021). Conflict is integral to human wildlife coexistence. *Frontiers of Conservation Science, 2*, 734314. https://doi.org/10.3389/fcosc.2021.734314

Hintzen, R. E., Papadopoulou, M., Mounce, R., Banks-Leite, C., Holt, R. D., Mills, M., Knight, T., Leroi, A. M., & Rosindell, J. (2020). Relationship between conservation biology and ecology shown through machine reading of 32,000 articles. *Conservation Biology, 34*(3), 721–732. https://doi.org/10.1111/cobi.13435

Hoang, N. T., & Kanemoto, K. (2021). Mapping the deforestation footprint of nations reveals growing threat to tropical forests. *Nature Ecology & Evolution, 5*, 845–853. https://doi.org/10.1038/s41559-021-01417-z

Hongbo, Y., Lupi, F., Zhang, J., & Liu, J. (2020). Hidden cost of conservation: A demonstration using losses from human-wildlife conflicts under a payments for ecosystem services program. *Ecological Economics, 169*, 106462.

Jha, A. K. (2021). *From hunters to ecological refugees-the Pardhi tribe of India.* BUUKS.

Jha, A. K. (2022). How fair-long arms of the law. In M. K. Misra (Ed.), *Wildlife India @ 50: Saving the wild, securing the future.* Rupa Publications.

Jha, K. K., & Campbell, M. (2021). *Critical research techniques in animal and habitat ecology: Examples from India.* Nova Science Publishers.

Johnsingh, A. J. T., & Negi, A. S. (2003). Status of tiger and leopard in Rajaji–Corbett Conservation Unit, northern India. *Biological Conservation, 111*, 385–393.

Kala, C. P., & Kothari, K. K. (2013). Livestock predation by common leopard in Binsar wildlife sanctuary, India: Human wildlife conflicts and conservation issues. *Human Wildlife Interactions, 7*(2), 325–333.

Karanth, K. K., & Anubhav, V. (2020). Wild seve: A novel conservation intervention to monitor and address human-wildlife conflict. *Frontiers in Ecology and Evolution, 8*, 198.

Kelly, D. A., & Mehta, T. (2019). The experiments with right of nature in India. In C. La Follette (Ed.), *Sustainability and the rights of nature in practice.* CRC Press.

Kinnebrew, E., Ochoa-Brito, J. I., French, M., Mills-Novoa, M., Shoffner, E., & Siegel, K. (2022). Biases and limitations of global forest change and author-generated land cover maps in detecting deforestation in the Amazon. *PLoS One, 17*(7), e0268970. https://doi.org/10.1371/journal.pone.0268970

König, H. J., Kiffner, C., Kramer-Schadt, S., Fürst, C., Keuling, O., & Ford, A. T. (2020). Human–wildlife coexistence in a changing world. Special section: Challenges of and solutions to human–wildlife conflicts in agricultural landscapes. *Conservation Biology, 34*(4), 786–794.

Kumar, A., Bargali, H. S., David, A., & Edgaonkar, A. (2017). Patterns of crop raiding by wild ungulates and elephants in Ramnagar Forest division, Uttarakhand. *Human-Wildlife Interactions, 11*(1), 8. https://doi.org/10.26077/j9j8-8e74

Levesque, R. J. R. (2011). In loco parentis. In R. J. R. Levesque (Ed.), *Encyclopedia of adolescence*. Springer. https://doi.org/10.1007/978-1-4419-1695-2_680

Lischka, S. A., Teel, T. L., Johnson, H. E., Reed, S. E., Breck, S., Don Carlos, A., & Crooks, K. R. (2018). A conceptual model for the integration of social and ecological information to understand human wildlife interactions. *Biological Conservation, 225*, 80–87.

Lodh, S. (2020). *Portrayal of 'hunting' in environmental history of India*. Academia.

Luetkemeier, R., Kraus, R., Mbidzo, M., Hauptfleisch, M., Liehr, S., & Blaum, N. (2023). A qualitative exploration of conflicts in human wildlife interactions in Namibia's Kunene region. *Diversity, 15*(3), 440. https://doi.org/10.3390/d15030440

Marques, A. (2021). Distant drivers of deforestation. *Nature Ecology & Evolution, 5*(6), 713–714. https://doi.org/10.1038/s41559-021-01420-4

Mehring, M., Bernard, B., Hummel, D., Liehr, S., & Lux, A. (2017). Halting biodiversity loss: How social–ecological biodiversity research makes a difference. *International Journal of Biodiversity Science, Ecosystem Services & Management, 13*(1), 172–180. https://doi.org/10.1080/21513732.2017.1289246

Milks, A. (2020). *Did prehistoric women hunt? New research suggests so*. The Conversation, November 4, 2020.

Munoz, L. (2023). *Bolivia's mother earth laws: Is the ecocentric legislation misleading?* Revista, Harvard Review of Latin America.

Mwamidi, D. M., Mwasi, S. M., & Nunow, A. A. (2012). The use of indigenous knowledge in minimizing human wildlife conflict: The case of Taita community, Kenya. *International Journal of Current Research, 4*(2), 026–030.

Nair, R., Athreya, V., John, D. C. L., Andheria, A., Surve, N., & Omkar, P. D. (2021). *Frontiers in conservation science*. Human Wildlife Interactions.

Nasi, R., Brown, D., Wilkie, D., Bennett, E., Tutin, C., van Tol, G., & Christophersen, T. (2008). *Conservation and use of wildlife-based resources: The bushmeat crisis*. Secretariat of the Convention on Biological Diversity.

Nishank. (2022). *Infrastructure projects in and around protected areas need to be held financially accountable*. The Bastion Development in Depth. September 8, 2022.

Nuntikorn, K., & Takeuchi, W. (2022). Human-elephant conflict risk assessment under coupled climatic and anthropogenic changes in Thailand. *Science of the Total Environment*. https://doi.org/10.1016/j.scitotenv.2022.155174

Odonne, G., Tareau, M. A., & van Andel, T. (2021). Geopolitics of bitterness: Deciphering the history and cultural biogeography of Quassia Amara L. *Journal of Ethnopharmacology, 267*, 113546. https://doi.org/10.1016/j.jep.2020.113546

Olivia, A. (2020). Humans aren't apex predators.

Pande, S. (2022). *Flora fauna and nature conservation values in Indian culture*. Ila Foundation.

Pande, S., & Satish, P. (2015). *Flora, fauna and nature in Buddhist thought*. Ila Foundation and Social Forestry Directorate, Government of Maharashtra.

Pande, S., & Satish, P. (2016). *Fauna and Deities: A perspective on nature conservation*. Social Forestry Directorate, Government of Maharashtra.

Panwar, H. S. (1979). Population dynamics and land tenures of tigers in Kanha National Park. *Indian Forester, 105*, 35–47.

Pendrill, F., & Persson, U. M. (2017). Combining global land cover datasets to quantify agricultural expansion into forests in Latin America: Limitations and challenges. *PLoS One, 12*(7), e0181202. https://doi.org/10.1371/journal.pone.0181202

Peterson, D. (2014). *Hunting as humanizer then and now*. Humans Nature, Centre for Humans and Nature.

Pinjarkar, V. (2014). In 4 years, tigers lost 12,000 sq. km territory. Times of India. January 2, 2014.

Rana, A. K., & Kumar, N. (2023). Current wildlife crime (Indian scenario): Major challenges and prevention approaches. *Biodiversity and Conservation, 32*(5), 1473–1491. https://doi.org/10.1007/s10531-023-02577-z

Rangarajan, M. (1998). The raj and the natural world: The war against "dangerous beasts" in colonial India. *Studies in History, 14*(1998), 265–299.

Reddy, U. S. (2021). *New policy will trigger more encroachments*. Times of India.

Ripley, S. D. (1964). *The land and wildlife of tropical Asia*. TIME Incorporated.

Rishi, V. (2005). Mitigating man–wildlife conflict identifying aberrant predators: A case study of the man-eaters of Sonaripur. *Indian Forester, 131*, 1255–1266.

Roy, K. (2022). *1,700 Indians killed by tigers, elephants in last 3 years: Environment ministry*. Deccan Herald News Service.

Saccò, M., Blyth, A., Bateman, P. W., Hua, Q., Mazumder, D., White, N., Humphreys, W. F., Laini, A., Griebler, C., & Grice, K. (2019). New light in the dark - A proposed multidisciplinary framework for studying functional ecology of groundwater fauna. *The Science of the Total Environment, 662*, 963–977. https://doi.org/10.1016/j.scitotenv.2019.01.296

Sahana, G. (2023). Helping wildlife navigate road and railway infrastructure. Mongabay, January 3, 2023.

Salerno, J., Bailey, K., Gaughan, A. E., Stevens, F. R., Hilton, T., Cassidy, L., Drake, M. D., Pricope, N. G., & Hartter, J. (2020). Wildlife impacts and vulnerable livelihoods in a transfrontier conservation landscape. *Conservation Biology, 34*(4), 891–902. https://doi.org/10.1111/cobi.13480

Sharma, J., Upgupta, S., Jayaraman, M., Chaturvedi, R. K., Bala, G., & Ravindranath, N. H. (2017). Vulnerability of forests in India: A national scale assessment. *Environmental Management, 60*(3), 544–553. https://doi.org/10.1007/s00267-017-0894-4

Shaurabh, A., & Radhakrishna, S. (2017). Investigating trends in human-wildlife conflict: Is conflict escalation real or imagined? *Journal of Asia-Pacific Biodiversity, 10*(2), 154–161.

Siegel, K., Farah Perez, A., Kinnebrew, E., Mills-Novoa, M., Ochoa, J., & Shoffner, E. (2022). Integration of qualitative and quantitative methods for land-use-change modeling in a deforestation frontier. *Conservation Biology, 36*(6), 13924. https://doi.org/10.1111/cobi.13924

Singh, N., Sonone, S., & Dharaiya, N. (2008). Sloth bear attacks on humans in Central India: Implications for species conservation. *Human-Wildlife Interactions, 12*(3), 5. https://doi.org/10.26077/2mgq-fs29

Sinha, B. K. P., & Jha, A. K. (2023). *Human-tiger clash: Beyond the stripes*. The Pioneer.

Sinha, B. K. P., & Jha, A. K. (2024). *Saving the great Indian bustard*. The Pioneer.

Smyntyna, O. V. (2003). The environmental approach to prehistoric studies: Concepts and theories. *History and Theory, 42*(4), 44–59. https://doi.org/10.1046/j.1468-2303.2003.00256.x

Stan, K., Sanchez-Azofeifa, A., Espírito-Santo, M., & Portillo-Quintero, C. (2015). Simulating deforestation in Minas Gerais, Brazil, under changing government policies and socioeconomic conditions. *PLoS One, 10*(9), e0137911. https://doi.org/10.1371/journal.pone.0137911

Tampakis, S., Andrea, V., Panagopoulos, T., Karanikola, P., Gkarmiri, R., & Georgoula, T. (2023). Managing the conflict of human–wildlife coexistence: A community-based approach. *Land, 12*(4), 832. https://doi.org/10.3390/land12040832

Taubert, F., Fischer, R., Groeneveld, J., Lehmann, S., Müller, M. S., Rödig, E., Wiegand, T., & Huth, A. (2018). Global patterns of tropical forest fragmentation. *Nature, 554*(7693), 519–522. https://doi.org/10.1038/nature25508

The Economic Times News, India. (2023). Why are man-animal battles rising?

Turvey, S., & Crees, J. (2019). Extinction in the anthropocene. *Current Biology, 29*(19), R982–R986. https://doi.org/10.1016/j.cub.2019.07.040

Uddin, M., Dutta, S., Kolipakam, V., Sharma, H., Usmani, F., & Jhala, Y. (2021). High bird mortality due to power lines invokes urgent environmental mitigation in a tropical desert. *Biological Conservation, 261*, 109262. https://doi.org/10.1016/j.biocon.2021.109262

Velmurugan, D. M. (2017). Historical development of wildlife protection in India. *International Journal of Current Research and Modern Education, 2*(2), 386–390.

Verghese, A., & Mendonce, V. (2016). Decline of bird diversity in an urban layout in residential Bengaluru, India—A human-bird conflict (1974–2014). In *Human animal conflict in agro-pastoral context: Issues & policies* (pp. 41–43). ICAR.

VIDHI Centre for Legal Policy, New Delhi. (2023). Report of the high level working group on the forest (conservation) amendment bill 2023.

Yap, H. T. (2019). One-ecosystem analysis for environmental conservation and sustainable livelihood. *F1000Research, 8*, 328. https://doi.org/10.12688/f1000research.13999.1

Chapter 6
Ecosystem Management and Trophy Hunting

Shahbaz Ahmad

Abstract Only during the last few decades has the concept of ecosystem services become important. The value of the ecosystem services of the entire biosphere is possibly greater than the global Gross-Domestic Product (GDP). This realization has led to a shift from unplanned exploitation to conservation and sustainable utilization of natural resources. Wild animals also enjoy protection through various laws and conventions. Tourist trophy hunting began in Kenya in the twentieth century and later spread throughout Africa. There is no evidence to suggest that it is essential as a management strategy or that it reduces human-wildlife conflict. With increasing awareness, public opinion is generally against trophy hunting. But since trophy hunting is a hobby of the rich, many justifications are put forward by those involved in this industry. However, these justifications are not supported by facts and scientific studies. The governments of various countries need to develop an exit policy to put an end to this practice of trophy hunting which is based on a colonial mindset. For the well-being of humanity, existing forests and natural ecosystems must be managed with a global vision to combat climate change and ensure the continuity of ecosystem services. An ecosystem manager has a multifaceted task—preserving and maintaining ecosystem services, involving people in the sustainable use of usufruct rights, managing human-wildlife conflict through an integrated strategy, creating awareness of nature through a non-lethal ecotourism program, running programs and activities for generating livelihood and other benefits to local population, creating a sense of serenity in the wild animals, etc. Trophy hunting has no place in a responsible ecosystem management.

Keywords Ecosystem · Ecosystem management · Ecosystem services · Ecotourism · Human-wildlife conflict · Trophy hunting · Wild animals

---

Shahbaz Ahmad is a retired Indian Forest Service Officer who has worked in Madhya Pradesh in various positions and has also been Principal Chief Conservator of Forests and Chief Wildlife Warden of the State. He has been concerned about the welfare of the people and the sustainable management of the ecosystems and natural resources. He has written several books on Wildlife and Socio-economic issues. He may be contacted through email at shahbaz_in@rediffmail.com.

S. Ahmad (✉)
Indian Institute of Forest Management, Bhopal, India

K. K. Jha and M. O. Campbell (eds.), *Intersections of Conservation Biogeography and Wildlife Management*, https://doi.org/10.1007/978-981-95-0121-2_6

Ecosystem Services and Objectives of Management

For a long time, forests and natural ecosystems were not considered economically beneficial because they did not generate sufficient revenue. Only in the last few decades has the concept of ecosystem services become more important[1]. Considering 17 ecosystem services for 16 biomes (Box 6.1), Costanza et al. (1997) estimated that in 2007, for the entire biosphere the value of ecosystem services was US$33 trillion while the Global Domestic Product was US$18 trillion. Again in 2011, the estimate for the total global ecosystem services was US$125 trillion while the Global Domestic Product was US$75 trillion (Costanza et al., 2014). It may also be noted that in 2022, the world's GDP was only US$95 trillion (https://ceoworld.biz/2022/03/31/economy-rankings-largest-countries-by-gdp-2022/), much less than the global value of ecosystem services. Although estimates for ecosystem services have been created to simplify decision-making, the fact is that natural ecosystems are irreplaceable, i.e., their services cannot be produced by money. Therefore, in the interest of human existence and human well-being, it is incumbent upon governments to protect and conserve the forests and natural ecosystems and to manage them in a sustainable manner.

Box 6.1 What Are Biomes?

Earth's biosphere consists of five main types of biomes: aquatic, grassland, forest, desert, and tundra. These biomes can be further divided according to location, elevation, or other definitions. Biomes are generally defined by the types of vegetation, soil, climate, and wildlife within them. Note that a biome is different from an ecosystem. A biome is a specific geographic area notable for the species living there, while an ecosystem is defined as the interaction of living and non-living things within an environment. Each biome can be home to several different ecosystems.

(*Source* University Corporation for Atmospheric Research https://scied.ucar.edu/learning-zone/earth-system/earths-changing-biomes)

[1] Ecosystem Services are outputs, conditions, or processes of natural systems that directly or indirectly benefit humans or enhance social welfare. Ecosystem services can benefit people in many ways, either directly or as inputs into the production of other goods and services. For example, the pollination of crops provided by bees and other organisms contributes to food production and is thus considered an ecosystem service. Another example is the attenuation of flooding in residential areas provided by riparian buffers and wetlands. (https://www.britannica.com/science/ecosystem-services).

Keeping in view the management objectives, the IUCN[2] has broadly classified the Protected Areas of the world into six categories, i.e. (1) Strict protection e.g., natural reserve and wilderness area, (2) Ecosystem Conservation and Protection, (3) Conservation of Natural Features, (4) Conservation through Active Management, i.e., Habitat/Species Management Area, (5) Landscape/Seascape Conservation and Recreation, and (6) Sustainable Use of Land Resource (IUCN, 2004). The objectives of the management of ecosystems are also incorporated into the UN Sustainable Development Goals (SDGs) (Box 6.2). The SDG 14 is "Life below Water," i.e., to conserve and sustainably use the oceans, sea, and marine resources for sustainable development. SDG 15 is "Life on Land," i.e., to protect, restore and promote sustainable use of terrestrial ecosystems, sustainably manage forests, combat desertification, halt and reverse land degradation, and halt biodiversity loss (Bertelsmann, Stiftung, & SDSN, 2018). Whatever be the objectives of management, the bottom line is that natural ecosystems are essential for the welfare of mankind and their benefits cannot be discounted by purely economic considerations. Forests account for a little over a third (38%) of the world's habitable land area. This is about a quarter (26%) of total land area (*ourworldindata.org/forest-area*). Interventions in forest and natural ecosystems should be carried out with the aim of judicious and sustainable management[3]. This concern has been an integral part of the forest policy of India and was aptly expressed more than a century ago by the then Inspector General of Forests Sainthill Eardley-Wilmot in his book *Forest Life and Sports in India*—"The object of State forestry in India is therefore to supply the requirements of its population in forest products, to protect the water supply of the country and to afford help to its industrial development. As in the case of European countries, the forest management should, as it always has, result in profit, but the profit should be subsidiary to the main object in view; it should be a consequence of, and not a reason for, a forest policy" (Eardley-Wilmot, 1910 p. 317).

[2] The International Union for Conservation of Nature (IUCN) is a membership Union uniquely composed of both government and civil society organizations. By harnessing the experience, resources and reach of its more than 1400 Member organizations and the input of some 15,000 experts, IUCN is the global authority on the status of the natural world and the measures needed to safeguard it.

[3] Sustainable management is a resource management technique that seeks to make any harvesting or consumption of natural resources as sustainable as possible. Thus, the main goal is to replenish any resources as fast as they are depleted. While this goal may not be practical, sustainable management can often help prolong the natural resource for as long as possible, such as with fossil fuels. It may be easier to sustain resources that are considered renewable, such as forests and fisheries (https://www.smartcapitalmind.com/what-is-sustainable-management.htm).

Box 6.2 Sustainable Development Goals
In September 2015, the UN General Assembly adopted the 2030 Agenda for Sustainable Development that includes 17 Sustainable Development Goals (SDGs). Building on the principle of "leaving no one behind," the new Agenda emphasizes a holistic approach to achieving sustainable development for all.

Ecotourism[4] is an integral part of ecosystem management, particularly wildlife management. Ecotourism or nature-based tourism provides the public with access to natural and wilderness areas with the aim of creating awareness about nature and conservation (Box 6.3). The global ecotourism market size was valued US$185–187 billion in 2021 and is expected to grow at the annual rate of 15.2% from 2022 to 2030. The African continent is the second fastest growing tourism region in the world after the Asia-Pacific region, and tourism accounts for approximately 8.5% of the continent's total GDP and roughly 24 million jobs (Sebunya, 2020). Trophy hunting is also projected as an ecotourism activity.

Box 6.3 Wildlife Tourism
Wildlife tourism refers to the observation and interaction with local animal and plant life in their natural habitats. It encompasses segments such as eco-tourism, safari tours and mountain tourism among others. Wildlife watching tourism occurs mainly in protected areas. Nature, national parks, and wildlife are considered the most important tourism assets for tourists travelling for instance to Africa.

According to the UNWTO Report '*Towards Measuring the Economic Value of Wildlife Watching Tourism in Africa*', 7% of world tourism relates to wildlife tourism, a segment growing annually at about 3%. The same document shows that a total of 14 countries in Africa are generating an estimated US$ 142 million in entrance fees for protected areas. A WWF report shows that 93% of all natural heritage sites support recreation and tourism and 91% of them provide jobs. For instance, in Belize, more

[4] Ecotourism is defined as "responsible travel to natural areas that conserves the environment, sustains the well-being of the local people, and involves interpretation and education" (International Ecotourism Society, 2015 https://ecotourism.org/what-is-ecotourism/).

than 50% of the population are said to be supported by income generated through reef-related tourism and fisheries.

Wildlife represents biodiversity, essential for our health and the well-being of the whole planet. We live in an interconnected ecological system, where each macro and microorganism, whether animal, plant or fish affects the other. Alteration of the natural habitat of any organism will trigger a dynamo effect, so non-equilibrium in the ecological system as a whole endangers the life cycle of many species. Around 40,000 species of animals, fungi, and plants benefit humans. More than the third of our pharmaceuticals originate from wild plants.

(https://www.unwto.org/asia/unwto-chimelong-why-wildlife)

Pros and Cons of Trophy Hunting

Trophy hunting is broadly defined as the killing of animals for recreation with the purpose of collecting trophies such as horns, antlers, skulls, tusks, or teeth for display (Sheikh, 2019). Trophy hunting is generally recognized as a key economic contributor to generating revenue for nature conservation and the well-being of the local communities. It is also known as "conservation hunting." Hunting and fences are also said to be the only means of keeping animals away from human habitats and reducing losses and difficulties (Pabla, 2020). Despite claims by hunters that shooting an animal is more humane than death by starvation, many wild animals are left wounded by unskilled hunters, and many others are said to have been killed accidentally. Untended traps are a possible source of injury to humans, as is the possibility of bites from trapped animals. An important reason for the overpopulation of certain animals is the destruction of their natural predators by humans. This upsets a delicate ecological balance in nature (UIA, 2021).

Safari Club International (SCI) is an international trophy hunting organization founded in 1970 and headquartered in Washington D.C. with the aim of protecting the freedom of hunting and to promoting wildlife conservation worldwide. A study was conducted by Southwick Associates on behalf of SCI for the period 2012–2014 covering eight African countries: Botswana, Ethiopia, Mozambique, Namibia, South Africa, Tanzania, Zambia, and Zimbabwe. It found that in the eight countries studied, US$ 326.5 million was spent by hunters annually and hunting supported 53,400 jobs. The United States accounts for the largest proportion of visiting hunters (74%) followed by Europe (16%) (Southwick Associates, 2015). The findings of this report were later challenged by another report from a study commissioned by Humane Society International. This study mentions that the economic benefits have been greatly overstated, and that the job creation is not 53,000 jobs but 7500–15,500 jobs. While the total contribution of tourism in the eight study centers ranges between 2.8% and 5.5% of GDP, the total economic contribution of trophy hunters is at most about 0.03% of GDP. In the countries studied, revenue from trophy hunting accounted for only 1.8% of the tourism revenue. Employment in trophy hunting represents only

Fig. 6.1 Animal trophies at the entrance of a taxidermy studio in Pretoria (https://www.cbc.ca/news/science/humans-are-superpredators-like-no-other-species-1.3197697)

0.76% or less of average direct tourism employment in the study countries (Murray, 2017). The fact is that the economic benefits of big game hunting are often overstated and dwarfed by the economic opportunities of environmentally friendly tourism. Studies have indicated that a live elephant can be worth up to US$1.6 million over its lifetime in photo tourism revenue. The fee a hunter pays to shoot a bull is typically US $40,000–50,000 (Kluck, 2020).

Trophy hunting originated, in part, during the colonial settlement in Africa (Sheikh, 2019). The establishment of the Dutch East India Company in 1652 attracted foreign hunters to Africa. With support from the British authorities, the number increased in the nineteenth century. Tourist trophy hunting began in Kenya in the twentieth century and later spread throughout Africa (Sheikh, 2019). Currently, Africa is the most popular destination for the trophy hunting of rare and threatened species (Fig. 6.1). South Africa and Namibia export the most CITES-listed mammal trophies.[5] Trophy hunting is also plagued by the problem of "canned hunting" by commercial operations that breed lions to be shot in captivity. Most lion hunts in Africa are "canned," leaving no means for the animal to escape from a fenced-in enclosure (Fig. 6.2). It goes without saying that game farms are not conservation centers (http://irafischer.com/trophy-hunting). Between 2014 and 2018 almost 125,000 trophies of CITES-protected species (Box 6.4) were imported globally, with the US and EU featuring as the biggest importers. US citizens import more wildlife trophies than citizens of any other country—over 65,000 trophies in 2017 alone.

[5] CITES (the Convention on International Trade in Endangered Species of Wild Fauna and Flora) is an international agreement between governments. Its aim is to ensure that international trade in specimens of wild animals and plants does not threaten the survival of the species. CITES was drafted as a result of a resolution adopted in 1963 at a meeting of members of IUCN (The World Conservation Union). The text of the Convention was finally agreed at a meeting of representatives of 80 countries in Washington, D.C., United States, on 3 March 1973, and on 1 July 1975 CITES entered in force.

Fig. 6.2 Depiction of the cruelty of canned hunting (https://mfoa.net/wildlife/join-our-campaign-end-canned-hunting-maine)

Box 6.4 The Cites Appendices

Over 38,700 species—including roughly 5950 species of animals and 32,800 species of plants—are protected by CITES against over-exploitation through international trade. They are listed in the three CITES Appendices. Appendix I lists species that are the most endangered among CITES-listed animals and plants. They are threatened with extinction and CITES prohibits international trade in specimens of these species except when the purpose of the import is not commercial, for instance for scientific research. Appendix II lists species that are not necessarily now threatened with extinction but that may become so unless trade is closely controlled. It also includes so-called "look-alike species", i.e., species whose specimens in trade look like those of species listed for conservation reasons. International trade in specimens of Appendix-II species may be authorized by the granting of an export permit or re-export certificate. Appendix III is a list of species included at the request of a Party that already regulates trade in the species and that needs the cooperation of other countries to prevent unsustainable or illegal exploitation. International trade in specimens of species listed in this Appendix is allowed only on presentation of the appropriate permits or certificates.

(https://cites.org/eng/app/index.php)

Trophy hunting is currently banned in 13 countries: Angola, Botswana, Congo, Gabon, Ghana, India, Kenya, Malawi, Mauritania, Niger, Nigeria, Rwanda, and South Sudan. It is generally legal in the rest of the world. Pakistan and Iran also allow trophy hunting. In Pakistan there are about 50 hunting areas covering nearly one-third of Gilgit-Baltistan[6] (Hale, 2022; Khan et al., 2023): every year, Iran's Environmental Protection Agency issues about 500 licenses to foreign visitors to hunt rare and protected breeds (Ardalen, 2015). The trophy hunters are mostly from US,

[6] Since 1947, parts of Jammu and Kashmir have been under Pakistan's occupation, with India referring to the area as 'Pakistan-occupied Kashmir (PoK)'. The occupied region comprises two ethnically and linguistically different regions: what Pakistan calls, 'Azad Jammu and Kashmir' or AJK, which includes parts of Kashmir and Jammu; and Gilgit Baltistan, which is the northernmost tip of Kashmir and covers parts of Ladakh.

EU, and wealthy people from the UAE and other countries. Although both Pakistan and Iran claim to be Islamic countries, for commercial reasons they find it convenient to overlook the saying of Prophet Muhammad: "Whoever kills a sparrow or anything bigger than that without a just cause, God will hold him accountable on the Day of Judgement." The listener asked: "O, Messenger, what is a just cause?" The Prophet replied: "That he will kill it to eat, not simply to chop off its head and then throw it away"[7] (https://islamreligion.com/pdf/en/prophet-muhammads-kindness-to-animals-8493-en.pdf).

Studies show that wildlife conservation is increasingly being used as an argument for evicting indigenous communities from their residential lands. This trend has been observed since the 1990s, particularly in South and Southeast Asia, North America, and Africa[8]. One reason for this change in discourse is that in the eyes of the international community economic reasons alone have lost their power to justify the violation of indigenous people's rights. Maasai people in northern Tanzania have experienced violent displacement from their ancestral homeland to make way for hunting companies. In Ngorongoro district people were displaced to enable the Ortello Business Corporation (OBC), based in the UAE, to conduct game hunts. In 1992, the Tanzanian government granted OBC an exclusive private hunting license. The long-term plan calls for the construction of a 1500 km^2 wildlife corridor exclusively for OBC hunters (Moko & Bens, 2022).

Trophy hunting is also justified as a tool to empower the local communities and examples are cited from Pakistan and neighboring countries (Adhikari, 2020). In Pakistan, a large portion of the trophy hunting fee goes directly to local communities where the government pays 80% of the fee to the affected Community Conserved Area (CCA) within 45 days after the hunting season is over (Adhikari, 2020). The irony is that *Markhor* (Box 6.5), a wild goat, is the national animal of Pakistan and people are paying to kill Pakistan's pride (Fig. 6.3). A holistic overview reveals that revenue generated by trophy hunting programs is only 1–2% of the tourism sector's total revenue. A study in Khunjerab National Park in northern Pakistan found that trophy hunting has strongly changed local residents' attitudes toward wildlife—a positive attitude toward wild ungulates and a strongly negative attitude toward wild carnivores (Rashid et al., 2020). This, in turn, has increased conflict between humans and snow leopards, as negative attitudes toward carnivores result in snow leopards being killed in retaliation. The study recommended that trophy hunting be stopped and that mass tourism be converted to ecotourism, given the limited benefits and significant problems associated with it. Many studies have also shown that placing

[7] The Prophet is also reported to have said that the act of man imprisoning animals under his control is a great evil (https://kitchenicious.com/hadith-about-kindness-to-animals/#0-quranic-verses-and-hadiths-about-kindness-to-animals-explain).

[8] Forced removals are drastic. Eviction is the most violent act a law-abiding State can inflict on its law-abiding citizens. Ultimately it is inimical to conservation's cause (Brockington & Igoe, 2006).

Fig. 6.3 A hunter with a killed Markhor, posing as a victor (https://i0.wp.com/khabarnaamaa.com/wp-content/uploads/2019/11/markhor-news.jpg?resize=496%2C347&ssl=1)

monetary values on wildlife can lead to conservation conflicts, whereas emphasizing the intrinsic value of wildlife contributes to conservation (Gomez-Baggethun & Ruiz-Perez, 2011). In addition, some studies have also indicated that animals in the hunted population showed clear signs of disturbance, i.e., a smaller group size, lower calf recruitment rates, and higher nervousness than conspecifics without trophy hunting (Hariohay et al., 2018).

Box 6.5 Markhor

The Markhor (*Capra falconeri*) is the largest of the goat family and is found in the rugged mountains of central Asia, from Southern Russia to the sparse woodland of the West Himalayas. The markhor is largely found in the Northern Areas of Pakistan especially in Chitral, Ghizar, and Hunza regions. Markhors live at altitudes of 500–3500 m (1600–11,000 feet) depending on season, summer is spent at higher altitudes and winter at lower altitudes. The markhor is the national animal of Pakistan.

The markhor is classed as an endangered species by the IUCN, meaning it is in danger of facing extinction in the near future if conservation efforts are not maintained. There are five subspecies of markhor:

Kashmir Markhor (*Capra falconeri cashmiriensis*)
Astor Markhor (*Capra falconeri falconeri*)
Bukharan Markhor (*Capra falconeri heptneri*)
Sulaiman Markhor (*Capra falconeri jerdoni*)
Kabul Markhor (*Capra falconeri megaceros*)
(https://animalcorner.org/animals/markhor/)

A Pragmatic Concept of People's Participation: Participation as a Mosaic of Roles

Whether it is trophy hunting or any other ecosystem management intervention, local community participation is essential. The World Bank's concept of "Participation" envisages full ownership and management of forests or other ecosystems by local village communities (Ahmad, 2020). Apparently, this concept seems very appealing and may have emotional value. However, it has serious disadvantages and inconvenient to implement. Ecosystems are generally networked and inter-linked, and they cannot be fragmented into many smaller units for the sake of theoretical convenience. The managing authority of an ecosystem entity must be the government or similar entity having control over all branches of governance and sources of revenue to manage the affairs of the government or entity. When a village-level entity owns or manages a small forest area, it is forced to generate revenue mostly from that area, ultimately leading to unsustainable management and degradation of the natural resource (Ahmad, 2020). An example is that of forest management in Papua New Guinea (PNG).[9] The land tenure system in PNG is very conducive for the so-called Community Forest Management (CFM). The communities there own 97% of the land making it difficult for the government to obtain land for conservation. PNG also derives all of its revenue from natural resource extraction. The result is that environmentally destructive projects take precedence over the protection and conservation of natural resources. The consequence of such a land tenure system is that over the last four to five decades, around a third of PNG's rainforest has been destroyed or degraded by deforestation. PNG is an example and an eye opener for everyone involved in natural resource management (Ahmad, 2020). Practical participatory management of forests or ecosystems is the mosaic of complementary roles of government, local communities, and other stakeholders. Local communities at the village level are not, in themselves, viable units for the ownership and management of forests or ecosystems. Therefore, justifying trophy hunting simplify because it brings some income to local communities is not a strong or valid argument. Policymaking requires a broader picture and a holistic vision.

[9] Papua New Guinea is an island country in the southwestern Pacific Ocean. It encompasses the eastern half of New Guinea, the world's second largest island (the western half is made up of the Indonesian provinces of Papua and West Papua); the Bismarck Archipelago (New Britain, New Ireland, the Admiralty Islands, and several others); Bougainville and Buka (part of the Solomon Islands chain); and small offshore islands and atolls. Since the country achieved independence in 1975, one of its principal challenges has been the difficulty of governing many hundreds of diverse, once-isolated local societies as a viable single nation (https://www.britannica.com/place/Papua-New-Guinea).

The Issue of Human-Wildlife Conflict

Another justification for trophy hunting is that it reduces human-wildlife conflict (HWC).[10] No positive association was found between trophy hunting and reductions in HWC. In the past, the human response to HWC was generally to kill the suspected wild animal. With the development of civilization, increasing awareness of conservation, and recognition of the importance of ecosystem services, traditional lethal retaliation against wildlife is now either illegal, regulated, or socially unacceptable in some places (Gallagher, 2021).

Globally, at least 24 species of terrestrial carnivores regularly hunt nine common domesticated species[11] (Gross et al., 2021). There are also several wild animals that regularly raid the crops of farmers.[12] An integrated approach is essential to dealing with HWC. The six elements of HWC Management are: understanding the conflict, mitigation, response, prevention, policy, and monitoring. Of these six elements, the first four are generally used in HWC management while the other two are ignored. Policies and monitoring are also essential elements which must be integrated into an appropriate HWC management strategy. According to a WWF Report in 2021—*A Future for All: Human Wildlife Co-existence*, India is one of the few countries in the world that have explicitly addressed co-existence and HWC in their national laws. The Wildlife (Protection) Act, 1972 empowers the Chief Wildlife Warden of the States to enable measures for the peaceful co-existence of humans and wildlife within and outside national parks and protected areas (Box 6.6).

---

[10] Human-wildlife conflict occurs when encounters between humans and wildlife lead to negative results, such as loss of property, livelihoods, and even life. Defensive and retaliatory killing may eventually drive these species to extinction. These encounters not only result in suffering for both people and wildlife immediately impacted by the conflict; they can also have a global reach, with groups such as sustainable development agencies and businesses feeling its residual effects.

[11] Species like wolves in Europe; leopards, hyenas, and lions in Africa; tigers, leopards, and snow leopards in Asia; jaguars in Latin America; and cougars and coyotes in North America hunt livestock, mostly at dusk or dawn. Marine carnivores, such as seals, dolphins, and otters, and avian species such as cormorants prey on fish stocks in fish farms and other fisheries, such as those that employ gill netting.

[12] Herbivores cause damage to crops by directly consuming or trampling them. In Asian and African countries, elephants cause significant damage to farms by feeding on or moving through a large variety of crops and crop fields, including maize, rice, oil palms, and fruit trees. Deer and antelopes are known to feed on the young leaves of plants, while monkeys, despite their small size, can cause massive damage, particularly to freshly sown fields. The wild boar is a species known globally to cause extensive damage to farms adjacent to forests or wooded areas.

A 2018 study in the International Journal of Agricultural Sciences looks at managing crop damage by wild animals in the State of Himachal Pradesh in India (Mehta et al., 2018). According to the study, people's suggestions to overcome the crop damage problem in order of preference are: (1) subsidies/funds/support, (2) fences, (3) compensation for crop damage, (4) use of poisonous agents in food crop, and (5) permission to kill wild animals. The study shows that people are more interested in financial assistance and protecting their crops and are least interested in killing or harming wild animals. Therefore, trophy hunting does not seem to be an acceptable strategy to address the problem of crop damage by wildlife. Internationally, governments are grappling with the problem of crop damage caused by wildlife. There have been isolated efforts to address the problem using indigenous strategies. But a comprehensive effort with an integrated strategy appears to be missing. In India, a national workshop on "Dealing with Crop Damage by Wildlife" was organized by the Madhya Pradesh State Forest Department in 2015 (Govt. of MP, 2015). The main recommendations of the workshop were:

1. Simplification of the compensation payment procedure.
2. Facilitation for various types of fences suitable for the area e.g., vegetative fences including medicinal plant cultivation, mechanical fencing, solar fencing, lac cultivation, early warning systems etc.
3. Change in cultivation pattern, e.g., growing crops that are inedible to wild animals.
4. Cultivation of fruit crops in peripheral areas.
5. Selected culling based on scientific studies.
6. Involvement of the people and eco-development activities
7. Improvement and integration of traditional crop protection strategies with other supportive interventions

Box 6.6 Wildlife (Protection Act), 1972

The Government of India enacted Wildlife (Protection) Act, 1972 with the objective of effectively protecting the wildlife of the country and to control poaching, smuggling and illegal trade in wildlife and its derivatives. Section 11 of the Act empowers the Chief Wildlife Warden to take decisions on issues related to human wildlife conflict

Hunting of wild animals to be permitted in certain cases.

1. Notwithstanding anything contained in any other law for the time being in force and subject to the provisions of Chapter IV

   (a) the Chief Wildlife Warden may, if he is satisfied that any wild animal specified in Schedule I has become dangerous to human life or is so disabled or diseased as to be beyond recovery, by Order in writing and stating the reasons therefor, permit any person to hunt such animal or cause such animal to be hunted;

   Provided that no wild animal shall be ordered to be killed unless the Chief Wildlife Warden is satisfied that such animal cannot be captured, tranquilized or translocated:

> Provided further that no such captured animal shall be kept in captivity unless the Chief Wildlife Warden is satisfied that such animal cannot be rehabilitated in the wild and the reasons for the same are recorded in writing.
>
> *Explanation*—For the purposes of clause (a), the process of capture or translocation, as the case may be, of such animal shall be made in such manner as to cause minimum trauma to the said animal];
>
> (b) the Chief Wild Life Warden or the authorized officer may, if he is satisfied that any wild animal specified in Schedule II has become dangerous to human life or to property (including standing crops on any land) or is so disabled or diseased as to be beyond recovery, by order in writing and stating the reasons therefor, permit any person to hunt such animal or group of animals in a specified area or cause such animal or group of animals in that specified area to be hunted.
>
> 2. The killing or wounding in good faith of any wild animal in defense of oneself or of any other person shall not be an offence:
>
>   Provided that nothing in this sub-section shall exonerate any person who, when such defense becomes necessary, was committing any act in contravention of any provision of this Act or any rule or order made thereunder.
> 3. Any wild animal killed or wounded in defense of any person shall be Government property.

It is incumbent on governments to use these recommendations and develop a comprehensive strategy for dealing with human-wildlife conflicts including minimizing crop damage and compensating for crop damage caused by wild animals.

Public Opinion on Trophy Hunting

There are therefore, only a few economic justifications for trophy hunting in the areas where it is still going on. There is no evidence to suggest that it is essential as a management strategy or reduces human-wildlife conflict. With increasing awareness, people's opinion is generally against trophy hunting. Surveys in the EU, Switzerland, and the US confirm that between 75% and 96% of respondents oppose trophy hunting and support a ban on imports of wildlife trophies. In South Africa, the main African exporter of trophies of protected species, a majority of 64% oppose trophy hunting (Hameleers, 2022). In August 2017, the British Columbia provincial government banned the trophy hunting of grizzly bear, but killing bears for meat was still permitted. It is now no longer possible to shoot at grizzly bears, with the exception of First Nations, Canada's indigenous people who hunt for food, social, and ceremonial reasons (Brenna, 2017). In 2017 a legal opinion of the World Commission on Environmental Law (WCEL) Ethics Specialist Group (ESG) stated that trophy hunting is not consistent with sustainable use

Fig. 6.4 Slaughter of baby seals in Canada (https://can2-prod.s3.amazonaws.com/petitions/photos/000/085/365/original/Seal_Hunt_2.jpg)

> Trophy hunting is not consistent with "sustainable use". And even if it were, "sustainable use" is not the sole criterion for the decision on eligibility of organizations seeking IUCN membership. The critical question is whether trophy hunting as it is practiced by individuals and promoted by certain hunting organizations may be consistent with IUCN's general objectives as expressed in Articles 2 and 7. This is clearly not the case. Any other view would threaten IUCN's credibility for providing moral and ethical leadership in conservation policies. It would certainly undermine the many efforts of IUCN members to promote a just and sustainable world. (Wiggins, 2020)

Trophy Hunting Expresses a Mindset

Trophy hunters are generally people from Europe and North America. Most of them are wealthy people who have the time and money to undertake this expensive and tedious endeavor. Trophy hunting is an expression of a state of mind. There is something ethically twisted when humans rise above the "lesser" animal kingdom and display their severed body parts as collectibles, souvenirs, and oddities to show a hunter's status (Rodgers, 2018). This colonial mindset is not just limited to trophy hunting. Seal pups aged 3 weeks to 3 months are shot with a rifle or killed using a spiked club known as *hakapik* (Fig. 6.4). This cruel killing in Canada is the largest authorized slaughter of marine mammals in the world (http://ifaw.org/press-release/2022-commercial-seal-hunt-resumes) (Box 6.7). The Asian houbara bustard or the MacQueen's bustard (*Chlamydotis macqueenii*, Gray, JE, 1832) (Fig. 6.5) is on the "vulnerable list" of the International Union for Conservation of Nature (IUCN) (Box 6.8). Its hunting is prohibited under Pakistan's wildlife protection laws. But every year, the Pakistan government grants UAE royalty permission to kill MacQueen's bustard (ANI, 2020). Hunting used to be a status symbol for the luxurious lifestyle of royalty, which is a current reality. This author thinks it appropriate to stand with the empathetic environmentalists by expressing the plight of the MacQueen's bustard and has expressed his concern in letters e-mailed to the Ruler of Dubai and the Prime Minister of Pakistan. These letters are attached as Annexures 1 and 2, respectively.

Fig. 6.5 The magnificent Asian Houbara Bustard (MacQueen's Bustard) (https://www.pakistankakhudahafiz.com/pkkhnew/wp-content/uploads/2016/01/HB-4-Fuert-11-Feb-10_990.jpg)

Box 6.7 Seal Slaughter

Every year, about a half million seals worldwide are subjected to painful and often lingering deaths, largely for the sake of fashion. Despite a worldwide outcry over the cruelty of the slaughter and a ban on seal products in many countries, the massacre continues.

Seals are slaughtered in Canada, Greenland, Namibia, Norway, and Russia. During Canada's annual commercial seal slaughter, as many as 400,000 seals can be shot or bludgeoned. Canada halted the slaughter of so-called "whitecoats" in 1987, but harp and grey seals lose their white fur at around 2 weeks of age, so most seals can still be shot or clubbed to death before they reach their first month of life.

Sealers use a variety of weapons, including clubs, "hakapiks" (heavy wooden clubs topped by a barbed metal hammer head), and rifles. Sealers often hook baby seals in the eye, cheek, or mouth to avoid damaging their fur, then drag them across the ice to skin them.

(https://www.peta.org/issues/wildlife/wildlife-factsheets/seal-slaughter/)

Box 6.8 Houbara Bustard

Bustards are medium-to-large game birds of the family Otididae, related to the cranes and rails in the order Gruiformes. There are about 23 species, confined to Africa, southern Europe, Asia, Australia, and part of New Guinea. Bustards have rather long legs, adapted to running. They have only three toes, lacking the hind toe (hallux). The body is compact, carried in a rather horizontal position, and the neck stands erect, forward of the legs, in the manner of other tall running birds.

The Asian houbara bustard, also known as MacQueen's bustard (Chlamydotis macqueenii), is a vulnerable species and its hunting is banned in Pakistan, but an exception is made for Arab royals who are allowed to do so via special permits. It is native to Central Asia and migrates to the Indian subcontinent, including Pakistan, during winter months. It is one of multiple bustard species and similar to the Great Indian Bustard, which is native to India. The Arab royals hunt the bird as a sport and also because its meat is believed to be an aphrodisiac. The IUCN has described bustard hunting in Pakistan as "unsustainable".

(https://www.britannica.com/animal/bustard; https://theprint.in/theprint-essential/aphrodisiac-meat-sport-why-arab-royals-hunt-bustard-in-pakistan-why-india-banned-it/584173/)

Conclusion

Forests and ecosystems were once considered natural resources to be exploited for human use. Wild animals were also considered as entities that could be hunted for different purposes. Gradually, trophy hunting became a status symbol for rich and influential people. In recent decades, the importance of ecosystem services and the intrinsic value of wildlife have been recognized. There has been a shift from unplanned exploitation to conservation and sustainable use of natural resources. Wild animals are also protected under various laws and conventions.

Trophy hunting as a hobby of the rich continued in various forms. Scientific studies do not show that trophy hunting is necessary or overall beneficial, and many countries are doing well by promoting non-lethal ecotourism. Since trophy hunting is a large commercial activity, many justifications are put forward by the players in this industry, such as Income generation, minimization of human-wildlife conflicts, welfare of local communities, conservation of wildlife. Furthermore, in the name of people's participation, there are concepts of total ownership and management of forests and ecosystems by village communities without any role of government. Such strategies aim to create a chaotic state with the vacillating attitude of the governments so that ultimately it can be claimed that governments or similar viable bodies are incapable of protecting and managing the forests and ecosystems; so they should be handed over to small local bodies where it will be beneficial for commercial forces and large business houses to exercise their influence. The general global trend suggests that the economic policies of most countries were designed to favor the rich. In the name of privatization, public assets were converted into private assets and governments became poorer. While national wealth has increased significantly, public wealth in rich countries is now negative or close to zero (Ahmad, 2022). If world's ecosystems become private properties, they would be managed for profit, not for ecosystem services and human well-being. Governments must promote and develop policies and strategies so that forests and natural ecosystems are managed with people's participation for the long and short-term well-being of the people.

The conflict between humans and animals is a problem that every ecosystem manager faces. A clear policy and an integrated strategy is essential to address this problem to reduce public hostility toward wildlife. Trophy hunting may generate some revenue, but does not resolve human-wildlife conflicts. In addition, the game farms are not wildlife sanctuaries. These are mostly private areas where wild animals are treated cruelly for profit. Owners of such areas and people who enjoy killing animals want to maintain the status quo as long as it makes a profit. However, with increasing public opinion against trophy hunting, the governments of various countries need to develop an exit policy to put an end to this practice, which is influenced by a colonial mindset. For the well-being of humanity, existing forests and natural ecosystems must be managed sustainably with regard to their intrinsic values. They must be scientifically managed with a global vision to combat climate change and ensure the continuity of ecosystem services that are essential for the well-being of humanity. An ecosystem manager must have a holistic view of the entire area being managed. He or she has a diverse role—conserving and sustaining ecosystem services, involving people in the sustainable utilization of the usufruct rights, managing human-wildlife conflict through an integrated strategy, raising awareness about nature through a non-lethal ecotourism program, running programs and activities to create livelihoods and other benefits for local populations, create a sense of serenity among wildlife, etc. Trophy hunting has no place in a responsible and responsive ecosystem management.

Annexures

Annexure-1

Shahbaz Ahmad IFS Retd.
Ex-Principal Chief Conservator of Forests &
Chief Wildlife Warden, MP India
Editor: Environmental Justice

To,

Mr. Sheikh Mohammed bin Rashid Al-Maktoum
Hon'ble Emir of Dubai Emirate

Subject: Hunting of Houbara Bustard in Pakistan

Respected Sir, Assalam Alaikum

It has come to notice that the Government of Pakistan has granted permission to you and your team for the killing of Houbara Bustard during the 2020-21 hunting season. By the grace of Allah you have been bestowed with all the luxuries of life and you need not go out for hunting for the sake of food or any other need. Technically, hunting may be permitted in Islam, but the killing of animals just for the sake of fun is against the very spirit of Islam.

Allah is Ar-Rahman and Ar-Rahim. Prophet Muhammad (Peace and Blessings of Allah be upon him) is also a mercy for this world. Just as Allah has been merciful to you, please be merciful to the creatures of Allah. Taking a vow not to kill any living being for the sake of fun or recreation would be a gesture compatible to the teachings of Islam and worth emulating by the leaders of the world.

I would request you to set an example by declaring that you would not be hunting any wild animal for fun or recreation. This declaration of yours, besides bringing honour to your country will also be honouring the sentiments of environmentalists all over the world.

With Regards

Date: 14-01-2021

Yours sincerely

Shahbaz Ahmad IFS Retd.
Mob: +91 9425121958

Annexure-2

Shahbaz Ahmad **IFS Retd.**
Ex-Principal Chief Conservator of Forests &
Chief Wildlife Warden, MP, India
Editor: Environmental Justice

To,
Mr. Imran Khan
Hon'ble Prime Minister of Pakistan

Subject: Permission to the Ruler of Dubai for hunting of Houbara Bustard

Respected Sir, Assalam Alaikum

It has come to notice that you have granted permission to the Ruler of Dubai and his team for the killing of Houbara Bustard during the hunting season 2020-2021. Technically there may nothing be wrong with this permission, but the killing of any creature just for fun or recreation is against the very spirit of Islam. If you claim to be an Islamic country, it is essential that the spirit of Islam is honoured both in letter and spirit.

As the Prime Minister of Pakistan you are the custodian of the natural wealth of your country. Houbara Bustard may not be able to directly communicate with you, but they would be expecting their custodian to be concerned about their welfare and safety. Obliging the Ruler of Dubai may have political significance for you, but what is more important is the treatment you give to the wild and helpless creatures of your country.

Honouring the sentiments of environmentalists all over the world, it would be prudent on your part to cancel the permits of the hunting of Houbara Bustard. This will not only raise your stature in the eyes of the people but will also add feathers to your statesmanship by setting an example and direction for the conservation of the natural wealth of the earth.

With Regards

Date: 14-01-2021

Yours sincerely

Shahbaz Ahmad IFS Retd.

Mob: +91 9425121958

References

Adhikari, L. (2020). Community based trophy hunting programs secure biodiversity and livelihoods: Learnings from Asia's high mountain communities and landscapes. *Environmental Challenges, 4*, 100175.
Ahmad, S. (2020). *The dilemma of people's participation*. Print World.
Ahmad, S. (2022). *Economics and development—An alternative to conventional economic thoughts and practices*. Print World.
ANI. (2020). *Pakistan gives "license to kill" houbara bustard, a potential bird species*. ANI.
Ardalen, S. (2015). *The big business of trophy hunting in Iran*. BBC.
Bertelsmann, Stiftung & SDSN. (2018). *SDG index and dashboards report 2018—Global responsibilities: Implementing the goals*. Pica Publishing Ltd.
Brenna, L. (2017). *British Columbia has banned grizzly bear hunting, permanently*. Life Gate.
Brockington, J., & Igoe, J. (2006). Eviction for conservation: A global overview. *Conservation and Society, 4*(3), 424–470.
Costanza, R., D'Arge, R., Groot, D., et al. (1997). The value of world's ecosystem services and natural capital. *Nature, 387*, 15.
Costanza, R., Groot, R., Paul, S., et al. (2014). Changes in the global value of ecosystem services. *Global Environmental Change, 26*, 152–158.
Eardley-Wilmot, S. (1910). *Forest life and sports in India*. Cornell University Library.
Gallagher, K. (2021). Human wildlife conflict; implications and solutions; sustainability for all.
Gomez-Baggethun, E., & Ruiz-Perez, M. (2011). Economic valuation and the commodification of ecosystem services. *Progress in Physical Geography, 35*(5), 613–628.
Govt. of MP. (2015). *Proceedings of national workshop on "Managing crop damage by wild animals"*. Forest Department.
Gross, E. J., et al. (2021). *A future for all—The need for human wildlife co-existence*. WWF.
Hale, G. (2022). *Community based conservation in Pakistan—Here, sustainability means "If it pays, it stays"*. Conservation Frontline.
Hameleers, R. (2022). *136 NGOs around the world call for a ban on hunting trophy imports*. Europe Group for Animals.
Hariohay, K. M., Jackson, C. R., Fyumagwa, R., & Røskaft, E. (2018). Trophy hunting versus ecotourism as a conservation model: Assessing the impacts on ungulate behaviour and demographics in the Ruaha-Rungwa ecosystem, Central Tanzania. *Environment and Natural Resources Research, 8*(2), 33.
IUCN. (2004). Protected areas category 2004. *Parks, 14*, 3.
Khan, B., Abubakar, S. M., & Shreshtha, K. (2023). *The story behind why Gilgit-Baltistan allows trophy hunting*. The Express Tribune.
Kluck, J. (2020). *Follow the money: Trophy hunting does not help communities*. Blog of Born Free.
Mehta, P., Negi, A., Chaudhary, R., & Janjhua, Y. (2018). A study on managing crop damage by wild animals in Himachal Pradesh. *International Journal of Agricultural Sciences, 10*(12), 6438–6442.
Moko, L. K., & Bens, J. (2022). *Colonial conservationism, wildlife tourism and land expropriation in Africa*. Review of African Potential Economy. Retrieved from https://roape.net/2022/07/07/colonial-conservationism-wildlife-tourism-and-land-expropriation-in-africa/
Murray, C. K. (2017). *The lion's share? On the economic benefits of trophy hunting*. Humane Society International, Melbourne.
Pabla, H. S. (2020). *Besides loving the beasts*. H. S. Pabla.
Rashid, W., Shi, J., Rahim, I., & Sultan, H. (2020). Issues and opportunities associated with trophy hunting and tourism in Khunjerab National Park, Northern Pakistan. *Animals, 10*(4), 597.
Rodgers, G. (2018). *Wildlife trophy hunting—The economics, ethics and emotions*. Retrieved from http://dyingwords.net/wildlife-trophy-hunting-the-economics-ethics-and-emotions
Sebunya K. (2020). Wildlife economics: What are they and how do they impact conservation?. African Wildlife Foundation

Sheikh, P. A. (2019). *International trophy hunting*. Congressional Research Service.
Southwick Associates. (2015). *The economic contribution of hunting related tourism in Eastern and Southern Africa*. Safari Club International.
UIA. (2021). *Hunting of animals*. World Problems & Human Potential.
Wiggins, S. (2020). *Informing the debate on trophy hunting*. International Wildlife Bond.

Chapter 7
Human Relations with Mesopredators

Michael O'Neal Campbell

Abstract In contrast to large carnivores, smaller carnivores, sometimes called mesopredators or mesocarnivores, are considered less dangerous to humans and livestock and are less charismatic, popular, and targeted in conservation research. However, small carnivores are important participators in ecosystems and have important relationships with people, human habitats, livestock, and domestic animals. In some cases, mesopredators can become apex predators through the regional or local extinction of larger species, but there is some disagreement on the size range of mammalian mesopredators. Avian mesopredators may be medium-sized birds of prey, but in terms of prey taken, the largest birds of prey, the eagles, are similar to the mesopredators of mammals such as medium-sized dogs and cats. Human attitudes toward mesopredators tend to be less intense because these smaller animals are generally less dangerous to humans and associated animals, have smaller ranges, and appear to be more amenable to semi-domestication as pets. This chapter examines case studies of human attitudes toward small carnivores, including mammals, birds, and reptiles in the context of human habitats and land use. The evidence suggests that smaller predators have been neglected in the conservation literature and the human dimension of wildlife literature in favor of larger, more charismatic and dangerous carnivores. However, this oversight is important for the study of both wildlife ecology and the human dimension of wildlife. The chapter takes a social science perspective, particularly that of conservation psychology, and presents evidence that this approach is important for the conservation biogeography of wildlife. It is concluded that conservation psychology is important for the study of conservation biogeography and human-wildlife relationships, even when focused on smaller predators. This contributes to nature conservation policy and protected area management.

Keywords Mesopredator · Reptile · Hawk · Raptor · Eagle · Lynx

M. O. Campbell (✉)
Simon Fraser University, Burnaby, BC, Canada
e-mail: ecol55@hotmail.com

K. K. Jha and M. O. Campbell (eds.), *Intersections of Conservation Biogeography and Wildlife Management*, https://doi.org/10.1007/978-981-95-0121-2_7

Introduction

Small or medium-sized predators or mesopredators may be mammals, reptiles, or birds, and this size or relationship category is defined variously, but the term generally describes predators below the apex predators in terms of size, dominance, and/or prey captured (Campbell, 2017a, 2017b). The term mesopredator is necessarily subjective, as mesopredators are associated with the size or status of apex predators and can be promoted to regional or local apex predators if the original apex predators are locally extirpated (Campbell, 2017b; Kilgo et al., 2012). For example, Prugh et al. (2009, p. 779) ask: "What exactly is a 'meso' or middle-rank predator?" and cite a hypothetical ecosystem in which wolves are the apex or top predators, "followed by a chain of potential mesopredators (coyote, cat, rat, lizard, spider, dragonfly)". Hodge (2013) notes that the term "mesopredator"has commonly been used to describe small or medium-sized carnivores such as foxes or coyotes, as opposed to large, apex carnivores such as bears and lions, but such size-based classifications are relative terms. A given species may be "medium-sized" in one community and yet be the largest in another community if the top predators have been extirpated or were always absent. An example is given of the coyote as the top predator as a result of the local extinction of wolves and bears (Hodge, 2013).

The terms mesopredator and mesocarnivore, as well as small predators and small carnivores, can be used interchangeably. Because some small members of the predominantly carnivorous genera, particularly mammals, may also eat plant material, it is argued that the term "mesopredator" is preferable to the term "mesocarnivore," which may imply an exclusively carnivorous diet (Hodge, 2013). This may also apply to birds, as pure avian carnivores are called birds of prey or avian predators, which may also occasionally eat carrion as part-time or facultative scavengers, and the classification may exclude other species such as many fish and insectivores such as herons, gulls, skuas, shrikes, and nightjars (Campbell, 2015, 2022). Because this point is disputed by various commentators, the terms mesopredator, mesocarnivore, and small carnivore are used interchangeably in this chapter. This is particularly relevant as the main issue is the relationship between these animals and humans, the key relevance being the animals' ability to hunt small livestock and companions animals and, unlike apex predators, pose few threats to people, associated animals, and economic activities (Campbell, 2017a; Campbell & Jha, 2021; Jha & Campbell, 2021).

Do Linh San et al. (2022) define mammalian small carnivores (which may be termed mesopredators) as the 232 species of the Order Carnivora with an average body mass of less than 21.5 kg, which may dominate areas where people have eradicated large carnivores. Similar to larger carnivores, many smaller carnivore species are Threatened or Near Threatened with extinction due to "regionally burgeoning human populations' industrial and agricultural activities, causing habitat reduction, destruction, fragmentation, and pollution. Overexploitation, persecution, and the impacts of introduced predators, competitors, and pathogens have also negatively affected many small carnivore species" and in addition "although small carnivores have been intensively studied over the past decades, bibliometric studies showed that they have not received the same attention given to large carnivores" (Do Linh San et al., 2022). Marneweck et al. (2021) define small mammalian carnivores as weighing less than 16 kg, with similar current conservation requirements as larger carnivores (based on IUCN Red List species-level assessments), but with fewer assessments than large carnivores, this showing the lower attention from conservation researchers for small carnivores (see also Brashares et al., 2010; Le Roux et al., 2019; Prugh et al., 2009; Ray et al., 2013; Ripple et al., 2014; Ripple & Beschta, 2012; Ritchie & Johnson, 2009).

Avian mesopredators were also defined separately from mammalian mesopredators because the above weight classifications for small mammalian carnivores (<21.5 kg) would include the largest avian predators, including apex predators such as the largest eagles: the Harpy (*Harpia harpyja* Linnaeus, 1758), Philippine (*Pithecophaga jefferyi* Ogilvie-Grant, 1896) and Steller's sea eagles (*Haliaeetus pelagicus* Pällas, 1811) (Campbell, 2022). Such avian apex predators may also perform similar ecological roles to mammal mesopredators by killing medium-sized to smaller prey (Lyly et al., 2015), but they may be of minimal threat to adult humans. Therefore, eagles and other large raptors, which may be apex predators among birds, may also be mesopredators in human life spaces, taking livestock of the same size as those taken by mammal mesopredators such as bobcats (*Lynx rufus* Schreber, 1777) coyotes (*Canis latrans* Say, 1823) and foxes (subfamily *Caninae* Fischer de Waldheim, 1817) (Campbell, 2022).

For example, Chakarov and Krüger (2010) describe an ecosystem with a three tiered guild in which the great eagle owl (*Bubo bubo*, Linnaeus, 1758) is the dominant apex predator, with two mesopredators, the northern goshawk (Accipiter gentilis, Linnaeus, 1758) and the Common buzzard (*Buteo buteo*, Linnaeus, 1758), with the goshawk being the dominant mesopredator (Chakarov & Krüger, 2010). In this classification, the great eagle owl would probably be at about the same level as a mammalian mespredator such as the bobcat (*Lynx rufus* Schreber, 1777) or European wildcat (*Felis silvestris* Schreber, 1777), in terms of prey selection and predatory ability (Breitenmoser et al., 2019; Feldhamer et al., 2004). O'Neil et al. (2023) also define common ravens (*Corvus corax* Linnaeus, 1758), as generalist avian mesopredators, which have thrived in landscapes modified by human actions.

In another example of predation, Davies (1999) documented that Verreaux's eagles killed small livestock, although much less frequently than mammalian mesopredators did. Black-backed jackals (*Lupulella mesomelas* or *Canis mesomela*s Schreber, 1775) and caracals (*Caracal caracal* Schreber, 1776), both important African mesopredators, were also documented as predators of small livestock in the Karoo, South Africa (Nattrass et al., 2020, p. 1222). These predators were highly adaptive, as the study showed that "irrespective of measurement approach, lethal control of mesopredators in this fragmented socio-economic landscape was associated with increased livestock losses the following year. Terrain ruggedness was positively, and number of farmworkers negatively, associated with livestock losses" (Nattrass et al., 2020, p. 1222). Small livestock killing by golden and bald eagles in North America and

Europe was also common, similar to coyotes, bobcats, and foxes (Seguin et al., 2001; Collins & Latta, 2009; Campbell, 2022).

Reptile mesopredators have also been defined, studied and compared with reptile apex predators such as crocodiles and aligators (*Crocodilia* Owen, 1842), but are generally less studied than mammalian or avian predators (Campbell, 2009). For example, Read and Scoleri (2015) describe the sand goanna (*Varanus gouldii*, Gray, 1838), a medium-sized reptile as a mesopredator that competes with other more dominant mammalian mesopredators, such as cats and foxes. Sutherland et al. (2011) also note that "varanid lizards are likely to be apex- or mesopredators, but being reptiles are rarely considered by managers and researchers when investigating the impacts of mammalian carnivore management", and there is "extensive dietary niche overlap between varanids, foxes and cats..." (Sutherland et al., 2011). Importantly, research on the management of carnivores has mostly ignored carnivorous reptiles, and has instead concentrated on mammalian apex predators and mesopredators. Among the Eurasian and African most significant terrestrial reptilian predators, varanid lizards seem to have a function similar to that of small to medium-sized carnivorous mammals (Sutherland et al., 2011; see also Pough, 1973; Ray et al., 2013; Nowak et al., 2008).

Prugh et al. (2009, p. 779) argue for the increasing importance of mesopredators in contemporary ecosystems through the concept or phenomenon of mesopredator release (see Soulé et al., 1988), which occurs with the decline or extinction of apex predators: "Apex predators have experienced catastrophic declines throughout the world as a result of human persecution and habitat loss. These collapses in top predator populations are commonly associated with dramatic increases in the abundance of smaller predators. Known as 'mesopredator release,' this trophic interaction has been recorded across a range of communities and ecosystems" (Prugh et al. (2009, p. 779). Mesopredator release is defined as "the expansion in density or distribution, or the change in behavior of a middle-rank predator, resulting from a decline in the density or distribution of an apex predator" (Prugh et al., 2009, p. 780, see also Brashares et al., 2010). This phenomenon is disputed by some researchers, who claim that small carnivores are affected by human activities in the same way as large carnivores (Allen et al., 2017; Haswell et al., 2017; Jachowski et al., 2020). Marneweck et al. (2021) contend that while some smaller, non-apex carnivores have spread across their ranges in the past century, the degree to which smaller carnivore species profit from competition from large carnivores and their "release" from larger predators' presence varies greatly depending on the circumstances and occurs in very few instances. As a result, opinions on the status and requirements for small carnivore conservation vary; some of these animals are categorized as needing protection, while others require control (Marneweck et al., 2021).

An example of mesopredator release occurred in North America, where the ranges of all apex predators have declined (due to human exploitation), while those of mesopredators have expanded (Prugh et al., 2009, p. 779; see also Roemer et al., 2002). Prugh et al. (2009, p. 779) further note that "the need to understand how best to predict and manage mesopredator release is urgent—mesopredator outbreaks are causing high ecological, economic, and social costs around the world." In North

America, wolf extinction may lead increased coyote populations; likewise in the African savanna, the decline of lions and leopards may contribute to increased baboon populations, both events affecting people's landuse activities (Prugh et al., 2009, p. 781). Prugh et al. (2009, p. 789) conclude that returning apex predators to areas where they have been exterminated could significantly reduce the negative effects of mesopredator releases. The extirpation of apex predators from direct persecution, killing of prey species, and habitat conversion has resulted in outbreaks of mesopredator populations worldwide. However, apex predator protection and/or reintroduction is a difficult undertaking, requiring compromise among those whose lives are most directly impacted by these predators, as well as thorough habitat restoration and increased public acceptance of large predators (Prugh et al., 2009, p. 789).

The remainder of this chapter examines the relationships between humans and mesopredators, including mammals, reptiles, and birds. These relationships are based on human social structures and actor actions as well as religious, sociocultural, political, and economic factors. It is noted that mesopredators are generally judged differently than large carnivores because they are less feared as a threat to human life, predation, and human occupation of habitat. Mesopredators also require less space for a sustained ecological presence and are therefore less likely to compete with humans for important land areas. Conversely, mesopredators are less likely to attract conservation management attention.

Mesopredator-Sized Mammals and People

Mammalian mesopredators, primarily cats, dogs, and mustelines, have important interactions with humans (Brashares et al., 2010; Elmhagen & Rushton, 2007). These interactions include real or imagined killings of small livestock and/or domestic animals, attacks on humans, home invasions, domestication, conservation, and social insights that create perspectives on these animals (Hosey & Melfi, 2014; MacGregor, 2021). Due to their smaller size compared with apex predators, these interactions may be generally ignored in research and less prioritized in policy actions. However, the greater numbers of such species compared with larger species and their generally elusive nature more increase the importance of such conflicts with humans. Common mammalian mesopredators from the Americas, Eurasia, and Africa are shown in Figs. 7.1, 7.2, 7.3, 7.4, 7.5, 7.6, 7.7, 7.8, 7.9, 7.10, 7.11, 7.12, 7.13, 7.14, 7.15, and 7.16. Figures 7.1, 7.2, 7.3, and 7.4 show mammal species from the Americas. Figures 7.5, 7.6, 7.7, and 7.8 show mammal species from Europe. Figures 7.9, 7.10, 7.11, and 7.12 show mammal species from Asia, and Figs. 7.13, 7.14, 7.15, and 7.16 show species from Africa.

Scasta et al., 2017 cite a study of rancher opinions of mesopredator impacts on livestock in Wyoming, USA, based on predation mitigation strategies for foxes (gray foxes (*Urocyon cinereoargenteus* Schreber, 1775, red foxes *Vulpes vulpes* Linnaeus, 1758, swift foxes (*Vulpes velox* Say, 1823), coyotes (*Canis latrans* Say, 1823), and bobcats (*Lynx rufus* Schreber, 1777), with larger predators comprising grizzly bears

Fig. 7.1 North American bobcat. *Source* Mazin Omron-Pexels

Fig. 7.2 North American coyote. *Source* YS-Unsplash

(*Ursus arctos* Linnaeus, 1758), black bears (*Ursus americanus* Pallas, 1780), cougars (*Puma concolor*, Linnaeus, 1771), wolves (*Canis lupus* Linnaeus, 1758) and dogs (*Canis familiaris* Linnaeus, 1758) (Scasta et al., 2017). Avian predators and scavengers were also cited, including turkey vultures (*Cathartes aura* Linnaeus, 1758), golden eagles (*Aquila chrysaetos* Linnaeus, 1758) and bald eagles (*Haliaeetus leucocephalus* Linnaeus, 1766), hawks (including ferruginous hawks *Buteo regalis* Gray, 1844, red-tailed hawk *Buteo jamaicensis* Gmelin, 1788, and Swainson's hawks *Buteo swainsoni* Bonaparte, 1838), and ravens (*Corvus corax* Linnaeus, 1758) (Scasta et al., 2017).

Scasta et al. (2017) found, based on ranchers testimony, that "efficacy of mitigation varied by predator species, mitigation strategy, and lethality of strategies, but not livestock type" (Scasta et al., 2017). In total, the livestock losses were of 5476 animals

Fig. 7.3 North American (and Eurasian) wolverine. *Source* Hans Veth-Unsplash

Fig. 7.4 North and South American ocelot. *Source* Victor Landaeta-Pexels

killed by various predators (including 25 adult cattle, 656 calves, 913 adult sheep, and 3882 lambs), with coyotes identified as the killers of 2867 animals, then predatory birds (1034 losses), grizzly bears (665 losses) and wolves (582) (Scasta et al., 2017). Wolves killed the most adult cattle (44), and coyotes killed the most calves (354), adult sheep (429), and lambs (2079) (Scasta et al., 2017). Lethal methods, such as trapping and shooting were effective against predation, and non-lethal methods such as fencing, stalling, and herding were less effective, for both the felid and canid predation on both cattle and sheep (Scasta et al., 2017). Ranchers noted that lethal methods were most effective at mitigating predation by foxes and coyotes, moderately effective at mitigating the impacts of large carnivores, and the least effective at mitigating birds. The use of guard animals was moderately or minimally effective.

Fig. 7.5 Eurasian lynx. *Source* Kenny Goossen-Unsplash

Fig. 7.6 Spanish lynx. *Source* Anne and Saturnino Miranda-Pixabay

Ranchers also reported that avian predators seem to be the "most challenging predator type" (Scasta et al., 2017).

Scasta et al. (2017) argued that the finding of the greater effectiveness of lethal methods over non-lethal methods, "presents a social difficulty given the greater public preference for non-lethal methods" and "currently, it is also difficult then for ranchers to shift from the predator control paradigm to the coexistence paradigm when the suite of non-lethal methods lack the same level of perceived efficacy." The findings of the study indicated that the use of guard animals was the most effective non-lethal technique, with possibilities for the use of more powerful guard animals, for larger carnivores such as wolves or bears, rather than the mesopredators. The findings indicated the greater difficulty in control avian predators: "While the function of avian predators is important, especially in reducing and removing carcasses, balancing

Fig. 7.7 European badger. *Source* Hans Veth-Unsplash

Fig. 7.8 Eurasian otter. *Source* Haberdoedas-Unsplash

livestock production and avian conservation is challenging" as the avian predators mentioned in this study are protected by the Migratory Bird Treaty Act, and "it is difficult to protect livestock that are predated by wildlife that are also protected yet have the advantage of flight" (Scasta et al. (2017).

In another example, medium-sized wild canids are known predators of livestock (Appleby & Smith, 2018). These authors raise the concept of surplus killing, which can "evoke particularly emotional responses in this regard." Kruuk (1972) is cited as defining surplus killing as the "killing of prey by a predator, without the killing individual, or members of its social unit, consuming any part of the kill" (Appleby & Smith, 2018). Kruuk (1972, p. 233) states that there are cases where predators "killed considerably more prey animals than they could possibly eat, and causal and functional aspects of this behavior are discussed. The species concerned were especially

Fig. 7.9 Indian dhole. *Source* Rohit Varma-Unsplash

Fig. 7.10 Asian clouded leopard. *Source* Lars Van Poucke-unsplash

foxes and spotted hyenas, and references are quoted about surplus killing by other Canidae, Felidae, and Ursidae species. It is argued that satiation in carnivores does not inhibit further catching and killing, but it probably does inhibit searching and hunting. Thus carnivores are able to procure an "easy prey" but normally satiation limits numbers killed. According to Appleby and Smith (2018), this is true even though the predator typically consumes the specific type of prey and access to the carcass is unrestricted. Some argue that because partial consumption of prey is frequently a crucial component of such events, it is preferable to use the phrase partial prey consumption instead of the term "surplus killing" due to its negative connotations. Keeping livestock and predators apart and preventing or changing what predators learn about the susceptibility of domestic prey may be the most obvious ways to resolve such conflicts (Appleby & Smith, 2018).

Fig. 7.11 Large Indian civet (*Viverra zibetha* Linnaeus, 1758). *Source* Tontantravel, Kaeng Krachan national park tours, CC BY-SA 2.0

Fig. 7.12 Asian golden jackal. *Source* Saleh Bakshiev-Pexels

Bickley et al. (2020, p. 546) give a related argument, noting that "local attitudes toward carnivores often reflect the degree of damage they are perceived to cause. Consequently, understanding the interactions between people and these species is essential to conservation efforts." Their study examined the impact of three canid species: Hoary fox (*Lycalopex vetulus* Lund, 1842), crab-eating fox (*Cerdocyon thous* Linnaeus, 1766), and the larger maned wolf (*Chrysocyon brachyurus* Illiger, 1815) on chicken farms in the Cerrado of Goiás, Brazil (Bickley et al., 2020, p. 546). The findings indicated that local respondents had greater knowledge about the maned wolf, and that although "chicken coops were identified as the most effective method for preventing predation" less than half of the respondents used coops. All three Cerrado canids were reportedly responsible for predating chickens, according to respondents, though the majority of these incidents allegedly happened during the

Fig. 7.13 African caracal. (*Caracal caracal* Schreber, 1776). *Source* Joshua J. Cotten-Unsplash

Fig. 7.14 African serval. (*Leptailurus serval* Schreber, 1776). *Source* Jolijn Rosendaal-Pexels

day, outside of the species' active seasons. Opinions about the canids were significantly predicted by reported cases of chicken predation. Respondents who had been affected by attacks on their chickens expressed distaste for having a wild canid on their land. Nevertheless, according to the majority of respondents, Cerrado canids should still be protected (Bickley et al., 2020, p. 546).

Mowbray (2023) argues that "small cat species can come into conflict with people across the globe, and though this plays out differently than big cat conflict, it can be devastating for farmers' livelihoods. When these cats are seen as pests, they can become targets for retaliatory killings, which threatens their conservation." It is acknowledged that small cat conflict may be less serious than big cat conflict, but there are still serious implications for rural land users. The main difference is that small cats and other mesopredators do not normally threaten human life, but

Fig. 7.15 African black-backed jackal. *Source* Leon Pauleikhoff-Unsplash

Fig. 7.16 African honey badger. *Source* Vincent Van Zalinge-Unsplash

may attack smaller livestock such as birds and young sheep and goats (Mowbray, 2023). Examples are cited of the leopard cat (*Prionailurus bengalensis*) or jungle cat (*Felis chaus*), the fishing cat (*Prionailurus viverrinus*, a fish-eater than can steal fish from fishing people), and non-felines such as jackals (Canis aureus) and Indian civets (Viverra zibetha) in tropical Asia, and pampas cats (*Leopardus colocolo*), Geoffroy's cat (*Leopardus geoffroyi*), ocelots, and margays in South America, which can affect local people through small livestock kills, and this behavior may invite retaliatory killing (Mowbray, 2023).

The study conducted by Campbell (2017b) examines the interactions between humans and small mammals in El Salvador, Central America. The country is overlooked in current research in terms of the relationships between humans and small wildlife, due to social conflicts in the past, high population density raptors, unusual large animal extinctions, and drastic environmental changes. People's attitudes toward small animals in rural La Union, in southeast El Salavdor varied greatly by age and gender, with older people more likely than younger people to have observed and killed larger wild animals. They also agreed that people were afraid of small wild animals and children were threatened by wild animals. Older people were less likely to believe that small wild animals improved human quality of life. Men were more likely than women to kill wild animals and to report that animals caused problems on farms. However, men were more likely to support conservation than women (Campbell, 2017b).

In this study, respondents attributed generational differences in attitudes toward wildlife to changing sociocultural norms over the past three decades (Campbell, 2017b). In the past, social networks were mainly focused on local communities and extended families, parents' opinions were more considered, age was more respected, and parents were more likely to be seen as role models by youth. There was a clear gender separation in the division of labor: trading in raw materials was predominantly a female occupation, while agriculture was reserved for both sexes and hunting was predominantly a male occupation. All respondents indicated that family and local sources as well as urbanization, access to internet, radio, television, and local and international travel have increased the level of environmental knowledge of local populations (Campbell, 2017b).

Although jaguars (*Panthera onca* L. 1758) and pumas (*Puma concolor* L. 1771) were extinct, small predators and omnivores reported on farmland included ocelot (*Leopardus pardalis* L. 1758), coyote (*Canis latrans* Say, 1823), jaguarondi (*Puma jaguarondi* Geoffroy Saint-Hilaire, 1803), margay (*Leopardus wiedii* Schinz, 1821), Oncilla (*Leopardus tigrinus* Schreber, 1775), White-nosed coati (*Nasua narica* L. 1766) and Common opossum (*Didelphis marsupialis* L. 1758) (Campbell, 2017b). Slight majorities of respondents irrespective of age and gender believed that there were too many, rather than too few, such mammals on the farms, and approximately equal numbers thought that there were more of fewer animals than in the past few decades. Mesopredators such as jaguarondis and coyotes were killed to protect livestock and people, especially children. Regarding conservation, respondents based their support or opposition on the species (e.g., differences in species size), the transient behavior of a species (e.g., rabid and non-rabid coyotes), or the human context (e.g., chicken farms would be more vulnerable than cattle ranches) (Campbell, 2017b). The documentation of increased numbers of small carnivores in cultivated land was significant research finding. It is probable that the overemphasis in the literature on large animals and the general disregard for smaller, more adaptable wildlife have understated the likelihood of population growth in the latter. The transition from dense canopy forests to a patchwork mosaic of farms, grasslands, and secondary mixed forests may have less of an impact on these smaller animals (Campbell, 2017b).

More younger people than older people were in favor of animal protection (Campbell, 2017b). Both men and women in their younger years found inspiration for and dedication to conservation in educational programs, nature documentaries, and internet resources like Facebook and information websites. Older adults were more likely to find stories with a shared cultural basis, information about long-term drops in animal populations, and traditional cultural guidance regarding whether to support conservation appealing. Men were more likely than women to focus on environmental issues in the media and link them to forests, wildlife, and tourism (Campbell, 2017b).

Regarding the fear of small carnivores in the La Union case study, Campbell (2017b) found that human fear of small predators was based on possible animal aggression such as biting, killing pets, stealing food, or entering homes. Some animals, such as the small cats and the opossum, were considered more dangerous than others on these counts, although none were considered capable of killing adult men or women, or even children. Older people cited previous encounters involving similar or larger animals, while younger people cited information from modern books, films, and music as reasons for their opinions about their own and other people's fear of wild animals. Gender and age were generally unrelated to the fear of animals. For the ocelot, margay, jaguarundi, coati, oncilla, coyote, and opossum, the primary contact was based on avoidance, throwing stones, defending with a cutlass, protecting small livestock and domestic animals, and all animals were considered clever and evasive. Most men and women believed that these species were dangerous to children, but less so to dogs and cats (Campbell, 2017b).

Older respondents related anecdotes from a time when older men and women were more afraid of larger animals because they were more common (Campbell, 2017b). Both men and women agreed that women were more afraid of wild animals than men, but they also noted that men would not completely acknowledge their fear. People explained their fear for similar reasons, for example being bitten, animals jumping on them, unexpected, frightening animal appearances when clearing farms, attacks on household dogs and cats, and stories of imaginary attacks. Just about half of men and women believed that wild animals were harmful to pets, compared to the majority who believed that wild animals were dangerous to children. According to every respondent, children are generally powerless to defend themselves against dogs, cats, or wild animals. Moreover, children could not be replaced, but pets could. Additionally, children were considered to be daring, oblivious to danger, and prone to copy adult behavior that is beyond their developmental level. These claims were supported by media films, shared narratives, eyewitness reports, and cultural norms regarding gender and child relationships (such as high regard for parents, experience and seniority, youthful exuberance) (Campbell, 2017b).

Another study by Campbell (2009, pp. 220–231) examined the importance of gender and age in the responses of rural coastal villages in Ghana to the presence of African palm civets. The case study was of the African palm civet, also known as the "bush cat," (*Nandinia binotata*; body length 87–97 cm; shoulder height 22 cm; tail 45–50 cm) is one of the more common mesopredator species in Ghana's coastal savanna, which is a mosaic of dry forests and shrub/grass areas. However, this species is currently little studied. In the nineteenth and early twentieth centuries, larger predatory mammals like the leopard Panthera pardus were common in the area; however, as human hunting increased, they became extinct due to a lack of protection (Carson, 1985; MacDonald, 1969). Because of its small size and flexible behavior, the civet managed to survive. It exploited the close proximity to human habitation provided by gutters, thick undergrowth on the edges of farms and villages, wood piles, old dead trees, leaf pits, thatched roofs, and overgrown bushes in landfill sites (Campbell, 2009). Although civets are normally nocturnal, human activity has disturbed their resting places, leading to daily interactions between humans and civets. Typical encounter locations were the edges of farms and villages, encircled by a mixture of farmland and bush, and in dense groups within the forest, as well as overgrown gutters and bushes. Fear of attack and the potential loss of food and livestock has formed the basis of human relationships with civets (Campbell, 2009).

The findings demonstrated that despite the fact that no one had ever been attacked by a civet cat, people were nonetheless afraid of them and that their movements were not greatly impacted by their knowledge of their existence (Campbell, 2009). Because they believed that civet cats were livestock robbers, food thieves, and village raiders, women and young people were noticeably more terrified of them. Women were also more likely to hit dense vegetation with sticks and scare off civets by carrying them out of fear for their homes and children (Campbell, 2009). This conclusion was pertinent to sub-Saharan Africa, where local actors' interactions with small mammals and their evaluation have received scant attention. In rural communities, larger mammals—especially those that are predatory—generally elicit stronger fear-related responses (Campbell, 2009). These responses can include fear of losing small livestock, being surprised or threatened when moving in familiar or unfamiliar surroundings, or even fearing that animals have spiritual powers (Campbell, 2009; see also Campbell & Jha, 2021; Jha & Campbell, 2021, 2022, 2023).

Campbell (2009) found that human-civet contacts happened mostly during hunting, gathering of firewood, harvesting, and burning vegetation for agriculture, and traveling through the unmanaged dense forests and bushes. Civets were killed by hunters using open-range shooting and bush clearance, using rifles, muskets, catapults, knives, dogs, and torches during nighttime hunts (Campbell, 2009). The civet cat was regarded as a match for a dog, but older hunters were less inclined to describe it as dangerous, citing their greater knowledge and previous experience with typical civet behavior (including encounters in gutters and hollow trees), compared with previously encountered larger wild mammals such as wild cats, wild boars, and monkeys. Some people were afraid to visit denser vegetation, because of the possibility of civets nearby (Campbell, 2009). Gender was more significant than age in the fear of civets, with women and younger people being more afraid of them. Civets were generally less feared than venomous snakes like cobras in terms of perceived fatal or injurious attacks; however, most women and younger people thought that the civet could bite if startled at close range (Campbell, 2009).

Encounters with civet cats were frequent in the village margins, especially with females handling the majority of the waste disposal in these areas. Small rodents were drawn to the village edges, which were covered in dense bushes, stands of trees, wooded cemeteries, rest areas, and garbage dumps (containing food waste, including feed and fish parts) (Campbell, 2009). Men and women alike reported losing livestock to civets, including chickens. In these regions, people saw civets as food thieves, hazards to children, and rodent predators; younger women were more worried about their children's safety than food theft. Food theft and the predatory role were the two things that worried men the most (Campbell, 2009).

In the rain-fed farms with seasonal farming during the rainy season (April to November), human-civet contact occurred mainly on the forested farm edges, particularly in the farms further from settlements, when farm clearance required the removal possible habitats for civets and other wildlife, including dense scrub, mixed tree-shrub stands, which frequently exposed burrows of civets, rats, snakes, and lizards (Campbell, 2009). Some people, especially younger women, refrained from participating in the management of remote areas due to the possibility of encounters causing stress, especially in the evening. Because of the more open terrain, the increased use of machinery (tractors, plows, water pumps), which could scare off wildlife, and the perception that civets are not as dangerous as large civet reptiles, there were fewer opportunities for human-civet interactions in irrigated agricultural areas. Some women, though, were put off farming in irrigated areas due to potential encounters with civet cats (Campbell, 2009).

Mesopredator-Sized Reptiles and People

Human relations with reptiles are complex, and in some cases reptiles have similar roles to birds and mammals (Campbell, 2009). Reptiles are generally understudied compared with other species in ecological functions (Campbell, 2009; Cox et al., 2022; Hamilton et al., 2022; Hoekstra et al., 2020; Kemp, 2023; Roll et al., 2017). Cox et al. (2022, p. 285) report that "reptiles are threatened by the same major factors that threaten other tetrapods-agriculture, logging, urban development and invasive species-although the threat posed by climate change remains uncertain. Reptiles inhabiting forests, where these threats are strongest, are more threatened than those in arid habitats, contrary to our prediction." Roll et al. (2017, p. 1677) add that "the distributions of amphibians, birds and mammals have underpinned global and local conservation priorities, and have been fundamental to our understanding of the determinants of global biodiversity. In contrast, the global distributions of reptiles, representing a third of terrestrial vertebrate diversity, have been unavailable. This prevented the incorporation of reptiles into conservation planning and biased our understanding of the underlying processes governing global vertebrate biodiversity." Common reptilian mesopredators from the America, Eurasia, and Africa are shown in Figs. 7.17, 7.18, 7.19, 7.20, 7.21, and 7.22.

In terms of reptilian mesopredator relations with people, and human social assessments of this guild and its conservation, there is also the need for more studies (Anan'eva et al., 2015; Miles, 2020; Roll et al., 2017). Janovcová et al. (2019) explain that "although there are many articles about reptiles, no one has ever studied the human perception of reptiles as a whole, a group that would include representatives of different taxonomic clades." Nevertheless, Janovcová et al. (2019) note that reptiles are important for human environment relations and their contribution

Fig. 7.17 Central American imperator. *Source* Nikola Tomasic-Pexels Boa

Fig. 7.18 North American diamond back rattle snake. *Source* Pexels-Pixabay

was to design a "study of human perception of all reptiles focusing on the relationship between perceived fear, disgust, and aesthetic preferences." The findings indicated that people classified reptiles into two groups, legless (snakes) and legged (lizards), with the snakes generally perceived as more as more "disgusting" and "in both groups, species perceived as the least beautiful were the same as those rated as the most disgusting" (Janovcová et al., 2019).

In terms of fear, it is noted that "of all reptiles, only certain groups, especially snakes and crocodiles, are really dangerous for humans" due to the size of the former and the venom (e.g., cobras and adders) and size (e.g., the boa constrictor, python, and anacondas) of the latter (Janovcová et al., 2019; see also Murphy & Henderson, 1997; Öhman & Mineka, 2001; Caldicott et al., 2005; McGregor, 2005). Writing on the human fear of reptiles, termed herpetophobia, Seladi-Schulman (2021)

Fig. 7.19 Indian cobra. *Source* Anil Sharma-Pexels

Fig. 7.20 Sri Lankan pit viper. *Source* Andreas-Pexels

notes that herpetophobia "is a fear of reptiles. People with herpetophobia are most often afraid of reptiles like snakes and lizards. However, they may also be afraid of other reptiles, such as turtles, alligators, and crocodiles." Ophidiophobia, or the fear of snakes, is thought to be the most common symptom of this anxiety disorder that is relatively common in Western countries. Though their exact nature is still unknown, contributing factors could include bad interactions with reptiles, other people's communications, and cultural elements found in religious texts, mythology, and folklore (Seladi-Schulman, 2021).

Janovcová et al. (2019) note that due to human fear of snakes, "it is rational to believe that snakes, although taxonomically a part of lizards, represent a different cognitive category for humans." The results confirmed that people tend to separate snakes and legless lizards from other lizards and other reptiles such as turtles, tuatara,

Fig. 7.21 African nile monitor. *Source* Petrus Bester-Pexels

Fig. 7.22 African black mamba. *Source* Tim Vickers Public Domain

and crocodiles (Janovcová et al., 2019). Respondents distinguished between snakes and other reptiles based on their ratings of beauty, disgust, and fear. It has been suggested that people confuse reptiles with other species primarily because of their outward similarities to snakes and legless lizards (Janovcová et al., 2019). Consequently, snakes are seen by humans as belonging to a different class of animals. People are also terrified of other creatures that resemble snakes but are not actually snakes (Janovcová et al., 2019).

Janovcová et al. (2019) note that prior research has demonstrated that human attitudes and perceptions, particularly with regard to reptiles, have a significant impact on the efficacy of local protection initiatives. Unlike other vertebrates like mammals or birds, reptiles are not very popular, and the dislike they arouse may prevent them from being protected (Janovcová et al., 2019). Because snake-like species are linked to strong feelings of disgust and fear, they may be more susceptible to attacks by humans and a decrease in conservation efforts (Janovcová et al., 2019). Previous studies have demonstrated that a variety of educational initiatives targeted at individuals living in the relevant area can help people overcome these negative emotions, which are mirrored in aversive attitudes (Janovcová et al., 2019).

Another case study by Ceríaco (2012) focused on the role of human values and wildlife folklore in conservation efforts, with variations based on demographic characteristics such as gender, age, or education. The authors argued that reptiles and amphibians are among the most underappreciated vertebrates and are victims of many negative values and misconceptions that result from the direct interpretation of folklore (Ceríaco, 2012). Furthermore, "although reptiles and amphibians are not responsible for major economic losses and most are harmless, they are feared and persecuted" (Ceríaco, 2012). Numerous studies indicate that fear of snakes is a common phobia in Western countries, including folklore and modern cultures (Ceríaco, 2012).

For example, Alves, Vieira, et al. (2012) there are a "variety of interactions between human cultures and herpetofauna", the "subject matter of ethnoherpetology, a subdivision of ethnozoology" and present a case study of Brazil, where "many reptiles interact with human communities because of their utility or because of the risks they represent." Here human-reptile interactions included snakes, lizards, turtles, and crocodilians, and were used for medicinal purposes, food, ornamental purposes, religious practices, and pets. Some species were killed as potentially dangerous.

Ceríaco (2012) also mentions ethnozoology, the study of the interaction between people and animals. Alves and Souto (2011) define this subject as interdisciplinary in nature, developed from studies in academic fields such as zoology, human ecology, sociology, and anthropology. Ethnoherpetology is a subfield of this study that is defined as "the study of human relationships with and knowledge about reptiles and amphibians" (Ceríaco, 2012). According to Ceríaco (2012), few studies have shown how this kind of knowledge negatively affects conservation, and nearly none have ever established a direct link between the existence of folklore, unfavorable values and preferences, and persecution as well as anti-conservation attitudes toward reptiles. The persecution of reptiles and amphibians is a serious issue that can affect their conservation. These myths, folklore, false beliefs, and negative values all play a part in this. The information in the literature on ethnozoology, conservation biology, and herpetology can be regarded as anecdotal. Very little research has been done to determine whether or not these problems affect people's perceptions of amphibians and reptiles (Ceríaco, 2012).

According to Ceríaco (2012), the findings indicate that prejudice and anti-conservation sentiments regarding amphibians and reptiles are specifically predicted by the existence of folklore and negative values. Folklore varies in sociodemographic ways as well, but people generally have negative attitudes toward these animals (Ceríaco, 2012). These findings imply that folklore and negative values, which are also influenced by sociodemographic factors, account for human persecution of reptiles and amphibians and anti-conservation attitudes toward them. Negative attitudes toward reptiles and amphibians were also significantly correlated with folklore about them. Citations are made to other studies (Coursey, 1998; Czech et al., 1998). Reptiles may have fewer supportive groups than other animals, as generally people prefer mammals and birds to reptiles and amphibians (Czech & Krausman, 2001; Gibbons, 1988). Nolan et al. (2006) discovered that, in comparison with other taxa like fish, birds, or mammals, some reptiles were the least valued creatures. More often than not, people mistook reptiles for amphibians, but amphibians were perceived more favorably than reptiles, and reptiles pose more threats than amphibians (Nolan et al., 2006; Ohman & Mineka, 2003; Ohman & Soares, 1994).

Font et al. (2023) present evidence that refutes the common perception of reptiles as stupid and behavioral and cognitively subpar, and the "persistent myth of the sluggish, primitive, stupid reptile can be traced, at least in part, to a long tradition of misconceptions about brain size and organization." It is argued that "reptilian brains were historically considered small and lacked the neural structures that support complex cognition in other vertebrates" and the idea that reptiles are animals that do not have a cerebral cortex has led to the expectation that their cognition should be simple compared to birds and mammals (Font et al., 2023). Current scientific evidence suggests that reptiles have highly developed brain structures similar to those of mammals and birds. The authors conclude that "reptiles are capable of spatial, social learning, reversal, problem solving, and many other types of learning and cognitively demanding behaviors, provided that experimental designs take into account some peculiarities of their biology involving their morphology, physiology and ecology" (Font et al., 2023).

A comparable study on the relationship between human perceptions of reptiles and social support for the conservation of reptile species is also presented by Tisdell (2011). The findings revealed, for instance, that "human attitudes toward reptile species appear to be more varied than those toward other vertebrates" and turtle species are more popular than crocodile and snake species (Tisdell, 2011). Factors for perceptions of reptile species include positive feelings, cultural attachments, reptile species survival status, the threat to humans, their economic use value, and perception of pest status (Tisdell, 2011). People's attitudes toward various wildlife species, such as their empathy and support for their survival, are said to both influence and be influenced by societal attitudes (Tisdell, 2011). Certain dangerous reptiles, like large crocodiles and venomous snakes, are often associated with social disgust. Conversely, most species of turtles are positively portrayed in society due to their benevolent relationship with humans (Tisdell, 2011). In addition to providing strong evidence for the survival of all wildlife species—including the least popular reptile species—the results also provided some evidence in favor of the phylogenetic similarity principle, which holds that there is a tendency for reptiles to be generally less popular than birds and, conversely, for birds to be less popular than mammals. Lower likability of these species is linked to a decrease in the proportion of individuals who support their survival. According to Tisdell (2011), this indicates that reptile species are probably overrepresented among extinct species and underrepresented among saved species in a scenario where some species inevitably go extinct owing to resource scarcity (Tisdell, 2011).

For a case study, based on the significant influence that reptiles have on West African social structures and the fact that reptiles receive less attention than mammals in environmental and social studies, Campbell (2009, 2013) investigated human-reptile interactions in Ghana's coastal savanna (Luiselli & Angelici, 2000; Luiselli et al., 2001; Tolley et al., 2016). Large reptiles are seen as bushmeat, sources of leather, predators, hazardous pests, barriers to land use, and symbols of evil in West Africa, where they are feared, revered, and even engaged in combat by humans (Bennet, 2002; Bennett & Takoordyal, 2003; Gorzula & Oduro, 1997; Jenkins & Broad, 1994). In the West African region, most studies focus on the ecology of small mammals, bushmeat production, other protein sources such as fish and livestock, and/or general environmental degradation, while human-reptile relationships are generally ignored, although gender differences in the superstitious fear of small animals in West Africa have been documented for snakes (Luiselli & Angelici, 2000; Luiselli et al., 2001; Mendelson et al., 2003).

Several species of snakes and lizards were studied, including mussurana (*Clelia clelia* Daudin, 1803; pit vipers (*Crotalus durissus* (L.1758, *Cerrophidion godmani* (Günther, 1863 or *Porthidium ophryomegas* (Bocourt, 1868), boa constrictor (*Boa constrictor* L. 1758), and green iguanas (*Iguana iguana* L. 1758) (Campbell, 2009, 2013). Human reactions to these animals included avoidance, defense with cutlasses, and avoidance altogether, with the exception of the avoided iguana. According to Campbell (2009), reptiles—aside from iguanas—were thought to be elusive, rare, dangerous, and capable of attacking humans and small animals like dogs, sheep, and goats. Rainfed farms, tree cover areas, and the edges of the village were all home to reptiles. While they were less common in village areas than in other areas, reptiles could occasionally be seen inside houses and were mostly found in gutters, garbage dumps, and edges with dense vegetation. Additionally, sightings on roofs, food stalls, window ledges, doors, and under beds have been reported (mostly of monitors) (Campbell, 2009, 2013). Given that women were more likely to use the locations listed and engage in related activities, the majority of sightings of this type were reported by females (Campbell, 2009, 2013). The sociocultural effects of human-reptile interaction included people avoiding certain risk areas, believing that large snakes and lizards could kill children, and young domestic fowl to reptiles. Some people linked spirits to reptiles, which were thought to be the spirits of the dead or gods from traditional religion, which made the particular reptile that possessed them evil. However, younger people were generally skeptical of these ideas, due to the influence of "Western" education and Christianity (Campbell, 2009).

Cobras were the commonest reptile in this study outside of the village areas on the rain-fed farms. People observed that cobra sightings could cause a variety of reactions, such as walking or running away, whereas pythons and monitor lizards elicited fewer reactions, primarily through loud behavior. When farming or logging, more women than men covered their legs with thick cloth (often in layers of denim or khaki) to protect against thorns and snakebites. These people also thought it was "possible" for spirits to possess reptiles (Campbell, 2009). About 80% of the large reptiles on the irrigated farms were pythons and monitor lizards; cobras were a little less common than on rain-fed farms, and rodents were the main prey. Farmers were approached and attacked by reptiles more frequently on irrigated farms, particularly by pythons and monitor lizards. The increased reptile presence was attributed to wet soils, denser vegetation on the farm's edges, puddles of water in the gutters, frequent waterlogging during sporadic rains, and excessive irrigation. Farmers increased their precautions because they anticipated higher densities of reptiles (Campbell, 2009). There were fewer reptiles in tree-covered areas than in irrigated areas but more than in rain-fed areas. Cobras made up around two thirds of all reptiles, especially where hunting caused disturbance due to manual clearing of shrubs, logging, and bush burning. The process of extracting wood led to increased interactions between humans and reptiles, as wood cutters would move in parallel rows which caused the animals to move forward and led to 50% more sightings (Campbell, 2009).

Long-term human behavior in logging and clearing new farms was impacted by reptiles, as evidenced by interactions such as people stepping on snakes, snakes biting pants, reptiles moving close to human legs, and monitor lizard tails striking human legs (Campbell, 2009). When a reptile was spotted, women were more likely than men to react by fleeing the scene. Before harvesting timber or clearing the farm, people observed that they would thrash the bush. Men were more likely to react to reptiles rather than run away, whereas young women were culturally more sensitive to contact with them. There were differing views on the spirituality of reptiles. Some, especially snakes, were thought to possess spiritual abilities similar to those of humans. Hunting of mammals such as antelopes, civets, and rodents was affected by the presence of reptiles, as the bushwhacking and fire required to drive mammals from cover also disturbed reptiles. Additionally, reptiles appeared adapted to human behavior and invaded cultural areas like farms, village dumps, homes, and marketplaces (Campbell, 2009).

Alves, Pereira Filho, et al. (2012, p. 6877) report a study from Brazil that examines how people's attitudes toward reptiles vary depending on their culture, environment, and individual experiences. The authors contend that there are at least 719 species of reptiles known to exist in Brazil, and that 11% of this fauna has been utilized for a range of purposes, such as bushmeat, leather, ornamentation, magical and religious uses, and traditional medicine. For a deeper understanding of how human societies use reptiles, Brazil can therefore be a useful case study (Alves, Pereira Filho, et al., 2012, p. 6877). In Brazil, there are over 80 species of reptiles that are significant to culture, which are exploited for a variety of therapeutic applications, culinary applications, decorative or ornamental uses, magical or religious rituals, pets (Alves, Pereira Filho, et al., 2012, p. 6877). Although most of these interactions are detrimental to the conservation of wildlife, the authors contend that there are numerous ways in which humans and reptiles interact, underscoring the significance of comprehending these uses and interactions in the context of reptile conservation in Brazil (Alves, Pereira Filho, et al., 2012, p. 6877). Alves, Pereira Filho, et al. (2012, p. 6877) conclude that "there are many forms of interaction between reptiles and humans in Brazil—although most of them are quite negative in terms of wildlife conservation—which reinforces the importance of understanding such uses and interactions in the context of protecting reptiles in Brazil. A better understanding of the cultural, social, and traditional roles of these reptiles is fundamental to establishing management plans for their sustainable use."

Pasmans et al. (2017) discuss how political actions are progressively restricting the ownership of reptiles as exotic pets. The authors argue that keeping reptiles and amphibians, like keeping other, more "traditional" pets, benefits society but raises concerns about animal welfare, human health and environmental sustainability. Unskilled pet owners' inadequate care and diet are issues with such intimate relationships between humans and reptiles. On the other hand, the authors note that they do not believe there are any good reasons to limit access to veterinary care, herpetofauna, or the keeping of reptiles and amphibians. Instead, they believe that specialist training should be enhanced and that veterinary curricula should further diversify herpetological medicine (Pasmans et al., 2017).

In comparison with reptiles, amphibians may appear less menacing to humans due to their lack of large, intimidating species like crocodiles and poisonous species like snakes (Ceríaco, 2012). Brom et al. (2020) for instance examined the relationships between amphibians and humans and found that amphibians are a class of animals that are both loved and loathed, making them a suitable subject to compare and decipher the factors that influence preferences and influencing attitudes toward animals. Based on cultural norms, early experiences, and the attitudes of primary caregivers, people's opinions and attitudes toward amphibians are largely formed in childhood and maintained into adulthood, according to the study's findings. It is also mentioned that anthropogenic global changes are posing a threat to amphibian species worldwide (Brom et al., 2020). Examples from urban settings are given, and data from North America indicates that even though amphibian populations in urban areas are smaller than those in the wild, they seem to be declining similarly or more slowly, indicating that urban environments may be able to serve as a haven for certain amphibian species. Ponds and natural urban wetlands are examples of amphibian habitats in urban settings. In many of these places, people have a negative attitude that is shaped by education, folklore, and fiction. This may have implications for amphibian conservation, which may receive less funding than that of mammals, birds, reptile, or fish (Brom et al., 2020).

Azevedo et al. (2022, p. 755) cite the cases of reptiles as pets, and note that "owning reptiles (e.g., chelonians, lizards, and snakes) poses a great deal of challenges in terms of welfare and care, as well as conservation and environmental concerns" and "despite the large scale of the reptile trade, little is known about the motives for acquiring and keeping these animals." The findings indicated that "human–reptile relations are driven by the same feelings as those previously reported for mammalian pets, although they are more diverse and nuanced by a sense of admiration and fascination for their mysterious nature and unusual behaviors. Building on these strong emotional bonds, promoting early education about wildlife conservation, and responsible pet keeping could play a crucial role in improving captive-reptile welfare" (Azevedo et al., 2022, p. 755).

This study was justified as globally, millions of reptiles are traded and kept as household pets, with more than 500 species of reptiles traded in the European Union, and eight million of them are thought to be kept as pets (Azevedo et al., 2022, p. 756). Although there are obstacles in the form of expensive enclosures and a lack of information about reptile care for prospective owners, the size of the reptile trade indicates that people are highly motivated to obtain and maintain reptiles as pets (Azevedo et al., 2022, p. 756). The authors nevertheless argue that despite a wealth of research on the relationship between humans and common mammals, it is unclear if the findings of these studies apply to reptiles, which seem less able to satisfy the human need for compassionate behavior (Azevedo et al., 2022, p. 756). With turtles being the most popular animals and snakes being the least, it appears that people's attitudes toward different species of reptiles vary. It may be more common to keep companion animals like parrots or primates that behave like humans. Reptiles may be devoid of these behavioral characteristics (Azevedo et al., 2022, p. 756; see also Almeida et al., 2014; George et al., 2016; Reuter & Schaefer, 2017; Warwick et al., 2018; Rodríguez et al., 2020; Toland et al., 2020).

Mesopredator-Sized Birds and People

Apex avian predators such as large eagles can be considered mesopredators in their relationship to humans because they take similar sized prey (including small livestock and domestic animals) as mammalian and reptile mesopredators, but unlike apex mammal predators such as big cats, these birds do not pose a comparable threat to humans and large livestock (Campbell, 2022; Jha & Campbell, 2022, 2023). These large eagles fulfill similar roles to medium-sized and small cats such as bobcats, caracals, servals, and ocelots, as well as reptiles such as medium-sized snakes and monitor lizards (Campbell, 2022). Examples are shown in Tables 7.1, 7.2, and 7.3. Table 7.3 shows human interactions with such raptors. Tables 7.4, 7.5, 7.6, 7.7 and 7.8 show descriptions of some of the world's largest eagles. Figures 7.23, 7.24, 7.25, 7.26, and 7.27 show examples of the largest eagle species. Figure 7.28 shows a domestic goat, which is a typical prey species for such eagles.

Giraldo-Amaya et al. (2021, p. 281) note that "among birds of prey, the largest species are usually considered more sensitive to human disturbance and threats. Large raptors are ecologically unique due to their strong effects on prey populations; for some species, even just the raptor's presence is perceived as a potential predation risk, affecting the distribution and behavior of prey." This chapter selects as examples four of the largest eagle species within the family Accipitridae (Vieillot, 1816), which hunt in different biogeographic realms, defined as large spatial regions in which ecosystems share broadly similar biota, such as the Afrotropical (mainly sub-Saharan Africa), the Nearctic (most of North America), the Indo-Malayan (India and Southeast Asia), the Neotropical (South America and parts of Central America), and the Palearctic (Eurasia) biogeographic areas (Hassan et al., 2005).

Table 7.1 Large tropical eagles according to size

Species	Scientific name	Mass (kg)	Length (cm)	Wingspan (cm)
Harpy eagle	*Harpia harpyja* (Linnaeus, 1758)	4–10	86.5–107	1.8–2.2
Philippine eagle	*Pithecophaga jefferyi* (Ogilvie-Grant, 2019)	4.7–8.0	86–102	1.8–2.2
Golden eagle	*Aquila chrysaetos* (Linnaeus, 1758)	2.5–6.4	66–102	1.8–2.3
Crowned eagle	*Stephanoaetus coronatus* (Linnaeus 1766)	2.6–4.1	80–95	1.5–1.8
Martial eagle	*Polemaetus bellicosus* (Daudin, 1800)	3.5–4.7	78–96	1.9–2.3

Sources Campbell (2022)

Table 7.2 Large raptors according to realm

Realm	Eagle species
Neotropical	Harpy eagle, crested eagle
Indomalayan	Philippine eagle,
Nearctic, palearctic	Golden eagle
Afrotropical	Crowned eagle, martial eagle

Source Campbell (2022)

Table 7.3 Social issues relevant to large raptors, by discipline

Discipline	Issues
Economics	Forestry, non-timber forest production, agriculture, urbanization, transport, hunting, industrial production
Sociology	Social perceptions of wildlife, conservation actions, social actions in environmental extraction, resource extraction
Political science	Stakeholder power in conservation, exploitation, urbanization, industrialization, decentralization, implementation
Psychology	Human perceptions of wildlife and habitats, human thinking shaped by the environment, background for social perceptions of conservation and exploitation
Anthropology	Cultural patterns of resource use, local knowledge of habitats, primal land cover, local participation in exploitation, conservation
History	Changing patterns of land use, modernization of resource use, trends in land use, human population growth, social, and environmental coevolution
Human geography	Spatial relations of human settlement, land use, exploitation and conservation impacts, scaled habitat change, patterns of socioenvironmental interactions

Sources Campbell (2015, 2022)

Fig. 7.23 Harpy eagle. *Source* Tee Zett-Pixabay

Fig. 7.24 Philippine eagle. *Source* Janssen Panizales-Pexels

The Harpy Eagle

The Harpy eagle is one of the largest eagles in the world, competing in this regard with the Steller's Sea eagle, and larger than the Philippine eagle, with no close avian competitor throughout its largely forested range (Campbell, 2022). Bates (1964, p. 99, 100) writes "the giant harpy, most powerful of the world's eagles, terrorizes mammals and birds of the forest canopy…the heaviest of all eagles" and is "the largest and fiercest inhabitant of the upper canopy… a most unusual bird of prey…Weighing some 20 pounds and with short, powerful wings" (104). Aguiar-Silva, Luz, et al. (2014, p. 24) writes that this species is "the heaviest and the most powerful bird of prey in the canopy of the Neotropical rainforests." The harpy eagle has a wide prey range, with up to 69 prey species reported by one source (Aguiar-Silva, Sanaiotti, & Luz,

Fig. 7.25 Golden eagle. *Source* Jesus Esteban San José-Pexels

Fig. 7.26 Martial eagle. *Source* Veterin TF-Pexels

2014, p. 24) and 102 prey species by another source (Miranda, 2018, p. 1). Such prey species include monkeys (Platyrrhini spp., E. Geoffroy, 1812), coatis (Nasua spp., Storr, 1780; Nasuella spp., Hollister, 1915), sloths (Bradypus spp., Linnaeus, 1758; Megalonychidae spp., Gervais 1855), agoutis (Dasyprocta., Illiger, 1811), opossum (Didelphidae., Gray, 1821) and New World porcupines (Erethizontidae.,Bonaparte 1845), and large birds and reptiles (Seymour et al., 2010; Rotenberg et al., 2012; Schulenberg, 2020; Campbell, 2022).

The harpy eagle is described as a threatened species due to habitat loss and direct persecution and as "a flagship species for Neotropical conservation" (Miranda, Peres, Carvalho-Rocha, et al., 2021, p. 110; see also Aguiar-Silva, Sanaiotti, & Luz, 2014). The Harpy eagle is commonest in Brazil and rarer in Mexico and Central America (Campbell, 2022). Miranda (2015, p. 70) writes that the species is threatened by

Fig. 7.27 Crowned eagle. *Source* Picfair Licence Purchase

Fig. 7.28 Young goat, typical domestic prey for eagles. *Source* Gundula Vogel-Pixabay

retaliatory killings due to possible predation on domestic animals, habitat loss, trophy shooting, deforestation, capture of birds (particularly juveniles) for social use and hunting and trapping of terrestrial prey animals (Ferguson-Lees & Christie, 2001; Trinca et al., 2008; Godoi et al., 2012; BirdLife International, 2021a; Campbell, 2022) (Table 7.4).

In terms of direct killings of harpy eagles, the main perpetrators are hunters (Gusmão et al., 2020; Giraldo-Amaya et al., 2021). In addition, indigenous groups kill harpy eagles in their Amazonian territory, to use its feathers for traditional headdresses, to make arrows, as souvenirs, or to keep or trade as "mascots" (Alvarez-Cordero, 1996; BirdLife International (2021b); Piana (2018). Harpy eagles may also be killing by people entering habitat areas for farming and mining operations (BirdLife International, 2021b). For example, in Venezuela, at least 45 harpy eagles

Table 7.4 Conservation status of the harpy eagle

Conservation status	Vulnerable
Adult numbers	100,000–250,000
Population	Decreasing
Habitat	Forest, Savanna
Decline in habitat	Yes
Threats	Deforestation, livestock farming and ranching, killing, hunting of prey species, industrialization, infrastructure development
Conservation actions	In Brazil, it is included in the National Action Plans for avian conservation (ICMBio 2014), the Cerrado and Pantanal (ICMBio 2019), and the Atlantic Forest (ICMBio 2017). Protected area development, research on ecology threats in Brazil, Ecuador, and Panama, captive breeding program, reintroductions in Belize, Panama, Colombia. Education programs in Peru, Panama, Venezuela, and Brazil

Sources BirdLife International (2021a)

were shot or poached from the 1980s to 1990s (Alvarez-Cordero, 1996). In Panama and Belize, eight of the 18 eagles released for reintroduction were shot (Watson et al., 2016).

Two main factors have been cited for human killings of harpy eagles: more importantly possible threats to livestock and to a lesser extent general curiosity on the part of people (Campbell, 2022). On the theme of threat to livestock, Miranda, Peres, Carvalho-Rocha, et al. (2021, p. 110) found that "livestock abundance and livestock husbandry were the best positive predictors of levels of self-reported livestock predation by harpy eagles", and the predated animals included chickens (48%), goats (22%), pigs (18%), sheep (8%) and pets (3%) out of a total of 192 animals, and small landholders were more likely than large landholders to kill eagles (Miranda, Peres, Carvalho-Rocha, et al., 2021, p. 110). Eagles were also killed out of "simple curiosity, and many interviewees reported later regretting their acts" (Miranda, Peres, Carvalho-Rocha, et al., 2021, p. 110; see also Wordley, 2021). These authors contend that ameliorative actions may include communication with smallholders regarding "education, compensation and tourism activities" (Miranda, Peres, Carvalho-Rocha, et al., 2021, p. 110).

In another study, Giraldo-Amaya et al. (2021, p. 281) concurred that the harpy eagles perceived threat to livestock and consequent human persecution. Over 130 cases of Harpy Eagle persecution were recorded between 1950 and 2020, but the information for some locations in Central and northern South America. Possible motivations for eagle persecution included ignorance, fear, perceived and real risk of livestock predation, curiosity, illegal trafficking, desire to keep an eagle in captivity, or food sources (Giraldo-Amaya et al., 2021). The size of the harpy eagle is cited as an important fact in human assessments, which is relevant even for the smaller eagles (Giraldo-Amaya et al., 2021; Campbell, 2022). For example, a study of the smaller Black-and-chestnut Eagle (Spizaetus isidori) in Andean communities in Colombia found that "the perception of S. isidori by the inhabitants was largely negative and influenced by different sociodemographic factors such as gender, chicken (Gallus gallus) ownership, and chicken management" and recorded killings were due to shootings and electrocution on high tension wires (Restrepo-Cardona et al., 2020). The authors concluded that "the persecution of *S. isidori* occurs as retaliation or as a preventative measure against chicken predation and is a significant cause of mortality of this species in Colombia (Restrepo-Cardona et al., 2020).

Because harpy eagle habitat may be within indigenous areas, native Amazonians kill harpy eagles for their primary wing and tail feathers for headdresses and fletching arrows, and trade headdresses commercially, and capture young eagles for sale on the black market. Therefore, cooperation with local human populations is important, including enaction of laws for eagle protection and supporting indigenous cultural practices which may indigenous communities will encourage sustainable use while fortifying the legal framework that supports conservation (Miranda, Peres, & Downs, 2021). These positions are echoed by BirdLife International (2021b) as there may be collaboration with indigenous groups to reduce eagle killings to at least a sustainable level. Eagle reintroductions, for example in the Brazilian Atlantic Rainforest may also be considered (see also Vargas et al., 2006; Piana, 2007, 2018; Trinca et al., 2008; Miranda et al., 2019; Miranda, Peres, Carvalho-Rocha, et al., 2021; Miranda, Peres, & Downs, 2021; Campbell, 2022).

Aguiar-Silva, Luz, et al. (2014, pp. 24–35) however contest the thesis of substantial livestock predation by the harpy eagle, citing evidence that the diet of some Harpy Eagles in the central Amazonian rainforest was mostly two species of sloths (Bradypus variegatus and Choloepus didactylus), and primates, rodents, carnivores, and birds, but "there was no evidence of Harpy Eagle predation on livestock or domestic animals" (Aguiar-Silva, Luz, et al., 2014, pp. 24–35). Aguiar-Silva, Luz, et al. (2014, p. 30) conclude that "in contrast to local beliefs, there was no evidence in the prey remains that Harpy Eagles fed on livestock or on domestic animals such as chickens or ducks, even where Harpy Eagle nesting territories overlapped human settlements… The biomass of wild animals in the study area is probably sufficient to support the population of Harpy Eagles, suggesting that there is enough forest cover to allow Harpy Eagles and arboreal mammals to coexist." However, some adult Harpy Eagles were killed due to livestock risk and curiosity (Aguiar-Silva, Luz, et al., 2014, p. 30; see also Concepcion et al., 2006).

Philippine Eagle

The Philippine eagle is another very large eagle, possibly the largest in the world by length and wing area, and only slightly smaller by weight and mass than the harpy eagle and Steller's sea eagle (Ferguson-Lees & Christie, 2001; Campbell, 2022). This species occurs in primary forests, forest-edge forests, and logged forests (Kennedy et al., 2000). Due to the absence of other large predators in its range, it has the status of a dominant apex predator with a specialized foraging ecology vulnerable to deforestation (Campbell, 2022). The main prey species include monkeys (the most common monkey species in its range is the crab-eating macaque (Macaca fascicularis, Raffles, 1821), and lemurs, civets, flying squirrels, flying foxes, and macaques; large snakes, monitor lizards, and even large birds such as hornbills (Brown & Amadon, 1968, see also Kennedy, 1985, Ibañez et al., 2003). Kennedy (1985) recorded 218 prey items. Of the 158 species found, mammals made up the majority (84.3%), followed by birds (7.8%) and reptiles (7.8%). Mammals that were most prevalent included flying lemurs (54.2%), palm civets (12.0%), flying squirrels (7.8%), rousette fruit bats (54.4%), long-tailed macaques (30.0%), and small mammals such as rats and deer (11.8%) percentage. Birds included a small juvenile owl and rufous hornbills. Reptile prey included Philippine cobras (1%), an unidentified snake, and a monitor lizard (0.6%) (Kennedy, 1985).

The Philippine eagle is "critically endangered" due to habitat losses and human persecution (BirdLife International, 2018a: see also Salvador & Ibanez, 2006, p. 171). Recent conservation efforts include reintroductions as the decline in the Philippine eagle population was previously attributed to killings and inadequate recruitment to the breeding population (Kennedy, 1977) Miranda, 2006). A recent example (2006) was a rescued bird (injured by gunshot) that was rehabilitated by the Philippine Eagle Foundation, then rehabilitated and released with a radio tracking device. This bird was found dead, shot and eaten by villagers (Salvador & Quiblatt, 2009). Subsequent research blamed poor dispersal and survival of juveniles and sub-adults for reintroduction failures (Miranda et al., 2008) (Table 7.5).

The factors for habitat destruction commercial logging and shifting agriculture (BirdLife International, 2018b). Lowland forest may be used by logging and mining concessions. Short-term threats include unregulated food hunting, zoo exhibits, and trade (Miranda et al., 2008). Adult eagles nesting close to forest edges are also easily shot or captured (BirdLife International, 2018b). Eagles may also be killed by snares while hunting on the forest floor and may be caught in deer and wild boar traps (BirdLife International, 2018b). Another possible but unproven threat that might impede already slow reproductive rates is pesticide accumulation in the environment

Table 7.5 Conservation status of the Philippine eagle

Conservation status	Critically endangered
Adult numbers	180–500
Population	Decreasing
Habitat	Forest
Decline in habitat	Yes
Threats	Annual and perennial non-timber crops, mining and quarrying, hunting and trapping terrestrial animals, logging and wood harvesting, agricultural and forestry effluents, storms and flooding
Conservation actions	CITES Appendix I. The species occurs in several protected areas including the northern Sierra Madre Natural Park on Luzon, and Mt. Kitanglad and Mt. Apo natural parks on Mindanao

Sources BirdLife International (2018b)

and in the bodies of eagles. The Philippine eagle population may also be impacted by extreme weather events, such as Typhoon Haiyan in 2013 (BirdLife International, 2018b; De Win, 2013).

Conservation measures include expansion of protected areas to include known eagle nests and habitats and conducting additional research on distribution, population, ecological needs, and threats. At the federal level, there could be political campaigns to save eagles and their habitats (BirdLife International, 2018b). Local governments may also consider the development of the Philippine Eagle Conservation Plan (Haribon Foundation, 2013). The release of captive-bred and rehabilitated birds can consider genetic differences among eagles in the dispersed ranges of Luzon, Mindanao, Samar, and Leyte (Miranda et al., 2008). It would also be beneficial to expand current research and development of conservation strategies for eagles in the Sierra Madre mountains (Haribon Foundation, 2013).

Golden Eagle

The golden eagle is arguably the best-known eagle or bird of prey in the world, due in part to its enormous range, hunting abilities, size, and the respect of the peoples with which it lives. This giant eagle is an avian apex predator throughout its range (Fernández-Gil et al., 2022). According to Katzner et al. (2020), golden eagles prey mostly on mammals. These mostly small mammals may include leporids (hares and rabbits), sciurids (squirrels, prairie dogs, marmots), fish, birds, reptiles, and even carrion (Collins & Latta, 2009; Kochert et al., 2002). Larger mammals such as red (*Cervus elaphus* Linnaeus, 1758) and mule deer (*Odocoileus hemionus*, Rafinesque, 1817) and pronghorn (*Antilocapra americana* Ord, 1815), which can reach 1.2–2.1 m in length and weigh over 50 kg, can also be killed and eaten (United States Fish and Wildlife Service, 2011; Watson, 2010).

Golden eagles also kill mammals that are mesopredators, especially foxes (Jackson et al., 2023; Lyly et al., 2015). These include the island fox (*Urocyon littoralis*, Baird, 1857) in California (Roemer et al., 2002), the gray fox (Urocyon cinereoargenteus), the swift fox (Vulpes velox), coyote (Canis latrans, Say, 1823), American badger (Taxidea taxus, Schreber, 1778) and fisher (Pekania pennanti, Erxleben, 1777) in North America, the corsac fox (Vulpes corsac, Linnaeus, 1768) in Mongolia, the Eurasian otter (Lutra lutra, Linnaeus, 1758) and badgers (Melesmeles, Linnaeus, 1758) in Scotland, and martens (Martes foina, Erxleben, 1777 and Martes martes, Linnaeus, 1758) in the French Alps.

These mesopredators can account for more than 5% of the prey in certain situations (Hoffmann et al., 2004; Campbell, 2022). For example, red foxes are commonly responsible for nest predation (Collins & Latta, 2009; Ellis et al., 2000; Olendorff, 1976; Ralls & White, 1995; Watson, 2010). The larger wolverines, coyotes, and bobcats, and even wild boars (*Sus scrofa domesticus* Erxleben, 1777) are also recorded as prey in North America and Europe, according to a few reports

(Olendorff, 1976; Bowen, 1980; Jordheim, 1980; Mason, 2000; Jung et al., 2009; Campbell, 2022).

Domestic animals, such as goats and lambs, may be killed by golden eagles (O'Gara, 1998). For example, O'Gara (1998) points out that "Golden eagles are more likely to prey on livestock than are bald eagles. Both species readily feed on livestock carrion and carcasses left by foxes and coyotes, although some individuals prefer live prey to carrion. Eagles are efficient predators and can cause severe losses of young livestock, particularly where concentrations of eagles exist. Generally, they prey on young animals, primarily lambs and kids, although they are capable of killing adults." Acorn & Dorrance (1990, p. 31) also cite instances of golden eagles killing lambs and kids in the United States and Canada. The golden eagle's diet in Scotland includes a large amount of dead lambs (Brown, 1976; Houston & Maddox, 1974). Additionally, dead pigs have been found in Corsica and on the island of Santa Cruz in California (Collins & Latta, 2009; Seguin et al., 2001). Roemer et al. (2002) cite the case of golden eagle predation on pigs in the California Channel Islands: Here "introduced pigs, by providing abundant food, enabled golden eagles to colonize the California Channel Islands. Eagles preyed heavily on the island fox, whose resulting decline toward extinction released populations of the competitively inferior island skunk." Geese, cranes, and wild birds like grouse are among the large captive birds that golden eagles have killed. In areas where they are common, wild birds can make up as much as 40% of the prey mass of golden eagles (Brown & Amadon, 1968). Carrion is an occasional food source for golden eagles (Bodio, 2012; Campbell, 2015).

According to Watson (2010), carrion accounts for a higher percentage of an eagle's diet during winter because there are fewer prey species available during this time of year than other seasons (Boag, 1977; Sánchez-Zapata et al., 2010; Wilmers et al., 2003). In America, at carrion food sources, golden eagles dominate the larger Andean and California condors, the smaller turkey and black vultures and occasionally the griffon vultures in Eurasia and Africa (Campbell, 2015). Along their shared range, larger sea eagles (sea eagles in Europe, bald eagles in North America) and golden eagles contend over carrion. Due to the white-tailed eagle's slightly larger size and the golden eagle's more aggressive nature, interspecies dominance is unclear (Halley & Gjershaug, 1996; Pettersson, 1977) (Table 7.6).

The relationship between people and golden eagles is based on the perception of its power and beauty (sometimes from a quasi-religious position) and also on the possibility that it may kill livestock or even attack people (hence its persecution) (Campbell, 2022). On the first point, the golden eagle one of the most significant religious and regal symbols, embodying majesty, and splendor. Eagles in this representation were portrayed with either one of two heads. The double-headed eagle was frequently portrayed in various Eurasian cultures and eras as a heraldic emblem of numerous royal houses and remains a state symbol in certain nations. The eagle was one of the representations of absolute power and even military prowess for the ancient Romans. The pre-Roman Greeks initially acknowledged the eagle's physical prowess as the foundation for its glorification. These included strong claws, swift reflexes, strength, power, and a strong sense of vision. A divine message (the eagle

Table 7.6 Conservation status of the golden eagle

Conservation status	Least concern
Adult numbers	100,000–200,000
Population	Stable
Habitat	Forest, Shrubland, grassland, rocky areas (e.g., inland cliffs, mountain peaks), desert
Decline in habitat	Yes
Threats	Wood and pulp plantations, utility and service lines, diseases, habitat shifting and alteration, renewable energy, hunting and trapping terrestrial animals, agricultural and forestry effluents
Conservation actions	Covered by North American Christmas bird count and breeding Bird survey. Systematic breeding surveys are carried out in at least 18 European countries. Retrofitting of power poles reduces electrocution risk

Sources BirdLife International (2021a)

is associated with Zeus, the king of the gods, the Roman Jupiter), victory in battle, freedom (exemplified by free, soaring flight), a god or goddess in bird form (e.g., goddess Athena, the Roman Minerva (who, in Homer's Odyssey, assumed the form of an eagle), and the intermediary between the gods and humanity. In ancient Rome, eagles were also depicted on coins and were believed to have served Jupiter. They were linked to Minerva and Mars, the warrior goddess of wisdom and the god of war, respectively. The eagle's contribution to the rise of the Roman military class was arguably its most significant function for the Romans (Carr-Gomm, 1994; Chariton, 2011; Collins, 2010; Greet, 2015; Matthews, 1995; Puplett, 2020; Wellman, 2015).

On the second point, that the golden eagle is perceived as a dangerous bird for small livestock, and therefore one of the threats to the golden eagle is direct persecution (Campbell, 2022). The species experienced severe persecution in the nineteenth century, and although the threat has significantly reduced and populations are currently stable, deliberate poisoning, shooting, and trapping continue, and species numbers are declining in North America and Spain (Campbell, 2022). Golden eagles were once affected by the use of organochlorine pesticides, but this is no longer a major problem. Deaths from electrocution while perching on power lines have been reported. Wind turbines are a direct cause of death, particularly in California, where they primarily harm young migratory birds (Watson, 2010). Adult migratory eagles may be affected by developments along flight routes, resulting in local collisions and actual loss of habitat (Katzner et al., 2012). In addition, reforestation, long-term changes in food supply, such as reductions in livestock numbers through changing management practices, and climate change may pose a threat to the species (Watson, 2010). A decline in rabbit populations due to viral hemorrhagic pneumonia may have a negative impact on productivity in the Mediterranean region (Fernández-Gil et al., 2022). BirdLife International (2021a) notes that retrofitting utility poles can reduce the risk of electrocution of eagles, particularly when prioritized around equipment poles and in high-risk areas (see Bedrosian et al., 2020; Dwyer et al., 2020).

In North America, Nelson (2024, p. 97) notes "until recent decades the Golden Eagle… was literally shot on sight by predator control agents of the federal government, animal damage control personnel from state and local governments, caretakers of private hunting clubs, and even the public at large. The persecution complex of certain segments of American society even extended to our national emblem, the Bald Eagle." Furthermore, "eagle electrocutions, shootings, poisonings, and other such killings were considered beneficial by many-long after the Bald Eagle Protection Act was passed in 1940. Ironically, bounties were paid in Alaska for years after the enactment of protective legislation. And for some time after the Bald Eagle Protection Act was amended in 1962 to afford Golden Eagles similar protection, it was equally apparent that persecution of that species might continue indefinitely. Evidence abounds in the literature for considerable shooting and poisoning of Golden Eagles through the early 1970s" (Nelson, 2024, p. 97).

The negative perception of eagles as livestock killers was the major factor for killings of Golden eagles throughout their range, with similarities to perspectives on bald eagles (Campbell, 2022). For example, O'Gara (1978, p. 206) document a "limited study on two ranches near Dillon, Montana, near the end of the lambing season in 1974" which found that 44 lambs were killed by golden eagles, which was 76% of all deaths recorded and respondent ranchers estimated an eagle kill of 1092 lambs were killed over the wide range and time, these valued around $38,000. The U.S. Fish and Wildlife Service responded by trapping and removing 249 golden eagles from the ranches, with a consequent decline in livestock predation.

O'Gara (1978, p. 206) adds that "the extent, or even existence, of predation by golden eagles on livestock is controversial. In fact, for many admirers of this majestic bird, simply discussing it at a pest conference is sheer heresy." It is further added that hunting of golden eagles was frequent in the western United States "until the Bald Eagle Act of 1940 was amended in 1962, providing protection for the golden eagle. The Act allows the shooting or trapping of golden eagles when serious depredations to the livestock industry are determined but forbids shooting from aircraft and poisoning, except for research purposes" (O'Gara (1978, p. 206). The author is however skeptical of the effectiveness of these laws, as ranchers have become frustrated, with "essentially had no recourse if faced by serious eagle predation", while "public pressure, abated by animal protection groups, is directed at 'saving' every golden eagle, no matter how much damage that eagle may do to personal property" and some argue more eagles were being killed after than before the laws were applied: "Knowing that no one has received a kill permit since 1970, many ranchers will continue to take the law into their own hands, and occasionally one will be caught. The ethics of forcing a rancher to break the law or go out of business must be considered" (O'Gara, 1978, p. 212).

This study mirrors a similar, more recent study of bald eagle predation on lambs in the western United States (Campbell, 2022). For example, Duvall et al. (2023) in a study titled "a win–win between farmers and an apex predator: investigating the relationship between bald eagles and dairy farms" argue that "among apex predators, bald eagles have received increased media attention in recent years due to conflicts with farmers across the United States", which "raises challenges for both farmers and wildlife managers as eagle abundance continues to increase and natural prey resources decline." The justification for this study in northwestern Washington State was that there was "high eagle activity on dairy farms in response to declines in salmon carcass availability, an important resource for wintering eagles across western North America" (Duvall et al., 2023). The relationship between eagles and dairy farmers was surprisingly positive, as the findings indicated that "eagles were attracted to dairy farms to feed primarily on cow afterbirth and calf carcasses, responding farmers had no issue with the presence of eagles on their farms, and many dairy farmers felt that eagles provided services to their farms" (Duvall et al., 2023). The eagles' presence proved beneficial to farmers as the eagles scavenged dairy byproducts and hunted unwanted pest species. The growing population of eagles on dairy farms played an important role in their ecological role as vital predators in agroecosystems. The farmers' actions to provide additional resources helped reduce conflict between humans and eagles and may also reduce pressure on other wild prey species. Collaboration between farmers and wildlife managers can therefore improve the complex relationship between dairy farms and eagles (Duvall et al., 2023).

Recent developments in conservation have supported golden eagle conservation (Campbell, 2022). Nelson (2024, p. 97) contends that the "decade of the 1970s was truly a transitional period in American conservation. Public attitudes toward eagles and other raptors changed, as did the understanding of possible long-term effects of man's extensive use of key components of natural ecosystems. Positive changes began in the Fifties and Sixties, culminated in the Seventies, and many of them are still with us in the Eighties." Such changes in favor of eagle conservation included the cessation of Native American tribes killing eagles for feathers (obtaining feathers instead from the U.S. Fish and Wildlife Service), the shooting and trapping of eagles by Federal and state predator control agents (with serious penalties for shooting eagles from aircraft and rancher and public violation of protective laws) and the use of some rodent and coyote control poisons, as well as DDT and other persistent organochlorine pesticides. There is also the more effective management of the design of power lines, to reduce eagle electrocution, and land management responsive to eagle habitats, the establishment of protected areas, examples of the latter being Eagle Rock Audubon Sanctuary in Colorado, the Seymour Eagle Management Area in Alaska, the Snake River Birds of Prey Natural Area, Idaho, and the Skagit River Bald Eagle Natural Area in Washington State. In the last few decades, there has also been increased research on the biological, environmental, and management needs of golden and bald eagles (Nelson, 2024).

Therefore, in North America, the golden eagle is currently similar to the bald eagle in conservation policy, as both are protected (Campbell, 2022). The United States Fish and Wildlife Service (2024) noted that "the Bald and Golden Eagle Protection Act and the Migratory Bird Treaty Act continue to provide protections not only for bald eagles, but golden eagles, too. The U.S. Fish and Wildlife Service works to assure the preservation of both species of eagles." However, the status is different as the bald eagle currently has a larger population. Brown (2024) writes that US law prohibits the killing, wounding, or disturbance of eagles, or removal of nests or eggs without a permit. Although bald eagles were exterminated over most of the

United States during the twentieth century, largely due to the pesticide DDT, eagle populations later recovered under federal protections and were removed from the federal endangered species list in 2007. The golden eagle is more threatened than the bald eagle, largely due to illegal shootings, energy infrastructure development, lead poisoning (Brown, 2024). Wagner (2014) notes that the golden eagle's usual prey is small to medium-sized wild animals such as geese, rabbits, and reptiles. However, American ranchers believed that the bird was devastating their livestock because it sporadically attacks larger animals such as lambs, deer, and even young brown bears. Between 1941 and 1961, around 20,000 golden eagles were shot from aircraft. Studies debunked claims that the eagles threatened ranchers' livelihoods, prompting the government to declare them a federally protected species in 1962 (Wagner, 2014).

In North America, Nelson, 2024 concludes with cautious optimism for the Golden eagle, noting that despite the serious problems from the past, including the historic use of eagle features by native cultures, serious persecution by control agents and ranchers, poisoning and shooting of eagles by farmers and the general public was prevalent, electrocution of eagles by power lines was also commonplace, use of persistent pesticides was common, and land use methods ignored the requirements of eagle habitation, with few protected areas, "when all of these things prevailed, there were still relatively high numbers of Golden Eagles throughout the West" (Nelson (2024). Nevertheless, Bedrosian et al. (2018) argue that "there has been increasing concern for Golden Eagle (Aquila chrysaetos) populations in North America due to current and future projections of mortality risk and habitat loss from anthropogenic sources."

Crowned Eagle

The eagles of the genus *Stephanoaetus* (Sclater, 1922), a single species called the crowned eagle, African crowned eagle or crowned hawk eagle, are very powerful eagles characterized by compressed bills, short wings, long, broad tails and feathered legs, short, powerful toes with huge claws and large crests. Along with the martial eagle, the crowned eagle is one of the largest and strongest eagles in Africa and is the largest bird of prey in the African forest. Although generally slightly smaller than the martial eagle, it is considered stronger due to its larger prey species (Campbell, 2022). The crowned eagle has been described as an African version of the South American harpy eagle and the Asian Philippine eagle due to its crested head, shorter wings (compared to the longer wings of savanna flying eagles such as the Martial and Verreaux eagles), and extremely powerful talons and prey species (including medium to large monkeys). Brown and Amadon (1968, p. 708) write that "this magnificent bird is possibly the finest of all African birds of prey and probably the most powerful, though not actually the largest." Kemp et al. (2020) describe the Crowned eagle's "long tail and short, rounded wings like a giant goshawk, with densely feathered legs, very large toes, heavy bill and comparatively small eyes."

The crowned eagle's habitat varies from dense montane and lowland forest to forest patches on rocky cliffs and bushy savanna, dense woodland, and forested ravines and gorges in open savannas and thornveld (Dowsett-Lemaire & Dowsett, 2006; Dowsett et al., 2008; Monadjem & Rasmussen, 2008; Campbell, 2022). The crowned eagle hunts from perches on tall trees near clearings or waterholes, or from flight over the forest floor, or soaring over the canopy (Brown & Amadon, 1968). Prey species are almost all mammals, especially monkeys and antelopes. Species taken include the small antelopes suni (*Nesotragus moschatus* Von Dueben, 1846), Harvey's duiker (*Cephalophus harveyi*, Thomas, 1893), Blue Duikers (Philantomba monticola Thunberg, 1789) klipspringers (*Oreotragus oreotragus*, Zimmermann, 1783), and Sharpe's Grysbok (*Raphicerus sharpie*, Thomas, 1897) (Irwin, 1981; Deacon & Rockingham-Gill, 2009; Campbell, 2022). Other prey taken includes monkeys, including Blue monkeys (*Cercopithecus mitis*, Wolf, 1822) and Vervet monkeys (*Chlorocebus pygerythrus*, F. Cuvier, 1821) (Deacon & Rockingham-Gill, 2009; Campbell, 2022), and domestic cats (Benson, 1942; Lowry, 2006). Two species of hyrax (Rock hyrax *Procavia capensis*, Pallas, 1766; and Yellow-spotted rock hyrax *Heterohyrax brucei*, Gray, 1868), have also been recorded. Domestic cats and game birds may be predated. Reptiles taken include the Rock monitor lizard (*Varanus albigularis*, Daudin, 1802) (Deacon & Rockingham-Gill, 2009; Dowsett et al., 2008; Irwin, 1981; Tarboton & Allan, 1984). There is also a record of a possible attack on a human child (Stjernstedt, 1975).

The crowned eagle is unique among eagles in the size of prey it commonly takes (Campbell, 2022). Steyn (1982) notes that the crowned eagle may take prey including antelopes up to six times the eagle's weight. Swatridge et al. (2014, p. 47) found that "the diet of African Crowned Eagles constituted almost entirely mammals (99%), which were predominantly antelopes (61%) and monkeys (25%)." Ferguson-Lees and Christie (2001) observed that the Crowned eagle had the highest weight range for prey among accipitrids, with the only match being the female Harpy eagle, and is among the few eagles to take prey weighing over 9 kg (McPherson et al., 2016a, 2016b). Crowned eagles, and in some cases the harpy and martial eagles, are among the few eagles that regularly kill primates (Mitani et al., 2001). For example, Mitani et al. (2001, 187), in a study of the diet of crowned eagles in Kibale National Park, Uganda, note that the results indicate that primates make up the vast majority of all prey (82%) and a nontrivial proportion of the total primate species population in Ngogo succumbs to predation by the crowned eagle every year. Eagles caught more male primates than females (although these may be larger) and killed more red-tailed monkeys (*Cercopithecus ascanius*, Audebert, 1799) than gray-cheeked mangabeys (*Lophocebus albugiena*, Gray, 1850) and did not distinguish between adults and adolescents (Mitani et al., 2001, p. 187).

Crowned eagles may forage in human-dominated landscapes (Campbell, 2022).

However, Maseko et al. (2023) found that in human-dominated landscapes, "occupancy of crowned eagles was positively influenced by forests", "occupancy was negatively influenced by disturbance and building density" yet "crowned eagles persist in the urban landscape mosaic due to natural and managed green spaces, especially forests."

McPherson et al. (2016a) report that in KwaZulu-Natal, South Africa, crowned eagles hunt in urban green spaces in Durban and Pietermaritzburg. This can lead to conflicts because the eagles may prey on domestic animals. The Hadeda ibis (Bostrychia hagedash, Latham, 1790) was a common prey species, according to results from time-lapse cameras at urban nesting sites used to document prey composition during the breeding season. The bird's abundance was many times higher than any previous diet description (McPherson et al., 2016a, 2016b, p. 383). Vervet monkeys and rock hyraxes were two other frequent prey species. The low percentage of native prey, which made up only 6% of the identifiable prey, was an important finding. The findings disproved the widespread notion that domestic cats and dogs were frequent prey. Only a small number of cats were killed by eagles, and no dogs were reported as prey. Yet, young eagles have been known to kill small dogs (McPherson et al., 2016b, p. 383).

The crowned eagle is an adaptable species, but is "threatened by persecution through trapping, shooting and nest destruction, competition for prey from humans, and habitat loss through deforestation" (BirdLife International, 2018b). BirdLife International (2018b) indicates that the numerous threats contribute to serious decline, influenced by the slow reproductive rate and variable tolerance of environmental change (see also Ferguson-Lees & Christie, 2001; Thomsett, 2011). The crowned eagle adapts to forests, woodland, savannas and shrublands, plantations and secondary growth. In particular, it can survive in small forest fragments, including urban greenspace forests, and in some areas shows high resilience to severe deforestation and degradation (BirdLife International, 2018a; see also Ferguson-Lees & Christie, 2001; Dowsett-Lemaire & Dowsett, 2006; McPherson et al., 2016a). BirdLife International (2018a) notes that factors for this adaptability include the "use of exotic invasive trees (especially Eucalyptus and Pinus spp.)" for nesting permits (see also McPherson et al., 2016a), "dietary plasticity" in terms of prey on new prey species, especially mammals (see also Swatridge et al., 2014) and the rarity of competitive mesopredators, "which may allow it to persist in certain areas…" and to "occur at higher densities than expected" (see also McPherson et al., 2016b; Symes & Antonites, 2014) (Table 7.7).

Crowned eagles are directly persecuted by people, with regional variations in this regard (Ferguson-Lees & Christie, 2001; Campbell, 2022). Global Raptor Information Network (2024a) reports that the population of this species has "declined greatly" in Liberia and Côte d'Ivoire, due to hunting pressure on the eagles and their prey species (see also Campbell, 2022) and in Malawi "the range of this species outside of protected areas is doubtless decreasing as the result of habitat loss, and it has been extirpated in some parts of the country" (see also Dowsett-Lemaire & Dowsett, 2006). Deforestation is also a factor for eagle species decline in southern Africa, where it is uncommon and considered Vulnerable, due to declines in prey species, direct persecution due to its predation on poultry, deforestation, and electrocution on powerlines (Monadjem & Rasmussen, 2008).

BirdLife International (2018b) lists human-induced impacts on crowned eagle populations as including deforestation, based on timber extraction, charcoal production, agricultural development (including plantations and shifting cultivation)

Table 7.7 Conservation status of the crowned eagle

Conservation status	Near threatened
Adult numbers	5000–50,000
Population	Decreasing
Habitat	Forest, savanna, Wetlands (inland), artificial/terrestrial
Decline in habitat	Yes
Threats	Annual and perennial non-timber crops, wood and pulp plantations, mining and quarrying, renewable energy, utility and service lines, hunting and trapping terrestrial animals, logging and wood harvesting
Conservation actions	Conservation actions underway The species occurs in a number of protected areas across its vast range

Sources BirdLife International (2018a)

mining, collisions with structures (wires, fences, vehicles, glass buildings, wind turbines), and electrocution on utility networks. There is also human competition for prey species with bushmeat exploitation, and crucially "direct persecution in an estimated 90% of its range" for food, arrow-fletching, traditional religious practices, ornaments, and threats to people and associated animals (see also Ferguson-Lees & Christie, 2001; Thomsett, 2011; McPherson et al., 2016b). McPherson et al. (2016b, p. 383) pose the question: "Diet of the crowned eagle (Stephanoaetus coronatus) in an urban landscape: potential for human-wildlife conflict?" and argue that conflict with humans can develop because crowned eagles may kill a few of livestock and pets, demonstrably fewer than commonly assumed (McPherson et al., 2016b). According to BirdLife International (2018b) to protect the crowned eagle, regular surveys must be conducted to monitor populations at selected sites throughout its range, monitor habitat loss and degradation, and research the effects of human competition for prey species. In addition, the total area of suitable protected habitat needs to be increased and educational measures implemented to reduce direct persecution and hunting pressure on prey animals. Power grids also need to be upgraded to bird-proof designs (BirdLife International, 2018b).

Martial Eagle

The martial eagle is the largest eagle in Africa and possibly one of the two or three strongest species, taking prey up to the size of small antelopes (Ferguson-Lees & Christie, 2001; Kemp et al., 2020; Campbell, 2022). Carr (1964, p. 108) notes that the martial eagle's large bill can tear a carcass in half if it is too heavy to be carried to the nest, and this eagle is the most powerful bird of prey on the open savanna. Martial eagles hunt from soaring heights, or perch on tree branches or pylons in treeless areas (BirdLife International, 2020; Simmons & Brown, 2009; Steyn, 1982; Verdoorn & Roth, 2000). Foraging habitats include open grassland, thornbush, light woodland, scattered tree savanna or more heavily wooded savannas, or riparian forests in arid landcover, and even semidesert in southern Africa, mostly lowlands, and generally excluding closed canopy forest and desert (Ferguson-Lees & Christie, 2001).

The martial eagle preys on mammals, birds, and reptiles, with regional variations in the prey taken (Campbell, 2022). Global Raptor Information Network (2024b) gives locations across Africa, were various prey types were taken, including birds such as guineafowl (Family *Numididae*, Longchamps, 1842) including helmeted guineafowl (*Numida meleagris*, Linnaeus, 1758), bustards (Family *Otididae*, Rafinesque, 1815) including Hartlaub's bustard (*Lissotis hartlaubii*, Heuglin, 1863) and Denham's or Stanley's bustard (*Neotis denhami*, Children & Vigors, 1826, francolins (subfamily *Perdicinae,* Horsfield, 1821), and even spotted eagle owls (*Bubo africanus*, Temminck, 1821), and possibly a hooded vulture (*Necrosyrtes monachus* Temminck, 1823) (Global Raptor Information Network, 2024b). Mammal species taken included dik-diks (Genus *Madoqua*, Ogilby, 1837), gerenuks (*Litocranius walleri*, Brooke, 1879), impala (*Aepyceros melampus*, Lichtenstein, 1812), Steenbok (*Raphicerus campestris*, Thunberg, 1811), Grey duiker (*Sylvicapra grimmia*, Linnaeus, 1758), Thompson's gazelle (*Eudorcas thomsonii*, Günther, 1884), common duiker (*Sylvicapra grimmia*, Linnaeus, 1758) and banded mongoose (*Mungos mungo*, Gmelin, 1788) (Global Raptor Information Network, 2024b). Reptiles taken included monitor lizards (Genus V*aranus*, Merrem, 1820), including the savannah monitor (*Varanus exanthematicus*, Bosc, 1792), the rock monitor (*Varanus albigularis*, Daudin, 1802), and the Nile monitor (*Varanus niloticus*, Linnaeus, 1766 (Naude et al., 2019). Prey varies regionally, with "mammalian prey being more common in eastern Africa and reptile prey being more common in southern Africa. Within South Africa, reptile prey proportion was greater in the east than in the west" (Naude et al., 2019) (Table 7.8).

The martial eagle occurs through the savanna land cover of West, Central, East, and Southern Africa. Simmons and Brown (2009, p. 5) write that although the martial

Table 7.8 Conservation status of the martial eagle

Conservation status	Endangered
Adult numbers	Unknown
Population	Decreasing
Habitat	Forest, savanna, Shrubland, grassland, Wetlands (inland)
Decline in habitat	Yes
Threats	Annual and perennial non-timber crops, utility and service lines, hunting and trapping terrestrial animals, dams and water management/ use, agricultural and forestry effluents
Conservation actions	Conservation actions underway A system to compensate farmers for stock losses has been initiated in South Africa. The species is listed as endangered in Namibia, South Africa, Lesotho, and Eswatini (Simmons, 2015, Taylor, 2015)

Sources BirdLife International (2020)

eagle probably occurs more widespread than any other eagle throughout southern Africa, it is becoming increasingly scarce in more heavily settled districts and may have undergone declines in much of its range, including West Africa (Ferguson-Lees & Christie, 2001; Cloete, 2013; Simmons, 2015; Amar & Cloete, 2018; BirdLife International, 2018a). Martial eagles have taken domesticated and semi-domesticated animals (Campbell, 2022). These include gamebirds and domestic poultry, which comprised 54% of prey recorded from one nest in Zimbabwe, with other animals being mammals (40.5%) and reptiles (5.4%) (Irwin, 1981). In northeastern South Africa, another study showed a focus on gamebirds (45%), followed by reptiles (38%) and mammals (17%) (Boshoff et al., 1990; Tarboton & Allan, 1984). Other domestic animals taken include young goats (*Capra aegagrus hircus*, Linnaeus, 1758) and lambs (*Ovis aries* Linnaeus, 1758) (Campbell, 2022).

Consequently, deliberate killings of eagles have increased, and with other factors, the martial eagle population has declined. BirdLife International (2018a) notes that the martial eagle population has declined over the past three generations (56 years) due to deliberate and incidental poisoning, habitat loss, reduction in available prey, pollution, and collisions with power lines, and that these threats may continue into the future. Threats include direct shooting and trapping by farmers, indirect poisoning through poisoned carcass consumption and possibly pesticides, drowning in reservoirs, collision and electrocution on electric power lines, the use of eagle parts in traditional medicine and habitat alteration and degradation, including local extirpation of prey species, and the loss of suitable trees for nesting (Verdoorn & Roth, 2000; Campbell, 2022). In Malawi and Namibia, human persecution from livestock predation and eagle persecution on farmlands have been mentioned (Dowsett-Lemaire & Dowsett, 2006; Simmons & Brown, 2009).

BirdLife International (2020) states that the martial eagle "suffers from direct persecution (shooting and trapping) by farmers and indirect poisoning (these two threats are by far the most important causes of loss)." Poisoning is carried out in retaliation for eagles killing livestock committed by large farmers and small-scale tribal communities. A reduction in natural prey may lead to an increase in predation on domestic animals, which in turn may lead to increased persecution by farmers. The species can use power poles as nesting sites, which can lead to power outages and thus create conflicts between humans and animals (Jenkins et al., 2013). BirdLife International (2020) lists conservation measures to protect the martial eagle, including programs that combine awareness campaigns and compensation to farmers for population losses due to eagle predation. Other measures include installing electric shock protection devices on electricity poles and conducting education and awareness campaigns to reduce the use of poison bait. Regular population surveys can also be carried out throughout the eagles' range (BirdLife International, 2020).

Davies (1999, p. 67) notes the high rates of killing of Martial eagles in the Karoo during the 1960s. The question is posed: "Has such persecution been based on any factual evidence of eagles killing livestock? Golden and Crowned eagles have killed prey (including domestic sheep) weighing 20–50 kg. Large eagles are, therefore, capable of killing domestic sheep up to about half adult size, so the issue appears to be not whether they kill domestic lambs, but to what extent does this occur and under what circumstances"(Davies, 1999, p. 67). Additionally, "Martial Eagles (plains), Black Eagles (mountains) and Crowned Eagles (forests) are regularly accused of killing lambs, with some confirmation that they do. There is little overlap between the ranges of large, forest-dwelling eagles and the major sheep farming regions of the world" (Davies, 1999, p. 67). The author nevertheless argues that eagle predation on livestock is generally lower than commonly supposed, and generally less frequent than that of mammalian mesopredators such as caracals. Additionally, "Black, Martial and Crowned Eagles are all highly territorial and are not inclined to scavenge. There are no records of these eagles being attracted in large numbers to an abundance of dead or dying lambs, unlike Golden and Wedge-tailed Eagles" (Davies, 1999, p. 69) and the author concludes that "Simply on the basis that lamb killing by eagles is very rare and can be dealt with, I would argue that large eagles are compatible with the open-range farming of small stock. Farmers must learn to tolerate compatible predators, especially in the case of vulnerable eagle populations" (Davies, 1999, p. 71).

Conclusions

Mesopredators can have an influence on humans' psyches and lifestyles that goes beyond their size, and in some cases it is as strong as the influence of large carnivores, as the smaller animals have greater adaptability, elusiveness of foraging, less detectable entries into human life spaces, lower levels of threat to humans and related animals, etc. In some cases, humans are unaware of their ecology or even their presence. Mesopredators lack the charisma and popularity assigned to large predators such as tigers or bears, but in some landscapes and habitats mesopredators are vital components of the ecological system, even assuming the role of apex predators with the extinction of large carnivores. Human psychology and observation, based on real and constructed issues, remain the key defining factor for mesopredator status and contact. This chapter has examined the perspective on conservation psychology as a factor for the biogeography of mesopredators, and argued that this approach is the determining factor for mesopredator conservation. The chapter examined the mesopredators from the perspective of the impacts on people, hence species that may be considered apex predators such as eagles may also be mesopredatrors, because of their similarity to mammalian mesopredators; they are generally unable to kill or harm people, and are threats mostly to small livestock rather than larger livestock such as cattle. Current evidence indicates that mesopredators are understudied, and from the human perspective remain in the shadow of the larger, more popular carnivores, partly because of their less extensive ecological requirements, and their low threat to people. Therefore, examination of the conservation psychology of mesopredators contributes to the conservation biogeography of these species.

References

Acorn, R. C., & Dorrance, M. J. (1990). *Methods of investigating predation of livestock*. Alberta Agriculture and Rural Development, Information Management Division.

Aguiar-Silva, F. H., Luz, B. B., Sanaiotti, T. M., Jaudoin, O., & Matos, S. P. (2014). Ameaça e conservação das aves de rapina de grande porte na Floresta Nacional do Tapajós e entorno. In: *Anais do II Seminário de Pesquisa Científica da Floresta Nacional do Tapajós, Santarém, Pará* (pp. 22–32).

Aguiar-Silva, F. H., Sanaiotti, T. M., & Luz, B. B. (2014). Food habits of the harpy eagle, a top predator from the Amazonian Rainforest Canopy. *Journal of Raptor Research, 48*(1), 24–35. https://doi.org/10.3356/JRR-13-00017.1

Allen, B. L., Allen, L. R., Andrén, H., Ballard, G., Boitani, L., Engeman, R. M., Fleming, P. J. S., Ford, A. T., Haswell, P. M., Kowalczyk, R., Linnell, J. D. C., David Mech, L., & Parker, D. M. (2017). Can we save large carnivores without losing large carnivore science? *Food Webs, 12*, 64–75. https://doi.org/10.1016/j.fooweb.2017.02.008

Almeida, A., Vasconcelos, C., & Strecht-Ribeiro, O. (2014). Attitudes toward animals: A study of Portuguese children. *Anthrozoös, 27*(2), 173–190. https://doi.org/10.2752/175303714X13903827487403

Alvarez-Cordero, E. (1996). *Biology and conservation of the harpy eagle in Venezuela and Panama*. University of Florida.

Alves, R. R., Pereira Filho, G. A., Silva Vieira, K., Silva Souto, W. M., Mendonça, L. E., Montenegro, P., Almeida, W. O., & Silva Vieira, W. L. (2012). A zoological catalogue of hunted reptiles in the semiarid region of Brazil. *Journal of Ethnobiology and Ethnomedicine, 8*, 27. https://doi.org/10.1186/1746-4269-8-27

Alves, R. R., & Souto, W. M. (2011). Ethnozoology in Brazil: Current status and perspectives. *Journal of Ethnobiology and Ethnomedicine, 7*, 22. https://doi.org/10.1186/1746-4269-7-22

Alves, R. R., Vieira, K. S., Santana, G. G., Vieira, W. L., Almeida, W. O., Souto, W. M., Montenegro, P. F., & Pezzuti, J. C. (2012). A review on human attitudes towards reptiles in Brazil. *Environmental Monitoring and Assessment, 184*(11), 6877–6901. https://doi.org/10.1007/s10661-011-2465-0

Amar, A., & Cloete, D. (2018). Quantifying the decline of the martial eagle Polemaetus bellicosus in South Africa. *Bird Conservation International, 28*(3), 363–374. https://doi.org/10.1017/S0959270917000314

Anan'eva, N. B., Uteshev, V. K., Orlova, N. L., & Gakhova, E. N. (2015). Izvestiia Akademii Nauk. *Seriia Biologicheskaia, 5*, 509–517.

Appleby, R. G., & Smith, B. P. (2018). Do wild canids kill for fun? In N. Carr & J. Young (Eds.), *Wild animals and leisure: Rights and wellbeing* (pp. 189–201). Routledge. https://doi.org/10.4324/9781315457413

Azevedo, A., Guimarães, L., Ferraz, J., Whiting, M., & Magalhães-Sant'Ana, M. (2022). Understanding the human–reptile bond: An exploratory mixed-methods study. *Anthrozoös, 35*(6), 755–772. https://doi.org/10.1080/08927936.2022.2051934

Bates, M. (1964). *The land and wildlife of South America*. Life Books.

Bedrosian, B. E., Domenech, R., Shreading, A., Hayes, M. M., Booms, T. L., & Barger, C. R. (2018). Migration corridors of adult Golden eagles originating in northwestern North America. *PLoS One, 13*(11), e0205204. https://doi.org/10.1371/journal.pone.0205204

Bedrosian, G., Carlisle, J. D., Woodbridge, B., Dunk, J. R., Wallace, Z. P., Dwyer, J. F., Harness, R. E., Mojica, E. K., Williams, G. E., & Jones, T. (2020). A spatially explicit model to predict the relative risk of Golden eagle electrocutions in the Northwestern Plains, USA. *Journal of Raptor Research, 54*(2), 110–125.

Bennet, D. (2002). Diet of juvenile Varanus niloticus (Sauria: Varanidae) on the black Volta River in Ghana. *Journal of Herpetology, 36*(1), 116–117.

Bennett, D., & Takoordyal, R. (2003). *The savanna monitor lizard: The truth about Varanus exanthematicus*. Viper Press.

Benson, C. W. (1942). Additional notes on Nyasaland birds. *Ibis, 6*, 299–337.
Bickley, S., Lemos, F., Gilmore, M., Azevedo, F., Freeman, E., & Songsasen, N. (2020). Human perceptions of and interactions with wild canids on cattle ranches in Central Brazil. *Oryx, 54*(4), 546–553. https://doi.org/10.1017/S0030605318000480
BirdLife International. (2018a). *Pithecophaga jefferyi* (amended version of 2017 assessment). The IUCN red list of threatened species 2018. https://doi.org/10.2305/IUCN.UK.2017-3.RLTS.T22696012A129595746.en
BirdLife International. (2018b). *Stephanoaetus coronatus*. The IUCN red list of threatened species. https://doi.org/10.2305/IUCN.UK.2018-2.RLTS.T22696201A129914678.en
BirdLife International. (2020). *Polemaetus bellicosus*. The IUCN red list of threatened species. https://doi.org/10.2305/IUCN.UK.2020-3.RLTS.T22696116A172287822.en
BirdLife International. (2021a). *Aquila chrysaetos*. The IUCN red list of threatened species 2021. https://doi.org/10.2305/IUCN.UK.2021-.RLTS.T22696060A202078899.en
BirdLife International. (2021b). *Harpia harpyja*. The IUCN red list of threatened species. https://doi.org/10.2305/IUCN.UK.2021-3.RLTS.T22695998A197957213.en
Boag, D. A. (1977). Summer food habits of Golden eagles in southwestern Alberta. *Canadian Field Naturalist, 91*, 296–298.
Bodio, S. J. (2012). *Eternity of eagles: The human history of the most fascinating bird in the world*. Rowman & Littlefield Publishing Group.
Boshoff, A. F., Palmer, N. G., & Avery, G. (1990). Regional variation in the diet of martial eagles in the Cape Province, South Africa. *South Africa Journal of Wildlife Research, 20*, 57–68.
Bowen, W. D. (1980). Coyote-Golden eagle interactions at an ungulate carcass. *Journal of Mammalogy, 61*(2), 376–377. https://doi.org/10.2307/1380075
Brashares, J., Prugh, L., Stoner, C. J., & Epps, C. (2010). Ecological and conservation implications of mesopredator release. In J. Terborgh & J. A. Estes (Eds.), *Trophic cascades: Predators, prey, and the changing dynamics of nature* (pp. 221–240). Island Press.
Breitenmoser, U., Lanz, T., & Breitenmoser-Würsten, C. (2019). *Conservation of the wildcat (Felis silvestris) in Scotland: Review of the conservation status and assessment of conservation activities (PDF) (report)*. IUCN Cat Specialist Group.
Brom, P., Anderson, P., Channing, A., & Underhill, L. G. (2020). The role of cultural norms in shaping attitudes towards amphibians in Cape Town, South Africa. *PLoS ONE, 15*(2), e0219331. https://doi.org/10.1371/journal.pone.0219331
Brown, L. H. (1976). *Eagles of the world*. David & Charles.
Brown, L. H., & Amadon, D. (1968). *Hawks, eagles, and falcons of the world* (Vol. 2). McGraw-Hill Book Company.
Brown, M. (2024). *Arrest warrant issued for Montana man accused of killing thousands of birds, including eagles*. The Associated Press. Retrieved from https://apnews.com/article/eagleskilled-indictment-black-market-b09d23ff49f338c9415240bd933642a0
Caldicott, D. G., Croser, D., Manolis, C., Webb, G., & Britton, A. (2005). Crocodile attack in Australia: An analysis of its incidence and review of the pathology and management of crocodilian attacks in general. *Wilderness & Environmental Medicine, 16*, 143–159.
Campbell, M. (2009). Repositioning zoogeography within the nature–culture borderlands: An animal geography of reptiles in southern Ghana. *Applied Geography, 29*, 260–268. https://doi.org/10.1016/j.apgeog.2008.09.001
Campbell, M. (2013). Reptile ecology, adaptive management and assessment techniques: Problems, developments and prospects. In W. I. Lutterschmidt (Ed.), *Reptiles in research investigations of ecology, physiology, and behavior from desert to sea* (pp. 387–406). Nova Science Publishers.
Campbell, M. (2015). *Vultures: Their evolution, ecology and conservation*. Taylor and Francis (CRC Press).
Campbell, M. (2017a). *Biological conservation in the 21st century: A conservation biology of large wildlife*. Nova Science Publishers.
Campbell, M. (2017b). Human perceptions of wild animals in La Union, El Salvador. In A. Columbus (Ed.), *Advances in psychology research* (Vol. 124). Nova Science Publishers.

Campbell, M. (2022). *The great eagles: Their evolution, ecology and conservation.* Taylor and Francis (CRC Press).

Campbell, M., & Jha, K. K. (2021). *Critical research techniques in animal and habitat ecology: International examples.* Nova Science Publishers.

Carr, A. (1964). *The land and wildlife of Africa.* Time Incorporated.

Carr-Gomm, P. S. (1994). *The druid animal oracle.* Fireside.

Carson, W. P. (1985). The ecology of the Accra and Winneba Plains with some aspects of related savanna ecosystems. In B. W. Garbrah (Ed.), *Impact of human activities on the structure and productivity of the savanna ecosystem in Ghana* (pp. 4–74). Ghana National Committee for the Man and Biosphere Program (MAB).

Ceríaco, L. M. (2012). Human attitudes towards herpetofauna: The influence of folklore and negative values on the conservation of amphibians and reptiles in Portugal. *Journal of Ethnobiology and Ethnomedicine, 8*, 8. https://doi.org/10.1186/17464269-8-8

Chakarov, N., & Krüger, O. (2010). Mesopredator release by an emergent superpredator: A natural experiment of predation in a three level guild. *PLoS One, 5*(12), e15229. https://doi.org/10.1371/journal.pone.0015229

Chariton, J. D. (2011). The mesopotamian origins of the Hittite double-headed eagle. *UWL Journal of Undergraduate Research, 14*, 1–13.

Cloete, D. (2013). *Investigating the decline of the martial eagle (Polemaetus bellicosus) in South Africa.* University of Cape Town.

Collins, B. J. (2010). Animal mastery in Hittite art and texts. In D. B. Counts & B. Arnold (Eds.), *The master of animals in old world iconography. Main series, No. 24* (pp. 59–74). Archaeolingua Foundation.

Collins, P. W., & Latta, B. C. (2009). Food habits of nesting Golden eagles (Aquila chrysaetos) on Santa Cruz and Santa Rosa islands, California. In C. C. Damiani & D. K. Garcelon (Eds.), *Proceedings of the 7th California Islands symposium* (pp. 255–268). Institute for Wildlife Studies.

Concepcion, C., Sulapas, M., & Ibañez, J. C. (2006). Notes on food habits and breeding and nestling behavior of Philippine eagles in mount Apo Natural Park, Mindanao, Philippines. *Banwa, 3*(1), 81–95.

Coursey, D. L. (1998). The revealed demand for a public good: Evidence from endangered and threatened species. *NYU Environmental Law Journal, 6*, 411–449.

Cox, N., Young, B. E., Bowles, P., Fernandez, M., Marin, J., Rapacciuolo, G., Böhm, M., Brooks, T. M., Hedges, S. B., Hilton-Taylor, C., Hoffmann, M., Jenkins, R. K. B., Tognelli, M. F., Alexander, G. J., Allison, A., Ananjeva, N. B., Auliya, M., Avila, L. J., Chapple, D. G., Cisneros-Heredia, D. F., & Xie, Y. (2022). A global reptile assessment highlights shared conservation needs of tetrapods. *Nature, 605*(7909), 285–290. https://doi.org/10.1038/s41586-022-04664-7

Czech, B., & Krausman, P. R. (2001). The endangered species act. In *History, conservation biology, and public policy.* Johns Hopkins University Press.

Czech, B., Krausman, P. R., & Borkhartaria, R. (1998). Social construction, political power, and allocation of benefits to endangered species. *Conservation Biology, 12*, 11031112. https://doi.org/10.1046/j.1523-1739.1998.97253.x

Davies, R. A. G. (1999). The extent, cost and control of livestock predation by eagles with a case study on black eagles (aquila Verreauxii) in the Karoo. *Journal of Raptor Research, 33*(1), 67–72. Retrieved from http://creativecommons.org/licenses/by-nc-sa/4.0/

De Win, S. (2013). *Typhoon Haiyan left parts of the Philippines behind as a wintery landscape typically found in temperate climates only!* Retrieved from http://www.birding2asia.com/articles/TyphoonHaiyan.html

Deacon, N. R., & Rockingham-Gill, D. V. (2009). Raptor survey in miombo woodland, Umfurudzi safari area, Shamva District, Zimbabwe: Preliminary results. *Gabar, 19*, 2–9.

Do Linh San, E., Sato, J., Belant, J., & Somers, M. (2022). *Small carnivores: Evolution, ecology, behaviour, and conservation.* Wiley. https://doi.org/10.1002/9781118943274

Dowsett, R. J., Aspinwall, D. R., & Dowsett-Lemaire, F. (2008). *The birds of Zambia: An atlas and handbook*. Tauraco Press.
Dowsett-Lemaire, F., & Dowsett, R. J. (2006). *The birds of Malawi: An atlas and handbook*. Tauraco Press & Aves.
Duvall, E. S., Schwabe, E. K., & Steensma, K. M. M. (2023). A win–win between farmers and an apex predator: Investigating the relationship between bald eagles and dairy farms. *Ecosphere, 14*(3), e4456. https://doi.org/10.1002/ecs2.4456
Dwyer, J. F., Murphy, R. K., Stahlecker, D. W., Dwyer, A. M., & Boal, C. W. (2020). Golden eagle perch-site use in the U.S. Southern Plains: Understanding electrocution risk. *Journal of Raptor Research, 54*(2), 126–135.
Ellis, D. H., Tsengeg, P., Whitlock, P., & Ellis, M. H. (2000). Predators as prey at a Golden eagle Aquila chrysaetos eyrie in Mongolia. *Ibis, 142*(1), 139–141. https://doi.org/10.1111/j.1474-919X.2000.tb07694.x
Elmhagen, B., & Rushton, S. P. (2007). Trophic control of mesopredators in terrestrial ecosystems: Top-down or bottom-up? *Ecology Letters, 10*(3), 197–206. https://doi.org/10.1111/j.1461-0248.2006.01010.x
Feldhamer, G. A., Thompson, B. C., & Chapman, J. A. (2004). *Wild mammals of North America*. Johns Hopkins University Press.
Ferguson-Lees, J., & Christie, D. A. (2001). *Raptors of the world*. Harcourt.
Fernández-Gil, A., Lamas, J. A., Ansola, L. M., Román, J., de Gabriel Hernando, M., & Revilla, E. (2022). Population dynamics of recovering apex predators: Golden eagles in a Mediterranean landscape. *Journal of Zoology, 319*(2), 99–111. https://doi.org/10.1111/jzo.13026
Font, E., Burghardt, G. M., & Leal, M. (2023). Brains, behaviour, and cognition: Multiple misconceptions. In C. Warwick, P. C. Arena, & G. M. Burghardt (Eds.), *Health and welfare of captive reptiles*. Springer. https://doi.org/10.1007/978-3-03086012-7_7
George, K. A., Slagle, K. M., Wilson, R. S., Moeller, S. J., & Bruskotter, J. T. (2016). Changes in attitudes toward animals in the United States from 1978 to 2014. *Biological Conservation, 201*, 237–242. https://doi.org/10.1016/j.biocon.2016.07.013
Gibbons, J. W. (1988). *The management of amphibians, reptiles and small mammals in North America: The need for an environmental attitude adjustment* (pp. 4–10). USDA Forest Service.
Giraldo-Amaya, M., Aguiar-Silva, F. H., Aparicio, U. K. M., & Zuluaga, S. (2021). Human persecution of the harpy eagle: A widespread threat? *Journal of Raptor Research, 55*(2), 281–286. https://doi.org/10.3356/0892-1016-55.2.28155
Global Raptor Information Network. (2024a). *Species account: Crowned hawk-eagle Stephanoaetus coronatus*. Retrieved from http://www.globalraptors.org
Global Raptor Information Network. (2024b). *Species account: Martial eagle Polemaetus bellicosus*. Retrieved from http://www.globalraptors.org
Godoi, M. N., Morante-Filho, J. C., Faxina, C., Modena, E. S., Pivatto, M. A., Manco, D. D., Bocchese, R., Teribele, R., Rosa, A. L., & Stavis, V. K. (2012). Aves de rapina raras no estado de Mato Grosso do Sul, Brasil. *Atualidades Ornitológicas, 170*, 41–47.
Gorzula, S., & Oduro, W. (1997). *Survey of the status and management of royal python Python regius in Ghana*. Cites Report.
Greet, B. J. R. (2015). *The Roman eagle: A symbol and its evolution*. University of Leeds.
Gusmão, A. C., Degra, D., da Silva, O. D., de Souza, L. S., da Frota, A. V. B., Tuyama, C. A., Tuyama, M. C., da Costa, T. M., Dalbem, A. P., Barnett, A. A., Aguiar-Silva, F. H., & dos Santos Filho, M. (2020). Power lines as a threat to a canopy predator: Electrocuted Harpy Eagle in southwestern Brazilian Amazon. *Journal of Threatened Taxa, 12*(13), 16904–16908. https://doi.org/10.11609/jott.6198.12.13.16904-16908
Halley, D. J., & Gjershaug, J. O. (1996). Inter- and intra-specific dominance relationships and feeding behaviour of Golden eagles (Aquila chrysaetos) and sea eagles (Haliaetus albicilla) at carcasses. *Ibis, 140*, 295–301.

Hamilton, J., Gartland, K. N., Jones, M., & Fuller, G. (2022). Behavioral assessment of six reptile species during a temporary zoo closure and reopening. *Animals, 12*(8), 1034. https://doi.org/10.3390/ani12081034
Haribon Foundation. (2013). *Preventing extinctions programme - Philippine eagle report*. Retrieved from https://speciesonthebrink.org/species/philippine-eagle/
Hassan, R., Scholes, R., & Ash, N. (2005). *Millenium ecosystem assessment: Ecosystems and human wellbeing, volume 1, current state and trends*. Island Press.
Haswell, P. M., Kusak, J., & Hayward, M. W. (2017). Large carnivore impacts are contextdependent. *Food Webs, 12*, 3–13. https://doi.org/10.1016/j.fooweb.2016.02.005
Hodge, A. M. (2013). *Laikipia plateau: What is a mesopredator?* Retrieved from https://blogs.scientificamerican.com/expeditions/laikipia-plateau-what-is-amesopredator/
Hoekstra, L. A., Schwartz, T. S., Sparkman, A. M., Miller, D. A. W., & Bronikowski, A. M. (2020). The untapped potential of reptile biodiversity for understanding how and why animals age. *Functional Ecology, 34*(1), 38–54. https://doi.org/10.1111/13652435.13450
Hoffmann, M., Mech, D., & Sillero-Zubiri, C. (2004). *Canids: Foxes, wolves, jackals and dogs: Status survey and conservation action plan*. International Union for Conservation of Nature.
Hosey, G., & Melfi, V. (2014). Human-animal interactions, relationships and bonds: A review and analysis of the literature. *International Journal of Comparative Psychology, 27*(1), 117–142. https://doi.org/10.46867/ijcp.2014.27.01.01
Houston, D. C., & Maddox, J. G. (1974). Causes of mortality among young Scottish blackface lambs. *Veterinary Record, 95*(25–26), 575. https://doi.org/10.1136/vr.95.25-26.575
Ibañez, J. C., Miranda, H. C., Balaquit-Ibañez, G., Afan, D. S., & Kennedy, R. S. (2003). Notes on the breeding behavior of a Philippine eagle pair at Mount Sinaka, Central Mindanao. *The Wilson Bulletin, 115*(3), 333–336. https://doi.org/10.1676/01-054
Irwin, M. P. S. (1981). *The birds of Zimbabwe*. Quest Publishing.
Jachowski, D. S., Butler, A., Eng, R. Y. Y., Gigliotti, L., Harris, S., & Williams, A. (2020). Identifying mesopredator release in multi-predator systems: A review of evidence from North America. *Mammal Review, 50*, 367–381. https://doi.org/10.1111/mam.12207
Jackson, C. R., Rød-Eriksen, L., Mattisson, J., Flagstad, Ø., Landa, A., Miller, A. L., Eide, N. E., & Ulvund, K. R. (2023). Predation of endangered Arctic foxes by Golden eagles: What do we know? *Ecology and Evolution, 13*(3), e9864. https://doi.org/10.1002/ece3.9864
Janovcová, M., Rádlová, S., Polák, J., Sedláčková, K., Peléšková, Š., Žampachová, B., Frynta, D., & Landová, E. (2019). Human attitude toward reptiles: A relationship between fear, disgust, and aesthetic preferences. *Animals, 9*(5), 238. https://doi.org/10.3390/ani9050238
Jenkins, A. R., De Goede, K. H., Sebele, L., & Diamond, M. (2013). Brokering a settlement between eagles and industry: Sustainable management of large raptors nesting on power infrastructure. *Bird Conservation International, 23*, 232–246.
Jenkins, M., & Broad, M. S. (1994). *International trade in reptile skins: A review and analysis of the main consumer markets 1983–91*. Traffic International.
Jha, K. K., & Campbell, M. (2021). *Critical research techniques in animal and habitat ecology: Examples from India*. Nova Science Publishers.
Jha, K. K., & Campbell, M. (2022). *Dynamics and interrelations between nature, science and society*. Nova Science Publishers.
Jha, K. K., & Campbell, M. (2023). *Vultures of India. Ecological developments, problems and prospects*. Nova Science Publishers.
Jordheim, S. (1980). Eagle-coyote incident. *Blue Jay, 38*, 47–48.
Jung, T. S., Everatt, K. T., & Andresen-Everatt, L. M. (2009). Kleptoparasitism of a coyote (Canis latrans) by a Golden eagle (Aquila chrysaetos)in northwestern Canada. *Northwestern Naturalist, 90*(1), 53–55. https://doi.org/10.1898/1051-1733-90.1.53
Katzner, T. E., Brandes, D., Miller, T., Lanzone, M., Maisonneuve, C., Tremblay, J. A., Mulvihill, R., & Merovich, G. T., Jr. (2012). Topography drives migratory flight altitude of golden eagles: Implications for on-shore wind energy development. *Journal of Applied Ecology, 49*, 1178–1186.

Katzner, T. E., Kochert, M. N., Steenhof, K., McIntyre, C. L., Craig, E. H., & Miller, T. A. (2020). Golden eagle (Aquila chrysaetos), version 2.0. In P. G. Rodewald & B. K. Keeney (Eds.), *Birds of the world*. Cornell Lab of Ornithology. https://doi.org/10.2173/bow.goleag.02

Kemp, A. C., Kirwan, G. M., & Marks, J. S. (2020). Crowned eagle (Stephanoaetus coronatus), version 1.0. In J. del Hoyo, A. Elliott, J. Sargatal, D. A. Christie, & E. de Juana (Eds.), *Birds of the world*. Cornell Lab of Ornithology. https://doi.org/10.2173/bow.crheag1.0

Kemp, M. E. (2023). Defaunation and species introductions alter long-term functional trait diversity in insular reptiles. *Proceedings of the National Academy of Sciences of the United States of America, 120*(7), e2201944119. https://doi.org/10.1073/pnas.2201944119

Kennedy, R. S. (1977). Notes on the biology and population status of the monkey-eating eagle of the Philippines. *The Wilson Bulletin, 89*(1), 1–20.

Kennedy, R. S. (1985). Conservation research of the Philippine eagle. *Research Reports, 18*, 401–414.

Kennedy, R. S., Gonzales, P. C., Dickinson, E. C., Miranda, H. C., Jr., & Fisher, T. H. (2000). *A guide to the birds of the Philippines*. Oxford University Press.

Kilgo, J. C., Ray, H. S., Vukovich, M., Goode, M. J., & Ruth, C. (2012). Predation by coyotes on white-tailed deer neonates in South Carolina. *The Journal of Wildlife Management, 76*(7), 1420–1430. https://doi.org/10.1002/jwmg.393

Kochert, M. N., Mcintyre, C. L., & Craig, E. H. (2002). Golden eagle (Aquila chrysaetos). In A. Poole & F. Gill (Eds.), *The birds of North America, 684*. Academy of Natural Sciences.

Kruuk, H. (1972). Surplus killing by carnivores. *Journal of Zoology, 166*(2), 233–244. https://doi.org/10.1111/j.1469-7998.1972.tb04087.x

Le Roux, E., Marneweck, D. G., Clinning, G., Druce, D. J., Kerley, G. I. H., & Cromsigt, J. P. G. M. (2019). Top–down limits on prey populations may be more severe in larger prey species, despite having fewer predators. *Ecography, 42*(6), 1115–1123. https://doi.org/10.1111/ecog.03791

Lowry, M. (2006). Courage and love. *Gabar, 17*, 29–30.

Luiselli, I., & Angelici, E. M. (2000). Ecological relationships in the Afrotropical cobra species (Naja melanoleuca and Naja nigricollis). *Canadian Journal of Zoology, 78*, 191–198.

Luiselli, I., Angelici, E. M., & Akani, G. C. (2001). Food habitats of Python sebae in suburban and natural habitats. *African Journal of Ecology, 39*, 116–118.

Lyly, M. S., Villers, A., Koivisto, E., Helle, P., Ollila, T., & Korpimäki, E. (2015). Avian top predator and the landscape of fear: Responses of mammalian mesopredators to risk imposed by the golden eagle. *Ecology and Evolution, 5*(2), 503–514. https://doi.org/10.1002/ece3.1370

MacDonald, G. (1969). *The gold coast past and present. A short description of the country and its people*. Greenwood Press.

MacGregor, L. B. (2021). Centring the animal: Re-evaluating the human–animal relationship since the middle ages. *The English Historical Review, 136*(578), 151–161. https://doi.org/10.1093/ehr/ceab023

Marneweck, C. J., Butler, A. R., Gigliotti, L. C., Harris, S. N., Jensen, A. J., Muthersbaugh, M., Newman, B. A., Saldo, E. A., Shute, K. E., Titus, K. L., Yu, S. W., & Jachowski, D. S. (2021). Shining the spotlight on small mammalian carnivores: Global status and threats. *Biological Conservation, 255*, 109005. https://doi.org/10.1016/j.biocon.2021.109005

Maseko, M. S. T., Zungu, M. M., Gumede, T. S., & Downs, C. T. (2023). Occupancy and detectability of African crowned eagles in an urban mosaic landscape: The importance of natural forest areas. *Urban Forestry & Urban Greening, 85*, 127976. https://doi.org/10.1016/j.ufug.2023.127976

Mason, J. R. (2000). Golden eagle attacks and kills adult male coyote. *Journal of Raptor Research, 34*(3), 244–245. Retrieved from https://sora.unm.edu/sites/default/files/journals/jrr/v034n03/p00244-p00245.pdf

Matthews, J. C. (1995). *British and Irish mythology*. Diamond Books.

McGregor, J. (2005). Crocodile crimes: People versus wildlife and the politics of postcolonial conservation on Lake Kariba, Zimbabwe. *Geoforum, 36*, 353–369.

McPherson, S. C., Brown, M., & Downs, C. T. (2016a). Crowned eagle nest sites in an urban landscape: Requirements of a large eagle in the Durban metropolitan open space system. *Landscape and Urban Planning, 146*, 43–50.
McPherson, S. C., Brown, M., & Downs, C. T. (2016b). Diet of the crowned eagle (Stephanoaetus coronatus) in an urban landscape: Potential for human-wildlife conflict? *Urban Ecosystems, 19*, 383–396. https://doi.org/10.1007/s11252-0150500-6
Mendelson, S., Cowlishaw, G., & Marcus Rowcliffe, J. (2003). Anatomy of a bushmeat commodity chain in Takoradi, Ghana. *Journal of Peasant Studies, 31*(1), 73–100.
Miles, D. B. (2020). Can morphology predict the conservation status of Iguanian lizards? *Integrative and Comparative Biology, 60*(2), 535–548. https://doi.org/10.1093/icb/icaa074
Miranda, E. B. (2018). Prey composition of harpy eagles (Harpia harpyja) in Raleighvallen, Suriname. *Conservation Science, 11*, 1940082918800789.
Miranda, E. B., Menezes, J. F. S., Farias, C. C., Munn, C., & Peres, C. A. (2019). Species distribution modeling reveals strongholds and potential reintroduction areas for the world's largest eagle. *PLoS One, 14*(5), e0216323.
Miranda, E. B., Peres, C. A., Carvalho-Rocha, V., Miguel, B. V., Lormand, N., Huizinga, N., Munn, C. A., Semedo, T. B., Ferreira, T. V., Pinho, J. B., & Piacentini, V. Q. (2021). Tropical deforestation induces thresholds of reproductive viability and habitat suitability in earth's largest eagles. *Scientific Reports, 11*(1), 1–17.
Miranda, E. B., Peres, C. A., & Downs, C. T. (2021). Landowner perceptions of livestock predation: Implications for persecution of an Amazonian apex predator. *Animal Conservation, 25*(1), 110–124. https://doi.org/10.1111/acv.12727
Miranda, E. B. P. (2015). Conservation implications of harpy eagle Harpia harpyja predation patterns. *Endangered Species Research, 29*(1), 69–79. https://doi.org/10.3354/esr00700
Miranda H., Salvador, D. I., & Bueser, G. L. (2008). *Updates on the nesting biology and population status of the Philippine eagle Pithecophaga jefferyi.*
Miranda, H. C. (2006). New insights on the population ecology and survival of the Philippine eagle based on radio-telemetry. In *Wings without borders: IV north American ornithological conference* (p. 225). American Ornithologists' Union.
Mitani, J. C., Sanders, W. J., Lwanga, J. S., & Windfelder, T. L. (2001). Predatory behavior of crowned hawk-eagles (Stephanoaetus coronatus) in Kibale National Park, Uganda. *Behavioral Ecology and Sociobiology, 49*(2-3), 187–195. https://doi.org/10.1007/s002650000283
Monadjem, A., & Rasmussen, M. W. (2008). Nest distribution and conservation status of eagles, selected hawks and owls in Swaziland. *Gabar, 19*, 1–22.
Mowbray, S. (2023). *Small wildcats pose big challenges, but coexistence is very much possible. Conservation News.* Retrieved from https://news.mongabay.com/2023/07/small-wildcats-posebig-challenges-but-coexistence-is-very-much-possible/
Murphy, J. C., & Henderson, R. W. (1997). *Tales of giant snakes: A historical natural history of anacondas and pythons.* Krieger Publishing Company.
Nattrass, N., Conradie, B., Stephens, J., & Drouilly, M. (2020). Culling recolonizing mesopredators increases livestock losses: Evidence from the south African Karoo. *Ambio, 49*(6), 1222–1231. https://doi.org/10.1007/s13280-019-01260-4
Naude, V. N., Smyth, L. K., Weideman, E. A., Krochuk, B. A., & Amar, A. (2019). Using web-sourced photography to explore the diet of a declining African raptor, the martial eagle (Polemaetus bellicosus). *Condor, 121*(1), 1–9. https://doi.org/10.1093/condor/duy015
Nelson, M. W. (2024). Human impacts on golden eagles: A positive outlook for the 1980s and 1990s. *Journal of Raptor Research, 16*(4), 97–103. Retrieved from https://sora.unm.edu/sites/default/files/journals/jrr/v016n04/p00097-p00103.pdf
Nolan, J. M., Jones, K. E., McDougal, K. W., & McFarlin, M. J. (2006). The lovable, the loathsome, and the liminal: Emotionality in ethnozoological cognition. *Journal of Ethnobiology, 26*, 126–138. https://doi.org/10.2993/0278-0771(2006)26[126:TLTLAT]2.0.CO;2

Nowak, E. M., Theimer, T. C., & Schuett, G. W. (2008). Functional and numerical responses of predators: Where do vipers fit in the traditional paradigms? *Biological Reviews of the Cambridge Philosophical Society, 83*(4), 601–620. https://doi.org/10.1111/j.1469-185X.2008.00056.x

O'Gara, B. W. (1978). *Sheep depredation by golden eagles in Montana.* Retrieved from https://escholarship.org/content/qt18p36935/qt18p36935_noSplash_9ba75cf7334d491ca70bbffda8663aa1.pdf?t=q82ah4

O'Gara, B. W. (1998). *Eagles.* Retrieved from https://pcwd.info/wpcontent/uploads/2016/12/1994Eagles.pdf

O'Neil, S. T., Coates, P. S., Webster, S. C., Brussee, B. E., Dettenmaier, S. J., Tull, J. C., Jackson, P. J., Casazza, M. L., & Espinosa, S. P. (2023). A spatially explicit modeling framework to guide management of subsidized avian predator densities. *Ecosphere, 14*(8). https://doi.org/10.1002/ecs2.4618

Öhman, A., & Mineka, S. (2001). Fears, phobias, and preparedness: Toward an evolved module of fear and fear learning. *Psychological Review, 108*, 483.

Ohman, A., & Mineka, S. (2003). The malicious serpent: Snakes as a prototypical stimulus for an evolved module of fear. *Current Directions in Psychological Science, 12*, 5–9. https://doi.org/10.1111/14678721.01211

Ohman, A., & Soares, J. J. F. (1994). Unconscious anxiety: Phobic responses to masked stimuli. *Journal of Abnormal Psychology, 103*, 231–240.

Olendorff, R. R. (1976). The food habits of north American golden eagles. *American Midland Naturalist, 95*(1), 231–236. https://doi.org/10.2307/2424254

Pasmans, F., Bogaerts, S., Braeckman, J., Cunningham, A. A., Hellebuyck, T., Griffiths, R. A., Sparreboom, M., Schmidt, B. R., & Martel, A. (2017). Future of keeping pet reptiles and amphibians: Towards integrating animal welfare, human health and environmental sustainability. *The Veterinary Record, 181*(17), 450. https://doi.org/10.1136/vr.104296

Pettersson, J. (1977). White-tailed eagle, Haliaeetus albicilla, robbing Golden eagle, Aquila chrysaetos, of its prey. *Vår Fågelvärld, 36*, 54.

Piana, R. P. (2007). Nesting and diet of Harpia harpyja Linnaeus in the native community of Infierno, Madre de Dios, Peru. *Revista Peruana de Biologıa, 14*, 135–138.

Piana, R. P. (2018). Harpia harpyja. In SERFOR (Ed.), *Libro Rojo de la Fauna Silvestre Amenazada del Perú* (pp. 203–204). Serfor (Servicio Nacional Forestal y de Fauna Silvestre).

Pough, F. H. (1973). Lizard energetics and diet. *Ecology, 54*, 837–844. https://doi.org/10.2307/1935678

Prugh, L. R., Stoner, C. J., Epps, C. W., Bean, W. T., Ripple, W. J., Laliberte, A. S., & Brashares, J. S. (2009). The rise of the mesopredator. *Bioscience, 59*(9), 779–791. https://doi.org/10.1525/bio.2009.59.9.9

Puplett, D. (2020). *Eagle mythology and folklore. Trees for life.* Retrieved from https://treesforlife.org.uk/into-the-forest/trees-plants-animals/birds/eagle/eaglemythology-and-folklore/

Ralls, K., & White, P. J. (1995). Predation on endangered San Joaquin kit foxes by larger canids. *Journal of Mammalogy, 76*(3), 723–729. https://doi.org/10.2307/1382743

Ray, J. C., Redford, K. H., Steneck, R., & Berger, J. (Eds.). (2013). *Large carnivores and the conservation of biodiversity*. Island Press.

Read, J. L., & Scoleri, V. (2015). Ecological implications of reptile mesopredator release in arid South Australia. *Journal of Herpetology, 49*(1), 64–69. https://doi.org/10.1670/13-208

Restrepo-Cardona, J. S., Echeverry-Galvis, M. Á., Maya, D. L., Vargas, F. H., Tapasco, O., & Renjifo, L. M. (2020). Human-raptor conflict in rural settlements of Colombia. *PLoS One, 15*(1), e0227704. https://doi.org/10.1371/journal.pone.0227704

Reuter, K. E., & Schaefer, M. S. (2017). Motivations for the ownership of captive lemurs in Madagascar. *Anthrozoös, 30*(1), 33–46. https://doi.org/10.1080/08927936.2017.1270589

Ripple, W. J., & Beschta, R. L. (2012). Large predators limit herbivore densities in northern forest ecosystems. *European Journal of Wildlife Research, 58*(4), 733–742. https://doi.org/10.1007/s10344-012-0623-5

Ripple, W. J., Estes, J. A., Beschta, R. L., Wilmers, C. C., Ritchie, E. G., Hebblewhite, M., Berger, J., Elmhagen, B., Letnic, M., Nelson, M. P., Schmitz, O. J., Smith, D. W., Wallach, A. D., & Wirsing, A. J. (2014). Status and ecological effects of the world's largest carnivores. *Science, 343*, 1241484.

Ritchie, E. G., & Johnson, C. N. (2009). Predator interactions, mesopredator release and biodiversity conservation. *Ecology Letters, 12*(9), 982–998. https://doi.org/10.1111/j.1461-0248.2009.01347.x

Rodríguez, L., Reuter, K. E., & Seabocy, M. S. (2020). Motivations of pet parrot ownership and captive conditions of the pets in Madagascar. *Madagascar Conservation & Development, 15*(1), 13–18. https://doi.org/10.4314/mcd.v15i1

Roemer, G. W., Donlan, C. J., & Courchamp, F. (2002). Golden eagles, feral pigs, and insular carnivores: How exotic species turn native predators into prey. *Proceedings of the National Academy of Sciences of the United States of America, 99*(2), 791–796. https://doi.org/10.1073/pnas.012422499

Roll, U., Feldman, A., Novosolov, M., Allison, A., Bauer, A. M., Bernard, R., Böhm, M., Castro-Herrera, F., Chirio, L., Collen, B., Colli, G. R., Dabool, L., Das, I., Doan, T. M., Grismer, L. L., Hoogmoed, M., Itescu, Y., Kraus, F., LeBreton, M., Lewin, A., & Meiri, S. (2017). The global distribution of tetrapods reveals a need for targeted reptile conservation. *Nature Ecology & Evolution, 1*(11), 1677–1682. https://doi.org/10.1038/s41559-017-0332-2

Rotenberg, J. A., Marlin, J. A., Pop, L., & Garcia, W. (2012). First record of a harpy eagle (Harpia harpyja) nest in Belize. *The Wilson Journal of Ornithology, 124*(2), 292–297.

Salvador, D., & Quiblatt, T. (2009). *Philippine eagle foundation annual report (2008).* Philippine Eagle Foundation.

Salvador, D. J. I., & Ibanez, J. C. (2006). Ecology and conservation of Philippine eagles. *Ornithological Science, 5*(2), 171–176.

Sánchez-Zapata, J. A., Eguía, S., Blázquez, M., Moleón, M., & Botella, F. (2010). Unexpected role of ungulate carcasses in the diet of Golden eagles Aquila chrysaetos in Mediterranean mountains. *Bird Study, 57*(3), 352–360. https://doi.org/10.1080/00063651003674946

Scasta, J. D., Stam, B., & Windh, J. L. (2017). Rancher-reported efficacy of lethal and nonlethal livestock predation mitigation strategies for a suite of carnivores. *Scientific Reports, 7*, 14105. https://doi.org/10.1038/s41598-017-14462-1

Schulenberg, T. S. (2020). Harpy eagle (Harpia harpyja), version 1.0. In T. S. Schulenberg (Ed.), *Birds of the world*. Cornell Lab of Ornithology.

Seguin, J. F., Thibault, J. C., Torre, J., Bayle, P., & Vigne, J. D. (2001). The diet of young Golden eagles Aquila chrysaetos in Corsica: Foraging in a man-made mammal fauna. *Ardea, 89*, 527–535.

Seladi-Schulman, J. (2021). *Herpetophobia, the fear of lizards and snakes*. Retrieved from https://www.healthline.com/health/herpetophobia

Seymour, A. S., Hatherley, G., Contreras, F. J., Aldred, J., & Beeley, F. (2010). Hatching synchrony, green branch collecting, and prey use by nesting harpy eagles (Harpia harpyja). *The Wilson Journal of Ornithology, 122*(4), 792–795. https://doi.org/10.1676/10-060.1

Simmons, R. E. (2015). Martial eagle Polemaetus bellicosus. In R. E. Simmons, C. J. Brown, & J. Kemper (Eds.), *Birds to watch in Namibia: Red, rare and endemic species* (pp. 135–137). Ministry of Environment and Tourism, Namibia Nature Foundation.

Simmons, R. E., & Brown, C. J. (2009). *Birds to watch in Namibia: Red, rare and endemic species*. National Biodiversity Programme.

Soulé, M. E., Bolger, D. T., Alberts, A. C., Wright, J., Sorice, M., & Hill, S. (1988). Reconstructed dynamics of rapid extinctions of chaparral-requiring birds in urban habitat islands. *Conservation Biology, 2*(1), 75–91. https://doi.org/10.1111/j.15231739.1988.tb00337.x

Steyn, P. (1982). *Birds of prey in southern Africa*. David Philip Publishers.

Stjernstedt, B. (1975). Eagle attack. *Black Lechwe, 12*, 18–22.

Sutherland, D. R., Glen, A. S., & de Tores, P. J. (2011). Could controlling mammalian carnivores lead to mesopredator release of carnivorous reptiles? *Biological Sciences, 278*(1706), 641–648. https://doi.org/10.1098/rspb.2010.2103

Swatridge, C. J., Monadjem, A., Steyn, D. J., Batchelor, G. R., & Hardy, I. C. (2014). Factors affecting diet, habitat selection and breeding success of the African crowned eagle Stephanoaetus coronatus in a fragmented landscape. *Ostrich, 85*(1), 47–55. https://doi.org/10.2989/00306525.2014.896832

Symes, C. T., & Antonites, A. R. (2014). Notes on African crowned eagle Stephanoaetus coronatus diet in savanna and forest in KwaZulu-Natal, South Africa. *Ostrich, 85*(1), 85–88. https://doi.org/10.2989/00306525.2014.892036

Tarboton, W. R., & Allan, D. G. (1984). The status and conservation of birds of prey in the transvaal. Transvaal Museum Monograph No. 3. Pretoria, 3.

Taylor, M. R. (2015). Martial Eagle Polemaetus bellicosus. In M. R. Taylor, F. Peacock, R. M. Wanless (Eds.), *The 2015 Eskom Red Data Book of Birds of South Africa, Lesotho and Swaziland* (pp. 113–115). BirdLife South Africa.

Thomsett, S. (2011). *Simon Thomsett on the African crowned eagle*. Retrieved from http://www.africanraptors.org/simon-thomsett-on-the-african-crowned-eagle-part-1/

Tisdell, C. (2011). The influence of public attitudes on policies for conserving reptiles. In *Reptiles, biology, behaviour and conservation*. Nova Science. https://doi.org/10.13140/2.1.4921.1523

Toland, E., Bando, M., Hamers, M., Cadenas, V., Laidlaw, R., Martínez-Silvestre, A., & van der Wielen, P. (2020). Turning negatives into positives for pet trading and keeping: A review of positive lists. *Animals, 10*(12), 2371. https://doi.org/10.3390/ani10122371

Tolley, K. A., Alexander, G. J., Branch, W. R., Bowles, P., & Maritz, B. (2016). Conservation status and threats for African reptiles. *Biological Conservation, 204*, 63–71. https://doi.org/10.1016/j.biocon.2016.04.006

Trinca, C. T., Ferrari, S. F., & Lees, A. C. (2008). Curiosity killed the bird: Arbitrary hunting of harpy eagles Harpia harpyja on an agricultural frontier in southern Brazilian Amazonia. *Cotinga, 15*, 12–15.

United States Fish and Wildlife Service. (2011). *Golden eagles status fact sheet*. Retrieved from https://www.fws.gov/migratorybirds/pdf/education/educationalactivities/golden-eagle-fact-sheet.pdf

United States Fish and Wildlife Service. (2024). *Eagle management*. Retrieved from https://www.fws.gov/program/eagle-management

Vargas, J., Whitacre, D., Mosquera, R., Albuquerque, J., Piana, R., Thiollay, J.-M., Márquez, C., Sánchez, J. E., Lezama-López, M., Midence, S., Matola, S., Aguilar, S., Rettig, N., & Sanaiotti, T. (2006). Status and current distribution of the harpy eagle (Harpia harpyja) in central and South America. *Ornitologia Neotropical, 17*, 39–55.

Verdoorn, G. H., & Roth, L. A. (2000). Polemaetus bellicosus martial eagle. In G. H. Verdoorn, K. L. Bildstein, & S. Ellis (Eds.), *Selected African Falconiformes conservation assessment and management plan* (pp. 68–69). International Union for Conservation of Nature and Natural Resources/SSC Conservation Breeding Specialist Group.

Wagner, E. (2014). *The deadly and devious golden eagle*. Smithsonian Magazine. Retrieved from https://smithsonianmag.com/science-nature/royal-raptor-180949437/

Warwick, C., Steedman, C., Jessop, M., Arena, P., Pilny, A., & Nicholas, E. (2018). Exotic pet suitability: Understanding some problems and using a labeling system to aid animal welfare, environment, and consumer protection. *Journal of Veterinary Behavior, 26*, 17–26. https://doi.org/10.1016/j.jveb.2018.03.015

Watson, J. (2010). *The golden eagle*. Poyser.

Watson, R. T., McClure, C. J., Vargas, F. H., & Jenny, J. P. (2016). Trial restoration of the harpy eagle, a large, long-lived, tropical forest raptor, in Panama and Belize. *Journal of Raptor Research, 50*(1), 3–22.

Wellman, J. (2015). *What does the eagle represent in the bible? A Christian study*. Retrieved from https://www.patheos.com/blogs/christiancrier/2015/07/17/whatdoes-the-eagle-represent-in-the-bible-a-christian-study/
Wilmers, C. C., Stahler, D. R., Crabtree, R. L., Smith, D. W., & Getz, W. M. (2003). Resource dispersion and consumer dominance: Scavenging at wolf- and hunter-killed carcasses in greater Yellowstone, USA. *Ecology Letters, 6*(11), 996–1003. https://doi.org/10.1046/j.1461-0248.2003.00522.x
Wordley, C. (2021). *Dying of curiosity: Why people shoot harpy eagles*. Mongabay. Retrieved from https://news.mongabay.com/2021/03/dying-of-curiosity-why-people-shoot-harpyeagles/

Chapter 8
Composition and Plant Species Diversity in the Diet of Some Wild Ungulates in India: A Systematic Review

Rohit Chaudhary, Kavya Shah, Taksh Kelaiya, Aadil Kazi, Prakhar Sharma, Mohmad Navaz Dahya, and Priyank Kapadi

Abstract Wild ungulates can influence the structure and function of their resident and transitory ecosystems through their feeding habits. Consequently, analysis of the feeding habits of ungulates can support efforts to protect ungulates, including the conservation of their wider habitats. We examined the dietary habits of nine species of wild ungulates in India by examining the literature published from 1990 to 2020. The species studied were chital (*Axis axis*), sambar (*Rusa unicolor*), nilgai (*Boselaphus tragocamelus*), blackbuck (*Antilope cervicapra*), gaur (*Bos gaurus*), swamp deer (*Rucervus duvaucelii*), chinkara (*Gazella bennetii*), Chowsingha (*Tetracerus quadricornis*), and hog deer (*Axis porcinus*). These species represent a wide range of body sizes and habitats and are therefore potential models for studying interspecific dietary differences and relationships. Our aim was to evaluate the methods used in studies of ungulate dietary habits to document ungulate diets and any interspecies-level similarities. The diet of ungulates was assessed by counting the number of plant species, life forms (trees, shrubs, herbs, grasses, sedges, and climbers), life form diversity, and tree/ grass species ratio. Cluster analysis was used to assess dietary similarities between species. A total of 21 studies were found, including six records for chital, five records for nilgai, seven for sambar, five for gaur, five for chowsingha, three for blackbuck, two for swamp deer, one for chinkara, and two for

R. Chaudhary (✉) · A. Kazi
Department of Wildlife Sciences, College of Forestry, Navsari Agricultural University, Navsari, India
e-mail: rchaudhary259@gmail.com

K. Shah · T. Kelaiya
Undergraduate Program; College of Forestry, Navsari Agricultural University, Navsari, India

P. Sharma
Department of Wildlife Sciences, Aligarh Muslim University, Aligarh, India

M. N. Dahya
Department of Zoology, B.P. Baria Science Institute, Navsari, India

P. Kapadi
Snake Research Institute, Valsad, India

K. K. Jha and M. O. Campbell (eds.), *Intersections of Conservation Biogeography and Wildlife Management*, https://doi.org/10.1007/978-981-95-0121-2_8

hog deer. Most plant species were found in the diet of chital (164), while the least plant species were found in the diet of hog deer (19). The highest tree/grass species ratio was recorded in the diet of sambar and the lowest in the diet of blackbuck. The diet of hog deer and swamp deer did not consist of tree species. The greatest similarity was between the diet of the hog deer and the swamp deer and between chital and nilgai. The results of this study suggest that as ungulate species become more specialized in using their habitat, the diversity of plant species included in their diet decreases. These findings support an integrated approach to ungulate conservation, including critical management of ungulate habitat and plant diversity in India.

Keywords Chital · Sambar · Nilgai · Blackbuck · Gaur · Swamp deer

Introduction

The increase in the global human population has led to declines in wildlife populations around the world (Cardinale et al., 2018). Key factors driving wildlife population and habitat declines are anthropogenic in nature, including infrastructure development, husbandry and farming practices, and demand for wildlife as resources such as meat and other products (Clavel et al., 2011). However, wildlife declines are not uniform across all species or groups of species, and the key factors driving wildlife declines, in addition to human density, are their body size and resource requirements (Lamprey & Reid, 2004; Gordon, 2009; Craigie et al., 2010; Ripple et al., 2015). For example, animals with large body size require larger living space, which makes them vulnerable to habitat destruction (Smith et al., 2018). Additionally, animals with large body size, such as elephants, are k-selective and produce a small number of offspring after long gestation periods, making them even more vulnerable to population decline (Pianka, 1970; Promislow & Harvey, 1990).

Wild ungulates are a group of mammals that show remarkable diversity in body size and ecological resource requirements (Nowak, 2018). The body size diversity of ungulates ranges from elephants, the largest land mammals on earth, to small mouse deer (Nowak, 2018). In addition, the diet of ungulates changes with body size, with smaller ungulates requiring high-quality food while larger ungulates can tolerate lower-quality food but require it in large quantities (Jarman, 1974). Habitat destruction on all continents impacts a range of wild ungulates as it can lead to loss of food and its diversity, affecting both diet generalist and specialist ungulates (Atwood et al., 2020). Conserving the remarkable diversity of ungulates is necessary for several reasons. For example, ungulates alter plant recruitment through browsing and antler/horn rubbing on trees (Beguin et al., 2016; Stokely & Betts, 2020), and through grazing, trampling, grubbing, and debarking, ungulates can influence the structure and composition of vegetation (Gill & Beardall, 2001). By changing the structure and composition of vegetation, ungulates can also create trophic cascades in the ecosystem in which they persist. For example, due to herbivory, ungulates suppress or eliminate nitrogen-fixing plants and reduce nitrogen accumulation (Boulanger et al., 2017; Knopps et al., 2000). The result of this reduced nutrient availability is a loss of forest productivity over time and a possible change in topstory composition (Etzold et al., 2020; Hobbs, 1996). The ungulates also provide important ecosystem services such as seed dispersal, as many of the ungulate species feed on seasonal fruits and disperse them over long distances. For example, Kitamura et al. (2007) found that elephants in Thailand disperse seeds of various plant species.

Successful species conservation requires information about vital ecological resources, such as food (Spitzer et al., 2020). This is particularly relevant to the ecology of ungulates, as ungulates2019 are important for the maintenance of the entire ecosystem due to their diverse diet and foraging behaviour (Hobbs, 1996; Ramirez,). Food also plays an important role in the different life histories of ungulates, as it is one of the most important determinants of ungulate density and carrying capacity, their social organization, and group size (Jarman, 1974; Takatsuti, 2022). Competition between ungulates is also reduced by food partitioning, making food a key dimension along which ungulates divide their resources and coexist (Schoener, 1974; Spitzer et al., 2020). Therefore, assessing the feeding habits of ungulates not only contributes to their conservation but also improves our ecological understanding of their life history traits (Ramirez, 2019; Spitzer et al., 2020).

India, is one of 17 mega diverse nations and is home to a diverse ungulate population ranging from Asian elephants (*Elephas maximus*, Linnaeus, 1758) to lesser mouse deer (*Tragulus* spp. Brisson, 1762) (UNEP, 2023). The conservation of ungulates in India is closely linked to the conservation of large carnivores such as tigers (*Panthera tigris tigris*, Linnaeus, 1758) and lions (*Panthera leo leo*, Linnaeus, 1758), which are among the most charismatic wild animals in the world (Srivathsa et al., 2022; Upadhyay et al., 2019). Most ecological information about ungulates in India relates to their density and biomass in protected areas as they serve as important prey for large carnivores such as tigers, leopards, and lions (Gupta & Krishnamurthy, 2023; Upadhyay et al., 2019). The All-India Tiger Assessment Report provides a tremendous amount of information on the abundance and density of the ungulate population on a regular basis (Jhala et al., 2020). However, there is a lack of information on the dietary habits of ungulates in India, and there is an urgent need to compile the current information on this subject in a review to develop a comprehensive understanding of the dietary habits of ungulates.

This chapter examines the dietary habits of ungulates in India using existing information from 1990 to 2020. The selected period covers the largest number of ungulate studies in India. Nine ungulate species are selected for this study: spotted deer or chital (*Axis axis*, Erxleben 1977), sambar (*Rusa unicolor*, Kerr 1792), blue bull or nilgai (*Boselaphus tragocamelus* Pallas 1766), blackbuck (*Antelope cervicapra*, Linnaeus 1758), Indian bison or gaur (*Bos gaurus* Smith 1827), swamp deer or barasingha (*Rucervus duvaucelii* G. Cauvier 1823), chinkara or Indian gazelle (*Gazella bennetii*, Sykes 1831), four-horned antelope or chowsingha (*Tetracerus quadricornis*, de Blainville 1816), and hog deer (*Hyelaphus porcinus* Zimmermann 1780). We include these species because they represent a wide variety of habitats and have been the subject of numerous studies (Tables 8.1, 8.2, and 8.3). The objectives of the present study are: (1) to evaluate the methods used in various studies investigating dietary habits, (2) to characterize the diet of ungulates, and (3) to assess similarities in the diet of ungulates.

Methodology

Since our was a review study, our data represent previous studies on the dietary habits of the selected ungulate species. Information for the study was derived from searches in Google Scholar using the keywords "food habits" and names of the ungulate species. The searches reviewed the first five pages of Google Scholar for the for the food habits studies using keyword like ungulates+diet, ungulates+food, species name+diet. If two studies used similar data, only one was included. We also obtained PhD theses data downloaded from "Shodhganga", a government archive of PhD theses in India (https://shodhganga.inflibnet.ac.in). The information included the ungulate species studied, the size of the study area, the sampling area, the methodology used (direct or indirect), the diet adequacy analysis performed, and the food plant species consumed by the ungulate species.

Analysis

We first described the number of plant species consumed by each ungulate species. We then grouped food species into six plant structural forms: tree, grass, herb, shrub, climber, and sedge, and assessed the relative contribution of each plant form in the ungulate diet. We tested whether there were significant differences in the consumption of different plant forms using the chi-square test of homogeneity (Zar, 2006). We also assessed the diversity of plant forms in the ungulate diet using the Shannon-Weiner diversity index (Magurran, 1988). Since the diet of ungulates varies primarily depending on the amount of tree and grass species consumed, we determined the ratio of tree species to grass species in the diet of the ungulates.

We also examined the similarity of plant species in the diet of ungulates, which can determine the partitioning strategy of ungulates in the diet. Because our data relate to several ungulate species and many plant species, we use hierarchical cluster analysis, a multivariate analysis that classifies species based on their similarity and dissimilarity (James & McCulloch, 1990). We use Ward's method in cluster analysis to assess the similarity in species consumption among ungulate species. Results of cluster analysis in different groups of clusters and species within the cluster show similarities in diet.

Table 8.1 Ecological characteristics of the ungulate's species considered in the study

Characteristics	Chital	Sambar	Nilgai	Chowsingha	Gaur	Chinkara	Blackbuck	Hog deer	Swamp deer
Adult body weight (average in kg)	45	180	200	22	800	23	42	35	180
Habitat preference	Open forest, meadows and grasslands	Dense forest	Open forest, grassland, scrubland, agriculture fields	Dry deciduous forest	Dense to moderate dense forest	Open forest, scrublands, scattered bush and sandy areas	Open habitat, grasslands, savanna, scrubland	Tall and moist grasslands	Swampy grasslands
Feeding habits	Mix feeder, preferably grazer	Mixed feeder, preferably browser	Mix feeder	Preferably browser	Mix feeder, preferably grazer	Preferably browser	Preferably grazer	Preferably grazer	Grazer
Social organization	Gregarious	Gregarious	Gregarious	Solitary or lives in pair	Gregarious	Solitary or in small groups	Gregarious	Solitary	Gregarious
Group size	Vary according to season, can reach up to >100 individuals	Small groups 1–5 generally, however, can reach up to 50 individuals	Vary according to habitat and season, but can reach up >100 individuals	Solitary or 1–3	Vary according to habitat, generally ranges between 1–16 individuals	Small groups 1–5 generally, however can reach up to 10 individuals	Small to very large groups which can ranges from 2 to 300 individuals	Solitary	Generally, ranges between 1–10 but can goes up to >100 individuals

Table 8.2 Details of study used in review to assess the diet

S. No.	Title of the study	PY	Species	SAN	PA/ Non PA	SASZ	SAS	D	S	M
1.	Study on the ecology of nilgai(*Boselaphus tragocamelus*) in Saurashtra	Mehta (2013)	Nilgai	Central Saurashtra, Gujarat	Non PA	NA	NA	16	3	Both
2.	Home range, habitat use and food habits of re-introduced Gaur in Bandhavgarh Tiger Reserve, Central India	Sankar et al. (2013)	Gaur	Bandhavgarh Tiger Reserve, Madhya Pradesh	PA	688	NA	12	3	Both
3.	Ecology of four-horned antelope in Pench Tiger Reserve, Seoni, M.P., with special prefrence to food & feeding habit	Haleem et al. (2014)	Chowsingha	Pench Tiger Reserve, Madhya Pradesh	PA	757.85	NA	NA	1	Both
4.	Studies on food and feeding habits of swamp deer in Jhilmil jheel conservation reserve, Haridwar, Uttarakhand, India	Tiwari and Rawat (2013)	Swamp deer	Jhilmil Jheel Conservation reserve, Uttarakhand	PA	37.83	6	24	6	Both

(continued)

Table 8.2 (continued)

S. No.	Title of the study	PY	Species	SAN	PA/ Non PA	SASZ	SAS	D	S	M
5.	Seasonal pattern of food habits of large herbivores in riverine alluvial grasslands of Brahmaputra floodplains, Assam	Devi et al. (2022)	Swamp deer, hog deer, sambar	Kaziranga National Park, Assam	PA	1090	NA	24	4	Both
6.	Browse plants of gaur in Mudumalai tiger reserve, southern India	Ashokkumar et al. (2010)	Gaur	Mudumalai tiger reserve, Tamil Nadu	PA	321	NA	20	NA	DO
7.	Ecology and management of chital and blackbuck in Guindy national park, madras	Raman et al. (1996)	Chital and blackbuck	Guindy National Park, Chennai	PA	2.7	2.7	24	6	DO
8.	Food habits of ungulates in dry tropical forests of Gir lion sanctuary, Gujarat, India	Khan et al. (1994)	Chital, sambar, Nilgai	Gir Lion sanctuary and National Park, Gujarat	PA	1412	NA	24	4	DO

(continued)

Table 8.2 (continued)

S. No.	Title of the study	PY	Species	SAN	PA/ Non PA	SASZ	SAS	D	S	M
9.	Conservation status and ecology of ungulates in Pench tiger reserve Madhya Pradesh with special reference to resource portioning	Haleem (2014)	Chowsingha, gaur, sambar	Pench Tiger Reserve, Madhya Pradesh	PA	757.85	NA	24	8	DO
10.	Ecology of chital in Gir	Dave (2008)	Chital	Gir Lion sanctuary and National Park, Gujarat	PA	1412	NA	24	6	DO
11.	Study on the ecology of wild ungulates of Keoladeo National Park, Bharatpur, Rajasthan	Haque (1990)	Chital, sambar, blackbuck, Nilgai, wild boar, feral cattle	Keoladeo National Park Bharatpur, Rajasthan	PA	29	NA	24	NA	DO
12.	Diet and habitat niche relationships within an assemblage of large herbivores in a seasonal tropical forest	Ahrestani et al. (2012)	Chital, sambar, Gaur and Asian Elephant	Bandipur and Mudumalai Tiger reserve	PA	400	NA	15	5	DO
13.	Food habits of nilgai in Van Vihar national park, Bhopal, Madhya Pradesh, India	Bhattarai (2019)	Nilgai	Van Vihar National Park, India	PA	4.45	NA	12	2	FMA

(continued)

Table 8.2 (continued)

S. No.	Title of the study	PY	Species	SAN	PA/ Non PA	SASZ	SAS	D	S	M
14.	Spatial and dietary overlap between blackbuck (Antilope cervicapra) and feral horse (Equus caballus) at point Calimere Wildlife sanctuary, southern India: Competition between native versus introduced species	Baskaran et al. (2016)	Blackbuck	Point Calimere Wildlife sanctuary, Tamil Nadu	PA	30	NA	7	2	FMA
15.	Seasonal food habits of the four-horned antelope (*Tetracerus quadricornis* de Blainville, 1816) in tropical deciduous forests of Aravalli Mountain range, Rajasthan, Western India	Meghwal et al. (2020)	Chowsingha	Sitamata, Kumbhalgarh, and Phulwari-ki-nal WLS, Rajasthan	PA		NA	11	3	FMA

(continued)

Table 8.2 (continued)

S. No.	Title of the study	PY	Species	SAN	PA/ Non PA	SASZ	SAS	D	S	M
16.	Behavioural ecology of four-horned antelope (Tetracerus quadricornis de Blainville, 1816) in the tropical forests of southern India	Baskaran et al. (2011)	Chowsingha	Mudumalai Wildlife sanctuary, Tamil Nadu	PA		NA	6	2	FMA
17.	Food plants and feeding habits of ungulates at Panna Tiger Reserve, Madhya Pradesh, India	Javid et al. (2022)	Chital, sambar, nilgai, chowsingha, chinkara	Panna Tiger Reserve, Madhya Pradesh	PA	542.7	NA	5	1	FMA
18.	Studies on the food and feeding habits of Gaur Bos gaurus H. Smith (Mammalia: Artiodactyla: Bovidae) in two protected areas of Goa	Gad and Shyama (2009)	Gaur	Bhagwan Mahavir Wildlife sanctuary and Mollem National Park, Goa	PA	133 and 107	NA	24	3	FMA
19.	Fruiting plants & wildlife	Kumar et al. (2020)	Gaur and barking deer	Bonai Forest division, Odisha	Non PA	NP	NA	12	3	FMA

(continued)

Table 8.2 (continued)

S. No.	Title of the study	PY	Species	SAN	PA/ Non PA	SASZ	SAS	D	S	M
20.	The habitat and food habits of hog deer (*Axis porcinus*) in protected areas of sub-Himalayan West Bengal	Bhowmik et al. (1999)	Barking deer	Jaldapara Wildlife sanctuary, West Bengal	PA		NA	NA	NA	FMA
21.	Factors affecting forage selection by the endangered Eld's deer and hog deer in the floating meadows of Barak-Chindwin Basin of north-East India	Tuboi and Hussain (2016)	Hog deer	Keibul Namjo, Manipur	PA	286	NA	23	3	Both

PY publishing year, *SAN* study area name, *PA* protected area, SASZ study area size (in km2), *SAS* sampling area size (in Km2), *D* duration, *S* number of season data was collected, *M* methodology, *DO* direct observation, *FMA* faecal matter analysis *Both* DO+FMA

Table 8.3 List of plant species found in the diet of ungulates (✓ = present; x = absent)

S. No.	Plant species	Plant form	Nilgai	Chowsingha	Blackbuck	Sambar	Gaur	Swamp deer	Chital	Chinkara	Hog deer
1.	*Asparagus racemosus*	Climber	✓	✓	x	✓	x	x	✓	x	x
2.	*Bauhinia vahlii*	Climber	x	✓	x	✓	✓	x	x	x	x
3.	*Combretum roxburghii*	Climber	x	x	x	✓	x	x	x	x	x
4.	*Cocculus hirsutus*	Climber	✓	x	x	x	x	x	✓	x	x
5.	*Cuscuta hyalina*	Climber	✓	x	x	x	x	x	x	x	x
6.	*Dolichandrona faleala*	Climber	✓	x	x	x	x	x	x	x	x
7.	*Hemidesmus indicus*	Climber	✓	x	x	x	x	x	✓	x	x
8.	*Ichnocarpus frutescens*	Climber	✓	x	x	✓	x	x	✓	x	x
9.	*Ipomoea fistulosa*	Climber	✓	x	x	x	x	x	✓	x	x
10.	*Ipomoea pes-tigridis*	Climber	✓	x	x	x	x	x	x	x	x
11.	*Ipomoea species*	Climber	✓	x	✓	x	x	x	x	x	x
12.	*Leptadenia spartium*	Climber	x	x	x	x	x	x	x	✓	x
13.	*Luffa acutangula*	Climber	✓	x	x	x	x	x	x	x	x
14.	*Mucuna pruriens*	Climber	✓	x	x	x	x	x	✓	x	x
15.	*Tinospora species*	Climber	x	✓	x	x	x	x	x	x	x
16.	*Vallaris solanacea*	Climber	x	✓	x	✓	✓	x	x	x	x

(continued)

Table 8.3 (continued)

S. No.	Plant species	Plant form	Nilgai	Chowsingha	Blackbuck	Sambar	Gaur	Swamp deer	Chital	Chinkara	Hog deer
17.	*Ventilago maderaspatana*	Climber	x	✓	x	x	x	x	x	x	x
18.	*Canavalia gladiata*	Climber	✓	x	x	✓	x	x	x	x	x
19.	*Cissampelos pareira*	Climber	x	x	x	x	x	x	x	x	x
20.	*Cryptolepis buchananii*	Climber	x	x	x	✓	x	x	x	✓	x
21.	*Dioscorea hispida*	Climber	x	x	x	x	x	x	x	x	x
22.	*Dioscorea oppositifolia*	Climber	x	x	x	x	x	x	x	x	x
23.	*Dioscorea pubera*	Climber	x	x	x	x	x	x	x	x	x
24.	*Dioscorea wallichi*	Climber	x	x	x	x	x	x	x	x	x
25.	*Dioscorea bulbifera*	Climbers	x	x	x	x	x	x	x	x	x
26.	*Acroceras zizanioides*	Grass	✓	x	x	x	x	x	✓	x	x
27.	*Andropogon contortus*	Grass	✓	✓	x	✓	x	x	✓	✓	x
28.	*Apluda mutica*	Grass	✓	✓	x	✓	✓	x	✓	✓	x
29.	*Apluda varia*	Grass	✓	x	x	x	x	x	✓	x	x
30.	*Aristida adscensionis*	Grass	x	✓	x	x	x	x	x	x	x
31.	*Aristida hystrix*	Grass	✓	✓	x	✓	x	x	✓	✓	x

(continued)

Table 8.3 (continued)

S. No.	Plant species	Plant form	Nilgai	Chowsingha	Blackbuck	Sambar	Gaur	Swamp deer	Chital	Chinkara	Hog deer
32.	*Aristida ischaemum*	Grass	x	x	x	x	x	x	✓	x	x
33.	*Aristida funiculata*	Grass	x	x	x	x	x	x	✓	x	x
34.	*Aristida setacea*	Grass	x	x	✓	x	x	x	✓	x	x
35.	*Arsitida mutabilis*	Grass	x	x	x	x	x	x	x	✓	x
36.	*Arundo donax*	Grass	x	x	x	x	x	x	x	x	✓
37.	*Arundinella leptochloa*	Grass	x	x	x	x	✓	x	x	x	x
38.	*Axonopus species*	Grass	x	✓	x	x	x	x	x	x	x
39.	*Bamboo species*	Grass	x	✓	x	✓	✓	x	x	x	x
40.	*Bambusa arundinacea*	Grass	x	x	x	x	✓	x	x	x	x
41.	*Bambusa species*	Grass	✓	x	x	✓	✓	x	✓	x	x
42.	*Bothriochloa pertusa*	Grass	✓	x	x	x	x	x	x	x	x
43.	*Brachiaria distachya*	Grass	✓	x	x	x	x	x	x	x	x
44.	*Brachiaria mutica*	Grass	x	x	x	x	x	x	✓	x	x
45.	*Brachiaria ramosa*	Grass	x	x	x	x	x	x	✓	x	x
46.	*Brachiaria reptans*	Grass	✓	x	x	x	x	x	✓	x	x
47.	*Capillipedium species*	Grass	x	x	x	x	x	x	x	x	✓
48.	*Cenchrus biflorus*	Grass	✓	x	x	x	x	x	x	x	x

(continued)

Table 8.3 (continued)

S. No.	Plant species	Plant form	Nilgai	Chowsingha	Blackbuck	Sambar	Gaur	Swamp deer	Chital	Chinkara	Hog deer
49.	*Cenchrus ciliaris*	Grass	x	✓	x	x	x	x	x	x	x
50.	*Cenchrus olitorius*	Grass	✓	x	x	x	x	x	x	x	x
51.	*Cenchrus pennisetiformis*	Grass	x	x	x	x	x	x	x	✓	x
52.	*Cenchrus species*	Grass	✓	x	x	x	x	x	x	x	x
53.	*Chionachne koenigii*	Grass	x	x	x	x	x	x	✓	x	x
54.	*Chloris barbata*	Grass	✓	✓	✓	✓	✓	x	✓	x	x
55.	*Chloris dolichostachya*	Grass	x	x	x	x	x	x	✓	x	x
56.	*Cholris dolichostachia*	Grass	x	x	x	x	x	x	x	x	x
57.	*Coix lacryma-jobi*	Grass	x	x	x	x	x	✓	x	x	✓
58.	*Cymbopogon flexuosus*	Grass	x	x	✓	x	x	x	✓	x	x
59.	*Cymbopogon martinii*	Grass	✓	x	x	x	x	x	x	x	x
60.	*Cenchrus biflorus*	Grass	x	x	x	x	x	x	✓	x	x
61.	*Cynodon barberi*	Grass	x	✓	✓	✓	x	x	x	x	x
62.	*Cynodon dactylon*	Grass	✓	✓	✓	✓	✓	✓	✓	✓	x
63.	*Cyrtococcum accrescens*	Grass	x	x	x	x	x	✓	x	x	x
64.	*Dendrocalamus strictus*	Grass	✓	✓	x	x	✓	x	x	x	x

(continued)

Table 8.3 (continued)

S. No.	Plant species	Plant form	Nilgai	Chowsingha	Blackbuck	Sambar	Gaur	Swamp deer	Chital	Chinkara	Hog deer
65.	*Desmostachya bipinnata*	Grass	x	x	✓	✓	x	x	x	x	x
66.	*Dichanthium annulatum*	Grass	✓	✓	✓	x	x	x	✓	x	x
67.	*Dicanthium species*	Grass	x	✓	x	x	✓	x	x	x	x
68.	*Digitaria ciliaris*	Grass	x	✓	x	x	x	x	✓	x	x
69.	*Digitaria longiflora*	Grass	x	✓	x	x	x	x	x	x	x
70.	*Echinochloa colona*	Grass	✓	x	x	x	x	✓	✓	x	x
71.	*Echinochloa species*	Grass	x	✓	✓	x	x	x	x	x	x
72.	*Echinochloa crus-galli*	Grass	x	x	x	x	x	✓	x	x	x
73.	*Eleusine coracana*	Grass	x	x	✓	x	x	x	x	x	x
74.	*Eleusine indica*	Grass	✓	x	x	x	x	x	✓	x	x
75.	*Eragrostis amabilis*	Grass	x	x	x	x	x	x	✓	x	x
76.	*Eragrotis species*	Grass	x	x	✓	x	x	x	✓	x	x
77.	*Eragrostis tenella*	Grass	x	✓	x	✓	✓	x	x	x	x
78.	*Eremopogon foveolatus*	Grass	x	✓	x	x	x	x	✓	x	x
79.	*Eulaliopsis binata*	Grass	x	✓	x	✓	✓	x	x	x	x

(continued)

Table 8.3 (continued)

S. No.	Plant species	Plant form	Nilgai	Chowsingha	Blackbuck	Sambar	Gaur	Swamp deer	Chital	Chinkara	Hog deer
80.	*Fimbristylis dichotoma*	Grass	x	x	x	x	x	✓	x	x	x
81.	*Fimbristylis miliaceae*	Grass	x	x	x	x	x	✓	x	x	x
82.	*Fimbristylis ovata*	Grass	x	x	✓	x	x	x	x	x	x
83.	*Hemarthria compressa*	Grass	✓	x	x	x	x	✓	✓	x	✓
84.	*Heteropogon contortus*	Grass	✓	✓	✓	✓	✓	x	✓	x	x
85.	*Imperata cylindrica*	Grass	x	✓	x	✓	✓	✓	x	x	✓
86.	*Iseilema laxum*	Grass	x	x	✓	x	x	x	x	x	x
87.	*Leersia hexandra*	Grass	✓	x	x	x	x	x	x	x	✓
88.	*Oplismenus compositus*	Grass	x	x	x	x	x	✓	x	x	x
89.	*Oplismenus undulatifolius*	Grass	x	x	x	x	✓	x	x	x	x
90.	*Oryza rufipogon*	Grass	x	x	x	x	x	x	x	x	✓
91.	*Oryza sativa*	Grass	x	x	✓	x	x	✓	x	x	x
92.	*Panicum species*	Grass	✓	✓	x	x	x	x	✓	x	x
93.	*Paspalidium* spp.	Grass	x	x	x	x	x	x	x	x	x
94.	*Paspalidium flavidum*	Grass	x	x	x	x	x	✓	x	x	x

(continued)

Table 8.3 (continued)

S. No.	Plant species	Plant form	Nilgai	Chowsingha	Blackbuck	Sambar	Gaur	Swamp deer	Chital	Chinkara	Hog deer
95.	*Paspalidium geminatum*	Grass	✓	x	x	x	x	x	✓	x	x
96.	*Paspalidium punctatum*	Grass	✓	x	x	x	x	x	✓	x	x
97.	*Phragmites karka*	Grass	x	x	x	x	x	✓	x	x	✓
98.	*Polypogon fugax*	Grass	x	x	x	x	x	✓	x	x	x
99.	*Pseudanthistiria umbellata*	Grass	x	✓	x	x	x	x	x	x	x
100.	*Saccharum officinarum*	Grass	x	x	✓	x	x	✓	x	x	x
101.	*Saccharum spontaenum*	Grass	✓	x	x	x	x	✓	✓	x	✓
102.	*Saccharum species*	Grass	x	x	x	x	x	✓	x	x	x
103.	*Sehima nervosum*	Grass	✓	✓	x	x	x	x	x	x	x
104.	*Sehima species*	Grass	x	✓	x	x	x	x	x	x	x
105.	*Setaria glauca*	Grass	x	x	x	x	x	✓	x	x	x
106.	*Setaria intermedia*	Grass	x	x	x	x	✓	x	x	x	x
107.	*Setaria species*	Grass	x	✓	x	x	x	x	x	x	✓
108.	*Setaria verticillata*	Grass	✓	x	x	x	x	x	x	x	x
109.	*Sporobolus species*	Grass	x	✓	✓	✓	x	x	x	x	x
110.	*Themeda cymbaria*	Grass	x	✓	x	x	✓	x		x	x
111.	*Themeda quadrivalvis*	Grass	x	✓	x	x	✓	x	✓	x	x

(continued)

Table 8.3 (continued)

S. No.	Plant species	Plant form	Nilgai	Chowsingha	Blackbuck	Sambar	Gaur	Swamp deer	Chital	Chinkara	Hog deer
112.	*Themeda triandra*	Grass	x	✓	x	x	✓	x	x	x	x
113.	*Thysanolaena maxima*	Grass	x	x	x	✓	x	x	x	x	x
114.	*Vetiveria zizanioides*	Grass	✓	✓	✓	✓	✓	✓	✓	✓	x
115.	*Zea mays*	Grass	✓	x	✓	x	x	x	x	x	x
116.	*Zizania latifolia*	Grass	x	x	x	x	x	x	x	x	✓
117.	*Acmella uliginosa*	Herb	x	x	x	x	x	✓	x	x	x
118.	*Agave ingens*	Herb	x	x	x	x	x	x	✓	x	x
119.	*Ageratum conyzoides*	Herb	✓	x	x	✓	x	x	x	✓	x
120.	*Allotropis cumiciana*	Herb	✓	x	x	x	x	x	x	x	x
121.	*Alysicarpus monilifera*	Herb	✓	x	x	x	x	x	x	x	x
122.	*Amaranthus species*	Herb	x	x	x	x	x	x	x	x	x
123.	*Anagallis arvensis*	Herb	✓	x	x	x	x	x	✓	x	x
124.	*Andrographis echioides*	Herb	✓	x	x	x	x	x	✓	x	x
125.	*Arachis hypogaea*	Herb	✓	x	x	x	x	x	x	x	x
126.	*Barleria cristata*	Herb	x	✓	x	x	x	x	x	x	x
127.	*Blainvillea acmella*	Herb	x	x	x	x	x	✓	x	x	x

(continued)

Table 8.3 (continued)

S. No.	Plant species	Plant form	Nilgai	Chowsingha	Blackbuck	Sambar	Gaur	Swamp deer	Chital	Chinkara	Hog deer
128.	*Boerhavia diffusa*	Herb	✓	x	x	x	x	x	x	x	x
129.	*Boechera stricta*	Herb	x	x	x	x	x	x	✓	x	x
130.	*Catharanthus pusillus*	Herb	✓	x	x	x	x	x	✓	x	x
131.	*Catharanthus pusillus*	Herb	x	x	x	x	x	x	x	x	x
132.	*Celosia argentea*	Herb	x	x	✓	x	x	x	x	x	x
133.	*Chlorophytum borivilianum*	Herb	x	x	x	x	x	x	x	x	x
134.	*Chromolaena odorata*	Herb	x	x	x	x	✓	x	x	x	x
135.	*Chrysopogon fulvus*	Herb	✓	x	x	x	x	x	✓	x	x
136.	*Cleome viscosa*	Herb	✓	x	x	x	x	x	x	x	x
137.	*Corchorus aestuans*	Herb	✓	x	x	x	x	x	x	x	x
138.	*Curculigo orchioides*	Herb	x	x	x	x	x	x	x	x	x
139.	*Cyanotis cristata*	Herb	✓	x	x	x	x	x	x	x	x
140.	*Cyanotis species*	Herb	✓	x	x	x	x	x	x	x	x
141.	*Cyanotis tuberosa*	Herb	x	x	x	x	x	x	x	x	x
142.	*Desmodium triflorum*	Herb	x	x	✓	x	x	x	x	x	x

(continued)

Table 8.3 (continued)

S. No.	Plant species	Plant form	Nilgai	Chowsingha	Blackbuck	Sambar	Gaur	Swamp deer	Chital	Chinkara	Hog deer
143.	*Elephantopus scaber*	Herb	x	✓	x	✓	x	x	x	x	x
144.	*Elytraria acaulis*	Herb	✓	x	x	x	x	x	✓	x	x
145.	*Eupatorium species*	Herb	x	x	✓	x	x	x	x	x	x
146.	*Euphorbia hirta*	Herb	✓	x	x	x	x	x	x	x	x
147.	*Euphorbia prostrata*	Herb	✓	x	x	x	x	x	x	x	x
148.	*Gossypium herbaceum*	Herb	✓	x	x	x	x	x	x	x	x
149.	*Grona triflora*	Herb	x	x	x	x	✓	x	x	x	x
150.	*Guizotia abyssinica*	Herb	x	✓	x	✓	✓	x	x	x	x
151.	*Hedychium coronarium*	Herb	x	x	x	x	x	x	x	x	✓
152.	*Helichrysm species*	Herb	x	✓	x	x	x	x	x	x	x
153.	*Hyptis suaveolens*	Herb	x	x	x	x	x	x	✓	x	x
154.	*Indigofera linnaei*	Herb	x	x	✓	x	x	x	x	x	x
155.	*Indigofera species*	Herb	x	✓	x	x	x	x	x	x	x
156.	*Indigofera tinctoria*	Herb	✓	x	x	x	x	x	x	x	x
157.	*Justicia prostrata*	Herb	x	x	✓	x	x	x	x	x	x
158.	*Lagascea mollis*	Herb	✓	x	x	x	x	x	✓	x	x

(continued)

Table 8.3 (continued)

S. No.	Plant species	Plant form	Nilgai	Chowsingha	Blackbuck	Sambar	Gaur	Swamp deer	Chital	Chinkara	Hog deer
159.	*Lathyrus sativus*	Herb	x	x	✓	x	x	x	x	x	x
160.	*Launaea nudicaulis*	Herb	x	x	x	x	x	x	✓	x	x
161.	*Launaea procumbens*	Herb	✓	x	x	x	x	x	x	x	x
162.	*Leucas aspera*	Herb	x	✓	x	x	x	x	x	x	x
163.	*Limnophila indica*	Herb	x	x	x	x	x	x	✓	x	x
164.	*Lindenbergia urticaefolia*	Herb	✓	x	x	x	x	x	✓	x	x
165.	*Marsilea quadrifolia*	Herb	x	✓	x	✓	✓	x	x	x	x
166.	*Medicago lupulina*	Herb	x	x	x	x	x	✓	x	x	x
167.	*Melilotus indica*	Herb	x	x	x	x	x	x	x	x	x
168.	*Melochia corchorifolia*	Herb	✓	x	x	x	x	x	✓	x	x
169.	*Ocimum basilicum*	Herb	x	x	x	✓	✓	x	x	x	x
170.	*Ocimum canum*	Herb	x	✓	x	✓	✓	x	x	x	x
171.	*Oenanthe javanica*	Herb	x	x	x	x	x	x	x	x	✓
172.	*Osbeckia stellata*	Herb	x	x	x	x	x	x	x	x	✓
173.	*Oxalis corniculata*	Herb	x	x	x	x	x	✓	x	x	x
174.	*Pennisetum typhoides*	Herb	x	x	x	x	x	x	x	✓	x

(continued)

Table 8.3 (continued)

S. No.	Plant species	Plant form	Nilgai	Chowsingha	Blackbuck	Sambar	Gaur	Swamp deer	Chital	Chinkara	Hog deer
175.	*Parthenium hysterophorus*	Herb	✓	x	x	✓	✓	✓	x	✓	x
176.	*Parthenium species*	Herb	x	x	x	x	x	x	✓	x	x
177.	*Paspalum distichum*	Herb	x	x	✓	✓	x	x	✓	x	x
178.	*Paspalum conjugatum*	Herb	✓	x	x	x	x	✓	✓	x	x
179.	*Paspalum scrobiculatum*	Herb	x	x	x	x	x	✓	x	x	x
180.	*Pavonia zeylanica*	Herb	✓	x	x	x	x	x	x	x	x
181.	*Pennisetum pedicellatum*	Herb	x	✓	x	✓	✓	x	x	x	x
182.	*Peristrophe bicalyculata*	Herb	x	x	x	x	x	x	✓	x	x
183.	*Phaseolus aconitifolius*	Herb	x	x	x	x	x	x	x	✓	x
184.	*Phaseolus mungo*	Herb	x	x	✓	x	x	x	x	x	x
185.	*Phyllanthus amarus*	Herb	x	✓	x	✓	✓	x	x	x	x
186.	*Plumbago zeylanica*	Herb	✓	x	x	x	x	x	x	x	x
187.	*Sesamum indicum*	Herb	x	x	✓	x	x	x	x	x	x
188.	*Sida acuta*	Herb	✓	✓	x	x	✓	x	✓	x	x

(continued)

Table 8.3 (continued)

S. No.	Plant species	Plant form	Nilgai	Chowsingha	Blackbuck	Sambar	Gaur	Swamp deer	Chital	Chinkara	Hog deer
189.	*Sida rhombifolia*	Herb	✓	x	x	x	x	x	✓	x	x
190.	*Sida species*	Herb	x	✓	x	✓	✓	x	x	x	x
191.	*Sida veronicifolia*	Herb	✓	x	x	x	x	x	✓	x	x
192.	*Silybum marianum*	Herb	x	x	x	x	x	x	x	x	x
193.	*Solanum americanum*	Herb	x	x	x	x	x	✓	x	x	x
194.	*Sonchus oleraceus*	Herb	x	x	x	x	x	x	✓	x	x
195.	*Spermacoce articularis*	Herb	x	x	x	x	✓	x	✓	x	x
196.	*Spirodela polyrhiza*	Herb	x	x	x	✓	✓	x	x	x	x
197.	*Stylosanthes species*	Herb	x	x	x	x	✓	x	x	x	x
198.	*Tribulus terrestris*	Herb	✓	x	x	x	✓	x	x	x	x
199.	*Tridax procumbens*	Herb	✓	x	x	x	x	x	✓	x	x
200.	*Trifolium alexandrinum*	Herb	x	x	x	x	x	x	x	x	x
201.	*Trifolium tomentosum*	Herb	x	x	x	x	x	✓	x	x	x
202.	*Typha angustifolia*	Herb	x	x	x	x	x	✓	x	x	x
203.	*Typha elephantina*	Herb	x	x	x	x	x	✓	x	x	x
204.	*Urena lobata*	Herb	x	x	x	x	✓	x	x	x	x
205.	*Xanthium indicum*	Herb	✓	x	x	x	x	x	✓	✓	x

(continued)

Table 8.3 (continued)

S. No.	Plant species	Plant form	Nilgai	Chowsingha	Blackbuck	Sambar	Gaur	Swamp deer	Chital	Chinkara	Hog deer
206.	*Xanthium strumarium*	Herb	x	x	x	✓	x	✓	x	x	x
207.	*Acanthospermum hispidum*	Herb	✓	x	x	✓	x	x	x	✓	x
208.	*Achyranthes aspera*	Herb	x	✓	x	x	x	x	x	x	x
209.	*Cassia tora*	Herb	✓	✓	x	✓	✓	x	✓	x	x
210.	*Convolvulus prostratus*	Herb	✓	✓	x	✓	x	x	✓	✓	x
211.	*Justicia simplex*	Herb	x	✓	x	x	x	x	x	x	x
212.	*Alternanthera philoxeroides*	Herb	x	x	x	x	x	x	x	x	✓
213.	*Carex cruciata*	Sedges	x	x	x	x	x	x	x	x	✓
214.	*Carex myosurus*	Sedges	x	x	x	x	x	x	x	x	x
215.	*Cyperus alopecuroides*	Sedges	x	x	x	x	x	x	x	x	x
216.	*Cyperus brevifolius*	Sedges	x	x	x	x	x	x	x	x	x
217.	*Cyperus cyperoides*	Sedges	x	x	x	x	x	✓	x	x	x
218.	*Cyperus rotundus*	Sedges	x	x	✓	x	✓	x	x	x	x
219.	*Cyperus scariosus*	Sedges	x	x	x	✓	✓	x	x	x	x
220.	*Cyperus species*	Sedges	x	x	x	✓	x	x	x	x	✓
221.	*Cyprus nutans*	Sedges	x	x	x	x	x	x	✓	x	x

(continued)

Table 8.3 (continued)

S. No.	Plant species	Plant form	Nilgai	Chowsingha	Blackbuck	Sambar	Gaur	Swamp deer	Chital	Chinkara	Hog deer
222.	*Scirpus lacustris*	Sedges	x	x	x	x	x	x	x	x	✓
223.	*Scirpus tuberosus*	Sedges	x	x	x	x	x	x	x	x	x
224.	*Scripus species*	Sedges	x	x	✓	✓	x	x	x	x	x
225.	*Abutilon indicum*	Shrub	✓	x	x	x	x	x	x	x	x
226.	*Barleria prionitis*	Shrub	x	x	x	x	x	x	✓	x	x
227.	*Cajanus cajan*	Shrub	x	x	x		x	x	x	✓	x
228.	*Calligonum polygonoides*	Shrub	x	x	x	x	x	x	x	✓	x
229.	*Calotropis gigantea*	Shrub	✓	x	x	x	x	x	✓	x	x
230.	*Calotropis procera*	Shrub	x	x	x	x	x	x	x	✓	x
231.	*Canthium parviflorum*	Shrub	x	x	x	x	x	x	✓	x	x
232.	*Capparis burhia*	Shrub	x	x	x	x	x	x	x	✓	x
233.	*Capparis decidua*	Shrub	✓	x	x	x	x	x	x	x	x
234.	*Capparis sepiara*	Shrub	✓	✓	x	✓	x	x	✓	x	x
235.	*Carissa carandas*	Shrub	✓	x	x	✓	x	x	✓	x	x
236.	*Carissa spinarum*	Shrub	✓	x	✓	✓	x	x	✓	x	x
237.	*Carrisa spinarum*	Shrub	x	x	x	x	x	x	x	x	x
238.	*Casearia tomentosa*	Shrub	x	x	x	✓	✓	x	x	x	x
239.	*Clerodendrum phlomidis*	Shrub	✓	✓	x	x	x	x	x	x	x

(continued)

Table 8.3 (continued)

S. No.	Plant species	Plant form	Nilgai	Chowsingha	Blackbuck	Sambar	Gaur	Swamp deer	Chital	Chinkara	Hog deer
240.	*Corchorus olitorius*	Shrub	✓	x	x	x	x	x	x	x	x
241.	*Dipterygium glaucumis*	Shrub	x	x	x	x	x	x	x	✓	x
242.	*Grewia hirsutovelutina*	Shrub	x	x	x	x	✓	x	x	x	x
243.	*Grewia flavescens*	Shrub	✓	x	x	x	x	x	x	x	x
244.	*Grewia hirsuta*	Shrub	x	✓	x	✓	x	x	x	x	x
245.	*Grewia latifolia*	Shrub	x	✓	x	✓	✓	x	x	x	x
246.	*Grewia ovalifolia*	Shrub	✓	x	x	x	x	x	x	x	x
247.	*Grewia species*	Shrub	x	x	x	x	✓	x	x	x	x
248.	*Grewia tiliifolia*	Shrub	✓	x	x	✓	✓	x	x	x	x
249.	*Grewia hirsuta*	Shrub	x	x	x	x	✓	x	x	x	x
250.	*Gymnosporia spinosa*	Shrub	x	✓	x	✓	✓	x	x	x	x
251.	*Helicteres isora*	Shrub	✓	x	x	✓	✓	x	✓	x	x
252.	*Ixora arborea*	Shrub	x	✓	x	✓	✓	x	✓	x	x
253.	*Ixora pavetta*	Shrub	✓	✓	x	✓	x	x	✓	✓	x
254.	*Jatropha species*	Shrub	x	x	✓	x	x	x	x	x	x
255.	*Lantana camara*	Shrub	✓	x	x	✓	x	x	✓	✓	x
256.	*Lantana indica*	Shrub	x	x	x	x	✓	x	✓	x	x
257.	*Nyctanthes arbor-tristis*	Shrub	x	✓	x	x	x	x	x	x	x

(continued)

Table 8.3 (continued)

S. No.	Plant species	Plant form	Nilgai	Chowsingha	Blackbuck	Sambar	Gaur	Swamp deer	Chital	Chinkara	Hog deer
258.	*Persicaria perfoliata*	Shrub	x	x	x	x	x	x	x	x	✓
259.	*Phoenix farinifera*	Shrub	x	x	x	x	x	x	✓	x	x
260.	*Randia dumetorum*	Shrub	✓	✓	✓	✓	x	x	✓	x	x
261.	*Randia malabarica*	Shrub	x	x	✓	x	x	x	✓	x	x
262.	*Rhus mysorensis*	Shrub	x	✓	x	x	x	x	x	x	x
263.	*Phyllanthus leucopyrus*	Shrub	x	x	x	x	x	x	✓	x	x
264.	*Solanum indicum*	Shrub	x	x	x	x	✓	x	x	x	x
265.	*Solanum species*	Shrub	✓	x	x	x	x	x	x	x	x
266.	*Solanum xanthocarpum*	Shrub	✓	x	x	x	x	x	x	x	x
267.	*Strobilanthes callosa*	Shrub	x	x	x	x	✓	x	x	x	x
268.	*Strobilanthes ixiocephalus*	Shrub	x	x	x	x	✓	x	x	x	x
269.	*Strobilanthes species*	Shrub	x	x	x	x	✓	x	x	x	x
270.	*Symplocos racemosa*	Shrub	x	x	x	x	✓	x	x	x	x
271.	*Tephrosia purpurea*	Shrub	✓	x	x	x	x	x	✓	x	x
272.	*Xeromphis spinosa*	Shrub	x	x	x	✓	x	x	✓	x	x

(continued)

Table 8.3 (continued)

S. No.	Plant species	Plant form	Nilgai	Chowsingha	Blackbuck	Sambar	Gaur	Swamp deer	Chital	Chinkara	Hog deer
273.	*Xeromphis uliginosa*	Shrub	x	x	x	✓	x	x	✓	x	x
274.	*Zizyphus glabarata*	Shrub	x	✓	x	x	x	x	x	x	x
275.	*Zizyphus jujuba*	Shrub	x	x	✓	x	x	✓	x	x	x
276.	*Zizyphus mauritiana*	Shrub	✓	✓	x	✓	x	x	✓	✓	x
277.	*Zizyphus nummularia*	Shrub	✓	✓	x	x	x	x	✓	✓	x
278.	*Zizyphus oenoplia*	Shrub	x	✓	x	✓	✓	x	✓	x	x
279.	*Zizyphus xylopyrus*	Shrub	✓	✓	x	✓	x	x	✓	✓	x
280.	*A. Funiculate*	Tree	x	x	x	x	x	x	x	✓	x
281.	*Acacia arabica*	Tree	x	x	✓	x	x	x	x	x	x
282.	*Acacia catechu*	Tree	✓	✓	x	✓	✓	x	✓	✓	x
283.	*Acacia jacquemontii*	Tree	✓	x	x	x	x	x	✓	✓	x
284.	*Acacia leucophloea*	Tree	✓	✓	x	✓	x	x	✓	x	x
285.	*Acacia nilotica*	Tree	✓	✓	x	✓	x	x	✓	✓	x
286.	*Acacia planiforns*	Tree	x	x	x	x	x	x	✓	x	x
287.	*Acacia Senegal*	Tree	✓	✓	x	x	x	x	✓	x	x
288.	*Acacia tortolis*	Tree	x	x	x	x	x	x	✓	x	x
289.	*Acacia species*	Tree	x	x	x	x	x	x	x	✓	x
290.	*Haldina cordifolia*	Tree	x	x	x	✓	✓	x	x	x	x

(continued)

Table 8.3 (continued)

S. No.	Plant species	Plant form	Nilgai	Chowsingha	Blackbuck	Sambar	Gaur	Swamp deer	Chital	Chinkara	Hog deer
291.	*Aegle marmelos*	Tree	✓	✓	✓	✓	✓	x	✓	✓	x
292.	*Ailanthus excelsa*	Tree	✓	x	x	x	x	x	✓	x	x
293.	*Alangium salvifolium*	Tree	x	x	x	✓	x	x	x	x	x
294.	*Albizia lebbeck*	Tree	✓	✓	x	x	x	x	✓	x	x
295.	*Albizia odoratissima*	Tree	x	✓	x	✓	✓	x	✓	x	x
296.	*Albizia procera*	Tree	x	x	x	x	x	x	✓	x	x
297.	*Anacardium occidentale*	Tree	x	x	✓	x	✓	x	x	x	x
298.	*Anogeissus acuminata*	Tree	x	✓	x	x	x	x	x	x	x
299.	*Anogeissus latifolia*	Tree	✓	✓	x	✓	✓	x	✓	✓	x
300.	*Anogeissus pendula*	Tree	✓	✓	x	✓	x	x	✓	✓	x
301.	*Annona species*	Tree	x	x	✓	x	x	x	x	x	x
302.	*Annona squamosa*	Tree	x	✓	x	✓	x	x	x	x	x
303.	*Azadirachta indica*	Tree	✓	✓	x	✓	x	x	✓	✓	x
304.	*Balanites roxburghii*	Tree	x	x	x	x	x	x	x	x	x
305.	*Balanites aegyptiaca*	Tree	✓	x	x	x	x	x	✓	x	x
306.	*Bauhinia variegata*	Tree	x	✓	x	✓	✓	x	x	x	x

(continued)

Table 8.3 (continued)

S. No.	Plant species	Plant form	Nilgai	Chowsingha	Blackbuck	Sambar	Gaur	Swamp deer	Chital	Chinkara	Hog deer
307.	*Bauhinia malabarica*	Tree	✓	x	x	x	x	x	x	x	x
308.	*Bauhinia racemosa*	Tree	x	✓	x	✓	✓	x	✓	x	x
309.	*Bombax ceiba*	Tree	x	✓	x	✓	x	x	✓	x	x
310.	*Borassus flabellifer*	Tree	x	x	x	x	x	x	✓	x	x
311.	*Boswellia serrata*	Tree	x	✓	x	x	x	x	x	x	x
312.	*Bridelia retusa*	Tree	x	✓	x	✓	x	x	✓	x	x
313.	*Buchanania cochinchinensis*	Tree	✓	✓	x	✓	x	x	✓	✓	x
314.	*Buchanania lanzan*	Tree	x	✓	x	✓	✓	x	✓	x	x
315.	*Butea monosperma*	Tree	✓	✓	x	✓	✓	x	✓	✓	x
316.	*Butea parviflora*	Tree	x	✓	x	✓	✓	x	x	x	x
317.	*Careya arborea*	Tree	x	x	x	✓	x	x	x	x	x
318.	*Cassia fistula*	Tree	x	✓	x	✓	✓	x	✓	x	x
319.	*Cassia pumila*	Tree	x	✓	x	x	x	x	x	x	x
320.	*Cassia roxburghii*	Tree	x	x	x	x	x	x	✓	x	x
321.	*Catunaregam spinosa*	Tree	✓	✓	x	✓	x	x	✓	x	x
322.	*Ceriscoides turgida*	Tree	✓	✓	x	✓	x	x	✓	✓	x

(continued)

Table 8.3 (continued)

S. No.	Plant species	Plant form	Nilgai	Chowsingha	Blackbuck	Sambar	Gaur	Swamp deer	Chital	Chinkara	Hog deer
323.	*Chloroxylon swietenia*	Tree	x	✓	x	✓	✓	x	✓	x	x
324.	*Cleistanthus collinus*	Tree	x	✓	x	✓	✓	x	x	x	x
325.	*Cordia dichotoma*	Tree	x	✓	x	x	x	x	✓	x	x
326.	*Cordia macleodii*	Tree	✓	x	x	✓	x	x	x	x	x
327.	*Cordia myxa*	Tree	x	x	x	x	✓	x	x	x	x
328.	*Cordia wallachi*	Tree	x	✓	x	x	x	x	x	x	x
329.	*Dalbergia sissoo*	Tree	✓	✓	x	x	x	x	✓	x	x
330.	*Dichrostachys cinerea*	Tree	✓	x	x	✓	x	x	x	x	x
331.	*Dillenia pentagyna*	Tree	x	x	x	x	✓	x	x	x	x
332.	*Diospyros melanoxylon*	Tree	✓	✓	x	✓	✓	x	✓	✓	x
333.	*Diospyros montana*	Tree	x	x	x	✓	✓	x	x	x	x
334.	*Ehretia laevis*	Tree	✓	x	x	x	x	x	✓	x	x
335.	*Elaeodendron glaucum*	Tree	✓	x	x	x	x	x	x	x	x
336.	*Emblica officinalis*	Tree	✓	✓	x	✓	✓	x	✓	x	x
337.	*Erythroxylum monogynum*	Tree	x	x	x	x	✓	x	x	x	x
338.	*Ehretia laevis*	Tree	x	x	x	x	x	x	x	x	x

(continued)

Table 8.3 (continued)

S. No.	Plant species	Plant form	Nilgai	Chowsingha	Blackbuck	Sambar	Gaur	Swamp deer	Chital	Chinkara	Hog deer
339.	*Eucalyptus species*	Tree	x	x	x	x	x	x	✓	x	x
340.	*Ficus amplissima*	Tree	x	x	x	✓	x	x	x	x	x
341.	*Ficus benghalensis*	Tree	✓	x	x	✓	x	x	✓	✓	x
342.	*Ficus glomerata*	Tree	✓	x	x	✓	✓	x	✓	x	x
343.	*Ficus hispidia*	Tree	x	x	x	✓	✓	x	✓	x	x
344.	*Ficus infectoria*	Tree	x	x	x	✓	x	x	x	x	x
345.	*Ficus racemosa*	Tree	✓	✓	x	✓	x	x	✓	x	x
346.	*Ficus religiosa*	Tree	✓	✓	x	✓	x	x	✓	✓	x
347.	*Ficus retusa*	Tree	✓	x	x	x	x	x	x	x	x
348.	*Ficus virens*	Tree	✓	✓	x	✓	x	x	✓	x	x
349.	*Flacourtia indica*	Tree	✓	✓	x	✓	x	x	✓	x	x
350.	*Gardenia latifolia*	Tree	✓	✓	x	✓	✓	x	✓	✓	x
351.	*Garuga pinnata*	Tree	x	x	x	✓	x	x	x	x	x
352.	*Glochidion heyneanum*	Tree	✓	✓	x	✓	x	x	✓	✓	x
353.	*Gmelina arborea*	Tree	✓	x	x	x	✓	x	x	x	x
354.	*Hardwickia binata*	Tree	x	x	x	x	✓	x	x	x	x
355.	*Haldina cordifolia*	Tree	x	✓	x	x	x	x	x	x	x
356.	*Holarrhena antidysenterica*	Tree	✓	x	x	✓	x	x	✓	x	x
357.	*Holoptelea integrifolia*	Tree	x	x	x	x	x	x	✓	x	x

(continued)

Table 8.3 (continued)

S. No.	Plant species	Plant form	Nilgai	Chowsingha	Blackbuck	Sambar	Gaur	Swamp deer	Chital	Chinkara	Hog deer
358.	*Hymenodictyon excelsum*	Tree	x	✓	x	x	x	x	x	x	x
359.	*Kydia calycina*	Tree	x	✓	x	x	x	x	x	x	x
360.	*Lagerstroemia parviflora*	Tree	✓	✓	x	✓	x	x	✓	x	x
361.	*Lannea coromandelica*	Tree	✓	✓	x	✓	x	x	✓	✓	x
362.	*Limonia acidissima*	Tree	✓	x	x	✓	x	x	✓	x	x
363.	*Madhuca emarginata*	Tree	x	x	x	x	x	x	x	✓	x
364.	*Madhuca indica*	Tree	✓	✓	x	✓	✓	x	✓	✓	x
365.	*Mangifera indica*	Tree	✓	✓	✓	✓	x	x	✓	x	x
366.	*Mallotus philippensis*	Tree	x	✓	x	x	x	x	x	x	x
367.	*Mayetenus emarginata*	Tree	x	x	x	x	x	x	✓	x	x
368.	*Miliusa tomentosa*	Tree	✓	✓	x	✓	✓	x	x	✓	x
369.	*Mitragyna parvifolia*	Tree	✓	✓	x	✓	✓	x	✓	x	x
370.	*Morinda tinctoria*	Tree	x	x	x	✓	x	x	x	x	x
371.	*Murraya paniculata*	Tree	✓	x	x	✓	x	x	x	x	x
372.	*Naringi crenulata*	Tree	✓	✓	x	✓	x	x	✓	✓	x

(continued)

Table 8.3 (continued)

S. No.	Plant species	Plant form	Nilgai	Chowsingha	Blackbuck	Sambar	Gaur	Swamp deer	Chital	Chinkara	Hog deer
373.	*Ougeinia oojeinense*	Tree	x	x	x	✓	x	x	✓	x	x
374.	*Pheonix aquilis*	Tree	x	✓	x	✓	✓	x	x	x	x
375.	*Phoenix acaulis*	Tree	✓	✓	x	✓	x	x	✓	x	x
376.	*Phoenix loureiroi*	Tree	x	x	x	x	x	x	✓	x	x
377.	*Phoenix sylvestris*	Tree	x	x	x	x	x	x	✓	x	x
378.	*Phyllanthus emblica*	Tree	✓	✓	x	✓	✓	x	✓	✓	x
379.	*Pithecellobium dulce*	Tree	x	✓	✓	x	x	x	x	x	x
380.	*Polyalthia longifolia*	Tree	x	x	x	x	x	x	✓	x	x
381.	*Pongamia pinnata*	Tree	✓	x	x	x	x	x	✓	x	x
382.	*Prosopis cineraria*	Tree	x	x	x	x	x	x	x	✓	x
383.	*Prosopis juliflora*	Tree	✓	✓	✓	x	x	x	✓	x	x
384.	*Pterocarpus marsupium*	Tree	✓	x	x	✓	x	x	✓	x	x
385.	*Sapindus emarginatus*	Tree	x	x	x	x	x	x	✓	x	x
386.	*Schleichera oleosa*	Tree	x	x	x	✓	✓	x	✓	x	x
387.	*Schrebera swietenioides*	Tree	x	x	x	✓	x	x	x	✓	x
388.	*Semecarpus anacardium*	Tree	x	x	x	✓	x	x	x	x	x

(continued)

Table 8.3 (continued)

S. No.	Plant species	Plant form	Nilgai	Chowsingha	Blackbuck	Sambar	Gaur	Swamp deer	Chital	Chinkara	Hog deer
389.	*Shorea robusta*	Tree	x	x	x	x	✓	x	x	x	x
390.	*Soymida febrifuga*	Tree	x	✓	x	x	x	x	x	x	x
391.	*Sterculia urens*	Tree	x	x	x	x	x	x	✓	x	x
392.	*Syzygium cumini*	Tree	✓	✓	x	✓	✓	x	✓	✓	x
393.	*Syzygium heyneanum*	Tree	x	x	x	x	x	x	✓	x	x
394.	*Syzygium rubicundum*	Tree	x	x	x	✓	x	x	✓	x	x
395.	*Syzygium salicifolium*	Tree	✓	x	x	✓	x	x	x	x	x
396.	*Tamarindus indica*	Tree	✓	✓	x	✓	x	x	✓	✓	x
397.	*Tamilnadia uliginosa*	Tree	✓	x	x	✓	✓	x	✓	✓	x
398.	*Tectona grandis*	Tree	x	✓	x	✓	✓	x	✓	x	x
399.	*Terminalia bellirica*	Tree	✓	✓	✓	✓	x	x	✓	✓	x
400.	*Terminalia crenulata*	Tree	x	✓	x	✓	x	x	✓	x	x
401.	*Terminalia elliptica*	Tree	✓	x	x	✓	x	x	x	✓	x
402.	*Terminalia paniculata*	Tree	x	x	x	x	✓	x	x	x	x
403.	*Terminalia tomentosa*	Tree	x	✓	x	✓	✓	x	✓	x	x

(continued)

Table 8.3 (continued)

S. No.	Plant species	Plant form	Nilgai	Chowsingha	Blackbuck	Sambar	Gaur	Swamp deer	Chital	Chinkara	Hog deer
404.	*Vitex negundo*	Tree	x	x	x	✓	✓	x	x	x	x
405.	*Wrightia tinctoria*	Tree	✓	✓	x	✓	x	x	✓	x	x
406.	*Digitaria species*		x	✓	x	x	✓	x	x	x	x

Results

Our literature search (Fig. 8.1) yielded a total of 21 studies that provided information on ungulate nutrition from 1990 to 2020 (Table 8.2). The review included six records for chital, five records for nilgai, seven for sambar, five for gaur, five for chowsingha, three for blackbuck, two for swamp deer, one for chinkara, and two for hog deer. The studies were conducted in 20 different protected areas, while two were conducted outside the protected areas. All the studies reported the size of the protected area, which ranged from 30 km^2 to 1412 km^2. The size of the effective sampling area within the study area (i.e. the area where faeces and stomachs were searched and direct foraging observations were reported) was reported in only two studies. The period of data collection was explicitly stated in all records and ranged from five to 23 months (mean = 20.8, median = 16, standard deviation = 16.3 months). Direct observation (8) and faecal analysis method (8) were the most used methods to assess ungulate food habits, while some studies used a combination of both methods. No study used the stomach contents analysis method to examine ungulate diet, and no study addressed the issue of sampling efficiency in ungulate diets.

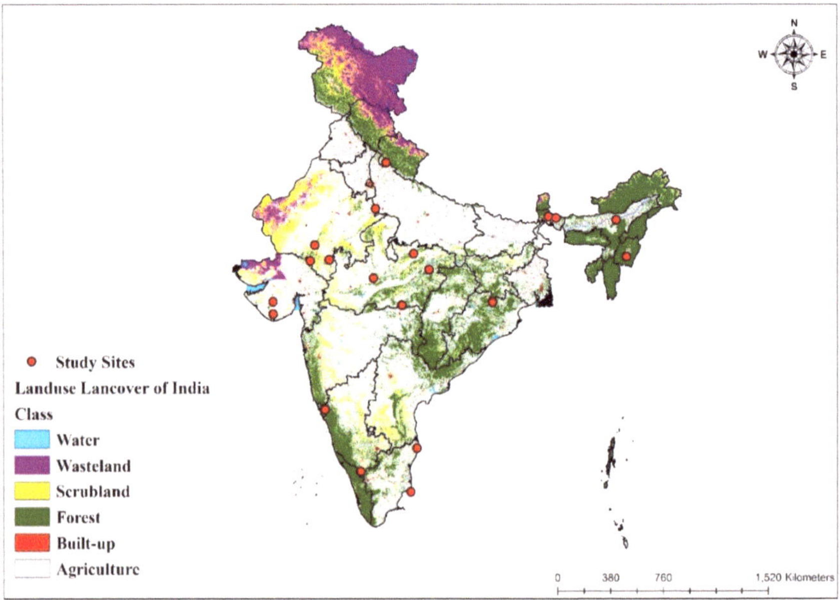

Fig. 8.1 Map showing location of study sites used in present review

Chital

The diet of Chital (Fig. 8.2) consists of 164 plant species (Fig. 8.3; Table 8.3) and six plant forms, namely trees, herbs, shrubs, climbers, sedges, and grasses (Fig. 8.4). Tree species contributed the most to the diet, i.e. 45.73%, followed by grass species 20.73%, herb species 15.85%, shrub species 13.41%, climber species (climbing plants) 3.66% and sedge species 0.60%. The chi-square test found a significant difference in the consumption of different plant forms by chital, indicating a high tendency for tree browse ($\chi^2 = 134$; df = 5; $p < 0.05$). Plant form diversity in case of chital was 1.39 (Fig. 8.5), while the browse-to-grass ratio was 2.20 (Fig. 8.6).

Fig. 8.2 A male chital captured while feeding on fallen seeds of *Ziziphus* species in Gir protected area, Gujarat, India (*Source* Rohit Chaudhary)

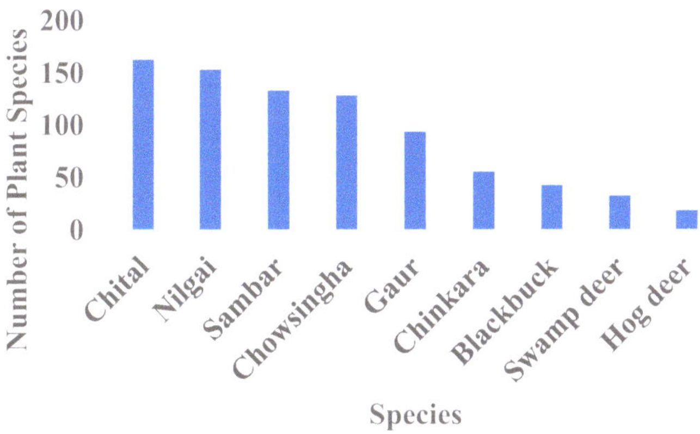

Fig. 8.3 Plant species number in the diet of ungulate species

Fig. 8.4 Percentage of plant form in the diet of chital

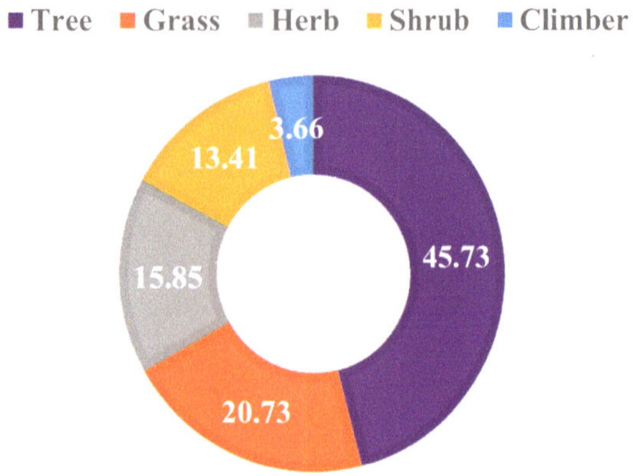

Fig. 8.5 Plant form diversity in the diet of ungulates (Box size is proportion to diversity)

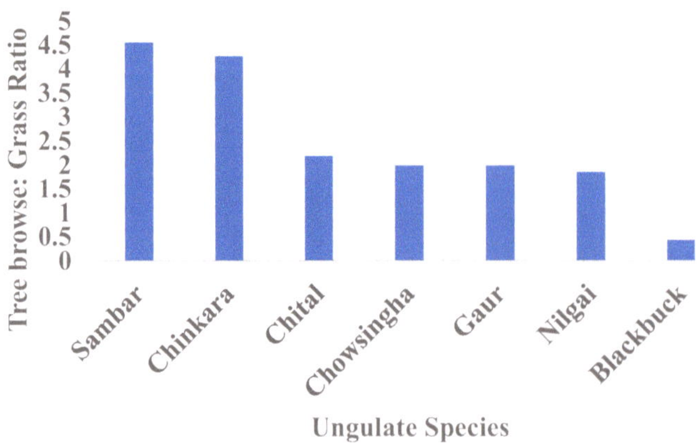

Fig. 8.6 Tree species to grass species ratio in the diet of ungulates

Sambar

The diet of sambar (Fig. 8.7) consisted of 134 species and six plant forms, namely trees, herbs, shrubs, climbers, sedges, and grasses. Tree species contributed the most to the sambar‘s diet i.e., 54.48%, followed by shrub species 14.18%, grass species 11.94%, herb species 11.94%, climbers 5.22% and sedges 2.23% (Fig. 8.8). A chi-square test showed significant differences in consumption of different plant forms, indicating a high propensity for tree browsing (χ^2 = 146; df = 5; $p < 0.05$). The plant form diversity in the sambar diet was 1.35, while the browsing to grass ratio was 4.56, indicating a high proportion of tree browsing in the diet of sambar.

Fig. 8.7 A female sambar captured from Gir protected area, Gujarat, India (*Source* Rohit Chaudhary)

Fig. 8.8 Percentage of plant form in the diet of sambar

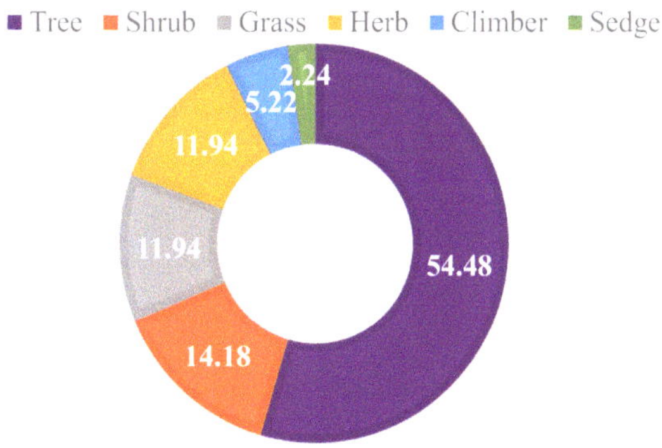

Nilgai

The diet of Nilgai (Fig. 8.9) consisted of 154 species and five plant forms, namely trees, shrubs, grasses, herbs, and climbers. Tree species contributed the most to the diet of Nilgai i.e., 36.36%, followed by herb species 22.73%, grass species 19.48%, shrub species 13.64% and climbers 7.79% (Fig. 8.10). A chi-square test showed significant differences in the consumption of different plant forms by nilgai, indicating a high propensity for tree browsing ($\chi^2 = 34$; df = 4; $p < 0.05$). The plant form diversity of the nilgai diet was 1.49 while the shrub to grass ratio was 1.86.

Fig. 8.9 A male nilgai captured in Gir protected area, Gujarat, India (*Source* Rohit Chaudhary)

Fig. 8.10 Percentage of plant form in the diet of nilgai

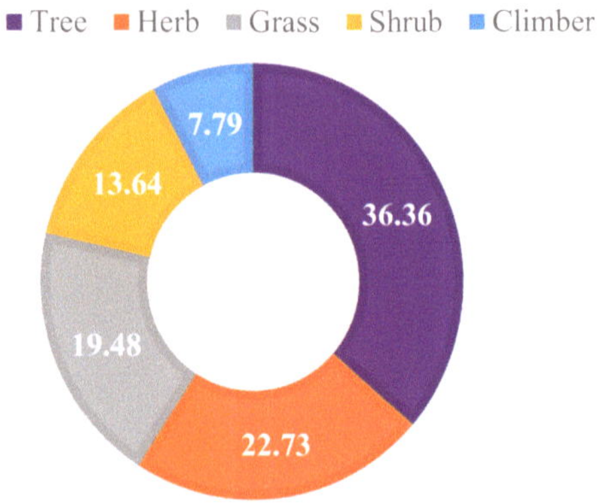

Gaur

The gaur diet (Fig. 8.11) consisted of 94 plant species and six plant forms, such as that of sambar and chital. Tree species contributed the highest proportion to gaur‘s diet i.e., 40.43%, followed by grass species 20.21%, herb species 18.09%, shrub species 17.02%, climbers 2.13%, and sedge species 2.13% (Fig. 8.12). The chi-square test showed a significant difference in consumption of different plant forms by gaur, indicating a high tendency for tree browsing ($\chi^2 = 56$; df = 5; $p < 0.05$). The diversity of plant forms at Gaur was 1.46 while the shrub to grass ratio was 2.

Fig. 8.11 A gaur captured from Rubber plantations of Northern Western Ghats (*Source* Rohit Chaudhary)

Fig. 8.12 Per cent of plant form in the diet of gaur

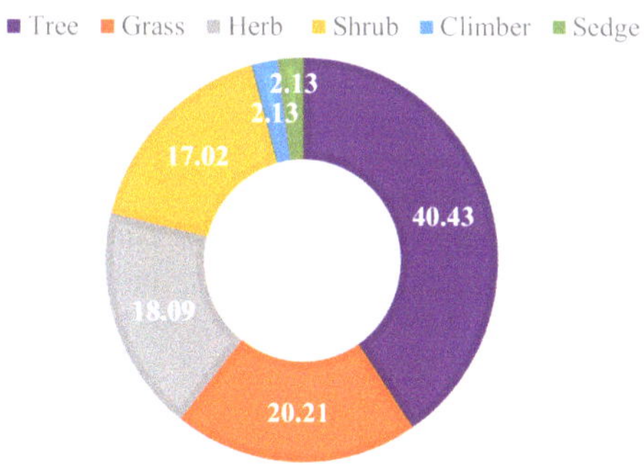

Fig. 8.13 A herd of swamp deer captured while feeding from Jhilmil Jheel conservation reserve, Uttarakhand, India (*Source* Prakhar Sharma)

Swamp Deer

The diet of swamp deer (Fig. 8.13) consisted of 33 plant species and four plant forms, namely grasses, herbs, sedges, and shrubs. Grass species contributed the most to the diet i.e., 57.58%, followed by herb species 36.36%, sedge species 3.03% and shrub species 3.03% (Fig. 8.14). A chi-square test showed significant differences in consumption of different plant forms by swamp deer, indicating a high tendency to feed on grass species ($\chi^2 = 28$; df = 3; $p < 0.05$). The diversity of plant forms in swamp deer was 0.89, while the pasture-to-grass ratio was not calculated because no tree species were detected in the diet of swamp deer.

Hog Deer

The diet of hog deer (Fig. 8.15) consisted of 19 plant species and four plant forms, namely grasses, herbs, sedges, and shrubs. Grass species contributed the most to the diet i.e., 57.89%, followed by herb species 21.05%, sedge species 15.79% and shrub species 5.26% (Fig. 8.16). A chi-square test revealed significant differences in consumption of different plant forms by hog deer, indicating a high tendency to feed on grass species ($\chi^2 = 11$; df = 3; $p < 0.05$). The diversity of plant forms in the hog deer diet was 1.09, while the ratio of pasture to grass was not calculated because no tree species were detected in the diet of the hog deer.

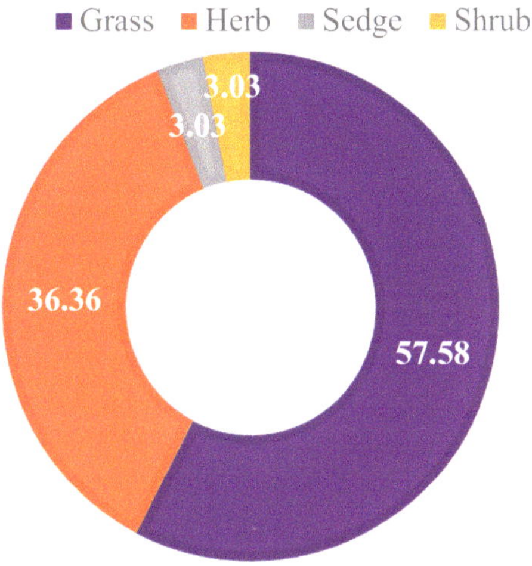

Fig. 8.14 Percentage of plant form in the diet of swamp deer

Fig. 8.15 A hog deer captured at Kaziranga National Park, Assam, India (*Source* Dhawal Mehta)

Chinkara

The diet of the Chinkara (Fig. 8.17) consists of 56 plant species and five plant forms, i.e. forms, namely grasses, herbs, sedges, climbers, and shrubs. Tree species contributed the highest proportion of diet of chinkara, i.e. 53.57% followed by shrub species 17.86%, grass species 12.50%, herb species 12.50%, and climber species climbing plants 3.57% (Fig. 8.18). Chi-square tests showed a significant difference in consumption of different plant forms by chinkara indicating a high tendency to feed upon on tree species ($\chi^2 = 26.65$; df = 5; $p < 0.05$). The diversity of plant forms

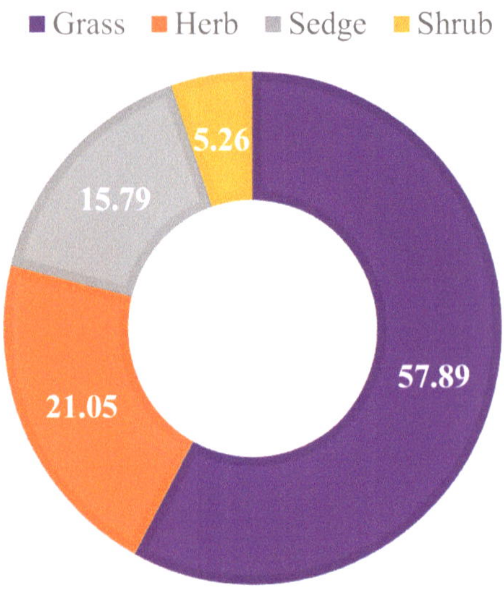

Fig. 8.16 Per cent of plant form in the diet of hog deer

Fig. 8.17 A Chinkara captured from Gir protected area, Gujarat, India (*Source* Rohit Chaudhary)

was 1.28, while the browse-to-grass ratio was 4.28 indicating a high contribution proportion of tree browse to grass in chinkara diet.

Blackbuck

The blackbuck's diet (Fig. 8.19) consists of 43 plant species and is similar to the chital, which includes six plant forms. Grass species contributed the highest proportion to the

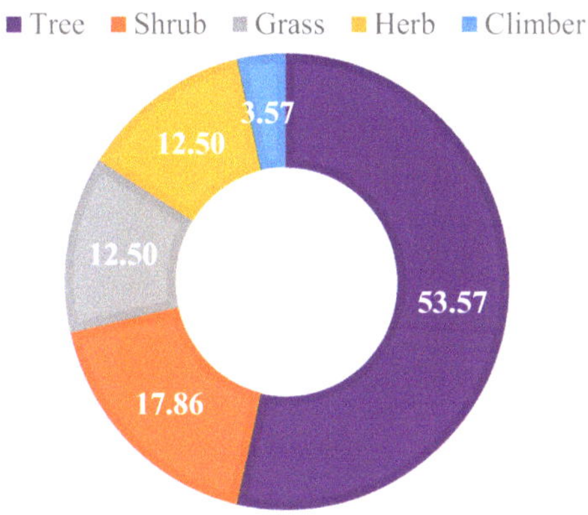

Fig. 8.18 Percentage of plant form in the diet of chinkara

Fig. 8.19 A blackbuck captured in Blackbuck National Park, Gujarat, India (*Source* Aadil Kazi)

blackbuck's diet i.e., 41.86%, followed by herb species20.93%, tree species 18.60%, shrub species 11.63%, sedge species 4.65%, and climber species 2.33% (Fig. 8.20). The chi-square test revealed a significant difference in the consumption of different plant forms by blackbucks, indicating a high propensity to consume grass species ($\chi^2 = 42.39$; df = 4; $p < 0.05$). The plant form diversity of blackbuck was 1.48, while the browse-to-grass ratio was 0.44, indicating that the proportion of tree browsing in the blackbuck's diet was low.

Chowsingha

The diet of chowsingha (Fig. 8.21) consists of 129 plant species and five plant forms including trees, grasses, herbs, shrubs, and climbers. Tree species contributed the highest proportion to chowsingha's diet 48.06%, followed by grass species 24.03%,

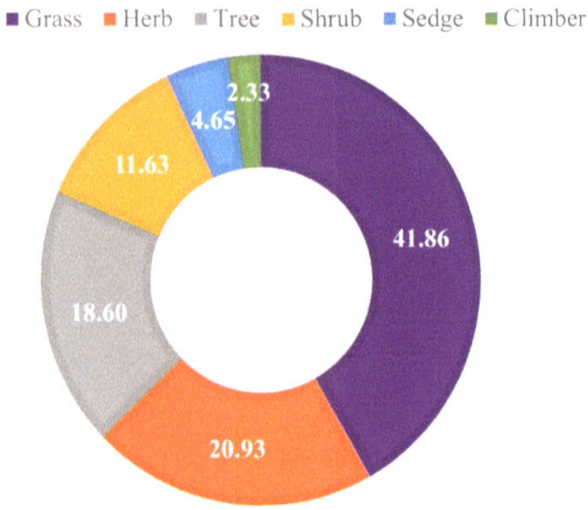

Fig. 8.20 Per cent of plant form in the diet of blackbuck

Fig. 8.21 A chowsingha captured in Kevadia zoo, Gujarat, India (*Source* Ravi Patel)

herb species 12.40%, shrub species 11.63%, and climbers at 3.88% (Fig. 8.22). The chi-square test revealed a significant difference in the consumption of different plant forms by chowsingha, indicating a high tendency towards consumption of tree browse ($\chi^2 = 76.85$; df = 4; $p < 0.05$). Plant form diversity at chowsingha was 1.30, while the browse-to-grass ratio was 2, indicating a high proportion of tree than grass in chowsingha's diet.

Diet Similarity in Ungulate Species

The cluster analysis shows two major clusters (Fig. 8.23), where one cluster includes swamp deer, hog deer, and blackbuck while the second cluster consists of chital,

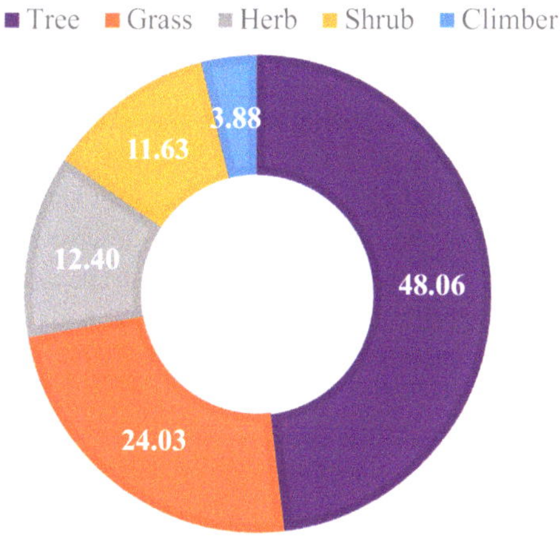

Fig. 8.22 Per cent of plant form in the diet of Chowsingha

nilgai, chinkara, sambar, chowsingha, and gaur. In the first cluster, the greatest similarity was found in the diet of swamp deer and hog deer, while in the second cluster, the greatest similarity was found in the diet of chowsingha, sambar, nilgai, and chital.

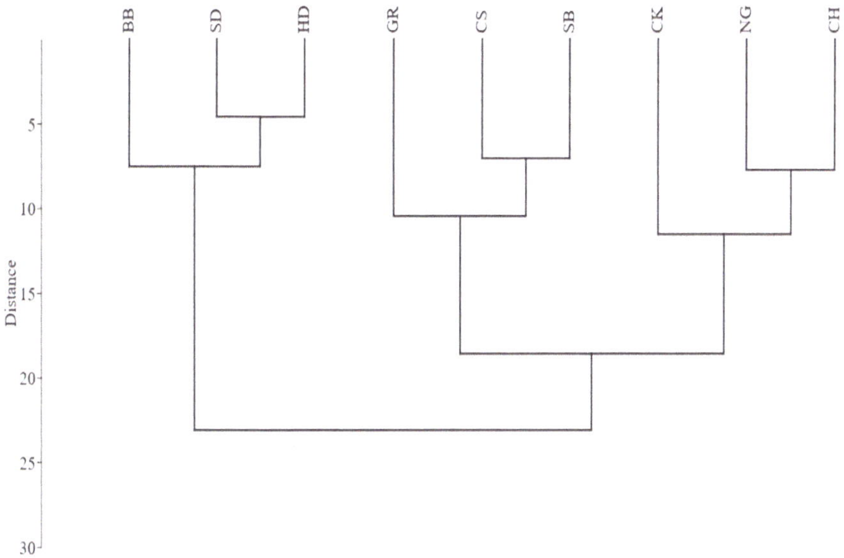

Fig. 8.23 Cluster analysis to assess similarity in consumption of plant species in ungulate diet (*BB* blackbuck, *SD* swamp deer, *HD* hog deer, *GR* gaur, *CS* chowsingha, *SB* sambar, *CK* chinkara, *NG* nilgai, *CH* Chital)

Discussion

Food plays a crucial role in the life history of ungulates, and therefore, the present study is important for the conservation of wild ungulates in India. The present study findings indicated that the chital diet had the highest number of dietary plant species, while the hog deer diet had the lowest number of plant species. The plant species richness in the diet of ungulates decreased as the species increasingly specialized in their habitat use. For example, chital and nilgai occur in a variety of habitats, while swamp deer and hog deer are grassland specialists (Dinerstein, 1979; Moe & Wegge, 1994; Dar et al., 2012; Berkel et al., 2022; Rastogi et al., 2022). However, studies from South African ecosystems have found that body size plays an important role in species richness in ungulate diets (Bukombe et al., 2019; Jarman, 1974). But, in the present case, it appears that habitat specialization has reduced the number of plant species available to ungulates, resulting in low food diversity for specialized species. Another possible reason is the lack of studies on the feeding habits of ungulates. There are fewer studies for hog deer and swamp deer than for chital and nilgai, and this may have led to assessments of low diet diversity in some ungulate species such as hog deer and swamp deer (Tiwari & Rawat, 2013; Tuboi & Hussain, 2016; Devi, 2022).

The highest number of plant species were found in the diet of the chital, suggesting that the chital is a generalist feeder. The chital‘s high dietary diversity also indicates its broad habitat preferences, ranging from forests to meadows (Schaller, 1967; Dinerstein, 1979; Bhattarai, 2019). The chital also shows seasonal changes in eating habits. For example, studies indicate that chital feed heavily on browse material during summer and winter, while during the monsoon season their diet consists mainly of grass material (Khan, 1994; Raman et al., 1996). The chital also feeds on fruits of some plant species. For example, in the Gir protected area in Gujarat, India, Khan et al. (1994) found that the chital feeds heavily on the fruits of *Ziziphus* species in winter. In the present review, the diet was found to consist of more tree species than grass species, which is also supported by the high browse-to-grass ratio. The high number of tree species could be due to the reason that grass is a temporary resource and is mainly available during the monsoon season—as in other seasons, grasses become dry and develop silica levels that make them inedible, while browse is a more permanent resource (Massey & Hartley, 2006). Therefore, the year-round availability of tree species contributes to a high proportion of tree browsing in the chital diet. Furthermore, in the present study, we only examined the number of plant species, and perhaps grass would dominate in the diet if their relative contribution in the chital diet had been quantified.

Nilgai, like the Chital, consist of many plant species in their diet. The high richness of plant species in Nilgai’s diet could have two main reasons. Firstly, the large body size of the nilgai, with average adult body weight of 200 kilograms (Khan et al., 1996), resulting in higher absolute energy costs and a high requirement of energy from plant species material (Jarman, 1974). Secondly, nilgai like chital inhabit a wide variety of habitats, which also contributes to the presence of many plant species in their diet (Khan et al., 1996; Meena et al., 2020; Sankar, 1994). The Nilgai, similar to the chital, is a seasonal feeder and feeds differently during the summer, monsoon, and winter seasons (Sankar, 1994). For example, nilgai in southern Texas are primarily grazers and feed exclusively on grass (Sheffield et al., 1983). Nilgai also feed on pods and fruits of various plant species, such as *Butea monosperma* and *Acacia catechu* (Meena et al., 2020; Sankar, 1994). The ratio of tree browsing to grass in the nilgai's diet also indicates its mixed foraging nature, unlike species such as sambar and chowsingha which rely heavily on tree browsing. Due to their adaptability to survive in a human-dominated landscape, nilgai also raid crops in agricultural fields, which may also contribute to their high number of plant species in their diet (Mehta, 2014; Meena et al., 2020).

The sambar is a large ungulate (average adult weight: 180 kg) (Khan et al., 1996) like nilgai, but its diet consists of fewer plant species than the diet of nilgai. Sambar, like nilgai, also has high absolute energy requirements due to their large body size, despite this smaller number of plant species were consumed by the sambar. Possibly, habitat choice plays an important role in defining the plant species in the sambar's diet. The sambar favours hilly terrain and dense forests (Chatterjee et al., 2014; Green, 1987; Haleem & Ilyas, 2022; Khan et al., 1996; Sankar, 1994). Therefore, restricted distribution in terms of habitat could lead to reduction in availability of plant species for sambar, resulting in lower intake of plant species than nilgai. The sambar's diet also features a high ratio of tree browsing to grass, possibly due to the sambar's heavy use of forest habitat. However, published research indicates that the diet of the sambar, like chital and nilgai, varied seasonally (Haleem & Ilyas, 2022; Khan, 1994; Sankar, 1994). Literature also suggests that sambar feeds on the bark of tree species during summers (Khan, 1994; Yen et al., 2015). Such debarking of trees can also lead to their death, which could further affect the structure and composition of vegetation (Khan, 1994; Yen et al., 2015). The chowsingha and sambar diets consist of a more or less similar number of plant species, possibly because of the similar habitat use of both species. Chowsingha, like sambar, is a habitat specialist with a strong preference for dry, open-canopy deciduous forests where tree growth is either reduced or stagnant (Baskaran et al., 2011; Sharma et al., 2013; Pokharel et al., 2015). However, the chowsingha also has a lower tree browseing to grass ratio than the sambar, suggesting that the chowsingha also feeds on grass species higher than sambar possibly during early monsoon season and has particular foraging preferences in other seasons (Rodgers & Panwar, 1988; Sharma et al., 2013).

The gaur diet consists of only 94 plant species, which is 1.2 times less than that of chital, and lesser than the number consumed by nilgai, sambar, and chowsingha. The possible reason for this limited number of plant species in the gaur's diet is its habitat preference. Sankar et al. (2013) found that gaur had a strong preference for grassland and bamboo forest and use the woodland habitat mainly for resting. Such limited use of the habitat might result in a limited number of plant species in the gaur's diet. The mouth structure of the gaur is similar to that of domestic cattle and is well adapted to grazing (Sankar et al., 2015). The gaur, unlike the chital and sambar, is also too heavy to stand on its hind legs for browsing, which might also lead to low plant richness in its diet. Gaur might also feed on the seedlings and saplings of tree species, leading to a high contribution of tree browse in its diet (Haleem & Ilyas, 2018; Nayak & Patra, 2015).

The Chinkara antelope had fewer plant species in its diet (56) than the gaur. The number of plant species in the chinkara's diet is approximately three times less than that of the chital and about 2.5 times less than the smaller chowsingha. The chinkara is highly dependent on open scrub or thorny area for habitat (Khan et al., 1996; Bagchi et al., 2003; Kumar et al., 2020). Due to its small body size, the chinkara may require a high-quality diet such as young shoots and leaves, which may be locally rare (Jarman, 1974). The habitat speciality and requirements for a high-quality diet may account for the small number of plant species in the chinkara's diet. The chinkara's habitat specialization was also evident through the high contribution of herb and shrub species, which comprise about 30% of its diet. The high tree browses to grass ratio also shows that chinkara are highly dependent on tree browse. The chinkara meets its moisture requirements through plant species, and therefore, the moisture content of plant species could play an important role in the Chinkara's feeding habits (Rahmani, 2015).

The blackbuck, swamp deer, and hog deer have a very low number of plant species in their diet, indicating their highly specialized diet. Of all three species, the blackbuck has a high proportion of plant species in its diet, while the hog deer has the lowest proportion of plant species. The blackbuck often lives in flat plains and favours open habitats such as grasslands, savannahs, and agricultural areas (Isvaran, 2015). The use of open habitats with low tree species numbers and tree diversity results in a mixed diet that includes both grass and tree browsing, unlike the grassland specialists swamp deer and hog deer. However, grass species dominate the blackbuck's diet, suggesting that it is primarily a grazer. Studies also suggest that blackbucks feed on the pods of *Acacia* and *Prosopis* tree species in summer when grass nutrient levels are low (Isvaran, 2005; Jhala, 1991). In summer, when there are no pods or tree bushes, the blackbuck raids the surrounding agricultural fields. Fresh, favoured food also leads to the formation of resource-based territories in blackbuck (Isvaran, 2005; Jhala, 1991). By defending resource-based territories, males can attract females. Therefore, food also plays an important role in the mating system of the blackbuck (Isvaran, 2005). Both swamp deer and hog deer feed on a minimal number of plant species, with more than half of their diet consisting of grass species and no tree species. Swamp deer are highly dependent on swamp habitats, while the hog deer is a tallgrass specialist, which may resulted its highly specialized diet (Rastogi et al., 2022; Tewari & Rawat, 2013).

Similarity in the Diet of Ungulates

Cluster analysis revealed that blackbuck, swamp deer, and hog deer have similar diets, with hog deer and swamp deer showing a strong association. All three species as previously stated preferred open habitats and fed primarily on grasses. However, the blackbuck's diet also includes tree material, placing it in a different row within the same group. As previously mentioned, both swamp and hog deer prefer grassland and grass for forage, which could lead to a close association in their diets.

Similarly, sambar and chowsingha show a close connection in their diet, while gaur occupies a different lineage in the same group. Both sambar and chowsingha have similar habitat preferences, which may result in them encountering the same plant species, resulting in a similar diet. Another lineage within the same cluster suggests that gaur may feed on some plant species consumed by both chowsingha and sambar. Nilgai and chital show a strong association in consumption of plant species, while chinkara occupies a different lineage within the same group, indicating a lower association with chital and nilgai. Both nilgai and chital are habitat generalists and can feed on a wide variety of plant species, which may result in similarity in their diet. The similarity of chinkara‘s diet with chital and nilgai could be due to their high consumption of tree, herb, and shrub species.

Conclusion

The current study found that ungulates are broadly divided into grazers (e.g. hog deer and swamp deer), mixed-feeders (e.g. blackbuck), and dominantly browsers (e.g. chital, nilgai, sambar, gaur, chowsingha, and chinkara). All ungulate species exhibit a gradient from high to low food niches, which may be due to the diversity of their habitat use. For example, chital and sambar used a diverse habitat that resulted in a larger number of plant species in their diet than the habitat specialists hog deer and swamp deer. Therefore, species' choice of habitat was more strongly correlated with diet than with body size and mouth structure. Some species such as hog deer and swamp deer are less studied than other species, such as chital and sambar. The current study found that ungulate conservation requires an integrated approach, where habitat and food plant restoration should support the conservation strategy.

References

Ahrestani, F. S., Heitkönig, I. M., & Prins, H. H. (2012). Diet and habitat-niche relationships within an assemblage of large herbivores in a seasonal tropical forest. *Journal of Tropical Ecology, 28*(4), 385–394.

Ashokkumar, M., Nagarajan, R., & Desai, A. A. (2010). Group size and age-sex composition of Asian elephant and gaur in Mudumalai Tiger Reserve, southern India. *Gajah, 32*, 27–34.

Atwood, T. B., Witt, A., Mayorga, J., Hammill, E., & Sala, E. (2020). Global patterns in marine sediment carbon stocks. *Frontiers in Marine Science, 7*, 165.

Bagchi, S., Goyal, S. P., & Sankar, K. (2003). Prey abundance and prey selection by tigers (Panthera tigris) in a semi-arid, dry deciduous forest in western India. *Journal of Zoology, 260*(3), 285–290.

Baskaran, N., Kannan, V., Thiyagesan, K., & Desai, A. A. (2011). Behavioural ecology of four-horned antelope (Tetracerus quadricornis de Blainville, 1816) in the tropical forests of southern India. *Mammalian Biology, 76*(6), 741–747.

Baskaran, N., Ramkumaran, K., & Karthikeyan, G. (2016). Spatial and dietary overlap between blackbuck (Antilope cervicapra) and feral horse (Equus caballus) at Point Calimere Wildlife Sanctuary, Southern India: Competition between native versus introduced species. *Mammalian Biology, 81*(3), 295–302.

Beguin, J., Tremblay, J. P., Thiffault, N., Pothier, D., & CôTé, S. D. (2016). Management of forest regeneration in boreal and temperate deer–forest systems: Challenges, guidelines, and research gaps. *Ecosphere, 7*(10), e01488.

Bhattarai, B. P. (2019). Factors associated with habitat segregation among the four species of cervids in the Chitwan National Park, Nepal. *Ekologia, 38*(1), 37–48. https://doi.org/10.2478/eko-2019-0004

Bhowmik, M. K., & Chakraborty, T. (1999). Status and distribution of hog deer (Axis porcinus) in protected areas of sub-Himalayan West Bengal. *ZOOS'PRINT JOURNAL, 14*(11), 151–152.

Boulanger, V., Dupouey, J. L., Archaux, F., Badeau, V., Baltzinger, C., et al. (2017). Ungulates increase forest plant species richness to the benefit of non-forest specialists. *Global Change Biology, 24*(2), 1–11. https://doi.org/10.1111/gcb.13899

Bukombe, J., Kittle, A., Senzota, R. B., Kija, K., Mduma, S., Fryxell, J. M., Magige, F., Mligo, C., & Sinclair, A. R. E. (2019). The influence of food availability, quality and body size on patch selection of coexisting grazer ungulates in western Serengeti National Park. *Wildlife Research, 46*(1), 54–63. https://doi.org/10.1071/WR18072

Cardinale, B. J., Gonzalez, A., Allington, G. R., & Loreau, M. (2018). Is local biodiversity declining or not? A summary of the debate over analysis of species richness time trends. *Biological Conservation, 219*, 175–183.

Chatterjee, D., Sankar, K., Qureshi, Q., Malik, P. K., & Nigam. P. (2014). Ranging pattern and habitat use of sambar (Rusa unicolor) in Sariska tiger reserve, Rajasthan, Western India. DSG Newsletter N°26.

Clavel, J., Julliard, R., & Devictor, V. (2011). Worldwide decline of specialist species: Toward a global functional homogenization? *Frontiers in Ecology and the Environment, 9*(4), 222–228.

Craigie, I. D., Baillie, J. E., Balmford, A., Carbone, C., Collen, B., Green, R. E., & Hutton, J. M. (2010). Large mammal population declines in Africa's protected areas. *Biological Conservation, 143*(9), 2221–2228.

Dave C. V. (2008). *Ecology of chital (Axis axis) in Gir* (Doctoral dissertation, Saurashtra University.

Dar, T. A., Habib, B., & Khan, J. A. (2012). Group size, habitat use and overlap analysis of four sympatric ungulate species in Shivalik Ecosystem, Uttarakhand, India. *Mammalia 76*(1), 31–41.

Demers, C., Long, A., Latt, C., & Willcox, E. (2020). *Establishing and maintaining wildlife food sources*. Retrieved from https://edis.ifas.ufl.edu

Devi, A., Hussain, S. A., Sharma, M., Gopi, G. V., & Badola, R. (2022). Seasonal pattern of food habits of large herbivores in riverine alluvial grasslands of Brahmaputra floodplains, Assam. *Scientific Reports, 12*(1), 482.

Dinerstein, E. (1979). An ecological survey of the Royal Karnali-Bardia wildlife reserve, Nepal. Part II: habitat/animal interactions. *Biological Conservation, 16*(4), 265–300.

Etzold, S., Ferretti, M., et al. (2020). Nitrogen deposition is the most important environmental driver of growth of pure, even-aged and managed European forests. *Forest Ecology Management, 458*, 1–13. https://doi.org/10.1016/j.foreco.2019.117762

Frank, S. J. D., Gopi, G. V., Sankar, K., Hussain, S. A., et al. (2021). Dry season resource selection among sympatric ungulates in a tropical coastal landscape: Implications for conservation and management. *Tropical Ecology, 62*, 418–426. https://doi.org/10.1007/s42965-021-00156-3

Gad, S. D., & Shyama, S. K. (2009). Studies on the food and feeding habits of gaur Bos taurus H. Smith (Mammalia: Artiodactyla: Bovidae) in two protected areas of Goa. *Journal of Threatened Taxa, 2*, 128–130. https://doi.org/10.11609/JoTT.o1589.128-30

Gill, R. M. A., & Beardall, V. (2001). The impact of deer on woodlands: The effects of browsing and seed dispersal on vegetation structure and composition. *Forestry, 74*, 209–218. https://doi.org/10.1093/forestry/74.3.209

Gordon, I. J. (2009). What is the future for wild, large herbivores in human-modified agricultural landscapes? *Wildlife Biology, 15*(1), 1–9.

Green, M. J. B. (1987). Ecological separation in Himalayan ungulates. *Journal of Zoology, 1*, 693–619. https://doi.org/10.1111/j.1096-3642.1987.tb00751.x

Gupta, D., & Krishnamurthy, R. (2023). Spatial density patterns of herbivore response to seasonal dynamics in the tropical deciduous forest of Central India. *Biotropica, 55*(2), 430–443. https://doi.org/10.1111/btp.13197

Haleem, A. (2014). *Conservation status and ecology of ungulates in Pench tiger reserve Madhya Pradesh with special reference to resource partitioning*. Ph.D. thesis submitted to Aligarh Muslim University, Aligarh, India.

Haleem, A., & Ilyas, O. (2018). Food and feeding habits of gaur (*Bos gaurus*) in highlands of Central India: A case study at Pench tiger reserve, Madhya Pradesh India. *Zoological Science, 35*(1), 57–67. https://doi.org/10.2108/zs170097

Haleem, A., & Ilyas, O. (2022). Status, abundance, and food habit of sambar deer (*Rusa unicolor*) in tropical deciduous forest of Central India. *Mammalia, 86*, 1–12. https://doi.org/10.1515/mammalia-2021-0109

Haleem, A., Ilyas, O., & Khan, S. (2014). Ecology of four-horned antelope in Pench tiger reserve, Seoni, M.P., with special preference to food & feeding habit. In *annual international conference on ecology, biodiversity and environment Singapore.*

Haque, N. M. D. (1990). *Study on the ecology of wild ungulates of Keoladeo National Park, Bharatpur, Rajasthan.* Aligarh Muslim University.

Hobbs, N. T. (1996). Modification of ecosystems by ungulates. *Journal of Wildlife Management, 60*, 695–713. https://doi.org/10.2307/3802368

Isvaran, K. (2005). Female grouping best predicts lekking in blackbuck. *Behavioural Ecology and Sociobiology, 57*, 283–294. https://doi.org/10.1007/s00265-004-0844-z

Isvaran, K. (2015). Blackbuck: *Antilope cervicapra.* In N. Manjrekar & A. J. T. Johnsingh (Eds.), *Mammals of South Asia.* University Press.

James, F. C., & McCulloch, C. E. (1990). Multivariate analysis in ecology and systematics: Pancea or Pandoras box? *Annual Review in Ecology and Systematics, 21*, 129–166. https://doi.org/10.1146/annurev.es.21.110190.001021

Jarman, P. J. (1974). The social organization of antelope in relation to their ecology. *Behaviour, 48*, 215–266. https://doi.org/10.1163/156853974X00345

Jhala, Y. V. (1991). *Habitat and population dynamics of wolves and blackbuck in Velavadar National Park, Gujarat, India.* Virginia Polytech Institute and State University.

Jhala, Y. V., Qureshi, Q., & Nayak, A. K. (2020). *Status of tigers, co-predators and prey in India, 2018.* National Tiger Conservation Authority, Government of India.

Khan, J. A. (1994). Food habits of ungulates in dry tropical forests of Gir lion sanctuary, Gujarat, India. *Acta Theriologica, 39*, 185–193.

Khan, J. A., Chellam, R., Rodgers, W. A., & Johnsingh, A. J. T. (1996). Ungulate densities and biomass in the dry deciduous forest of Gir, Gujarat, India. *Journal of Tropical Ecology, 12*, 149–162. https://doi.org/10.1017/S0266467400009366

Khan, J. A., Rodgers, W. A., Johnsingh, A. J. T., & Mathur, P. K. (1994). Tree and shrub mortality and debarking by sambar Cervus unicolor (kerr) in Gir after a drought in Gujarat, India. *Biological Conservation, 68*, 149–154. https://doi.org/10.1016/0006-3207(94)90346-8

Kitamura, S., Yumoto, T., Poonswad, P., & Wohandee, P. (2007). Frugivory and seed dispersal by Asian elephants, *Elephas maximus*, in a moist evergreen forest of Thailand. *Journal of Tropical Ecology, 23*(3), 373–376. https://doi.org/10.1017/S0266467407004026

Knopps, J. M. H., Ritichie, M. E., & Tilman, D. (2000). Selective herbivory on a nitrogen fixing legume (Lathyrus venosus) influences productivity and ecosystem nitrogen pools in an oak savanna. *Ecoscience, 7*, 166–174. https://doi.org/10.1080/11956860.2000.11682585

Kumar, D., Velankar, A. D., Rao, N. V. R., Kumara, H. N., Mishra, P. S., Bhattacharya, P., & Raj, V. M. (2020). *Ecological determinants of occupancy and abundance of chinkara (Gazella bennettii) in Yadahalli wildlife sanctuary, Karnataka, India.*

Lamprey, R. H., & Reid, R. S. (2004). Expansion of human settlement in Kenya's Maasai Mara: What future for pastoralism and wildlife. *Journal of Biogeography, 31*(6), 997–132. https://doi.org/10.1111/j.1365-2699.2004.01062.x

Magurran, A. (1988). *Ecological diversity and its measurement*. Princeton University Press.

Massey, F. P., & Hartley, S. E. (2006). Experimental demonstration of the antiherbivore effects of silica in grasses: Impacts on foliage digestibility and vole growth rates. *Proceedings Royal Scosiety B, 273*(1599), 299–304. https://doi.org/10.1098/rspb.2006.3586

Meena, M., & Choudhary, K. (2020). A perceptive study on feeding pattern of Nilgai, (*Boselaphus tragocamelus*) in Jodhpur region, Rajasthan, India. *Journal of Experimental Zoology India, 23*(2), 1951–1956.

Meena, M., Jaipal, A. K., & Garg, K. (2020). Crop damage by the blackbuck (*Antilope cervicapra*) and other ungulates in Aroun Rohat region, Rajasthan, India. *Alochana Chakra Journal, 10*(5), 2866.

Meghwal, R., Bhatnagar, C., & Koli, V. K. (2020). Seasonal food habits of the four-horned antelope (*Tetracerus quadricornis* de Blainville, 1816) in tropical deciduous forests of Aravalli mountain range, Rajasthan, Western India. *The Journal of Basic and Applied Zoology, 81*, 1–11. https://doi.org/10.1186/s41936-020-00193-9

Mehta, D., Soni, V. C., & Sankar, K. (2013). Food habits of Nilgai (Boselaphus tragocamelus) as determined by pellet analysis in a human-dominated landscape, Western Gujarat, India. *Scientific Transactions in Environment and Technovation, 7*(1), 7–13.

Mehta, D. (2014). *Study on the Ecology of Nilgai (Boselaphus tragocamelus) in Saurashtra*. Ph.D. thesis. Saurashtra University, Rajkot, India.

Moe, S. R., & Wegge, P. (1994). Spacing behaviour and habitat use of axis deer (*Axis axis*) in lowland Nepal. *Canadian Journal of Zoology, 72*, 1735–1744. https://doi.org/10.1139/z94-234

Javid, M., Parveen, T., & Ilyas, O. (2022). Food plants and feeding habits of ungulates at Panna Tiger Reserve, Madhya Pradesh, India.

Nayak, B. K., & Patra, A. K. (2015). Food and feeding habits of Indian bison, *Bos gaurus* (Smith, 1827) in Kuldiha wildlife sanctuary, Balasore, Odisha, India and its conservation. *International Research Journal of Biological Sciences, 4*(5), 73–79. Retrieved from http://www.isca.me/IJBS/Archive/v4/i5/14.ISCA-IRJBS-2015-043.pdf

Nowak, R. M. (2018). *Walker's mammals of the world: Monotremes, marsupials, Afrotherians, Xenarthrans, and Sundatherians*. Johns and Hopkins University Press.

Pianka, E. R. (1970). On R- and K-selection. *The American Naturalist, 104*(940), 592–597. https://doi.org/10.1086/282697

Pokharel, K. P., Ludwig, T., & Storch, I. (2015). Spatial niche partitioning in sub-tropical solitary ungulates: Four-horned antelope and barking deer in Nepal. *PLoS One, 10*(2), 1–16. https://doi.org/10.1371/journal.pone.0117917

Promislow, D. E. L., & Harvey, P. H. (1990). Living fast and dying young: A comparative analysis of life-history variation among mammals. *Journal of Zoology, 220*, 417–437. https://doi.org/10.1111/j.1469-7998.1990.tb04316.x

Rahmani, A. R. (2015). Chinkara or Indian Gazelle: *Gazella bennetti*. In N. Manjrekar & A. J. T. Johnsingh (Eds.), *Mammals of South Asia*. University Press.

Raman, T. R. S., Menon, R. K. G., & Sukumar, R. (1996). Ecology and management of chital and blackbuck in Guindy National Park, Madras. *Journal of Bombay Natural History Society, 93*, 178–192.

Ramirez, J. I., Jansen, P. A., den Ouden, J., Goudzwaard, L., & Poorter, L. (2019). Long-term effects of wild ungulates on the structure, composition and succession of temperate forests. *Forest Ecology and Management, 432*, 478–488.

Rastogi, S., Chanchani, P., Sankaran, M., & Warrier, R. (2022). Grasslands half-full: Investigating drivers of spatial heterogeneity in ungulate occurrence in Indian Terai. *Journal of Zoology, 316*, 139–153. https://doi.org/10.1111/jzo.12939

Ripple, W. J., Newsome, T. M., Wolf, C., Dirzo, R., Everatt, K. T., Galetti, M., ... & Van Valkenburgh, B. (2015). Collapse of the world's largest herbivores. *Science Advances, 1*(4), e1400103.

Rodgers, W. A., & Panwar, H. S. (1988). Planning a wildlife protected area network in India. *FAO Field Document Dehradun, 1*(9), 341.

Sankar, K. (1994). *Ecology of three sympatric herbivores (chital, sambar and nilgai) with special reference to reserve management in Sariska tiger reserve, Rajasthan, India.* University of Rajasthan.

Sankar, K., Pabla, H. S., Patil, C. K., Nigam, P., Qureshi, Q., Navaneethan, B., Manjreakar, M., Virkar, P., & Mondal. (2013). Home range, habitat use and food habits of re-introduced gaur (*Bos gaurus gaurus*) in Bandhavgarh Tiger Reserve, Central India. *Tropical Conservation Science, 6*, 50–69. https://doi.org/10.1177/194008291300600108

Sankar, K., Pasha, M. K. S., Areendran, G., & Qureshi, Q. (2015). Gaur: Bos gaurus. In N. Manjrekar & A. J. T. Johnsingh (Eds.), *Mammals of South Asia.* University Press.

Schaller, G. (1967). *The deer and the tiger.* University of Chicago Press.

Schoener, T. W. (1974). Resource partitioning in ecological communities. *Science, 185*, 27–39. https://doi.org/10.1126/science.185.4145.27

Sharma, K., Chundawat, R. S., Gruisen, J. V., & Rahmani, A. S. (2013). Understanding the patchy distribution of four-horned antelope *Tetracerus quadricornis* in a tropical dry deciduous forest in Central India. *Journal of Tropical Ecology, 30*, 45–54. https://doi.org/10.1017/S0266467413000722

Sheffield, W. J., Fall, B. A., & Brown, B. A. (1983). *The nilgai antelope. The Caesar kleberg program in wildlife ecology a department of wildlife and fisheries sciences.* The Texas A&M University.

Shodhganaga. (2023). *Repository of Indian Ph.D. thesis.* Retrieved from https://shodhganga.inflibnet.ac.in/

Smith, F. A., Smith, R. E. E., & Payne, J. L. (2018). Body size downgrading of mammals over the late quaternary. *Science, 360*, 310–313. https://doi.org/10.1126/science.aao5987

Spitzer, R., Felton, A., Landman, M., Singh, N. J., Widemo, F., & Cromsigt, J. P. G. M. (2020). Fifty years of European ungulate dietary studies: A synthesis. *Oikos, 10*, 1–13. https://doi.org/10.1111/oik.07435

Srivathsa, A., Banerjee, A., Banerjee, S., Chawla, M. M., Das, A., et al. (2022). Chasms in charismatic species research: Seventy years of carnivore science and its implications for conservation and policy in India. *Biological Conservation, 273*, 109694. https://doi.org/10.1016/j.biocon.2022.109694

Stokely, T. D., & Betts, M. G. (2020). Deer-mediated ecosystem service vs. disservice depends on forest management intensity. *Journal of Applied Ecology, 57*, 31–42. https://doi.org/10.1111/1365-2664.13532

Takatsuti, S. (2022). Long-term changes in food habits of deer and habitat vegetation: 25-year monitoring on a small Island. *Ecological Research, 37*, 92–101. https://doi.org/10.1111/1440-1703.12264

Tewari, R., & Rawat, G. S. (2013). Habitat use of swamp deer (*Rucervus duvaucelii duvaucelii*) in Jhilmil jheel conservation reserve, Haridwar, Uttarakhand, India. *International Journal of Ecology and Environmental Sciences, 39*, 79–85.

Tiwari, R., & Rawat, G. S. (2013). Studies on the food and feeding habits of swamp deer (*Rucervus duvaucelii duvaucelii*) in Jhilmil Jheel conservation reserve, Haridwar, Uttarakhand, India. *ISRN Zoology, 6*, 278213. https://doi.org/10.1155/2013/278213

Tuboi, C., & Hussain, S. A. (2016). Factors affecting forage selection by the endangered Eld's deer and hog deer in the floating meadows of Barak-Chindwin Basin of north-East India. *Mammalian Biology, 81*, 53–60. https://doi.org/10.1016/j.mambio.2014.10.006

UNEP. (2023). *Megadiverse countries.* Retrieved from https://www.biodiversitya-z.org/

Upadhyay, H. S., Behera, S., Dutta, S. K., Sahu, H. K., & Sethy, J. (2019). A viable tiger population in Similipal Tiger Reserve, India? Calculating if the ungulate prey base is limiting. *Wildlife Biology*. https://doi.org/10.2981/wlb.00474

Van Berkel, T., Emsens, W. J., Eam, S. U., Simoes, S., Puls, S., Rin, N., ... & Jocque, M. (2022). Population density, habitat use and activity patterns of endangered hog deer in Cambodia. *Mammal Research, 67*(3), 311–316.
Yen, S. C., Lin, C. Y., Hew, S. W., Yang, S. Y., Yeh, C. F., & Weng, G. J. (2015). Characterization of debarking behavior by sambar deer (*Rusa unicolor*) in Taiwan. *Mammal Study, 40*(3), 167–179. https://doi.org/10.3106/041.040.0304
Zar, J. H. (2006). *Biostatistical analysis*. Pearson Publication.

Chapter 9
Geomatics Applications to Conservation and Social Developments

Michael O'Neal Campbell

Abstract The computer-aided systems, which are usually referred to as geomatics in North America and to some extent geoinformatics in Europe and Asia, represent an important technological advance with widespread applicability to conservation biogeography as well as to the environmental and spatial aspects of economics, anthropology, sociology, human geography, psychology and political science. Geomatics, an extremely broad field or combination of fields, includes remote sensing, geographic information science or geographical information systems (GIS), global positioning systems (GPS), geodesy and cartography and mapping. Because conservation biogeography is a relatively new discipline or sub-discipline, applications are developing to some of the supporting disciplines such as anthropology or human geography, and the conservation context may be recognized or inferred. This chapter examines the current state of geomatics in conservation biogeography. The methodology consists of examining the research results in the social sciences, conservation biology and biogeography and selecting those applications and results that can be assigned to the field of conservation biogeography. Due to the multidisciplinary background of conservation biogeography, biogeography and conservation science, a certain degree of subjectivity may arise, particularly due to increasing social science involvement directly caused by increasing human interference with the environment, with increasingly complex environmental outcomes from this involvement. Current evidence suggests that geomatics-based systems enable more reliable and precise calculations for environmental research with more applicable results. Geomatics-based techniques easily measure land cover changes that can impact conservation and biogeography. However, more problematic may be the predictive modeling of human activities in the context of human decision-making and the measurement of the socioeconomic, political, cultural and even psychological impacts of biogeographical changes. Further difficulties arise when measuring questions that concern applied disciplines such as nature conservation, cultural and urban studies. Assessing these complications can contribute to the application of geomatic methods to conservation biogeography.

M. O. Campbell (✉)
Simon Fraser University, Burnaby, BC, Canada

K. K. Jha and M. O. Campbell (eds.), *Intersections of Conservation Biogeography and Wildlife Management*, https://doi.org/10.1007/978-981-95-0121-2_9

Keywords Geomatics · Aerial photography · Geographical information science · Remote sensing · Carnivore · Conservation

Introduction

Geomatics (derived from the French word Géomatique, coined by the French photogrammetrist Dubuisson) (Dubuisson, 1975) is an extensive scientific and technological field increasingly used for numerous environmental applications (Campbell, 2018). There are several mostly application-related definitions (Campbell, 2022). Gomarasca (2010, p. 137) writes that the term geomatics was designed at Laval University in Canada during the early 1980s, based on the increasing applications of computing software to the geographical sciences. According to Gomarasca (2010, p. 137) geomatics was "erected on the scientific framework of geodesy"...using "terrestrial, marine, airborne and satellite-based sensors to acquire spatial and other data" and became "a systemic, multidisciplinary, integrated approach to selecting the instruments and the appropriate techniques for collecting, storing, integrating, modeling, analyzing, retrieving at will, transforming, displaying and distributing spatially georeferenced data from different sources with well defined accuracy characteristics and continuity in a digital format" (Gomarasca, 2010, p. 137). Ehlers (2008, p. 20) adds that the terms geomatics and geoinformatics are largely synonymous, with the former mostly being a North American term and the latter used primarily in Europe. Natural Resources Canada (2018) defines geomatics as "the modern discipline which integrates the tasks of gathering, storing, processing, modeling, analyzing and delivering spatially referenced or location information... surveying, hydrography, mapping, remote sensing (often called earth observation) and geographic information processing, often called geographic information systems (or GIS)."

The breadth and depth of this subject are frequently highlighted, based on the utility of its branches, including geographical information systems, remote sensing, cartography and related statistical analyses in applications to environmental and spatial aspects of the social and natural science topics. For example, Duffy (2020) referring to geomatics and geomatic engineering notes that these terms comprise "an all-inclusive description of the science that comprises photogrammetry, GIS, surveying, remote sensing, satellite navigation, geodesy, GPS/GNSS, LiDAR, and mapping and cartography. The study of geomatics also incorporates mathematics, computer science, and engineering." The Journal of Remote Sensing and GIS (2023) gives a broad but technologically based definition, noting that geomatics, which is also termed "geospatial technology or geomatics engineering, or geomatic engineering" is..."the discipline of gathering, storing, processing, and delivering geographic information, or spatially referenced information. It consists of products, services and tools involved in the collection, integration and management of geographic data." Notably, the recent development "has been made possible by advances in computer hardware, computer science, and software engineering, as well as by airborne and space observation remote-sensing technologies" (Journal of Remote Sensing and GIS, 2023).

Other definitions have a greater focus on the tools of geomatics-based techniques and the multiple supports from various techniques (Geomatica, 2002). Abdalla (2021) defines "as a systematic, multidisciplinary, integrated science that is dealing with systems and technologies for collecting, storing, integrating, modeling, analyzing, retrieving, transforming, displaying and distributing spatially georeferenced data from different measurements with well-defined precision and accuracy characteristics and continuity and in a digital format." The British Columbia Institute of Technology (2018) defines geomatics as: "the art and science of measuring, storing, managing and presenting of geospatial data" and notes that there are important future possibilities "where rapidly changing technology is applied directly to the sustainable management of the world's resources". In addition, "geomatics consists of a wide range of topics, including GPS technology, field surveying, cartography, geodesy, land use planning, photogrammetry, hydrography and cadastral studies" (British Columbia Institute of Technology 2018).

Geomatics techniques are relevant to conservation biogeography because these methods can identify, record, analyze and present data and information that can improve understanding, facilitate improvement actions and support relevant policies (Campbell & Jha, 2021; Han et al., 2014). As the human-environment relationship becomes increasingly dynamic, geomatic techniques are being applied to increasingly complex areas at the interface between environmental and social sciences, encompassing simple descriptions, complex analysis and even predictive modeling. Advances can include technical, biological and social methods based on the social and environmental sciences. The remainder of this chapter examines the applications of geomatics-based methods to conservation biogeography. Some selectivity is required because geomatic methods can be applied to the spatial aspects of all natural and social sciences. Case studies are selected from those topics that have a focus on the spatial distribution of organisms (biogeography), the human dimensions of such change (cultural biogeography), and the parts of relevant social sciences that directly impact biogeographic changes (conservation studies and environmental studies). The two main branches of geomatics, remote sensing and geographical information systems, are examined one after the other.

Types of Geomatics

Remote Sensing

Remote sensing refers to recorded visual data, from primitive photography to more advanced space-based technology, currently supported by computerization. For conservation biogeography, relevant remote sensing applications include more detailed identification of life forms (smaller species, at greater distances, over larger areas), recording of animal behavior (for example, movements, breeding, mating and feeding patterns), land cover change (including vegetation, temperature, moisture and species change) and the analysis of such variables for ecosystem assessment. Horning et al. (2015) point out that "remote sensing, using satellite and aerial imaging to measure and map the environment, increasingly provides a vital tool for effective collection of the information needed to research and set policy for conservation priorities." In addition, "remote sensing data, and the tools that ecologists and conservation biologists use to analyse them, have become more accessible and plentiful than ever" (Pettorelli et al., 2024). For a definition, remote sensing is "the acquiring of data about an object without touching it" (Jensen, 2007, p. 4). A longer definition is that "remote sensing is the noncontact recording of information from the ultraviolet, visible, infrared and microwave regions of the electromagnetic spectrum by means of instruments such as cameras, scanners, lasers, linear arrays, and/or area arrays located on platforms such as aircraft or spacecraft, and the analysis of acquired information by means of visual an digital image processing" (Jensen, 2007, p. 4). Another definition is that of Natural Resources Canada (2015): "Remote sensing is the science (and to some extent, art) of acquiring information about the Earth's surface without actually being in contact with it. This is done by sensing and recording reflected or emitted energy and processing, analyzing, and applying that information." Technological devices include "cameras, optical-mechanical scanners, linear and area arrays, lasers, radar systems, sonar, seismographs, gravimeters, magnetometers, and scintillation counters" (Jensen, 2007, p. 4).

Important developments for remote sensing include increasingly refined techniques in aerial photography and satellite imagery (Kwok, 2018). Aerial photography developed mostly in North America after the Second World War, using modified combat aircraft, such as the North American Aviation P-51 Mustang (a fighter or fighter-bomber), the Lockheed P-38 Lightning (a fighter), the Douglas B-66 Destroyer (a light bomber) and the McDonnell Douglas F-4 Phantom II (a supersonic jet interceptor and fighter-bomber) (Campbell 2018). Airplanes designed for photographic espionage and reconnaissance included the Lockheed U-2, the Lockheed SR-71 "Blackbird," North American A-5 Vigilante and the Grumman OV-1 Mohawk, which was equipped with radar imaging capability, the Side-Looking Airborne Radar (SLAR) capable of penetrating foliage and mapping terrain features and moving targets (Campbell 2018; The Military Factory 2025). Satellite development occurred later, creating images which were more powerful than aerial photography, being taken from higher altitudes with larger coverage (Morgan & Falkner, 2001; Campbell 2018). The utility of satellite images depends on the coverage and the coarseness of the resolution, based on the ground area covered by one pixel (picture element) in the image (Jensen, 2007). The wavelengths of the electromagnetic spectrum were used for the creation of images (Table 9.1).

An important development in satellite imagery was the Landsat program, which began on July 23, 1972, with the launch of the Earth Resources Technology Satellite (later Landsat), the beginning of a series up to Landsat 8, launched on February 11, 2013 (National Aeronautics and Space Administration (NASA), 2018). Other satellite-based imaging systems are shown in the tables (Ramsay et al., 1998; Flett

Table 9.1 Wavelengths on the electromagnetic spectrum in meters

Name	Wavelength (m)
Radio	$>1 \times 10^{-7}$
Microwave	1×10^{-7} to 1×10^{-1}
Infrared	7×10^{-7} to 1×10^{-3}
Optical	4×10^{-7} to 7×10^{-7}
Ultraviolet UV	1×10^{-8} to 4×10^{-7}
X-ray	1×10^{-11} to 1×10^{-8}
Gamma-ray	$< 1 \times 10^{-11}$

Source: Adapted from National Aeronautics and Space Administration (NASA) (2018)

et al., 2001; Lindsey, 2005; National Aeronautics and Space Administration (NASA), 2018) (Tables 9.2 and 9.3). Some of these satellites produced multispectral images, with bands that represent a wider range of the electromagnetic spectrum, from the ultraviolet, to the visible and infrared bands (Figs. 9.1, 9.2, 9.3, 9.4, 9.5, and 9.6). Synthetic aperture radar (SAR) was more advanced, because it transmits microwave energy to the observed surface and collects the reflected energy, rather than the reflected solar energy from the earth's surface. This removes the effect of day or night, or atmospheric interference such as cloud cover, smoke, and haze in the image (Ramsay et al., 1998; Flett et al., 2001; Livingston, 2015, p. 16).

The United States Geological Service (2001) defined infrared imagery: "Color-infrared (CIR) aerial photography—often called "false color" photography because it renders the scene in colors not normally seen by the human eye—is widely used for interpretation of natural resources. Atmospheric haze does not interfere with the acquisition of the image." It is further added that "mages obtained by satellites and

Table 9.2 Recent satellite launched by the United States

Name	Nation	Active dates	Spatial resolution (m)
EarlyBird-1	USA	Dec. 24, 1997–Dec. 28, 1997	3 m PAN, 15 m MS
Ikonos	USA	Sept. 24, 1999–mar. 31, 2015	82 cm PAN 3.2 m MS
QuickBird	USA	Oct. 18, 2001–Dec.17, 2014	60 cm PAN 2.4 m MS
OrbView-3	USA	Jun. 2003–march 13, 2011	1 m PAN 4 m MS
WorldView-1	USA	Sept. 18, 2007–2018	0.5 m PAN
GeoEye-1	USA	Sep. 6, 2008–2018	0.41–0.46 m PAN, 1.65 m MS
WorldView-2	USA	Oct. 8, 2009–2018	46 cm PAN 1.84 m MS
WorldView-3	USA	Aug. 13, 2014–2018	31 cm PAN, 1.24 m MS, 3.7 m SWIR, 30 m CAVIS

PAN refers to panchromatic, MS refers to multispectral and SAR to synthetic aperture radar. CAVIS refers to Clouds, Aerosols, Vapors, Ice and Snow. SW refers to short wave, IR to infrared, NIR to near infrared and SWIR to shortwave infrared imagery. Sources: Livingston (2015, pp. 15–16), National Aeronautics and Space Administration (NASA) (2018), Mann (2019)

Table 9.3 Recent satellites launched by nations other than the United States

Name	Nation	Dates (launch to retirement)	Spatial resolution (m)
RADARSAT-1	Canada	Nov. 4, 1995–Mar. 28, 2013	10–100 SAR
IRS-1C	India	Dec. 28, 1995–Sept. 21, 2007	6 m PAN
EROS-A	Israel	Dec.5, 2000–May 2016	1.8 meter
Spot 5	France	May 4, 2002–Mar. 31, 2015	5 m PAN 10 m MS
Carosat-1	India	May 5, 2005–2018	2.5 m PAN
EROS-B	Israel	Apr. 25, 2006–2018	70 cm
Cartosat-2	India	Jan. 10, 2007–2018	<1 m
EROS-B	Israel	Apr. 25, 2006–2018	70 cm
Cartosat-2	India	Jan. 10, 2007–2018	<1 m
Cosmo SkyMed-1	Italy	Jun. 8, 2007–2018	SAR
TerraSar-X	Germany	Jun. 15, 2007	Variable, to 1 m SAR
SAR-Lupe 1–5	Germany	Dec. 19, 2006–Jul. 22, 2008+	<1 m, SAR
Cosmo SkyMed-2	Italy	Dec. 9, 2007–2018	SAR
RADARSAT-2	Canada	Dec. 14, 2007–2018	3–100, <1 m SAR
Cartosat-2A	India	April 28, 2008	<1 m PAN
Cosmo SkyMed-3	Italy	Oct. 25, 2008–2018	SAR
TanDEM-X	Germany	Jun. 21, 2010–2018	2—4 m
Cartosat-2B	India	Jul. 12, 2010–2018	<1 m PAN
Cosmo SkyMed-4	Italy	Nov. 6, 2010–2018	SAR
NigComSat-1R	Nigeria	Dec. 19, 2011–2018	2.5 m PAN 5 m MS
Spot 6	France	Sept. 9, 2012–2018	PAN-1.5 m, MS-6 m, BGR, NIR
Spot 7	France	Jun. 30, 2014–2018	PAN 1.5 m MS- 6 m (BGR, NIR)
Pleiades-1A	France	Dec. 16, 2011–2018	PAN 0.5 m, 2 m MS
Pleiades-1B	France	Dec. 2, 2012–2018	PAN 0.5 m 2-m MS

PAN refers to panchromatic, MS refers to multispectral and SAR to synthetic aperture radar. SW refers to short wave, IR to infrared and SWIR to shortwave infrared imagery. BGR refers to blue, green and red. Note for the Spot satellites, Spot 1–5 were launched 1986–2002, and deactivated 1990–2015. Sources: Livingston (2015, pp. 15–16), National Aeronautics and Space Administration (NASA) (2018), Mann (2019)

high-altitude aircraft give engineers and scientists a tool to study landforms, vegetation health patterns, environmental pollution, and other effects of human activities on the planet's surface" and "satellites and high-altitude aircraft equipped to record scenes of the Earth use both visible and invisible parts of the electromagnetic spectrum. Near-infrared light is invisible to the human eye, but adding it to these images allows scientists to 'see' the surface of the Earth in other than natural colors. The result is "color infrared" photography" (United States Geological Service, 2001).

Fig. 9.1 Landsat 8 satellite in orbit. Source: National Aeronautics and Space Administration (NASA) (n.d.). Public Domain. https://creativecommons.org/licenses/by/2.0/

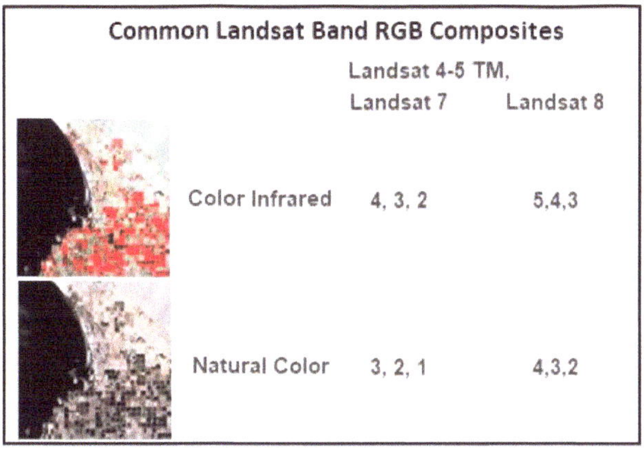

Fig. 9.2 Landsat bands. Source: United States Geological Service (2024b)

Light Detection and Ranging or LiDAR is a recent, highly efficient technology (Natural Resources Canada, 2018), which uses light pulses that generate detailed three-dimensional information on Earth surface features. LiDAR apparatus is usually based on an aircraft platform, and includes a laser, a scanner and a global positioning system (GPS) receiver, and uses ultraviolet, visible, or near infrared light to targets, such as vegetation, earth materials, precipitation and clouds, aerosols and chemical compounds. LIDAR, with high resolution, can generate more accurate models and maps than older remote sensing methods (Campbell, 2018). LIDAR products have been developed for applications in many earth and environmental sciences, including geology, archaeology, geography, seismology, forestry and atmospheric sciences, and these techniques may support related methods for incisive assessments (Campbell, 2018). For example, Kamoske et al. (2022, p. 1440) write that

Fig. 9.3 Landsat infrared image of Mt. Etna, Sicily (the mountain is in the lower center). Red colors signify vegetation. The sea is black. Note urban development and farms in the lower left corner. Source: United States Geological Service (2024b)

LiDAR is "an approach that jointly leverages spectral properties related to leaf and canopy functional traits and forest health" and that "LiDAR derived estimates of forest structure, fine-resolution topographic diversity, and careful consideration of biogeographical differences within and among biomes is needed to accurately map biodiversity variation from above."

Drone-based research is also a comparatively recent development, based on drones (also termed unmanned aerial vehicles or uncrewed aerial vehicles, UAV, or remotely piloted aircraft, RPA) (Figs. 9.7, 9.8, 9.9, and 9.10). More advanced models include global positioning systems (GPS) (Hammill et al., 2017; Johnston et al., 2017; Lucas, 2018). The drones are controlled by a ground-based human operator or by onboard computers. Drones were initially a military development, but recent civilian applications include commercial,, agricultural and ecological applications (Franke, 2015;

Fig. 9.4 Landsat infrared image of the Syrian desert. Note the light blue color signifying the desert, and the small, dark blue water body to the right middle. Sediment may have made it slightly lighter colored than the sea in the previous image (Fig. 9.3). Source: United States Geological Service (2024b)

Kennedy & Rogers, 2015; Federal Aviation Administration (FAA), 2018). Drone sizes range from insect-sized to small airplane-sized machines, and GPS drones (linked to satellites through GPS for navigation) and photography drones. Most advanced drones have an autopilot system, including a computer system, a GPS, a compass, a barometric altimeter and sensors, and may carry a video camera and/or photographic technology. Some may have optical sensors (infrared and/or ultraviolet capable), physical sensors (for measuring temperature, humidity, etc.) and sound sensors (Ditmer et al., 2015; Ivošević et al., 2015).

Drones used for biogeographical, ecological and conservation studies may be termed "eco-drones" or "conservation drones" (Ivošević et al., 2015, p. 113). Hodgson et al. (2018, p. 1160) note that "ecologists are increasingly using technologies, such as remotely piloted aircraft… unmanned aerial systems or unmanned aerial vehicles), for wildlife monitoring applications." Johnston et al. (2017, p. 178) add that "unoccupied aircraft systems (UAS) are now frequently used in wildlife research, including studies of marine species." Ivošević et al. (2015, p. 113) note that "a drone

Fig. 9.5 Landsat infrared image of southern coast of the Netherlands. Note the little squares representing cropland (pink), trees (red), black representing clear water, dark blue shades representing sediment-laden water and light blue shades representing soil and bare ground, including concrete and beaches. Source: United States Geological Service (2024b)

can provide a low-cost and low-impact solution to environmental managers working in a variety of ecosystems." The World Wildlife Fund (2020) contends that "drones help conservationists with surveying, monitoring, tracking and anti-poaching efforts. They are becoming increasingly popular among conservation practitioners because of their ability to collect information from hard-to-reach locations, while minimizing disturbances to wildlife." Jewitt (2018) adds that "there is a growing body of literature on the use of drones for monitoring species and their habitats. Drones have been used successfully, among others, to monitor a wide range of species from birds and reptiles to elephants and marine species such as turtles, whales, dolphins and dugongs."

In terms of the applications of remote sensing to conservation biogeography, the main issue concerns the assessment and monitoring of vegetation, which may be used for habitat, deforestation, agricultural, urbanization and wildlife studies (Campbell, 2018; Gong et al., 2024; Matyukira & Mhangara, 2024). In addition, tracking and observation tools such as drones may be used to record wildlife presence. Petso

Fig. 9.6 Landsat infrared image of Foxe Basin, Nunavut, Canada. Note the sediment-laden water, with possible ice. Source: United States Geological Service (2024b)

Fig. 9.7 Mavic Pro drone over the fog around Mt. Tamalpais, California, USA. Source: Ian Usher-Unsplash

Fig. 9.8 Mavic Pro drone over YuLong mountain, China. Source: Iewek Gnos-Unsplash

Fig. 9.9 Phantom 3 advanced drone in a forested area. Source: Jason Mavrommatis-Unsplash

and Jamisola (2023) point out that "the use of drones and artificial intelligence may offer more reliable methods of counting populations and monitoring wildlife." This is particularly true for sophisticated devices such as thermal drones: "Drones equipped with thermal sensors have shown ability to overcome some of the limitations often associated with traditional human-occupied aerial surveys (e.g., low detection, high operational cost, human safety risk)"... and "the method provides the ability to perform accurate and reliable population surveys in a safe and cost-effective manner

Fig. 9.10 GPS. Source: Mike Birdy-Pexels

compared with traditional aerial surveys and is only expected to continue to improve as sensor technology and machine learning analytics continue to advance" (Beaver et al., 2020, p. 434).

McDermid et al. (2010, p. 193) argues that such technologies allow "an impressive suite of wildlife ecology and habitat attributes to be modeled, predicted and monitored through time, including land cover physiognomy, vegetation structure and condition, forage characteristics, specific nutrient concentrations, overall productivity, and biomass." Additionally, "the diversity of remote sensing systems and their varying spectral, spatial, temporal, and radiometric resolutions allows for an impressive suite of wildlife ecology and habitat attributes to be modeled, predicted, and monitored through time" (McDermid et al., 2010, p. 202). For wildlife studies, Neumann et al. (2015, p. 8) note that "animal movement patterns in space and time are a central aspect of animal ecology... Remotely sensed environmental indices can play a key role in understanding movement patterns by providing contiguous, relatively fine-scale data that link animal movements to their environment," and "animal movement studies incorporate remotely sensed products that provide time series of environmental response variables." Environmental variables relevant to animal presence may be represented by such technology at varying levels of detail (Table 9.4). This can support conservation mangement efforts, and facilitate predictive modeling of movements (Neumann et al., 2015).

Table 9.4 Color associations of infrared imagery based on landscape features

Landcover	Description	Coloursrange		
Vegetation	Red hues are almost always associated with living vegetation. Living, dense vegetation is indicated by extremely intense red tones. Lighter pink, red, green tones, and brown colors show lower plant vigor.			
Bare earth	Shades of white, blue, or green are associated with bare soils. Each color's darker tones can suggest wetter soil.			
Built features	Constructed features are colored according to the compositional materials. Asphalt roads are dark blue or black; gravel or dirt roads are lighter colors and clean concrete roads are lighter colored.			
Water	Blue shades are associated with water from almost black (clean water) to paler shades (based on sediment). Colors of very shallow water may be determined by the bottom sediment materials.			

Source: United States Geological Service (2001)

Geographical Information Systems/Science

Geographical information science is the discipline that uses geographical information systems (GIS) to create computer-generated maps, images, analyses and scientific supports (Campbell, 2018). The United States Geological Service (2024a) defines GIS as "computer system that analyzes and displays geographically referenced information. It uses data that is attached to a unique location." Bechtel et al. (2015, p. 145) writes that this type of system "serves for capturing, storing, analysis, and visualization of data that describe a part of the Earth's surface, the technical and administrative entities, as well as findings of geoscience, economics, and ecological applications" (Bechtel et al., 2015, p. 148). Other definitions from the formative period of GIS may describe its components: "it is an information system with a database of observables of spatially distributed objects, activities, or events, which can be described by points, lines, or surfaces" (Dueker, 1979). Other such definitions look at its capabilities: "It is a comprehensive collection of tools for capturing, storing, retrieval, transformation, and visualization of spatial data of the real world for special applications" (Burrough, 1979). Applications are also important: "it is an information system containing all spatial data of the atmosphere, the Earth's surface, and the lithosphere, allowing the systematic capture, update, manipulation, and analysis of such data, based on a standardized reference frame" (Göpfert, 1991); it also integrates spatial data from several sources, with the results used to reinforce decison-making and problem solving (Campbell 2028; Cowen, 1988).

A key contribution of GIS to conservation requirements is its integrative ability to connect social and environmental research, as GIS is heavily used in social research (Campbell, 2018). For example, Ballas et al. (2017) demonstrate the utility of GIS in social applications including urban and transportation planning, emergency planning, retail locations and networks, education, health, crime, geodemographics and poverty/income inequality. These social parameters are related to conservation management through financial support, land use competition and the requirement for a balanced approach to social and conservation issues. Another important contribution of GIS is its ability to cover both socially derived and natural features and land cover (Campbell, 2022; Campbell & Jha, 2021). For example, Subak (2000) argues that biodiversity conservation requires support through methods that integrate the social activities and imprints of scaled actors (individuals, communities, NGOs, business groups and governments) with the goals and methods of environmental policy and management, combined with inventories of environmental variables.

Such diverse, integrated studies can be exemplified by the unprotected zones surrounding protected areas (see, e.g., the argument of Burroughs & Clark, 1995, in a study of Yellowstone and Georges Bank National Parks), and human activities must be examined as an essential part of this larger socio-ecological system (Salwasser, 1995; Quon et al., 2001; Campbell 2017, 2018). Such assessments therefore require the integration of the interpretive and analytical methods of the social sciences with the usually more rigorous methods of the natural sciences into GIS models. For example, Haberl et al. (2001) describe comparative studies that require interfacial studies of landscape ecology, economic issues and social history. In the narrower area of protected area management, new difficulties arise which, in addition to the interface problems of the natural and social sciences, primarily concern short-term and small-scale impact assessments and the flexibility of methods (Campbell, 2022).

The accuracy of analysis of spatial data is another important contribution of GIS to conservation science, particularly the spatial parameters of organisms' behavior and locations, as well as temporal and spatial variations in land cover, topography and land use (Campbell, 2018). Changes in landscape variation can be based on measurements of vegetation parameters such as canopy height, position, perimeter canopy dynamics, density and open spots, which can be assessed in field studies or image-based analysis using the parameters of tone, shape, texture, size and pattern of features over time to reveal the phytogeographic changes essential to the needs of conservation biogeography. The size, shape, density and pattern of urban features also provide information about the rate and patterns of urbanization, which can determine impacts on ecosystems (Jensen, 2007). By assessing combined cultural and natural land use patterns, animal habitats can be determined and differences between and within species can be assessed (Campbell, 2018).

For such land cover assessment, remote sensing images can be the starting point for GIS processes that help determine coordinates and enable location network-based studies (Campbell, 2018). Even when using image-based data, numerical data collected in the field (point, density, line, pattern and elevation data) is entered into GIS software as tabular data. Image-based data have limitations due to accuracy issues, even with high-resolution images when coverage includes areas of high species and population densities (Campbell, 2022). This suggests that image interpretation may be subjective and requires critical assessment (Campbell 2018). GIS methods can therefore, at least in part, take advantage of the most accurate images developed in the twenty-first century with high-resolution imagery and light detection and ranging (LIDAR), radio detection and ranging (RADAR), or interferometric synthetic aperture radar (IFSAR), which can create more sophisticated products such as digital elevation models, LIDAR-based images and other advanced outputs (Jensen, 2007).

Geomatics Applications to Conservation Biogeography

The application of remote sensing and GIS to conservation biogeography is mainly concerned with the assessment and monitoring of deforestation, reforestation and changes in human footprints such as agricultural and urban development, thus enabling an effective quantification of conservation outcomes, based on visible and measureable parameters (Campbell, 2018, 2022). Such applications are principally based on image analysis (using aerial photographs, satellite images, LiDAR), tracking (using global positioning systems GPS, drones) and field data analyzed statistically for map and image creation and analyses (using GIS software) (Campbell, 2022, Campbell & Jha, 2021; Campbell, 2022). This section presents case studies of such applications using remote sensing and GIS techniques, which are illustrative of applications relevant to conservation biogeography. Unlike such applications to biogeography, applications to conservation biogeography must focus on conservation themes, outcomes and imprints, which may increase the complexity of the studied features and processes. Human imprints and processes on the landscape may be extremely dynamic, with variable impacts and possibilities, requiring the inclusion of social science methods into technology-based approaches (Campbell, 2018). The investigation of the roles of remote sensing for conservation biogeography, and applications to the supporting, environmentally focused aspects of the social sciences must be acknowledged, as the environmental subdisciplines of economics, political science, sociology, psychology, anthropology and human geography all make important contributions to conservation biogeography (Campbell, 2018; Campbell & Jha, 2021).

Remote Sensing and Conservation Biogeography

Remote sensing applications to conservation biogeography create images and analyses of the factors and parameters of conservation, applied to spatial features of the biosphere (Campbell, 2018). Cavender-Bares et al. (2022) point out that "remote sensing has transformed the monitoring of life on Earth by revealing spatial and temporal dimensions of biological diversity through structural, compositional and functional measurements of ecosystems." This engagement is comparatively recent, as noted by Pettorelli et al., 2014: "Assessing and predicting ecosystem responses to global environmental change and its impacts on human well-being are high priority targets for the scientific community. The potential for synergies between remote sensing science and ecology, especially satellite remote sensing and conservation biology, has been highlighted by many in the past. Yet, the two research communities have only recently begun to coordinate their agendas." However, "such synchronization is the key to improving the potential for satellite data effectively to support future environmental management decision-making processes" and enhancing conservation management" (Pettorelli et al. (2014).

Rose et al. (2015, p. 350) pose ten questions crafted by "a group of remote sensing scientists affiliated with government and academic institutions and conservation organizations," which are argued to be the key topics needed to "increase conservation effectiveness through the use of Earth observation technologies." The objective was to "increase conservation practitioners' use of remote sensing to support their work, increase collaboration between the conservation science and remote sensing communities, identify and develop new and innovative uses of remote sensing for advancing conservation science, provide guidance to space agencies on how future satellite missions can support conservation science and generate support from the public and private sector in the use of remote sensing data to address the 10 conservation questions" (Rose et al., 2015, p. 350). The questions concerned how remotely sensed data may be used for modeling, analyzing, monitoring, predicting and understanding of species distributions and abundances, animal movements, ecosystem response and resilience to multiple stressor. In addition, topics included the impacts of climate on ecosystems, applications to threat reduction, governance and regulation compliance, and resource management decisions, configuration of protected area networks at relevant spatial extents, changes in ecosystem services, effectiveness of conservation efforts, impacts of agriculture and aquaculture on ecosystems, and the disturbance and degradation of changes on species and ecosystem functions (Rose et al., 2015, p. 350).

Remote sensing techniques may also be assessed for their role in the social science research that contributes to conservation biogeography (Campbell, 2018). For example, remote sensing is an important research tool for economics, especially considering socioeconomic and imprints on the Earth's surface (Zhao et al., 2023). Keola et al. (2015) describe the use of night time lights, to distinguish between areas of high human activity and income growth and those with less human activity, the latter including areas of agriculture and forestry. Keola et al. (2015) note that "nighttime lights alone may not explain value-added by agriculture and forestry. By adding land cover data, our framework can be used to estimate economic growth in administrative areas of virtually any size." The addition of other economic data allows assessments of land cover and economic activities, which may be relevant to land use and may have implications for conservation and other competing land uses.

Howitt et al. (2022, p. 183) "remote sensing technologies (RST) currently in use and on the cusp of development will vastly increase data availability" with important applications for economic development, as "agricultural and resource economics applications over the last half-century have taken advantage of and promoted the development of methods of dynamic analysis and optimal control." It is argued that "this technology offers more precise information-gathering with enhanced temporal and spatial resolution, and spectral capability that maps underground resources. These new technologies enable society to increase agricultural productivity while protecting natural resources for future use. These technologies will also enable economists to learn about behavioral responses to changing policy and environmental conditions" (Howitt et al., 2022, p. 183).

Donaldson and Storeygar (2016, p. 171) note that "the past decade or so has seen a dramatic change in the way that economists can learn by watching our planet from above. A revolution has taken place in remote sensing and allied fields such as computer science, engineering and geography. Petabytes of satellite imagery have become publicly accessible at increasing resolution, many algorithms for extracting meaningful social science information from these images are now routine, and modern cloud-based processing power allows these algorithms to be run at global scale." It is further noted that the main advantages of such remote sensing data to economists may be divided "into three categories: (1) access to information difficult to obtain by other means; (2) unusually high spatial resolution; and (3) wide geographic coverage" (Donaldson & Storeygar, 2016, p. 171).

Issues of interest to economists include "economic analysis that already uses remotely sensed data on night lights, precipitation, wind speed, flooding, topography, forest cover, crop choice, agricultural productivity, urban development, building type, roads, pollution, beach quality, and fish abundance." Other important factors are for example the the studies within the field of economics that examine "the impact of short-run weather fluctuations and longer-term climate trends on human activity" (Donaldson & Storeygar, 2016, p. 184), the application of digital elevation modeling to the role of topography in urban and large infrastructure development, and predictive economic modeling for agriculture, especially agronomic models, such as GAEZ, which "enable such predictions by using characteristics of the location (including its topography, from a remotely sourced digital elevation model) and crop as modeling inputs. The result is a prediction about the yield that could potentially be achieved for any crop at any location on Earth" (Donaldson & Storeygar, 2016, p. 187).

For urban development, spatial change can be monitored for expansion, with contrasts between "urban (residential, commercial, and industrial development and transportation networks) and nonurban (water, bare rock and sand, forest, range and grassland, agricultural land and wetlands) land cover categories over time" (Donaldson & Storeygar, 2016, p. 187). Forestry is another economic activity monitored with remote sensing and modeling (Campbell, 2018). Donaldson and Storeygar (2016, p. 189) note that remote sensing for measuring large-scale deforestation has been used since the early 1990s (citing Skole & Tucker, 1993), the first major application to an economic model being the work by Foster and Rosenzweig (2003). This study used 30 years of Landsat data to document changes in forest cover in India. The satellite data and village-level survey data found that the main factor for forest change was the increase in the demand for forest products due to income growth, rather than the supply-side factors (see, e.g., land and labor prices). Other forest-related activities monitored with satellite images include illegal logging, and conservation attempts, with cases from Brazil's 2006 anti-deforestation policy (Burgess et al., 2016) and farmer-tree interfaces in Uganda (Jayachandran et al., 2016).

He et al. (2014, p. 2886) give the example of the extensive, rapid urbanization of China, which "has resulted in a massive loss of natural habitat, which is threatening the nation's biodiversity and socioeconomic sustainability." The role of remotely sensed data lies in the fact that "the impact of urban expansion on natural habitats is not well-understood, primarily due to the lack of accurate spatial information regarding urban expansion across China" (He et al., 2014, p. 2886). The chosen technique was the "integration of data on nighttime light levels, a vegetation index, and land surface temperature" for the study of (He et al., 2014, p. 2886). The natural habitat loss was evaluated during study period 1992–2012. The findings indicated rapid urban growth, an average annual growth rate of 8.74% (above the global average of 3.20%), significant natural habitat loss in some areas, with extreme losses in the Pearl River Delta (25.79% or 1518 km^2 loss of natural habitat and 41.99% or 760 km^2 loss of the local wetlands).

Remote sensing techniques may be applied to the spatial assessments of economic forestry and impacts of economic activities at more detailed levels (Campbell, 2018). For example, Braun and Koch (2016) employ a geomatics-based assessment to measure issues with plantation forestry in southern Chile, using biodiversity as the main parameter. The authors argue that while using remote sensing techniques can create estimates for landscapes, to relate land use intensity to biodiversity conditions, to derive information that is required to ensure the development of biodiversity conservation plans, a new approach is needed. Hence, "instead of inferring biodiversity using environmental variables or the variability of spectral values, a hypothesis-based approach is applied... The proposed approach yields a coefficient of correlation of 0.73 and produces a convincing estimate of regional biodiversity."

Remote sensing plays an important, albeit recent role in anthropology, largely through studies of archaeology, and possible roles of ancient human cultures on historic biogeographical features and change (Campbell, 2018). The role of remote sensing methods lies in the capability of this technology in the detection of remnants of human imprints on the Earths' surface, and the possible utility of these findings in the cultural biogeography and consequently the conservation biogeography of the studied region (Campbell, 2022; Campbell & Jha, 2021). Landscape archaeology is one of the principal areas of remote sensing, which is defined as "the understanding of archaeological remains (artifacts, sites and site complexes) in terms of the wider spatial realms (both physical and meaningful) of past human experience" and interacts with term landscape, within which "the environmental aspects include the combined effects of climate, hydrology, landforms, vegetation, and fauna" and the "human imprint reflects the cumulative effects of cultural processes that have occurred through time, including: the histories of vegetation disturbance, clearance, and planting; earthmoving; settlement; pollution; and so on" (Denham, 2017, p. 464).

Parcak (2017) describes the role of ground-based remote sensing, photogrammetry, satellite remote sensing and LiDAR in landscape archaeology, noting that the ground-based remote sensing enables the derivation of otherwise difficult to detect data: "Unlike other sensors, ground-penetrating radar (GPR) offers a way for archaeologists to visualize subsurface data in three dimensions" and "features previously obscured by infilling or saturated clay-rich soils can be revealed by GPR, as can features obscured by reconstruction or tourism development." However, it is argued that "different landscape types can greatly affect GPR results because there can be many obstacles in the form of modern vegetation or topography" (Parcak

(2017). Leucci et al. (2016, p. 290) point out that "extensive urbanization may cover important archaeological structures that are still buried in urban areas", and "Ground Penetrating Radar is one of the most reliable geophysical techniques for resolving buried archaeological structures in urban territories."

Other studies have used LiDAR to derive data explaining past human activities, which may be environmentally relevant. Agapiou and Lysandrou (2015, p. 192) write on the fact that "time-stamped maps from 1999 until 2015 indicate on one hand that remote sensing technologies are widely accepted and applied by the archaeological community" and there is an existing gap between the different scientific fields of remote sensing archaeology." Li (2023) in a study of the Greater Khingan Range, in northeastern China, documents the use of LiDAR for the assessment of archaeological traces of nomadic people such as the Mongols, which penetrates the thick forest that previously impeded archaeological surveys and mapping. Despite the lack of such studies in China, especially of the Greater Khingan Range, the study examines an "application of UAV LiDAR and photogrammetry to detect and map the ancient sites hidden beneath the forest canopy in Sheshan Mountain, located southeast of the Greater Khingan Range" (Li, 2023). These LiDAR-laden UAVs showed with higher clarity, the archaeological structures under forest in the earth materials, giving "a holistic picture of the archaeological sites in the area, allowing defining a probable ancient ritual or worship landscape that may date back to the Liao period (907–1125 AD)" (Li, 2023). This study "demonstrated for the first time the immense promise of UAV LiDAR for archaeological surveys and landscape archaeology in the area, highlighting new opportunities for archaeological research in the Greater Khingan Range" (Li, 2023).

Remote sensing is also relevant to socioeconomics, psychology and mental health, with relevance to conservation biogeography. One example is the research on green spaces in urban areas, which is an important aspect of conservation biogeography, and may be related to social issues of access, social status, health and conservation policy. However, Honold et al. (2015) argue that "despite promising experimental findings, few studies have addressed the potential long-term health benefits of frequent contact with different kinds of urban nature" and the case study examined "the cross-sectional relations between two kinds of urban nature (neighborhood vegetation visible from the home, use of public green spaces) and health outcomes (life satisfaction, perceived general health, 2-months hair cortisol levels) in a sample population from Berlin." The findings indicated that residents living in homes with views of vegetation had significantly lower cortisol levels, and regular users of vegetated trails along canals had significantly lower cortisol levels and significantly higher life satisfaction than less frequent users. Such vegetated routes were important for activities of the residents (Honold et al. 2015).

Human exercise in nature may confer similar psychological benefits. For example, Barton and Pretty (2010, p. 3947) argue that "green exercise" which is human exertion in natural surroundings, may confer both short- and long-term positive health outcomes, based on the analysis of ten UK-based studies. The findings indicated that "dose responses for both intensity and duration showed large benefits from short engagements in green exercise, and then diminishing but still positive returns.

Every green environment improved both self-esteem and mood; the presence of water generated greater effects" (Barton & Pretty, 2010, p. 3947). Gender neutrality was indicated, as both men and women showed improvements in self-esteem after such exercise, but men showed greater differences in mood. For age, youth was positively correlated with increased self-esteem, and for mood, the least change was in the young and old. People with mental illness had strong self-esteem improvements (Barton & Pretty, 2010, p. 3947).

Some studies dispute these findings. For example, Lahart et al. (2019) cite a study which examined the "evidence for the proposed additive effect of exercise in the presence of nature (green exercise) by systematically reviewing studies that investigated the effects of outdoor or virtual green exercise compared with indoor exercise." The findings indicated that "compared with indoor exercise, acute bouts of outdoor green exercise may favorably influence affective valence and enjoyment, but not emotion, perceived exertion, exercise intensity, and biological markers. No other consistent statistical differences were observed, apart from a higher enjoyment of outdoor green versus virtual green exercise"; hence "there was limited evidence to support the view that green exercise offers superior benefits to exercise without exposure to nature. The low quality of evidence prohibits clear interpretation of trial findings. Future robust and rigorously designed trials are needed to evaluate the effects of long-term and multiple-bout exposure to nature during exercise compared with exercise indoors" (Lahart et al., 2019).

Other research, more broadly, examines the overall possibility for environmental, social benefits. For example, Weigand et al. (2019) examine the issues of the physical environment, people's physical condition, and inequalities based on residential place and space distribution, social segregation and social status/inequalities of exposure. Remote sensing is vital to these investigations, as "recent technological and methodological developments in satellite remote sensing have proven to provide highly detailed information on environmental conditions" (Weigand et al., 2019). It is further argued that this issue, which falls under both environmental justice and health, is a remote sensing opportunity, as "especially due to their high level of spatial detail and the provided large-area coverage, they allow for spatially continuous description of environmental characteristics, and "as a future perspective, ongoing earth observation missions, as well as processing architectures, ensure data availability and applicability of 'big earth data' for future environmental justice analyses" (Weigand et al., 2019).

Weigand et al. (2019) mention the use of geomatics for several environmental variables relevant to human health and economic status. Among the most important are applications to urban green spaces, air pollutants, noise, and heat (Campbell, 2022; Campbell & Jha, 2021; Solimini, 2016). Urban green spaces (urban parks, greenbelts, green hearts, treed avenues, even gardens) are very important for human health, especially mental and physical health issues, including stress relief, reduction of the risks of cardiovascular and respiratory diseases, and mortality reduction, especially in socioeconomically deprived areas (Campbell, 2018, 2020, see also Maas et al., 2006; Hansmann et al., 2007; Mitchell & Popham, 2008; van Eeden et al., 2018; Thompson et al., 2010; Villeneuve et al., 2012; Richardson et al., 2013). Urban green spaces may reduce surface temperatures and heat stress in surrounding areas, and create incentives for human exercise, although poorer people have sometimes have less access to green spaces (Du et al., 2017; Heynen et al., 2006; Kabisch & Haase, 2014; Milligan & Bingley, 2007; Mitchell & Popham, 2008; Pham et al., 2016; Wüstemann et al., 2017). Remote sensing of urban green spaces is mostly based on the use of multispectral data with the visible, infrared and near infrared parts of the electromagnetic spectrum. This data records the area of vegetation cover, and properties such as primary production and yields (Solimini, 2016; Campbell, 2018, 2022). The varying heterogeneity of landscapes, which is higher in complex urban environments than in more homogeneous rural land cover, usually requires a high spatial resolution for the satellite images. (e.g., Liu et al., 2016; Van de Voorde, 2016; Li et al., 2014; Bertram & Rehdanz, 2015; Santos et al., 2016; Banzhaf & de la Barrera, 2017; Du et al., 2017; Ekkel & de Vries, 2017).

Remote sensing tools may also be applied to urban air pollution studies. Weigand et al. (2019) note that "after adjusting for individual behavior, air pollutants like ozone, fine particulate matter, nitrogen oxide, blacksmoke, and sulfur have been shown to increase mortality rates in cities" (see also Dockery et al., 1993; Hoek et al., 2002), and additionally "environmental justice research studies which focused on air pollutants identified stronger exposures of low-income or racialized neighborhoods in Canada, Great Britain, and Germany" (Flacke et al., 2016; Kershaw et al., 2013; Mitchell & Dorling, 2003). Studies of these air pollution phenomena include particulate matter, sulfur, ozone, NOx, carbon monoxide, ammonia, which may be documented by ground-based sensors, which may be limited in effectiveness (Martin, 2008). Remote sensing tools may also be used for broader coverage (e.g., Sentinel-5(P) with higher resolution), but the resolution may be coarse (Brauer et al., 2016; Duncan et al., 2014; Gupta et al., 2006; van Donkelaar et al., 2015). More recent studies have attempted a high-resolution approach (Briggs et al., 1997; Guo et al., 2014; Hu, 2009; Li et al., 2017; Prud'homme et al., 2013; van Donkelaar et al., 2010; Voigtländer et al., 2011). The remote sensing data may also be used to document the distribution of "health burdens" (Weigand et al., 2019). Land use regression models (LUR), or "Regression Mapping," based on satellite imagery and ground-based measurements, may analyze such environmental health burdens (Bertazzon et al., 2015; Briggs et al., 1997; Habermann et al., 2015; Hoek et al., 2008; Lee et al., 2017; Passchier-Vermeer & Passchier, 2000; Pearce et al., 2006; Ryan & LeMasters, 2007; Vardoulakis et al., 2003; Wang et al., 2015).

Noise pollution is another important issue, as urban noise has been found to significantly affect urban and suburban wildlife, as for example Sordello et al. (2020) note that "ecological research now deals increasingly with the effects of noise pollution on biodiversity. Indeed, many studies have shown the impacts of anthropogenic noise and concluded that it is potentially a threat to the persistence of many species." The findings of this research were that "the impacts of anthropogenic noises on species and ecosystems have been researched for many years. In particular, some taxonomic groups (mammals, birds, fishes), types of noise (transportation, industrial, abstract) and outcomes (behavioural, biophysiological, communication) have been studied more than others. Conversely, less knowledge is available on certain species (amphibians, reptiles, invertebrates), noises (recreational, military, urban) and impacts (space use, reproduction, ecosystems) (Sordello et al., 2020). This study also notes that sound is not a real problem, as most species emit sounds: "The problem arises when sounds turn into "noise," which depends on each species (sensitivity threshold) and on the type of impact generated (e.g., disturbances, avoidance, damage). In this case, we may speak of "noise pollution" (Sordello et al., 2020).

Such noise can affect animals in terms of communication use of space and reproduction, and affects most species, including mammals, birds, amphibians, reptiles, fishes and invertebrates, terrestrial, coast and aquatic/marine ecosystems (Sordello et al., 2020; see also (Sun & Narins, 2005; (Amoser et al., 2004; Bernath-Plaisted & Koper, 2016; Bunkley et al., 2017; Cunnington & Fahrig, 2010; Drolet et al., 2016; Jacobsen et al., 2014; Koper et al., 2016; LaZerte et al., 2016; Mancera et al., 2017; Shannon et al., 2016; Thompson et al., 2010). Human noise pollution may emit from vehicular traffic, industrial developments, and marine and aerial transport (Brown, 1990; Lengagne, 2008; Mason et al., 2016; Sordello et al., 2020; Vasconcelos et al., 2007). Noise impacts on people, can include sleeping issues, long-term memory impairment and psychosocial symptoms, hypertension, and cardiovascular disease, which may be ameliorated with urban green spaces (Bunkley et al., 2017; Fyhri & Klaeboe, 2009; Stansfeld & Matheson, 2003; Von Szombathely et al., 2018). Remote sensing and ground sensor techniques can be used to estimate noise levels, through noise models measuring some parameters such as "traffic speed, vehicle counts, road type, surrounding built-up structures, topography, and land cover, etc. to approximate the local noise levels (De Souza & Giunta, 2011; Garg & Maji, 2014; Seto et al., 2007) and landscape metrics, including building type, orientation and height, green space distribution, geometric features and densities may be used to detect the links between urban structure and noise" (Weigand et al., 2019).

Urban heat may also affect urban wildlife and people (Campbell, 2018). For animals, evidence indicates that "urbanization appears to take a greater toll on wild mammals in hotter, less vegetated locales than in cooler, greener ones" (Anthes, 2023). Haight et al. (2023) describe a study in partnership with the Urban Wildlife Information Network, which used "occurrence data from 725 sites located across 20 North American cities and a multi-city, multi-species occupancy modeling approach," and the findings indicated that "Species occupancy and diversity were most negatively related to urbanization in the warmer, less vegetated cities. Additionally, larger-bodied species were most negatively impacted by urbanization across North America" (Haight et al., 2023, p. 1654). These findings suggest that "shifting climate conditions could worsen the effects of urbanization on native wildlife communities, such that conservation strategies should seek to mitigate the combined effects of a warming and urbanizing world" (Haight et al., 2023, p. 1654). Weigand et al. (2019) note the impact on people, including heat waves, which may especially result in deaths of vulnerable people (Kovats et al., 2006; Rey et al., 2009; Scherer et al., 2014; Urban et al., 2016; Vandentorren et al., 2006). Such heat may be dependent on local variations, in land use patterns, urban substrates and vegetation cover, especially vegetation deficiency, and water surfaces, building materials, surface roughness, wind speed, and the size of the urban area (Bornstein, 1968; Johnson et al., 2009; Oke, 1973; Weng et al., 2003).

Remote sensing techniques may be used for urban heat measurement, as ground temperatures may be measured from space, using thermal infrared sensors from satellite imagery with high spatial resolution, including Landsat TM/ETM+ or Landsat-8/TIRS (Ayanlade & Jegede, 2015; De Faria Peres et al., 2018; Ngie et al., 2014; Pearsall, 2017; Solimini, 2016; Stathopoulou & Cartalis, 2007). Intra-day temperature variations have also been detected by measuring daytime and nocturnal MODIS imagery, showing increased temperatures in urban areas (Meng et al., 2018). Local climate zones are also been identified using satellite imagery (e.g., Bechtel et al., 2015; Danylo et al., 2016; Tuia et al., 2017; Um & Um, 2015). Studies have used high-resolution SPOT (20 m ground resolution) images and high-resolution satellite imagery (Banzhaf et al., 2009; Gamba et al., 2000; Leichtle et al., 2017; Montanges et al., 2015; Wurm et al., 2011).

Drones are another important aspect of remote sensing applications to the social intersection with conservation biogeography (Campbell, 2018). However, limitations and even problems may emerge from their operation (Campbell, 2022). For example, Sandbrook (2015) point out that "unmanned aerial vehicles, or 'drones', appear to offer a flexible, accurate and affordable solution to some of the technical challenges of nature conservation monitoring and law enforcement. However, little attention has been given to their possible social impacts." Possible social impacts of drone operations for conservation management include the issues of "safety, privacy, psychological well-being, data security and the wider understanding of conservation problems" and the requirement that conservation must adhere to ethical standards, and that any such negative social impacts may affect the long-term sustainability of conservation applications (Sandbrook, 2015, p. 636). It is argued that there is requirement for "empirical research to establish whether the identified social risks of drones occur in reality and how they could be mitigated, and for self-regulation of drone use by the conservation sector to ensure good ethical practice and minimise the risk of unintended consequences" (Sandbrook, 2015, p. 636).

Psychological and social impacts from drone usage are thus important (Campbell, 2018). Several arguments have been advanced on this thesis. For example, Sandbrook (2015, p. 640) mentions psychological well-being, as drones usage "can be empowering for local groups if it provides them with the means to collect their own data, enforce rules and challenge the claims of others who may wish to mislead them," and the data may forest data for carbon assessment (see also Lewis & Nkuintchua, 2012; Paneque-Gálvez et al., 2014). However, possible negative consequences include privacy issues, as "drones used for conservation are very likely to collect information about people that could be used to identify them and what they are doing—indeed in the case of law enforcement applications, this is the deliberate intention"; as such, these issues have attracted attention including the "ethical implications and the ways in which they might infringe privacy and civil liberties" (Sandbrook, 2015, p. 640). Other problems include possible "fear, confusion and hostility among those on the ground" even the generation of "conspiracy theories, suspicions and fantasies, particularly when they are used in remote areas in developing countries which have little prior exposure to electronic devices" and "people may recognise a drone for what it is, but have misconceptions about its purpose, perhaps believing it to be sent by a private company, the military, a terrorist group or any number of others. Such perceptions could fuel existing conflicts or create new ones" (Sandbrook, 2015, p. 640).

There is also the psychological impact of distancing, which means that drones may be applied to different conservation actions than a physically present human operator (see Sparrow, 2009), and may also therefore create local social tensions. In this regard, Sandbrook (2015, p. 641) argues that "conservation organisations and agencies have a long history of conflict and disputes with local people, derived from evictions, exclusion, limitation of activities and other actions that are contrary to local interests" and that "It seems more likely that the use of drones for conservation purposes will be negatively received in places with a history of such conflict" sometimes with hostility (see also West et al., 2006). Drones may create the image of punishment for local stakeholders in conservation issues (see Mulero-Pázmány et al., 2014) for this issue in South Africa) and may even in result in drones being shot down, as has happened in the United States, where hunters may shoot at drones operated by animal welfare groups (Schroyer, 2012; Times & Democrat, 2014). Drone operators must overcome such issues, as recent conservation efforts have emphasized more sustainable, community-based inclusive approaches, and sought to avoid enforcement based on exclusion (Hulme & Murphree, 2001; Adams et al., 2004; Maffey et al., 2015; Saito et al., 2015; Campbell & Jha, 2021).

GIS and Conservation Biogeography

GIS techniques are increasingly applied to conservation biogeography topics (Böhm & Popescu, 2016). This application is partly due to the increasing role of GIS methods in both environmental and social science research, and the increasing recognition of the role of social issues and research in conservation studies (Campbell, 2018). For example, concerning GIS and social research, Higgins (2017) writes that "areas like geography, psychology, sociology, economics and political science can be better explored with GIS when spatial data is a part of the study. Examples include analyzing political elections, population distribution, land use, environmental degradation, along with how any of these (or other) variables interact with each other." Cioffi-Revilla (2010, p. 259) writes about computational social science, which involves the interfaces between computer science and the "traditional social science disciplines: social psychology, anthropology, economics, political science and sociology, each of which comprises several specialized branches five social sciences." Social GIS, one of the computational social science branches, is "closely related to the vast field of spatial statistical analysis, but with a greater emphasis on visualization of layers of social data" (Cioffi-Revilla (2010, p. 262). For application to the interfaces of social and physical sciences, and conservation biology, Klinkenberg (2020) writes that "there is growing significance of space, spatiality, location, and place in social science and physical science research. In particular, spatial analysis plays an increasing role in Conservation Biology as the spatial dimension of species and habitats gains recognition" and the "use of these technologies has dramatically changed how we conduct research and has led to some novel approaches in biodiversity studies and the related field of conservation biology" (Klinkenberg, 2020).

The applications of GIS to conservation biogeography are vast, because such numerical data may be derived from an array of land cover parameters, and also on the activities and other dynamics of human presence, within and between the topics of the social sciences (Campbell, 2018; Campbell & Jha, 2021). The main issues concern the utility of GIS techniques to the topics researched by the particular social science (e.g., economics), the applicability of this research to the field of conservation biogeography, and the beneficial results and outcomes for the latter discipline. As the application of GIS to social data may indicate the spatialization of the information, such studies may evolve into human geography, for example, for research clustering around environmental economics, or the economics of activities with environmental outcomes, focus on the parameters of economic issues that may affect the spatial environment, and cluster within the interface of economic geography, and possibly urban geography and biogeography and conservation science. Although there are numerous topics relevant to GIS applications to the economic side of conservation biogeography, the assessment of the economic and ecological value of urban green spaces is one of the most important current developments (Campbell, 2018, 2020).

Coroian et al. (2021) give the example of the multiple utility of urban green spaces, citing both economic and ecological concerns: "cooling effect, air and noise pollution mitigation, recreation, aesthetic pleasure, quality of life and public health, as well as economic benefits, such as raising the price of the real state in the vicinity of UGS." The contribution of GIS involves spatial statistics using GIS, which "have made possible the development of correct, coherent explanatory UGS variables such as distance, size, density, connectivity, sight and accessibility" (Coroian et al., 2021). The study developed analyses that made a contribution to the assessment of the economics of green spaces: "According to the spatial analyses from ArcGis, during the 10 years, an intense real estate market was developed, affecting the distribution of green spaces per capita." Such green spaces, sometimes influenced by economic considerations, may have major implications for urban wildlife, especially where connectivity with suburban and rural green land cover is emphasized (Coroian et al., 2021).

Other studies have used statistical analysis to determine the relationship between urban green space and other land cover developments and urban economic growth, which may serve as a background of other studies on the ecological merits of such developments (Rozenstein & Karnieli, 2011). For example, Zhou et al. (2022) examine "the impact of urban green space construction rate on urban economic growth from the perspective of spatial interaction." The study presented findings, indicating the outcome that "the level of green space construction can attract talents and investment by improving the environmental level of the city, and these financial expenditures, foreign investment, and talents are conducive to urbanization, thus having a significant positive impact on urban economic development. In addition, it also has a significant positive spatial spillover effect. In addition, the construction of urban green space will also stimulate the environmental protection of neighboring cities, which has a significant positive spatial dependence."

Another study of green spaces using GIS examined another important economic issue related to urban development: the relationship between real estate, green spaces and social development (Kim & Peiser, 2018). The study researches "how green spaces geographically influence property values in two different prototypes of urban growth: planned versus unplanned communities" (Kim & Peiser, 2018). The methodology used "hedonic regression and GIS analysis to assess the impact of green spaces on adjacent home values. The hedonic method deconstructs an observed home sale price into a set of inherent attributes associated with the property, thereby enabling the identification and quantification of various impacts on property sale prices." The findings of this quantitative (GIS and statistics) research indicate that "property values inconsistently reflect the five major attributes of green spaces (view, distance, size, type, character)" with differences between the planned and unplanned communities. In the planned community areas, "the active recreational green spaces have a negative impact on property values, while views, larger-sized green spaces, proximities to parks, greenways, and passive recreational green spaces have significant positive impacts. In contrast, in the communities characterized by unplanned growth, only proximity to park is directly reflected in housing values." The authors conclude that "the findings signify that different attributes of green spaces impact property values in unique ways in planned versus unplanned communities" (Kim & Peiser, 2018).

GIS methods may also be used for assessing the suitability of green space development, given the acknowledgment of their socioeconomic benefits (Rozenstein & Karnieli, 2011). For example, Gelan (2021) notes that "in the planning and management of urban centers, attention to the appropriate site selection of urban green spaces with regard to the importance that these spaces have from the perspectives of ecology, socioeconomic, mentality, etc., is an inevitable requirement." The study used GIS-based Multi-criteria analysis (MCA) to select locations for urban green spaces, with the parameters land use, proximity to settlements, roads, water bodies, population density, land ownership, topography and scenic attractiveness as the main factors determining urban green space suitability. The findings indicated that some areas were highly suitable, suitable, moderately suitable and poorly suitable, for urban green development, and a small section of the study area was unsuitable for urban green development. This study gave a good example of predictive, GIS-based modeling based on GIS-based MCA for conservation issues in biogeography (Gelan, 2021).

GIS applications have also contributed to sociology and psychology research in an environmental context (Campbell, 2018). Common research areas related to social issues include recreational landuse, in relation to natural and cultural landscapes, and possible social choices for the use of such land cover (Campbell, 2022). For example, Aklıbaşında and Bulut (2014) present a GIS-based study of tourism planning, examining the "effectiveness of GIS on holistic evaluation of natural and cultural resources in planning tourism terrains was analyzed." This study attempted, through GIS data analysis, "potential tourism and recreation terrains, level of suitability, and rate of coverage," and found that certain percentages of land area were quite suitable for tourism, and other parts were suitable, less suitable and not suitable. The findings indicated that "the water resources and valley landscapes were the basic sources of the tourism and recreation activities" in the study area (Aklıbaşında & Bulut, 2014).

Kuvan (2005) presents a similar study situated at the interface of tourist landuse and social development and the conservation of forests, the latter comprising one of the main attractants for tourism. Taking a Turkish case study, it is argued that "the use and conversion of forest lands is central to tourism development. Land use change is responsible for the majority of the negative environmental impacts of tourism on natural resources" and "the rapid emergence of mass tourism development in Turkey, encouraged without considering protection and sustainable use of natural and cultural resources, has resulted in serious problems in forest areas" (Kuvan, 2005).) It is concluded that in the case study of Belek, "a tourism center located in Antalya province's coastal areas and the fastest growing destination of the country, all tourist facilities have been established in forest lands under the status of conservation forest. Today, the Belek Conservation Forest is under severe pressure from tourism" (Kuvan, 2005). Such studies may form the background for the application of GIS methods, which may lend quantification to the consequent land cover changes in such contexts (Campbell, 2018).

For political issues, the main contribution of GIS may be to provide quantified support for decision-making and support the political empowerment of variously powerful stakeholders (Campbell 2018). For example, Jankowski (2009, p. 1966) "discuss the potential of Geographic Information Systems (GIS) to become an information technology enabling groups of people to participate in decisions shaping their communities and promoting sustainable use of natural resources" and explains "the concept of participation in the context of planning and decision making." It is argued that "Participatory GIS (PGIS) offers tools that can be used to help the public become meaningfully involved in decision making processes affecting their communities" (Jankowski 2009, p. 1966). Anderson et al., (2009) present a similar study on public participatory geographic information system (PPGIS), which focused on the creation of an "integrated GIS-based decision-support tool for community-focused communication and conservation assessment" among the stakeholders of the case study, which included Nova Forest Alliance (NFA) of Nova Scotia, private woodland owners, government agencies, forest companies, and non-government organizations and possible applications to other cases studies within the forests in Canada. The findings indicated that "the GIS-based tool, once set up and populated with the necessary data, was able to generate effective visual alternatives to support decision-making within the NFA community and elsewhere" (Anderson et al. (2009). Limitations of this application included data quality, use and sharing policies, problems with participation, and the lack of GIS expertise in the community. Suggested solutions were greater data accessibility, fewer bureaucratic and private sector limits, more community participation in the initial project design and objective determination, and latter developments, and the use of simpler participatory technical systems for participants (Anderson et al., 2009).

Geomatics Applications to Large Carnivore Conservation

This section takes the example of large carnivore conservation, as a case study that exemplifies the issues discussed above. Large carnivores comprise the most charismatic and problematic wildlife for conservation efforts, and as such represent most of the problems encountered in the conservation of smaller animals (Campbell, 2022; Durant et al., 2021). Large carnivore conservation is extremely problematic from economic, political, sociological, psychological, anthropological and geographical perspectives (Campbell, 2018). Wolf and Ripple (2018) argue that "Earth's terrestrial large carnivores form a highly endangered group of species with unique conservation challenges. The majority of these species have experienced major geographical range contractions, which puts many of them at high risk of extinction or of becoming ecologically ineffective." Hovardas et al. (2021) add that "large carnivores present multiple conservation challenges for various stakeholders across geographical scales and socio-cultural contexts. There have been many manifestations of human-carnivore conflict worldwide but also positive examples of human-carnivore coexistence." The large carnivores are mostly large felines (cats), ursine (bears) and canids (dogs), which may compete for land cover, and be seen as threats to people and companion animals. Human reactions such competing species include variably successful extermination efforts (Campbell & Torres Alvarado, 2011). Outcomes of such eradication efforts can be failure due to conservation opposition or animal adaptation, or success, which may indicate local extirpation or even regional extinction (Carbone & Gittleman, 2002; Karanth et al., 2004, 2010). In some cases, species extinction results (Naughton-Treves et al., 2005; Karanth & Chellam, 2009; Campbell 2027).

Geomatics techniques are particularly useful for the study of the interactions between people and large carnivores because these relations cover competition for preferred spaces, the spatial distribution and patterns of land use conflicts, social and psychological impacts on people (including fear, interests, conservation preferences), economic and political conflicts (including alternative landuses, controls on industrial and urban landuses for habitat conservation), and historical human-animal relations within anthropological systems (Hodgson et al., 2018; Ivošević et al., 2015; Johnston et al., 2017). Different types of geomatics-based techniques are used to measure the parameters associated with these varying and complex relationships, based on the requirements of time, areal extent, complexity of inter-relations and human social variables such as actor-oriented micro-scale studies, and large socio-cultural and economic structures (Bhattarai et al., 2019; Creel et al., 2015; Ingeman et al., 2022; van Eeden et al., 2018).

The main human dimensions of large carnivores concern the relationships between people and mammals (Durant et al., 2021; Ghosal et al., 2015). Durant et al. (2021) argues that "Coexistence with large carnivores poses challenges to human well-being, livelihoods, development, resource management and policy. Even where people and carnivores have historically coexisted, traditional patterns of behavior toward large carnivores may be disrupted by wider processes of economic, social, political, and climate change" and that within such spaces of cooperation and conflict, "conservation interventions have typically focused on changing behaviors of those living alongside large carnivores to promote sustainable practices." Ghosal et al. (2015) add that "people's reactions to large carnivores take many forms, ranging from support and coexistence to resistance and conflict" which result from many factors, and explore the "social construction of landscapes as a key concept to explore responses to large carnivores in the context of ecological, economic, social, and cultural changes" in the case study areas. Additionally, "the response to large carnivore presence is influenced by many different factors, of which the interpretation of change-particularly landscape change-plays a significant role" (Ghosal et al., 2015).

Issues that affect the relations between people and large carnivores largely concern the characteristics of the large carnivores (Campbell, 2018). Large carnivores have greater mobility and habitat requirements (large land areas, wide ranging, larger prey species, which can include people, and dispersed sometimes migratory populations) (Carbone & Gittleman, 2002; Cardillo et al., 2004; Berger et al., 2013; Wolf & Ripple, 2018). Large carnivore presence may induce human mental constructs, including potential and real physical attacks on people and/or companion animals and livestock, and competition for favored land cover, and the social and economic impacts of ameliorative actions (Inskip & Zimmermann, 2009; Kuijper et al., 2016). Common conflict incidents may enable the construction of legends, stories, half-truths and detailed studies, which in turn may enable or constrain conservation policies and impinge on human quality of life (QOL) issues (Sinha, 2019). Large carnivores, including birds, may prefer water sources, forest cover, fertile soils with grasses and shrubs to access prey species, and these may conflict with human activities such as farming, urban settlement, recreation activities and wildlife association (Carter et al., 2015; Campbell, 2018).

Human quality of life (QOL) is a major socio-economic, psychological and political, "multidimensional" concept that is related to such issues (Sinha, 2019). QOL has definitions based on human physical, material, social and emotional well-being, and human development and activity, and issues such as human age and gender (Felce & Perry, 1995). QOL is affected by economic issues, which may include competition for socioeconomically valuable land such as trails, rural and pristine forests, protected areas and urban green spaces (Campbell 2013). Psychological issues for QOL mainly concern human preferences for wild, pristine environments, and the fear of carnivore presence, regardless of the reality of the actual threat, which in some cases may minimal or non-existent, especially in the case of bird species, and some of the smaller and/or more elusive mammalian carnivores (Williamson, 2002; Karanth & Chellam, 2009; Fontúrbel & Simonetti, 2011).

Conservation actions to safeguard large carnivores include the creation and maintenance of protected areas, which may be studied using GIS methods (Gairola et al., 2013). For example, Altaweel (2017) writes about the role of GIS predictive modeling in wildlife conservation, habitat analysis, and environmental threats. One development concerns predictive modeling, using GIS for the assessment of land cover regions suitable for sustainable wildlife communities, and possible routes for migration (see, e.g.,, Perkl, 2016). Issues that may be examined using GIS methods include possibilities of wildlife-vehicle collisions, due to transport networks in natural habitats, extents of habitat area for wild species conservation (Shanee et al., 2013), and natural water sources for habitat sustenance (Altaweel, 2017; Yang et al., 2014). Social issues also assume prominence, termed "enhancing public participation in conservation planning with GIS" (Altaweel, 2017). Altaweel (2017) argues that "given that conservation and mitigation are needed to combat threats to wildlife, one novel approach has been to increase public involvement. In effect, this entails using the public as a stakeholder on where they would like to see species conservation to occur for particular animals and plants. These locations, selected by the public, are then compared with experts in wildlife" (see also Cox et al., 2014). Hence, "increasingly, the merger of public interest, including in water quality and wanting to see species conserved, has highlighted the need to develop complex plans that address stakeholder interests" (Altaweel, 2017).

Conclusions

As conservation biogeography advances and its contacts with supporting social sciences increase, quantitative, accurate and integrative measurements of land use change and human-environment networks are needed. Geomatics techniques, including remote sensing and GIS, are increasingly being used to meet this requirement. Geomatics-based applications in this complex socio-ecological domain are still in their infancy, although applications in ecology and natural biogeography may be established. This paper points to the potential of geomatics as a bridging discipline and methodology capable of creating syntheses that integrate both natural and social sciences by presenting more authoritative integrations of both sets of disciplines, thereby providing a new strand of socio-ecological understanding. Current relevant developments in geomatics include increased data sources and integration, sharper image resolution and increased collaboration between the social and environmental sciences, particularly in relation to the parameters measurable with geomatics techniques, such as percentage land cover change, human and non-human-animal movements and impressions on the landscape, and interactions between humans, non-human-animals and land cover. Research guided by these parameters is a major current dynamic with enormous possibilities.

References

Abdalla, R. (2021). Geomatics. In B. Daya Sagar, Q. Cheng, J. McKinley, & F. Agterberg (Eds.), *Encyclopedia of mathematical geosciences. Encyclopedia of earth sciences series*. Springer. https://doi.org/10.1007/978-3-030-26050-7_139-1

Adams, W., Aveling, R., Brockington, D., Dickson, B., Elliott, J., Hutton, J., Roe, D., Vira, B., et al. (2004). Biodiversity conservation and the eradication of poverty. *Science, 306*, 1146–1149. https://doi.org/10.1126/science.1097920

Agapiou, A., & Lysandrou, V. (2015). Remote sensing archaeology: Tracking and mapping evolution in European scientific literature from 1999 to 2015. *Journal of Archaeological Science: Reports, 4*, 192–200. https://doi.org/10.1016/j.jasrep.2015.09.010

Aklıbaşında, M., & Bulut, Y. (2014). Analysis of terrains suitable for tourism and recreation by using geographic information system (GIS). *Environmental Monitoring and Assessment, 186*(9), 5711–5719. https://doi.org/10.1007/s10661-014-3814-6

Altaweel, M. (2017). *The role of GIS in wildlife research and conservation.* GIS Lounge. Retrieved from https://www.gislounge.com/role-gis-wildlife-research-conservation/

Amoser, S., Wysocki, L. E., & Ladich, F. (2004). Noise emission during the first powerboat race in an Alpine lake and potential impact on fish communities. *The Journal of the Acoustical Society of America, 116*(6), 3789–3797. https://doi.org/10.1121/1.1808219

Anderson, C., Beazley, K., & Boxall, J. (2009). Lessons for PPGIS from the application of a decision-support tool in the Nova Forest Alliance of Nova Scotia, Canada. *Journal of Environmental Management, 90*(6), 2081–2089. https://doi.org/10.1016/j.jenvman.2007.08.031

Anthes, E. (2023). *Urban animals can't take the heat, study finds.* Retrieved from https://www.nytimes.com/2023/09/04/science/cities-wildlife-climate-change.html

Ayanlade, A., & Jegede, O. (2015). Evaluation of the intensity of the daytime surface urban heat Island: How can remote sensing help? *International Journal of Image and Data Fusion, 6*, 348–365.

Ballas, D., Clarke, G., Franklin, R., & Newing, A. (2017). *GIS and the social sciences: Theory and applications.* Routledge. https://doi.org/10.4324/9781315759326

Banzhaf, E., & de la Barrera, F. (2017). Evaluating public green spaces for the quality of life in cities by integrating RS mapping tools and social science techniques. In *Proceedings of the 2017 Joint Urban remote sensing event (JURSE), Dubai, UAE.*

Banzhaf, E., Grescho, V., & Kindler, A. (2009). Monitoring urban to peri-urban development with integrated remote sensing and GIS information: A Leipzig, Germany case study. *International Journal of Remote Sensing, 30*, 1675–1696.

Barton, J., & Pretty, J. (2010). What is the best dose of nature and green exercise for improving mental health? A multi-study analysis. *Environmental Science & Technology, 44*(10), 3947–3955. https://doi.org/10.1021/es903183r

Beaver, J. T., Baldwin, R. W., Messinger, M., Newbolt, C. H., Ditchkoff, S. S., & Silman, M. R. (2020). Evaluating the use of drones equipped with thermal sensors as an effective method for estimating wildlife. *Wildlife Society Bulletin, 44*(2), 434–443. https://doi.org/10.1002/wsb.1090

Bechtel, B., Alexander, P., Böhner, J., Ching, J., Conrad, O., Feddema, J., Mills, G., See, L., & Stewart, I. (2015). Mapping local climate zones for a worldwide database of the form and function of cities. *ISPRS International Journal of Geo-Information, 4*, 199–219.

Berger, J., Buuveibaatar, B., & Mishra, C. (2013). Globalization of the cashmere market and the decline of large mammals in Central Asia. *Conservation Biology, 27*(4), 679–689. https://doi.org/10.1111/cobi.12100

Bernath-Plaisted, J., & Koper, N. (2016). Physical footprint of oil and gas infrastructure, not anthropogenic noise, reduces nesting success of some grassland songbirds. *Biolological Conservation, 204*, 434–441. https://doi.org/10.1016/j.biocon.2016.11.002

Bertazzon, S., Johnson, M., Eccles, K., & Kaplan, G. G. (2015). Accounting for spatial effects in land use regression for urban air pollution modeling. *Spatial and Spatio-temporal Epidemiology, 14-15*, 9–21.

Bertram, C., & Rehdanz, K. (2015). The role of urban green space for human well-being. *Ecological Economics, 120*, 139–152.

Bhattarai, B. R., Wright, W., Morgan, D., Cook, S., & Baral, H. S. (2019). Managing human-tiger conflict: Lessons from Bardia and Chitwan National Parks, Nepal. *European Journal of Wildlife Research, 65*(3), 34. https://doi.org/10.1007/s10344-019-1270-x

Böhm, M., & Popescu, V. D. (2016). Landscape ecology, biogeography, and GIS methods. In C. K. Dodd (Ed.), *Reptile ecology and conservation: A handbook of techniques.* Oxford Academic. https://doi.org/10.1093/acprof:oso/9780198726135.003.0022

Bornstein, R. D. (1968). Observations of the urban heat Island effect in New York city. *Journal of Applied Meteorology, 7*, 575–582.

Brauer, M., Freedman, G., Frostad, J., van Donkelaar, A., Martin, R. V., Dentener, F., van Dingenen, R., Estep, K., Amini, H., Apte, J. S., Balakrishnan, K., Barregard, L., Broday, D., Feigin, V., Ghosh, S., Hopke, P. K., Knibbs, L. D., Kokubo, Y., Liu, Y., Ma, S., & Cohen, A. (2016). Ambient air pollution exposure estimation for the global burden of disease 2013. *Environmental Science & Technology, 50*(1), 79–88. https://doi.org/10.1021/acs.est.5b03709

Braun, A. C., & Koch, B. (2016). Estimating impacts of plantation forestry on plant biodiversity in southern Chile-a spatially explicit modelling approach. *Environmental Monitoring and Assessment, 188*(10), 564. https://doi.org/10.1007/s10661-016-5547-1

Briggs, D. J., Collins, S., Elliot, P., Fischer, P., Kingham, S., Lebert, E., Pryl, K., van Reeuwijk, H., Smallbone, K., & van der Veen, A. (1997). Mapping urban air pollution using GIS: A regression-based approach. *International Journal of Geographical Information Science, 11*(7), 699–718. https://doi.org/10.1080/136588197242158

British Columbia Institute of Technology. (2018). *Geomatics*. https://www.bcit.ca/study/programs/geomatics

Brown, A. L. (1990). Measuring the effect of aircraft noise on sea birds. *Environment International, 16*(4-6), 587–592. https://doi.org/10.1016/0160-4120(90)90029-6

Bunkley, J. P., McClure, C. J. W., Kawahara, A. Y., Francis, C. D., & Barber, J. R. (2017). Anthropogenic noise changes arthropod abundances. *Ecology and Evolution, 7*(9), 2977–2985. https://doi.org/10.1002/ece3.2698

Burgess, R., Costa, F., & Olken, B. A. (2016). *The power of the state: National Borders and the deforestation of the Amazon*. Retrieved from https://conference.nber.org/confer//2016/EEEs16/Burgess_Costa_Olken.Pdf

Burrough, P. A. (1979). *Principles of Geographical Information Systems for Land Resources Assessment*. Clarendon.

Burroughs, R. H., & Clark, T. W. (1995). Ecosystem management: A comparison of greater yellowstone and georges bank. *Environmental Management, 19*(5), 649–663. https://doi.org/10.1007/bf02471947

Campbell, M. (2017). *Biological Conservation in the 21st Century: A Conservation Biology of Large Wildlife*. Nova Science Publishers.

Campbell, M. (2018). *Geomatics and conservation biology*. Nova Science Publishers.

Campbell, M. (2020). Changing study methodologies and large carnivore/human interactions in British Columbia, Canada. In S. A. Lund (Ed.), *Canada past, present and future perspectives* (pp. 71–130). Nova Science Publishers.

Campbell, M. (2022). *The great eagles: Their evolution, ecology and conservation*. CRC Press.

Campbell, M. O. (2013). The relevance of age and gender for public attitudes to brown bears (Ursus arctos), black bears (Ursus americanus), and cougars (Puma concolor) in Kamloops, British Columbia. *Society & Animals, 21*(4), 341–359. https://doi.org/10.1163/15685306-12341260

Campbell, M., & Jha, K. K. (2021). *Critical research techniques in animal and habitat ecology: International examples*. Nova Science Publishers.

Campbell, M. O., & Torres Alvarado, M. E. (2011). Public perceptions of jaguars Panthera onca, pumas Puma concolor and coyotes Canis latrans in El Salvador. *Area, 43*(3), 250–256. https://doi.org/10.1111/j.1475-4762.2011.00996.x

Carbone, C., & Gittleman, J. L. (2002). A common rule for the scaling of carnivore density. *Science, 295*(5563), 2273–2276. https://doi.org/10.1126/science.1067994

Cardillo, M., Purvis, A., Sechrest, W., Gittleman, J. L., Bielby, J., & Mace, G. M. (2004). Human population density and extinction risk in the world's carnivores. *PLoS Biology, 2*(7), e197. https://doi.org/10.1371/journal.pbio.0020197

Carter, N., Levin, S., Barlow, A., & Grimm, V. (2015). Modeling tiger population and territory dynamics using an agent-based approach. *Ecological Modelling, 312*, 347–362. https://doi.org/10.1016/j.ecolmodel.2015.06.008

Cavender-Bares, J., Schneider, F. D., Santos, M. J., Armstrong, A., Carnaval, A., Dahlin, K. M., Fatoyinbo, L., Hurtt, G. C., Schimel, D., Townsend, P. A., Ustin, S. L., Wang, Z., & Wilson, A. M. (2022). Integrating remote sensing with ecology and evolution to advance biodiversity conservation. *Nature Ecology & Evolution*. https://doi.org/10.1038/s41559-022-01702-5

Cioffi-Revilla, C. (2010). Computational social science. *WIREs Computational Statistics, 2*(3), 259–271. https://doi.org/10.1002/wics.95

Coroian, I., Nap, M. E., Pop, I. D., Matei, F., Sălăgean, T., Deak, J., Chiorean, S., Șuba, E. E., Lupuţ, I., & Ficior, D. (2021). Using GIS analysis to assess urban green space in terms of real estate development. *Bulletin of University of Agricultural Sciences and Veterinary Medicine Cluj-Napoca. Horticulture, 78*(1), 61. https://doi.org/10.15835/buasvmcn-hort:2020.0061

Cox, C., Morse, W., Anderson, C., & Marzen, L. (2014). Applying public participation geographic information systems to wildlife management. *Human Dimensions of Wildlife, 19*(2), 200–214.

Creel, S., Becker, M., Christianson, D., Dröge, E., Hammerschlag, N., Hayward, M. W., Karanth, U., Loveridge, A., MacDonald, D. W., Matandiko, W., M'soka, J., Murray, D., Rosenblatt, E., & Schuette, P. (2015). Conservation policy. Questionable policy for large carnivore hunting. *Science, 350*(6267), 1473–1475. https://doi.org/10.1126/science.aac4768

Cunnington, G. M., & Fahrig, L. (2010). Plasticity in the vocalizations of anurans in response to traffic noise. *Acta Oecologica-International Journal of Ecology, 36*, 463–470. https://doi.org/10.1016/j.actao.2010.06.002

Danylo, O., See, L., Bechtel, B., Schepaschenko, D., & Fritz, S. (2016). Contributing to WUDAPT: A local climate zone classification of two cities in Ukraine. *IEEE Journal of Selected Topics in Applied Earth Observations and Remote Sensing, 9*, 1841–1853.

Denham, T. (2017). Landscape archaeology. In A. S. Gilbert (Ed.), *Encyclopedia of geoarchaeology. Encyclopedia of earth sciences series* (pp. 464–468). Springer. https://doi.org/10.1007/978-1-4020-4409-0_168

De Faria Peres, L., de Lucena, A. J., Filho, O. C. R., & de Almeida França, J. R. (2018). The urban heat Island in Rio de Janeiro, Brazil, in the last 30 years using remote sensing data. *International Journal of Applied Earth Observation and Geoinformation, 64*, 104–116. https://doi.org/10.1016/j.jag.2017.08.012

De Souza, L. C. L., & Giunta, M. B. (2011). Urban indices as environmental noise indicators. *Computers, Environment and Urban Systems, 35*(5), 421–430. https://doi.org/10.1016/j.compenvurbsys.2011.06.001

Ditmer, M., Vincent, J. B., Werden, L. K., Tanner, J. C., Laske, T. G., Iaizzo, P. A., Garshelis, D. L., & Fieberg, J. R. (2015). Bears show a physiological but limited behavioral response to unmanned aerial vehicles. *Current Biology, 25*(17), 2278–2283. https://doi.org/10.1016/j.cub.2015.07.024

Dockery, D. W., Pope, C. A., Xu, X., Spengler, J. D., Ware, J. H., Fay, M. E., Ferris, B. G., Jr., & Speizer, F. E. (1993). An association between air pollution and mortality in six U.S. cities. *The New England Journal of Medicine, 329*(24), 1753–1759. https://doi.org/10.1056/NEJM199312093292401

Donaldson, D., & Storeygar, A. (2016). The view from above: Applications of satellite data in economics. *Journal of Economic Perspectives, 30*(4), 171–198. https://doi.org/10.1257/jep.30.4.171

Drolet, A., Dussault, C., & Côté, S. D. (2016). Simulated drilling noise affects the space use of a large terrestrial mammal. *Wildlife Biology, 22*(6), 284–293. https://doi.org/10.2981/wlb.00225

Du, H., Cai, W., Xu, Y., Wang, Z., Wang, Y., & Cai, Y. (2017). Quantifying the cool Island effects of urban green spaces using remote sensing data. *Urban Forestry and Urban Greening, 27*, 24–31. https://doi.org/10.1016/j.ufug.2017.06.008

Dubuisson, B. (1975). *Practique de la photogrammétrie et de moyens cartographiques derivés des ordinateurs*. Editions Eyrolles. [Practical photogrammetry and cartographic means derived from computers].

Dueker, K. J. (1979). Land resource information systems: A review of fifteen years experience. *Geo-Processing (Netherlands), 1*(2).

Duffy, L. (2020). *Fundamentals of surveying: Land surveying, GIS and geomatics*. Retrieved from https://www.enr.com/articles/53273-fundamentals-of-surveying-land-surveying-gis-and-geomatics
Duncan, B. N., Prados, A. I., Lamsal, L., Liu, Y., Streets, D. G., Gupta, P., et al. (2014). Satellite data of atmospheric pollution for U.S. air quality applications: Examples of applications, summary of data end-user resources, answers to FAQs, and common mistakes to avoid. *Atmospheric Environment, 94*, 647–662. https://doi.org/10.1016/j.atmosenv.2014.05.061
Durant, S. M., Marino, A., Linnell, J. D. C., Oriol-Cotterill, A., Dloniak, S., Dolrenry, S., Funston, P., Groom, R. J., Hanssen, L., Horgan, J., Ikanda, D., Ipavec, A., Kissui, B., Lichtenfeld, L., McNutt, J. W., Mitchell, N., Naro, E., Samna, A., & Yirga, G. (2021). Fostering coexistence between people and large carnivores in Africa: Using a theory of change to identify pathways to impact and their underlying assumptions. *Frontiers in Conservation Science, 2*, 698631. https://doi.org/10.3389/fcosc.2021.698631
Ehlers, M. (2008). Geoinformatics and digital earth initiatives: A German perspective. *International Journal of Digital Earth, 1*(1), 17–30. https://doi.org/10.1080/17538940701781975
Ekkel, E. D., & de Vries, S. (2017). Nearby green space and human health: Evaluating accessibility metrics. *Landscape and Urban Planning, 157*, 214–220. https://doi.org/10.1016/j.landurbplan.2016.06.008
Federal Aviation Administration (FAA). (2018). *FAA aerospace forecast*. Retrieved from https://www.faa.gov/data_research/aviation/aerospace_forecasts/media/FY2019-39_FAA_Aerospace_Forecast.pdf
Felce, D., & Perry, J. (1995). Quality of life: Its definition and measurement. *Research in Developmental Disabilities, 16*(1), 51–74. https://doi.org/10.1016/0891-4222(94)00028-8
Flacke, J., Schüle, S. A., Köckler, H., & Bolte, G. (2016). Mapping environmental inequalities relevant for health for informing urban planning interventions-A case study in the City of Dortmund, Germany. *International Journal of Environmental Research and Public Health, 13*(7), 711. https://doi.org/10.3390/ijerph13070711
Flett, D. G., Manore, M. J., Ramsay, B. R., & Falkingham, J. C. (2001). *Preparing for operational use of RADARSAT-2 data at the Canadian ice service*. https://ieeexplore.ieee.org/document/976200
Fontúrbel, F. E., & Simonetti, J. A. (2011). Translocations and human-carnivore conflicts: Problem solving or problem creating? *Wildlife Biology, 17*(2), 217–224. https://doi.org/10.2981/10-091
Foster, A. D., & Rosenzweig, M. R. (2003). Economic growth and the rise of forests. *The Quarterly Journal of Economics, 118*(2), 601–637. https://doi.org/10.1162/003355303321675464
Franke, U. E. (2015). *Civilian drones: Fixing an image problem?* International Relations and Security Network. Retrieved from https://isnblog.ethz.ch/security/civilian-drones-fixing-an-imageproblem
Fyhri, A., & Klaeboe, R. (2009). Road traffic noise, sensitivity, annoyance and self-reported health--a structural equation model exercise. *Environment International, 35*(1), 91–97. https://doi.org/10.1016/j.envint.2008.08.006
Gairola, S., Procheş, Ş., & Rocchini, D. (2013). High-resolution satellite remote sensing: A new frontier for biodiversity exploration in Indian Himalayan forests. *International Journal of Remote Sensing, 34*(6), 2006–2022. https://doi.org/10.1080/01431161.2012.730161
Gamba, P., Houshmand, B., & Saccani, M. (2000). Detection and extraction of buildings from interferometric SAR data. *IEEE Transactions on Geoscience and Remote Sensing, 38*, 611–617. https://doi.org/10.1080/01431161.2012.730161
Garg, N., & Maji, S. (2014). A critical review of principal traffic noise models: Strategies and implications. *Environmental Impact Assessment Review, 46*, 68–81. https://doi.org/10.1016/j.eiar.2014.02.001
Gelan, E. (2021). GIS-based multi-criteria analysis for sustainable urban green spaces planning in emerging towns of Ethiopia: The case of Sululta town. *Environ Systems Research, 10*, 13. https://doi.org/10.1186/s40068-021-00220-w
Geomatica. (2002). *Geomatics*. Canadian Institute of Geomatics.

Ghosal, S., Skogen, K., & Krishnan, S. (2015). Locating human-wildlife interactions: Landscape constructions and responses to large carnivore conservation in India and Norway. *Conservation and Society, 13*(3), 265–274. https://doi.org/10.4103/0972-4923.170403

Gomarasca, M. (2010). Basics of geomatics. *Applied Geomatics, 2*, 137–146. https://doi.org/10.1007/s12518-010-0029-6

Gong, Z., Ge, W., Guo, J., & Liu, J. (2024). Satellite remote sensing of vegetation phenology: Progress, challenges, and opportunities. *ISPRS Journal of Photogrammetry and Remote Sensing, 217*, 149–164. https://doi.org/10.1016/j.isprsjprs.2024.08.011

Göpfert, W. (1991). Raumbezogene Informationssysteme. Wichmann.

Guo, Y., Feng, N., Christopher, S. A., Kang, P., Zhan, F. B., & Hong, S. (2014). Satellite remote sensing of fine particulate matter ($PM_{2.5}$) air quality over Beijing using MODIS. *International Journal of Remote Sensing, 35*(17), 6522–6544. https://doi.org/10.1080/01431161.2014.958245

Gupta, P., Christopher, S. A., Wang, J., Gehrig, R., Lee, Y., & Kumar, N. (2006). Satellite remote sensing of particulate matter and air quality assessment over global cities. *Atmospheric Environment, 40*(30), 5880–5892. https://doi.org/10.1016/j.atmosenv.2006.03.016

Haberl, H., Batterbury, S., & Moran, E. (2001). Using and shaping the land: A long-term perspective. *Land Use Policy, 18*(1), 1–8. https://doi.org/10.1016/s0264-8377(00)00040-5

Habermann, M., Billger, M., & Haeger-Eugensson, M. (2015). Land use regression as method to model air pollution. Previous results for Gothenburg/Sweden. *Procedia Engineering, 115*, 21–28. https://doi.org/10.1016/j.proeng.2015.07.350

Haight, J. D., Hall, S. J., Fidino, M., Adalsteinsson, S. A., Ahlers, A. A., Angstmann, J., Anthonysamy, W. J. B., Biro, E., Collins, M. K., Dugelby, B., Gallo, T., Green, A. M., Hartley, L., Jordan, M. J., Kay, C. A. M., Lehrer, E. W., Long, R. A., MacDougall, B., Magle, S. B., Minier, D. E., & Lewis, J. S. (2023). Urbanization, climate and species traits shape mammal communities from local to continental scales. *Nature Ecology & Evolution, 7*(10), 1654–1666. https://doi.org/10.1038/s41559-023-02166-x

Hammill, M. O., Dale, J., Stenson, G. B., den Heyer, C., Gosselin, J. F., Leblanc, P., & Johnston, D. W. (2017). *Comparison of methods to estimate grey seal pup production at different colonies*. DFO Canadian Science Advisory Secretariat (CSAS) Research Document 2017/041.

Han, R. B., Cao, J. K., & Ma, S. Q. (2014). Geomatics in ecosystem assessment and management. In B. Y. Cao, S. Q. Ma, & H. H. Cao (Eds.), *Ecosystem assessment and fuzzy systems management. Advances in intelligent systems and computing* (Vol. 254). Springer. https://doi.org/10.1007/978-3-319-03449-2_2

Hansmann, R., Hug, S. M., & Seeland, K. (2007). Restoration and stress relief through physical activities in forests and parks. *Urban Forestry and Urban Greening, 6*(4), 213–225. https://doi.org/10.1016/j.ufug.2007.08.004

He, C., Liu, Z., Tian, J., & Ma, Q. (2014). Urban expansion dynamics and natural habitat loss in China: A multiscale landscape perspective. *Global Change Biology, 20*(9), 2886–2902. https://doi.org/10.1111/gcb.12553

Heynen, N., Perkins, H. A., & Roy, P. (2006). The political ecology of uneven urban green space: The impact of political economy on race and ethnicity in producing environmental inequality in Milwaukee. *Urban Affairs Review, 42*, 3–25. https://doi.org/10.1177/1078087406290729

Higgins, A. K. (2017). *Using GIS in social science research. Rural sociology group.* Wageningen University.

Hodgson, J. C., Mott, R., Baylis, S. M., Pham, T. T., Wotherspoon, S., Kilpatrick, A. D., Segaran, R. R., Reid, I., Terauds, A., & Koh, L. P. (2018). Drones count wildlife more accurately and precisely than humans. *Methods in Ecology and Evolution, 9*(5), 1160–1167. https://doi.org/10.1111/2041-210X.12974

Hoek, G., Beelen, R., de Hoogh, K., Vienneau, D., Gulliver, J., Fischer, P., & Briggs, D. (2008). A review of land-use regression models to assess spatial variation of outdoor air pollution. *Atmospheric Environment, 42*(33), 7561–7578. https://doi.org/10.1016/j.atmosenv.2008.05.057

Hoek, G., Brunekreef, B., Goldbohm, S., Fischer, P., & van den Brandt, P. A. (2002). Association between mortality and indicators of traffic-related air pollution in The Netherlands: A cohort study. *Lancet, 360*(9341), 1203–1209. https://doi.org/10.1016/S0140-6736(02)11280-3

Honold, J., Lakes, T., Beyer, R., & van der Meer, E. (2015). Restoration in Urban spaces: Nature views from home, greenways, and public parks. *Environment and Behavior, 48*(6). https://doi.org/10.1177/0013916514568556

Horning, N., Robinson, J. A., Sterling, E. J., Turner, W., & Spector, S. (2015). *Remote sensing for ecology and conservation: A handbook of techniques*. Oxford University Press.

Hovardas, T., Penteriani, V., Trouwborst, A., & López-Bao, J. V. (2021). Editorial: Conservation and management of large carnivores—local insights for global challenges. *Frontiers in Ecology and Evolution, 9*, 682444. https://doi.org/10.3389/fevo.2021.68244

Howitt, R., Karp, L., & Rausser, G. (2022). Remote sensing technologies: Implications for agricultural and resource economics. In H. de Gorter, J. McCluskey, J. Swinnen, & D. Zilberman (Eds.), *Modern agricultural and resource economics and policy. Natural resource management and policy* (Vol. 55, pp. 183–217). Springer. https://doi.org/10.1007/978-3-030-77760-9_9

Hu, Z. (2009). Spatial analysis of MODIS aerosol optical depth, PM2.5, and chronic coronary heart disease. *International Journal of Health Geographics, 8*, 27. https://doi.org/10.1186/1476-072X-8-27

Hulme, D., & Murphree, M. W. (2001). *African wildlife and livelihoods: The promise and performance of community conservation*. James Currey.

Ingeman, K. E., Zhao, L. Z., Wolf, C., et al. (2022). Glimmers of hope in large carnivore recoveries. *Scientific Reports, 12*, 10005. https://doi.org/10.1038/s41598-022-13671-7

Inskip, C., & Zimmermann, A. (2009). Human-felid conflict: A review of patterns and priorities worldwide. *Oryx, 43*(01), 18. https://doi.org/10.1017/s003060530899030x

Ivošević, B., Han, Y., Cho, Y., & Kwon, O. (2015). The use of conservation drones in ecology and wildlife research. *Journal of Ecology and Environment, 38*(1), 113–118. https://doi.org/10.5141/Ecoenv.2015.012

Jacobsen, L., Baktoft, H., Jepsen, N., Aarestrup, K., Berg, S., & Skov, C. (2014). Effect of boat noise and angling on lake fish behaviour: Disturbance and lake fish behaviour. *Journal of Fish Biology, 84*(6), 1768–1780. https://doi.org/10.1111/jfb.12395

Jankowski, P. (2009). Towards participatory geographic information systems for community-based environmental decision making. *Journal of Environmental Management, 90*(6), 1966–1971. https://doi.org/10.1016/j.jenvman.2007.08.028

Jayachandran, S., de Laat, J., Lambin, E. F., & Stanton, C. Y. (2016). *Cash for carbon: A randomized controlled trial of payments for ecosystem services to reduce deforestation*. Working paper 22378. Retrieved from http://www.nber.org/papers/w22378

Jensen, J. R. (2007). *Remote Sensing of the Environment: An Earth Resource Perspective. 2nd Edition*. Pearson Prentice Hall.

Jewitt, D. (2018). The use of drones in conservation. *Geomatics, Surveying and Measurement*, 20–23.

Johnson, D. P., Wilson, J. S., & Luber, G. C. (2009). Socioeconomic indicators of heat-related health risk supplemented with remotely sensed data. *International Journal of Health Geographics, 8*, 57. https://doi.org/10.1186/1476-072X-8-57

Johnston, D. W., Dale, J., Murray, K. T., Josephson, E., Newton, E., & Wood, S. (2017). Comparing occupied and unoccupied aircraft surveys of wildlife populations: Assessing the gray seal (Halichoerus grypus) breeding colony on Muskeget Island, USA. *Journal of Unmanned Vehicle Systems, 5*(4), 178–191. https://doi.org/10.1139/juvs-2017-0012

Journal of Remote Sensing and GIS. (2023). *Geomatics*. Walsh Medical Media.

Kabisch, N., & Haase, D. (2014). Green justice or just green? Provision of urban green spaces in Berlin, Germany. *Landscape and Urban Planning, 122*, 129–139. https://doi.org/10.1016/j.landurbplan.2013.11.016

Kamoske, A. G., Dahlin, K. M., Read, Q. D., Record, S., Stark, S. C., Serbin, S. P., Zarnetske, P. L., & Dornelas, M. (2022). Towards mapping biodiversity from above: Can fusing lidar and hyperspectral remote sensing predict taxonomic, functional, and phylogenetic tree diversity in temperate forests? *Global Ecology and Biogeography, 31*(7), 1440–1460. https://doi.org/10.1111/geb.13516

Karanth, K. U., & Chellam, R. (2009). Carnivore conservation at the crossroads. *Oryx, 43*(01), 1. https://doi.org/10.1017/s003060530843106x

Karanth, K. K., Nichols, J. D., Karanth, K. U., Hines, J. E., & Christensen, N. L. (2010). The shrinking ark: Patterns of large mammal extinctions in India. *Proceedings of the Royal Society B: Biological Sciences, 277*(1690), 1971–1979. https://doi.org/10.1098/rspb.2010.0171

Karanth, K. U., Nichols, J. D., Kumar, N. S., Link, W. A., & Hines, J. E. (2004). Tigers and their prey: Predicting carnivore densities from prey abundance. *Proceedings of the National Academy of Sciences, 101*(14), 4854–4858. https://doi.org/10.1073/pnas.030621010

Kennedy, C., & Rogers, J. I. (2015). Virtuous drones? *The International Journal of Human Rights, 19*(2), 211–227. https://doi.org/10.1080/13642987.2014.991217

Keola, S., Andersson, M., & Hall, O. (2015). Monitoring economic development from space: Using nighttime light and land cover data to measure economic growth. *World Development, 66*, 322–334. https://doi.org/10.1016/j.worlddev.2014.08.017

Kershaw, S., Gower, S., Rinner, C., & Campbell, M. (2013). Identifying inequitable exposure to toxic air pollution in racialized and low-income neighbourhoods to support pollution prevention. *Geospatial Health, 7*(2), 265–278. https://doi.org/10.4081/gh.2013.85

Kim, S. K., & Peiser, R. B. (2018). The economic effects of green spaces in planned and unplanned communities. *Journal of Architectural and Planning Research, 35*(4), 323–342. Retrieved from https://www.jstor.org/stable/26893776

Klinkenberg, B. (Ed.). (2020). *Biodiversity of British Columbia. Lab for advanced spatial analysis*. Department of Geography, University of British Columbia. Retrieved from www.biodiversity.bc.ca

Koper, R. P., Erbe, C., du Preez, D. R., & Plön, S. (2016). Comparison of soundscape contributors between two neighboring southern right whale nursing areas along the South African coast. *Proceedings of Meetings on Acoustics, 27*, 070014. https://doi.org/10.1121/2.0000303

Kovats, R. S., Johnson, H., & Griffith, C. (2006). Mortality in southern England during the 2003 heat wave by place of death. *Health Statistics Quarterly, 29*, 6–8.

Kuijper, D. P., Sahlén, E., Elmhagen, B., Chamaillé-Jammes, S., Sand, H., Lone, K., & Cromsigt, J. P. (2016). Paws without claws? Ecological effects of large carnivores in anthropogenic landscapes. Proceedings. *Biological Sciences, 283*(1841), 20161625. https://doi.org/10.1098/rspb.2016.1625

Kuvan, Y. (2005). The use of forests for the purpose of tourism: The case of Belek tourism Center in Turkey. *Journal of Environmental Management, 75*(3), 263–274. https://doi.org/10.1016/j.jenvman.2005.01.003

Kwok, R. (2018). Ecology's remote-sensing revolution. *Nature, 556*, 137–138. https://doi.org/10.1038/d41586-018-03924-9

Lahart, I., Darcy, P., Gidlow, C., & Calogiuri, G. (2019). The effects of Green exercise on physical and mental wellbeing: A systematic review. *International Journal of Environmental Research and Public Health, 16*(8), 1352. https://doi.org/10.3390/ijerph16081352

LaZerte, S. E., Slabbekoorn, H., & Otter, K. A. (2016). Learning to cope: Vocal adjustment to urban noise is correlated with prior experience in black-capped chickadees. *Proceedings of the Royal Society B: Biological Sciences, 283*, 20161058. https://doi.org/10.1098/rspb.2016.1058

Lee, M., Brauer, M., Wong, P., Tang, R., Tsui, T. H., Choi, C., Cheng, W., Lai, P. C., Tian, L., Thach, T. Q., Allen, R., & Barratt, B. (2017). Land use regression modelling of air pollution in high density high rise cities: A case study in Hong Kong. *The Science of the Total Environment, 592*, 306–315. https://doi.org/10.1016/j.scitotenv.2017.03.094

Leichtle, T., Geiß, C., Wurm, M., Lakes, T., & Taubenböck, H. (2017). Unsupervised change detection in VHR remote sensing imagery—An object-based clustering approach in a dynamic urban environment. *International Journal of Applied Earth Observation and Geoinformation, 54*, 15–27. https://doi.org/10.1016/j.jag.2016.08.010

Lengagne, T. (2008). Traffic noise affects communication behaviour in a breeding anuran, Hyla arborea. *Biological Conservation, 141*(8), 2023–2031. https://doi.org/10.1016/j.biocon.2008.05.017

Leucci, G., Giorgi, L. D., Giacomo, G. D., Ditaranto, I., Miccoli, I., & Scardozzi, G. (2016). 3D GPR survey for the archaeological characterization of the ancient Messapian necropolis in Lecce, South Italy. *Journal of Archaeological Science: Reports, 7*, 290–302. https://doi.org/10.1016/j.jasrep.2016.05.027

Lewis, J. D., & Nkuintchua, T. (2012). Accessible technologies and FPIC: Independent monitoring with forest communities in Cameroon. *Participatory Learning and Action, 65*, 151–165.

Li, T., Shen, H., Yuan, Q., Zhang, X., & Zhang, L. (2017). Estimating ground-level $PM_{2.5}$ by fusing satellite and station observations: A geo-intelligent deep learning approach. *Geophysical Research Letters, 44*(23), 11985–11993. https://doi.org/10.1002/2017GL075710

Li, X., Meng, Q., Li, W., Zhang, C., Jancso, T., & Mavromatis, S. (2014). An explorative study on the proximity of buildings to green spaces in urban areas using remotely sensed imagery. *Annals of GIS, 20*(3), 193–203. https://doi.org/10.1080/19475683.2014.945482

Li, Z. (2023). New opportunities for archaeological research in the greater Ghingan range, China: Application of UAV LiDAR in the archaeological survey of the Shenshan Mountain. *Journal of Archaeological Science: Reports, 51*, 104182. https://doi.org/10.1016/j.jasrep.2023.104182

Lindsey, R. (2005). *Nimbus: 40th anniversary*. Retrieved from https://earthobservatory.nasa.gov/features/Nimbus

Liu, Y., Meng, Q., Zhang, J., Zhang, L., Jancso, T., & Vatseva, R. (2016). An effective building neighborhood Green index model for measuring urban green space. *International Journal of Digital Earth, 9*(4), 387–409. https://doi.org/10.1080/17538947.2015.1037870

Livingston, S. (2015). Commercial remote sensing satellites and the regulation of.

Lucas, T. (2018). *Drones prove reliable, cost-efficient tool for wildlife surveys*. Retrieved from https://nicholas.duke.edu/about/news/dronesprove-reliable-cost-efficient-tool-wildlife-surveys

Maas, J., Verheij, R. A., Groenewegen, P. P., de Vries, S., & Spreeuwenberg, P. (2006). Green space, urbanity, and health: How strong is the relation? *Journal of Epidemiology and Community Health, 60*(7), 587–592. https://doi.org/10.1136/jech.2005.043125

Maffey, G., Homans, H., Banks, K., & Arts, K. (2015). Digital technology and human development: A charter for nature conservation. *Ambio, 44*(Suppl 4), 527–537. https://doi.org/10.1007/s13280-015-0703-3

Mancera, K. F., Murray, P. J., Lisle, A., Dupont, C., Faucheux, F., & Phillips, C. J. C. (2017). The effects of acute exposure to mining machinery noise on the behaviour of eastern blue-tongued lizards (Tiliqua scincoides). *Animal Welfare, 26*(1), 11–24. https://doi.org/10.7120/09627286.26.1.011

Mann, A. (2019). *ISRO: The Indian Space research organization*. Retrieved from https://www.space.com/indian-space-research-organization.html

Martin, R. V. (2008). Satellite remote sensing of surface air quality. *Atmospheric Environment, 42*, 7823–7843. https://doi.org/10.1016/j.atmosenv.2008.07.018

Mason, J., McClure, C., & Barber, J. (2016). Anthropogenic noise impairs owl hunting behavior. *Biological Conservation, 199*, 29–32. https://doi.org/10.1016/j.biocon.2016.04.009

Matyukira, C., & Mhangara, P. (2024). Advances in vegetation mapping through remote sensing and machine learning techniques: A scientometric review. *European Journal of Remote Sensing*. https://doi.org/10.1080/22797254.2024.2422330

McDermid, G. J., Coops, N. C., Wulder, M. A., Franklin, S. E., & Seitz, N. E. (2010). Critical remote sensing contributions to spatial wildlife ecological knowledge and management. In S. A. Cushman & F. Huettmann (Eds.), *Spatial complexity, informatics and wildlife conservation* (pp. 193–221). Springer.

Meng, Q., Zhang, L., Sun, Z., Meng, F., Liang, W., & Sun, Y. (2018). Characterizing spatial and temporal trends of surface urban heat Island effect in an urban main built-up area: A 12-year case study in Beijing, China. *Remote Sensing of Environment, 204*, 826–837. https://doi.org/10.1016/j.rse.2017.09.019

Milligan, C., & Bingley, A. (2007). Restorative places or scary spaces? The impact of woodland on the mental well-being of young adults. *Health & Place, 13*(4), 799–811. https://doi.org/10.1016/j.healthplace.2007.01.005
Mitchell, G., & Dorling, D. (2003). An environmental justice analysis of British air quality. *Environment and Planning A, 35*, 909–929. https://doi.org/10.1068/a3524
Mitchell, R., & Popham, F. (2008). Effect of exposure to natural environment on health inequalities: An observational population study. *Lancet, 372*(9650), 1655–1660. https://doi.org/10.1016/S0140-6736(08)61689-X
Montanges, A. P., Moser, G., Taubenböck, H., Wurm, M., & Tuia, D. (2015). Classification of urban structural types with multisource data and structured models. In *Proceedings of the 2015 joint Urban remote sensing event (JURSE), Lausanne, Switzerland, 30 March–1 April 2015.*
Morgan, D., & Falkner, E. (2001). *Aerial Mapping*. CRC Press.
Mulero-Pázmány, M., Stolper, R., van Essen, L. D., Negro, J. J., & Sassen, T. (2014). Remotely piloted aircraft systems as a rhinoceros anti-poaching tool in Africa. *PLoS One, 9*(1), e83873. https://doi.org/10.1371/journal.pone.0083873
National Aeronautics and Space Administration (NASA). (2018). *Fundamentals of satellite remote sensing*. Retrieved from https://appliedsciences.nasa.gov/sites/default/files/D1P3_Fundamentals.pdf
National Aeronautics and Space Administration (NASA). (n.d.). *Landsat 8*. Retrieved from https://science.nasa.gov/mission/landsat-8-ldcm/
Natural Resources Canada. (2015). *Fundamentals of remote sensing—Introduction*. Retrieved from https://natural-resources.canada.ca/maps-tools-and-publications/satellite-imagery-elevation-data-and-air-photos/tutorial-fundamentals-remote-sensing/introduction/9363
Natural Resources Canada. (2018). *Geomatics*. Retrieved from https://www.nrcan.gc.ca/earth-sciences/geomatics/10776
Naughton-Treves, L., Holland, M. B., & Brandon, K. (2005). The role of protected areas in conserving biodiversity and sustaining local livelihoods. *Annual Review of Environment and Resources, 30*(1), 219–252. https://doi.org/10.1146/annurev.energy.30.050504.164507
Neumann, W., Martinuzzi, S., Estes, A. B., Pidgeon, A. M., Dettki, H., Ericsson, G., & Radeloff, V. C. (2015). Opportunities for the application of advanced remotely-sensed data in ecological studies of terrestrial animal movement. *Movement Ecology, 3*(1). https://doi.org/10.1186/s40462-015-0036-7
Ngie, A., Abutaleb, K., Ahmed, F. B., Darwish, A., & Ahmed, M. (2014). Assessment of urban heat Island using satellite remotely sensed imagery: A review. *South African Geographical Journal, 96*(2), 198–214. https://doi.org/10.1080/03736245.2014.924864
Oke, T. R. (1973). City size and the urban heat Island. *Atmospheric Environment, 7*, 769–779. https://doi.org/10.1016/0004-6981(73)90140-6
Paneque-Gálvez, J., McCall, M. K., Napoletano, B. M., Wich, S. A., & Koh, L. P. (2014). Small drones for community-based forest monitoring: An assessment of their feasibility and potential in tropical areas. *Forests, 5*(6), 1481–1507. https://doi.org/10.3390/f5061481
Parcak, S. H. (2017). GIS, remote sensing, and landscape archaeology. *The Oxford Handbook of Topics in Archaeology*. https://doi.org/10.1093/oxfordhb/9780199935413.013.11
Passchier-Vermeer, W., & Passchier, W. F. (2000). Noise exposure and public health. *Environmental Health Perspectives, 108*(suppl. 1), 123–131. https://doi.org/10.1289/ehp.00108s1123
Pearce, J., Kingham, S., & Zawar-Reza, P. (2006). Every breath you take? Environmental justice and air pollution in Christchurch, New Zealand. *Environment and Planning A: Economy and Space, 38*(5), 919–938. https://doi.org/10.1068/a37446
Pearsall, H. (2017). Staying cool in the compact city: Vacant land and urban heating in Philadelphia, Pennsylvania. *Applied Geography, 79*, 84–92. https://doi.org/10.1016/j.apgeog.2016.12.010
Perkl, R. M. (2016). Geodesigning landscape linkages: Coupling GIS with wildlife corridor design in conservation planning. *Landscape and Urban Planning, 156*, 44–58. https://doi.org/10.1016/j.landurbplan.2016.05.016

Petso, T., & Jamisola, R. S. (2023). Wildlife conservation using drones and artificial intelligence in Africa. *Science Robotics, 8*(85), 7008. https://doi.org/10.1126/scirobotics.adm7008

Pettorelli, N., Safi, K., & Turner, W. (2014). Satellite remote sensing, biodiversity research and conservation of the future. *Philosophical Transactions of the Royal Society of London Series B, Biological Sciences, 369*(1643), 20130190. https://doi.org/10.1098/rstb.2013.0190

Pettorelli, N., Williams, J., Schulte, H., & Crowson, M. (2024). Deep learning and satellite remote sensing for biodiversity monitoring and conservation. *Remote Sensing in Ecology and Conservation.* https://doi.org/10.1002/rse2.415

Pham, T. T. H., Apparicio, P., Séguin, A. M., Landry, S., & Gagnon, M. (2016). Spatial distribution of vegetation in Montreal: An uneven distribution or environmental inequity? *Landscape and Urban Planning, 107*(3), 214–224. https://doi.org/10.1016/j.apgeog.2016.12.010

Prud'homme, G., Dobbin, N. A., Sun, L., Burnett, R. T., Martin, R. V., Davidson, A., Cakmak, S., Villeneuve, P. J., Lamsal, L. N., van Donkelaar, A., et al. (2013). Comparison of remote sensing and fixed-site monitoring approaches for examining air pollution and health in a national study population. *Atmospheric Environment, 80*, 161–171. https://doi.org/10.1016/j.atmosenv.2013.07.020

Quon, S., Martin, L., & Murphy, S. (2001). Effective planning and implementation of ecological rehabilitation projects: A case study of the regional municipality of waterloo (Ontario, Canada). *Environmental Management, 27*(3), 421–433. https://doi.org/10.1007/s002670010159

Ramsay, B., Manore, M., Weir, L., Wilson, K., & Bradley, D. (1998). Use of RADARSAT data in the Canadian ice service. *Canadian Journal of Remote Sensing, 24*(1), 36–42.

Rey, G., Fouillet, A., Bessemoulin, P., Frayssinet, P., Dufour, A., Jougla, E., & Hémon, D. (2009). Heat exposure and socio-economic vulnerability as synergistic factors in heat-wave-related mortality. *European Journal of Epidemiology, 24*(9), 495–502. https://doi.org/10.1007/s10654-009-9374-3

Richardson, E. A., Pearce, J., Mitchell, R., & Kingham, S. (2013). Role of physical activity in the relationship between urban green space and health. *Public Health, 127*(4), 318–324. https://doi.org/10.1016/j.puhe.2013.01.004

Rose, R. A., Byler, D., Eastman, J. R., Fleishman, E., Geller, G., Goetz, S., Guild, L., Hamilton, H., Hansen, M., Headley, R., Hewson, J., Horning, N., Kaplin, B. A., Laporte, N., Leidner, A., Leimgruber, P., Morisette, J., Musinsky, J., Pintea, L., Prados, A., & Wilson, C. (2015). Ten ways remote sensing can contribute to conservation. *Conservation Biology, 29*(2), 350–359. https://doi.org/10.1111/cobi.12397

Rozenstein, O., & Karnieli, A. (2011). Comparison of methods for land-use classification incorporating remote sensing and GIS inputs. *Applied Geography, 31*(2), 533–544. https://doi.org/10.1016/j.apgeog.2010.11.006

Ryan, P. H., & LeMasters, G. K. (2007). A review of land-use regression models for characterizing intraurban air pollution exposure. *Inhalation Toxicology, 19*(Suppl 1), 127–133. https://doi.org/10.1080/08958370701495998

Saito, K., Nakamura, K., Ueta, M., Kurosawa, R., Fujiwara, A., Kobayashi, H. H., Nakayama, M., Toko, A., & Nagahama, K. (2015). Utilizing the Cyberforest live sound system with social media to remotely conduct woodland bird censuses in Central Japan. *Ambio, 44*(Suppl 4), 572–583. https://doi.org/10.1007/s13280-015-0708-y

Salwasser, H. (1995). Factors influencing the context and principles of ecosystem management. *Natural Resources and Environmental Issues, 5*(1), 2. https://digitalcommons.usu.edu/nrei/vol5/iss1/2

Sandbrook, C. (2015). The social implications of using drones for biodiversity conservation. *Ambio, 44*(Suppl 4), 636–647. https://doi.org/10.1007/s13280-015-0714-0

Santos, T., Tenedório, J. A., & Gonçalves, J. A. (2016). Quantifying the City's green area potential gain using remote sensing data. *Sustainability, 8*, 1247. https://doi.org/10.3390/su8121247

Scherer, D., Fehrenbach, U., Lakes, T., Lauf, S., Meier, F., & Schuster, C. (2014). Quantification of heat-stress related mortality hazard, vulnerability and risk in Berlin, Germany. *DIE ERDE—Journal of the Geographical Society of Berlin, 144*(3–4), 238–259. https://doi.org/10.12854/erde-144-17

Schroyer, M. (2012). *Activists' drone shot out of the sky for fourth time.* Retrieved from https://www.suasnews.com/2012/11/activists-drone-shot-out-of-the-sky-for-fourth-time/

Seto, E. Y., Holt, A., Rivard, T., & Bhatia, R. (2007). Spatial distribution of traffic induced noise exposures in a US city: An analytic tool for assessing the health impacts of urban planning decisions. *International Journal of Health Geographics, 6*, 24. https://doi.org/10.1186/1476-072X-6-24

Shanee, S., Tello-Alvarado, J. C., Vermeer, J., & Bóveda-Penalba, A. J. (2013). GIS risk assessment and GAP analysis for the Andean Titi monkey (Callicebus oenanthe). *Primate Conservation, 26*(1), 17–23. https://doi.org/10.1896/052.026.0111

Shannon, G., Crooks, K. R., Wittemyer, G., Fristrup, K. M., & Angeloni, L. M. (2016). Road noise causes earlier predator detection and flight response in a free-ranging mammal. *Behavioral Ecology, 27*(5), 1370–1375. https://doi.org/10.1093/beheco/arw058

Sinha, B. R. K. (2019). *Multidimensional approach to quality of life issues.* In Springer eBooks. https://doi.org/10.1007/978-981-13-6958-2

Skole, D., & Tucker, C. (1993). Tropical deforestation and habitat fragmentation in the Amazon: Satellite data from 1978 to 1988. *Science, 260*(5116), 1905–1910. https://doi.org/10.1126/science.260.5116.1905

Solimini, D. (2016). *Understanding earth observation.* Springer International Publishing.

Sordello, R., Ratel, O., Flamerie, F., Léger, C., Dambry, A., & Sylvie, V. (2020). Evidence of the impact of noise pollution on biodiversity: A systematic map. *Environmental Evidence, 9*, 20. https://doi.org/10.1186/s13750-020-00202-y

Sparrow, R. (2009). Building a better warbot: Ethical issues in the design of unmanned systems for military applications. *Science and Engineering Ethics, 15*(2), 169–187. https://doi.org/10.1007/s11948-008-9107-0

Stansfeld, S. A., & Matheson, M. P. (2003). Noise pollution: Non-auditory effects on health. *British Medical Bulletin, 68*, 243–257. https://doi.org/10.1093/bmb/ldg033

Stathopoulou, M. I., & Cartalis, C. (2007). Use of satellite remote sensing in support of Urban Heat Island studies. *Advances in Building Energy Research, 1*, 203–212. https://doi.org/10.1080/17512549.2007.9687275

Subak, S. (2000). Forest protection and reforestation in Costa Rica: Evaluation of a clean development mechanism prototype. *Environmental Management, 26*(3), 283–297. https://doi.org/10.1007/s002670010087

Sun, J. W. C., & Narins, P. M. (2005). Anthropogenic sounds differentially affect amphibian call rate. *Biological Conservation, 121*(3), 419–427. https://doi.org/10.1016/j.biocon.2004.05.017

The Military Factory. (2025). Reconnaissance and Scout Aircraft. https://www.militaryfactory.com/aircraft/reconnaissance-scout-aircraft.php

Thompson, P. M., Lusseau, D., Barton, T., Simmons, D., Rusin, J., & Bailey, H. (2010). Assessing the responses of coastal cetaceans to the construction of offshore wind turbines. *Marine Pollution Bulletin, 60*(8), 1200–1208. https://doi.org/10.1016/j.marpolbul.2010.03.030

Times and Democrat. (2014). *Animal rights group says drone shot down.* The Times and Democrat. Retrieved from http://thetandd.com/animal-rights-group-says-drone-shot-down/article_017a720a-56ce-11e1-afc4-001871e3ce6c.html

Tuia, D., Moser, G., Wurm, M., & Taubenböck, H. (2017). Land use modeling in North Rhine-Westphalia with interaction and scaling laws. In *proceedings of the 2017 joint Urban remote sensing event (JURSE)*, Dubai, UAE, 6–8 March 2017.

Um, S. B., & Um, J. S. (2015). Metropolitan urban hotspots of chronic sleep deprivation: Evidence from a community health survey in Gyeongbuk Province, South Korea. *Geospatial Health, 10*(2), 382. https://doi.org/10.4081/gh.2015.382

United States Geological Service. (2001). *Understanding color infrared photographs*. Retrieved from https://pubs.usgs.gov/fs/2001/0129/report.pdf

United States Geological Service. (2024a). *What is a geographic information system (GIS)?* Retrieved from https://www.usgs.gov/faqs/what-a-geographic-information-system-gis

United States Geological Service. (2024b). *Common landsat band RGB composites*. Retrieved from https://www.usgs.gov/media/images/common-landsat-band-rgb-composites

Urban, A., Burkart, K., Kyselý, J., Schuster, C., Plavcová, E., Hanzlíková, H., Štěpánek, P., & Lakes, T. (2016). Spatial patterns of heat-related cardiovascular mortality in The Czech Republic. *International Journal of Environmental Research and Public Health, 13*(3), 284. https://doi.org/10.3390/ijerph13030284

Van de Voorde, T. (2016). Spatially explicit urban green indicators for characterizing vegetation cover and public green space proximity: A case study on Brussels, Belgium. *International Journal of Digital Earth, 10*(8), 798–813. https://doi.org/10.1080/17538947.2016.1252434

van Donkelaar, A., Martin, R. V., Brauer, M., & Boys, B. L. (2015). Use of satellite observations for long-term exposure assessment of global concentrations of fine particulate matter. *Environmental Health Perspectives, 123*(2), 135–143. https://doi.org/10.1289/ehp.1408646

van Donkelaar, A., Martin, R. V., Brauer, M., Kahn, R., Levy, R., Verduzco, C., & Villeneuve, P. J. (2010). Global estimates of ambient fine particulate matter concentrations from satellite-based aerosol optical depth: Development and application. *Environmental Health Perspectives, 118*(6), 847–855. https://doi.org/10.1289/ehp.0901623

van Eeden, L. M., Crowther, M. S., Dickman, C. R., Macdonald, D. W., Ripple, W. J., Ritchie, E. G., & Newsome, T. M. (2018). Managing conflict between large carnivores and livestock. *Conservation Biology: The Journal of the Society for Conservation Biology, 32*(1), 26–34. https://doi.org/10.1111/cobi.12959

Vandentorren, S., Bretin, P., Zeghnoun, A., Mandereau-Bruno, L., Croisier, A., Cochet, C., Ribéron, J., Siberan, I., Declercq, B., & Ledrans, M. (2006). August 2003 heat wave in France: Risk factors for death of elderly people living at home. *European Journal of Public Health, 16*(6), 583–591. https://doi.org/10.1093/eurpub/ckl063

Vardoulakis, S., Fisher, B. E., Pericleous, K. A., & Gonzalez-Flesca, N. (2003). Modelling air quality in street canyons: A review. *Atmospheric Environment, 37*(2), 155–182. https://doi.org/10.1016/S1352-2310(02)00857-9

Vasconcelos, R. O., Amorim, M. C. P., & Ladich, F. (2007). Effects of ship noise on the detectability of communication signals in the Lusitanian toadfish. *Journal of Experimental Biology, 210*(12), 2104–2112. https://doi.org/10.1242/jeb.004317

Villeneuve, P. J., Jerrett, M., Su, J. G., Burnett, R. T., Chen, H., Wheeler, A. J., & Goldberg, M. S. (2012). A cohort study relating urban green space with mortality in Ontario, Canada. *Environmental Research, 115*, 51–58. https://doi.org/10.1016/j.envres.2012.03.003

Voigtländer, S., Goebel, J., Claßen, T., Wurm, M., Berger, U., Stunk, A., & Elbern, H. (2011). Using geographically referenced data on environmental exposures for public health research: A feasibility study based on the German socio-economic panel study (SOEP). *SSRN Electronic Journal*. https://doi.org/10.2139/ssrn.1884910

Von Szombathely, M., Albrecht, M., Augustin, J., Bechtel, B., Dwinger, I., Gaffron, P., Krefis, A., Oßenbrügge, J., & Strüver, A. (2018). Relation between observed and perceived traffic noise and socio-economic status in urban blocks of different characteristics. *Urban Science, 2*(1), 20. https://doi.org/10.3390/urbansci2010020

Wang, M., Gehring, U., Hoek, G., Keuken, M., Jonkers, S., Beelen, R., Eeftens, M., Postma, D. S., & Brunekreef, B. (2015). Air pollution and lung function in Dutch children: A comparison of exposure estimates and associations based on land use regression and dispersion exposure modeling approaches. *Environmental Health Perspectives, 123*(8), 847–851. https://doi.org/10.1289/ehp.1408541

Weigand, M., Wurm, M., Dech, S., & Taubenböck, H. (2019). Remote sensing in environmental justice research—A review. *ISPRS International Journal of Geo-Information, 8*(1), 20. https://doi.org/10.3390/ijgi8010020

Weng, Q., Lu, D., & Schubring, J. (2003). Estimation of land surface temperature–vegetation abundance relationship for urban heat island studies. *Remote Sensing of Environment, 89*, 467–483. https://doi.org/10.1016/j.rse.2003.11.005
West, P., Igoe, J., & Brockington, D. (2006). Parks and peoples: The social impact of protected areas. *Annual Review of Anthropology, 35*, 251–277. https://doi.org/10.1146/annurev.anthro.35.081705.123308
Williamson, D. F. (2002). *Status, management, and trade of the American black bear (Ursus americanus) in North America.* World Wildlife Fund.
Wolf, C., & Ripple, W. (2018). Rewilding the world's large carnivores. *Royal Society Open Science, 5*, 172235. https://doi.org/10.1098/rsos.172235
World Wildlife Fund. (2020). *Explore the power of conservation drones.* Retrieved from https://techhub.wwf.ca/technology/drones-uav-monitoring-surveying/
Wurm, M., Taubenböck, H., Roth, A., & Dech, S. (2011). Urban structuring using multisensoral remote sensing data: By the example of the German cities Cologne and Dresden. In *Proceedings of the 2009 joint urban remote sensing event*. IEEE. Retrieved from https://elib.dlr.de/58970/1/paper_wurm_final.pdf
Wüstemann, H., Kalisch, D., & Kolbe, J. (2017). Access to urban green space and environmental inequalities in Germany. *Landscape and Urban Planning, 164*, 124–131. https://doi.org/10.1016/j.landurbplan.2017.04.002
Yang, W., Liu, W., Liu, Y., Corry, R. C., & Kreutzwiser, R. D. (2014). Cost-effective targeting of riparian buffers to achieve water quality and wildlife habitat benefits. *International Journal of River Basin Management, 12*(1), 43–55. https://doi.org/10.1080/15715124.2014.880710
Zhao, N., Cao, G., Ghosh, T., & Zheng, Q. (2023). Remote sensing imagery for mapping economic activities. *Remote Sensing*. Retrieved from https://www.mdpi.com/journal/remotesensing
Zhou, H., Liu, Y., & He, M. (2022). The spatial interaction effect of green spaces on urban economic growth: Empirical evidence from China. *International Journal of Environmental Research and Public Health, 19*(16), 10360. https://doi.org/10.3390/ijerph191610360

Index

K. K. Jha and M. O. Campbell (eds.), *Intersections of Conservation Biogeography and Wildlife Management*, https://doi.org/10.1007/978-981-95-0121-2

J

K

L

M

N

P

Q

R

MIX
Papier aus verantwortungsvollen Quellen
Paper from responsible sources
FSC® C105338

If you have any concerns about our products,
you can contact us on
ProductSafety@springernature.com

In case Publisher is established outside the EU,
the EU authorized representative is:
Springer Nature Customer Service Center GmbH
Europaplatz 3, 69115 Heidelberg, Germany

Printed by Libri Plureos GmbH
in Hamburg, Germany